TELENOTHIANS

TELENOTHIANS

An Inquiry into the Limits of Hybridization

Volume I: Heterotherians

Eugene M. McCarthy, Ph.D. Genetics

Truly, experience is superior to reason.

—JEAN RIOLAN
Opera Anatomica

NOTHOS PRESS

2025

Third edition 2025

NOTHOS PRESS

ISBN 979-8-3492-8610-0

Published by Nothos Press
P.O. Box 142
Athens, GA 30603
USA

Cover: *The Minotaur* (George Frederic Watts, 1885).

CONTENTS

Acknowledgments

When you try to do something good and honest, as I have tried to do in this book, it often isn't possible without the help of others. I would like to thank two friends for their help with this book. John Pelter, with his editorial recommendations, his many suggestions and his open mind, was a huge help. Ed Falkowski, too, aided greatly with the production of this book. And above all, I thank my wife Rebecca, who gifted me not only endless hours of editing, but also what has been far more important—limitless patience and love.

In the ninth rank of prerogative instances, we will place bordering instances, ... They are such as exhibit those species of bodies which appear to be composed of two species, or to be the rudiments between the one and the other. ... from their dignity, they must be treated of and classed separately, for they point out admirably the order and constitution of things and suggest the causes of the number and quality of the more common species in the universe, leading the understanding from that which is, to that which is possible.

—FRANCIS BACON, *Novum Organum*[1]

INTRODUCTION

The true things rarely get into circulation. It's usually the false things. It's hard to know where to start, if you don't start with the truth.

—MARILYN MONROE

This book is about *telenothians* (/TEL-ə-NŌ-thē-ənz/), creatures combining traits ordinarily found only separately in two or more highly distinct kinds of organisms.† A well-known example is the platypus, which has a bill like that of a duck but a body much like that of an otter. Platypuses combine other seemingly contradictory traits. For example, they both lay eggs (a trait of birds) and produce milk (a trait of mammals).

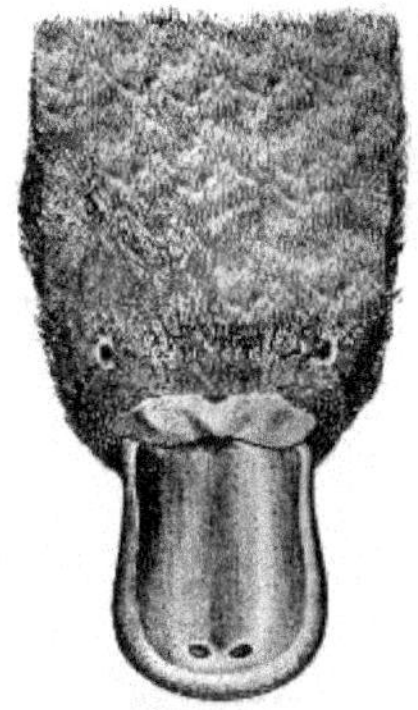

The platypus, classified as a mammal, has a bill like a bird's, specifically, a duck's

Platypuses are well-known, not only to scientists, but also to the general public. When they come up in discussion, no one is likely to doubt they exist. Such is not the case, however, with certain other kinds of telenothians. For example, suppose someone asserted a bird had the traits of both a chicken and a duck. In this book, such a bird is called a *galanatid* (*gal-* for chicken + *-anatid* for duck). Many would dismiss such a claim, even though such a combination (bird with bird) is less disparate than the composite represented by a platypus (bird with mammal).

†. The degree of distinctness of two organisms can be defined in terms of taxonomic difference. Thus, in terms of categories of biological classification, platypuses are *interclass* telenothians because they mix traits characteristic of two different classes of the same phylum (Chordata): those usually seen only in Class Mammalia with ones usually found only in Class Aves (birds). An *interordinal* telenothian would be one that combined traits characteristic of two different orders of the same class. Here, we will be concerned mostly with interordinal and interclass telenothians, usually ones having the characteristics of either mammals or birds (or of both, as in the case of the platypus).

Nevertheless, opinions that are not accepted today, may well be tomorrow. Even the platypus, when it was first discovered, was viewed with suspicion. Many scholars wondered whether the specimen might be a fraud. In the first scientific description of the platypus, zoologist George Kearsley Shaw (1799, p. 228) voiced the same kinds of doubts that many scientists would express today about a galanatid.

George Kearsley Shaw
1751-1813

> On a subject so extraordinary as the present [i.e., the platypus], a degree of skepticism is not only pardonable, but laudable; and I ought perhaps to acknowledge that I almost doubt the testimony of my own eyes with respect to the structure of this animal's beak; yet must confess that I can see no appearance of any deceptive preparation … nor can the most accurate examination of expert anatomists discover any deception in this particular.

And today, "a degree of skepticism is not only pardonable, but laudable" in connection, too, with the question of the existence of galanatids. With respect to such birds, we stand today at the same threshold of knowledge where Shaw once stood with respect to the platypus. But as information accumulates, it may be that galanatids, too, will turn out to be every bit as real as the platypus. In the meantime, my approach has been what I consider scientific: to collect as much evidence as possible about such birds so that a well-informed assessment of their status can be made. That evidence is presented in the chapter on galanatids.

There is another reason galanatids have been viewed with suspicion: The nature of their origin is unclear. Everyone knows where platypuses come from. They arise from parents of their own kind. So no questions arise. But reports about galanatids say they hatch from eggs laid either by chickens or, sometimes, by ducks. Such claims seem to entail one of two events, either of which many people would deem impossible:

- that a chicken and a duck mated and produced offspring; *or*
- that the non-hybrid offspring of a mating between a hen and a rooster must somehow have been affected by a mutation

causing it to take on various traits of a duck (or that the offspring of a drake and duck took on those of a chicken).

Anyone who deems such events impossible will view galanatids in the same light. And those with such beliefs will likely judge that the easiest and most sensible approach is simply to reject all such reports in advance as false and not worth the reading. By this means, a myriad of witnesses can be dismissed at once.

Nevertheless, as we shall see, reports confirm, quite beyond any reasonable doubt, the existence of such birds. For example, a video shows a living galanatid with what appears to be the head of a chicken and the body of a duck. This bird, "Chuck the Chicken-duck" (shown at right), hatched in Texas in 2012 from an egg laid by a chicken who was being kept in a pen with ducks. That the mother, a hen, was penned with ducks suggests this bird was a chicken-duck hybrid, as does its mixed traits (and the fact that it is well known that chickens and ducks sometimes mate). Details about "Chuck" appear in the chapter on galanatids.

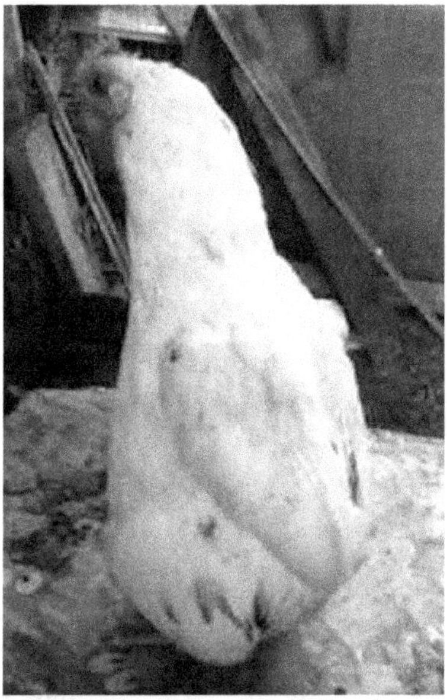

"Chuck the Chicken-duck," a Texan galanatid (source: ||y3m8gxa6).

In this book, each different kind of telenothian is assigned a name, and a section is devoted to each, so that all relevant reports and information for each particular kind of composite will be collected under a single heading. This arrangement is intended to facilitate evaluation of four issues:

- Whether telenothians of the sort under consideration exist.
- What they are like, if they do exist.
- How much variation exists between individuals.
- The nature of their origins.

How well these points are resolved depends on the type of telenothian in question. But two facts emerge from the review provided in these pages:

- Many telenothians, not generally recognized by science, do exist;
- Organisms assigned to different orders, or even to different classes, do sometimes hybridize.

Judging from the platypus, one can see that the tags "possible" and "impossible" are not static categories but, instead, potentially transient judgments based on what science currently accepts as correct. Indeed, each individual human being on earth is forced to make judgments about what is correct. Such assessments can, do and should change over time, as we improve our knowledge about what does, or can, exist.

But such advances in knowledge take place through investigation. Through investigation we sometimes learn that things we once believed imaginary do in fact exist. Progress toward a more accurate knowledge of the exact position of this boundary between the real and the fabulous is one of science's goals. But we are allowed only to approach the true line of demarcation between fact and fancy, never to arrive there. To imagine that we have achieved such knowledge is merely to close our minds and embrace a dogma.

There is a widespread faith in the existence of natural laws, immutable rules that govern reality. Such laws may indeed exist. But we mortals have no definite assurance that we have correctly determined them. What might now seem to be a violation of natural law might later, with a better understanding of nature's rules, turn out to be possible. In this sense, the age of miracles has not passed, and never can. As Castaneda put it, "We are men, and our lot is to learn and to be hurled into inconceivable new worlds."

But let us return to the subject of telenothians, especially to their mode of origin. This is a troublesome question. To understand why, consider the platypus. Everyone agrees that platypuses do exist, and most people, even most biologists, agree platypuses have traits of both mammals and birds. Studies have even shown that despite their being classified as mammals, many of their genes are otherwise seen only in birds. The explanation, accepted among biologists, of these mingled features and birdlike genes is that the platypus is a mammal of ancient origin that has retained certain ancestral traits. There seems to be only one other potential explanation, in terms of natural phenomena, of how platypuses might have come into being: an otter-like mammal mated with a duck-like bird to produce a hybrid that had the traits of both mammals and birds. But among biologists this idea is widely and strongly disavowed.

This attitude is inconsistent with ordinary scientific practice. Ordinarily, when faced with two alternative hypotheses, scientists are—or at least should be—indifferent to which of the two gains acceptance. In science the criterion should always be whether the hypothesis is consistent with observation. But many scientists have a strong tendency not only to deny that reported telenothians might be hybrids, but even to reject the idea that such hybrids might be possible. And such opinions seem not to be based on experimental findings. For instance, there have been virtually no published studies reporting the results of efforts—whether failed or otherwise—to produce telenothians experimentally with hybrid crosses. Likewise, the question of whether the various peculiar telenothian specimens available in museums might be of hybrid origin receives little or no attention, even though it could be approached via biomolecular analysis. And yet, even those who say merely that they are interested in considering whether such crosses might be possible are likely to encounter disapproval or even mockery from their colleagues. In short, there is a strong and widespread bias with regard to hybrids.

Why does this prejudice exist? In part, it seems to be a custom arising out of superstition. Just as buildings customarily lack a thirteenth floor, the topic of telenothians is avoided because they, too, elicit irrational fears. And just as building owners who are not superstitious omit thirteenth floors because they have negative effects (e.g., lower occupancy rates), biologists who are not superstitious omit investigation of telenothians because such investigation has negative effects (e.g., damage to reputation). But this kowtowing to ignorance on the part of the very people who should be guarding and advancing biological knowledge tends to prevent any progress in our knowledge of telenothians.

The reader may wonder what superstitious fears telenothians might inspire. But composite beings have long elicited irrational awe. Among the ancient Babylonians and Egyptians, creatures that combined, or seemed to combine, the traits of very distinct kinds of animals—especially those of humans with those of non-humans—were regarded as portents or even worshiped as gods. The Greeks not only believed in a wide variety of such *tertium quids* (centaurs, satyrs, mermaids, etc.), but also reverenced them. Among the Romans, any report of such a birth was investigated by government officials, and when it became certified as fact, augurs would evaluate the omen and pronounce it good or bad.

The notion that such births were omens persisted for centuries. Indeed, the notion that composite animals are harbingers of evil persists even today, as evidenced by the many contemporary reports about telenothians cited in this book that mention people being dismayed to see such creatures and fearing them as signs of misfortunes to come. For example, in 2014, when a mare in the Romanian town of Zărand foaled a colt with reported camel-like features, townspeople were shocked and flocked to the church to pray (see the detailed account of this event in the chapter on HIPPOCAMELIDS).

Many people connect telenothians with witchcraft. During the Middle Ages, the idea arose that telenothians were the creation of demons, witches and sorcerers, working to overturn God's natural order (Barnes 1991, Huizinga 1922). Telenothians with seeming human characteristics, especially, were seen in an evil light. Women who bore such composites were frequently executed, and many men, too, if convicted of bestiality, suffered the same fate. In a legal context such things became "that horrible sin not to be mentioned among Christians" ("peccatum illud horribile, inter christianos non nominandum"). If such behavior was not to be mentioned even in courts of law, it certainly was not mentioned in polite conversation. Nor were telenothians, whose very existence seemed to imply the occurrence of such behavior

Over the last three centuries, notions about telenothians have slowly shifted. In the seventeenth century scholarly opinion generally accepted the idea that very distant hybridization could occur, including hybridization between humans and animals, though such things were dreaded and abhorred. But during the eighteenth century, scholars gradually came to reject the possibility of such crosses and began using the term *lusus naturae* (freak or sport of nature) to describe reported telenothians. This term had the happy advantage of removing all uncomfortable mention of mixed copulation from discussion. The idea was that an all-powerful Nature was entertaining herself ("making sport") by creating such composites on a whim or—as *on a whim* was expressed in those days—"on a freak." This sleight of tongue was early embraced by the conservative side of society. In the popular press, however, telenothians continued to be reported, and such reports sometimes included explicit claims that they had had a hybrid origin. But among scholars, *lusus naturae* and *freak of nature* were themselves eventually replaced by the word *mutation,* which in the context of telenothians is no more informative than the terms it superseded. It tells one nothing, because to say a trait arose by mutation, without identifying a corresponding change in the mutated individual's DNA, is no more enlightening than to say that an individual is different from its parents. The word does, however, have the advantage of sounding scientific.

To drive this point home, consider one of the telenothians documented in this book, the *canibovid.* Canibovids are offspring with canine characteristics calved by cattle. Among the innumerable reports about such animals, the early ones use the terms *lusus naturae* and *freak of nature,* whereas in later years they speak of "mutation." But merely using the word *mutation* does nothing to explain how canibovids come into being. To explain such a phenomenon, we would have to understand how cattle DNA could be altered so that the offspring of a cow would take on many of the traits of a dog. Such an alteration at the genetic level might conceivably be feasible. But the exact nature of what that change might be has not, to date, been revealed. On the other hand, male dogs and recumbent cattle have been observed *in coitu* (a fact documented in the chapter on canibovids). The question then arises whether such copulations occasionally produce hybrids. This issue could certainly be addressed via experimentation, say, via artificial insemination of cattle with dog semen. But even though canibovids are obvious composites commingling the traits of dogs with those of cattle, it would seem that no such breeding experiments have been conducted, nor has any DNA study been done to investigate whether these animals are hybrids.

And what can be said of canibovids can be said, too, of telenothians in general: They do exist—beyond reasonable doubt—but the nature of their origins needs to be better established, at least in the case of certain kinds of telenothians. Though in some reports the clearly composite nature of the specimens does seem to shriek hybrid, further investigation is needed in many cases to establish the fact of hybrid parentage beyond doubt.

The nature of telenothian origins is not a matter of mere idle curiosity. It is in fact a vital issue for biology. Distant hybridization would allow the rapid production of radically new forms of life. A platypus, for example, is radically different from an otter or a duck. And the potential significance of such large and rapid changes looms even larger since it is now well known that many hybrids can produce offspring and therefore pass on their traits.

Let the reader be forewarned that some of the evidence collected in this book may be disturbing. There are many reports about telenothians mingling the traits of humans with those of nonhuman animals. Indeed, this book contains many illustrations and descriptions (as well as links to videos) of living beings that might well be described as monsters—things that some may find shocking even to contemplate. But gaining knowledge is often a disturbing experience. And even distressing facts can be informative. As the early English anatomist William Lawrence (1814, p. 170) once commented in connection with the study of stillborn infants with major deformities,

> Should any doubt be entertained, whether the examination of these productions can lead to interesting or useful results, we may observe, that our conceptions of nature will be more correct in proportion as they are drawn from a more extensive survey of her works; and, that we cannot fail to learn something of her mode of proceeding by attending even to the imperfect sketches, which she seems to have abandoned as incapable or unworthy of being completed. The causes of these failures, and the conditions necessary to the production of perfect beings, cannot be understood without carefully noting in what the imperfections consist, and the circumstances under which they are produced.

So, too, it is with hybridization. To have any hope of attaining a full understanding of that phenomenon and, hence, of its role in the evolutionary process, we must "cast a cold eye on life, on death," and pass on to consider the implications of the fullest possible range of observations.

Observations have a wonderful way of dispelling murky thinking, particularly when the reports are numerous and generally tend to support the same conclusion. But a downside of observation is that it is such hard work collecting all the facts relevant to a topic, especially one as broad as the present one, or even to absorb that information once assembled. It is far and away easier to simply make pronouncements based on hearsay. But to those who prefer fact to fantasy, this book permits many witnesses to speak.

Yet, if there are some so biased as to claim that certain plain truths should nowhere be discussed, let them content themselves with this book's introduction, or even with its title. For I have written it for those who enjoy exploring Nature's games. "Let no Cato enter my theater, or if he enters, let him look."

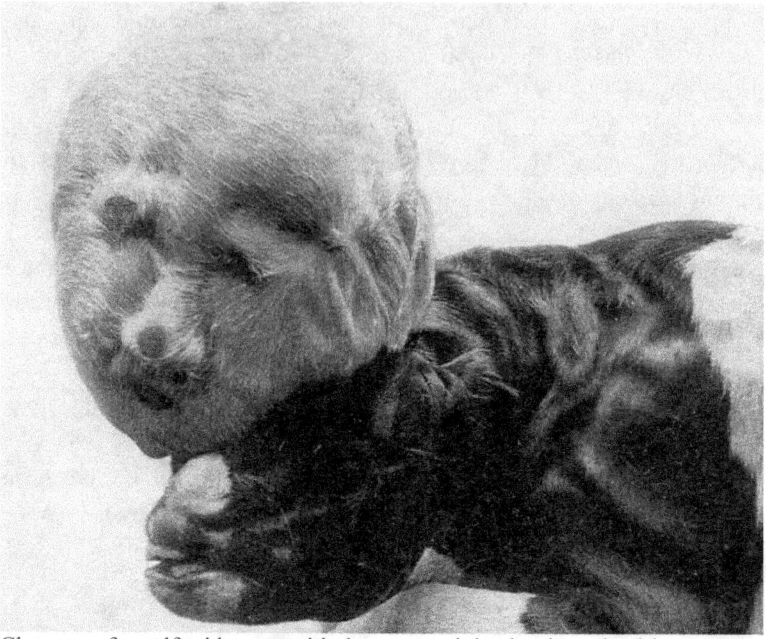

Close-up of a calf with a parasitic human conjoined twin, calved by a cow at Beppu, in Oita Prefecture, Japan, in 1926 (for more information about the Beppu birth, see p. 76). This cattle-human (*boshominid*) is one of the many different kinds of telenothians discussed in this book.

Almost all persons remind us of, and do really resemble some one of the lower animals. Some men remind us of the lion, some look like a bulldog or a mastiff, or a rat terrier. And some even resemble a cat or a rat. Some look like a crane, others like an eagle, while others have the features of a monkey or baboon. Others are proud of their likeness to the gorilla. Again some have the exact expression of a sheep, and alas, some remind us forcibly of the ass.

—Robert Lee Payne, *Valedictory Address* (1879, p. 92)

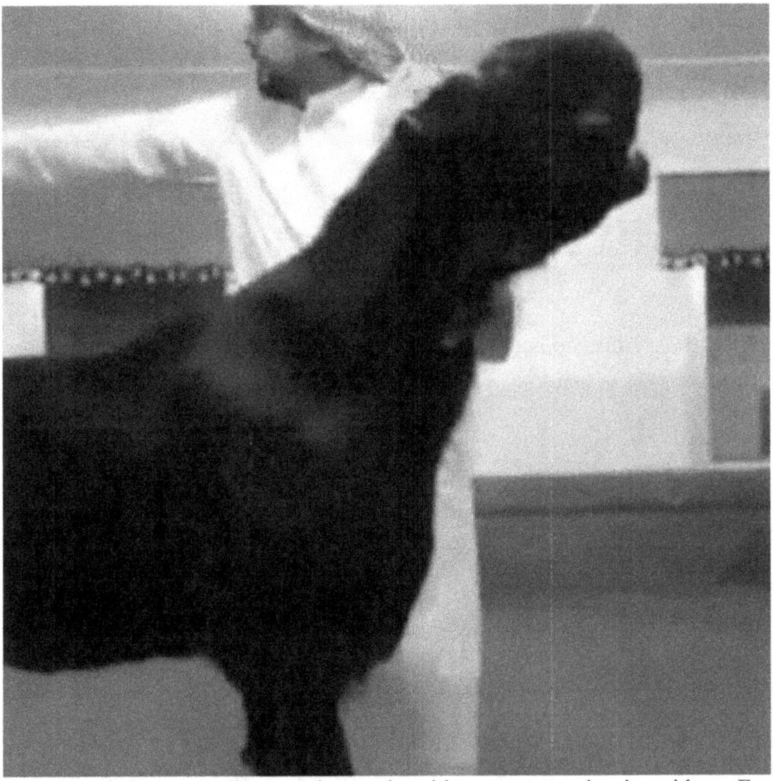

An alleged dog-goat. Extreme heterosis with respect to size is evident. For further information, see the chapter on CANICAPRIDS.

Heterotherians

> *All that we see could be other than it seems. Indeed, anything we describe could be other than it seems. There is no a priori order of things.*
>
> —LUDWIG WITTGENSTEIN
> *Tractatus Logico-Philosophicus*

A heterotherian is an animal combining traits typical of two distinct mammalian orders. In this book, more than half the cited reports about heterotherians refer to composites involving humans, a preponderance likely due to the greater attention such births receive, that is, there seems to be a reporting bias.

For he who is acquainted with the paths of nature, will more readily observe her deviations; and vice versa, he who has learnt her deviations, will be able more accurately to describe her paths.

—FRANCIS BACON, *Novum Organum*

Homungulates

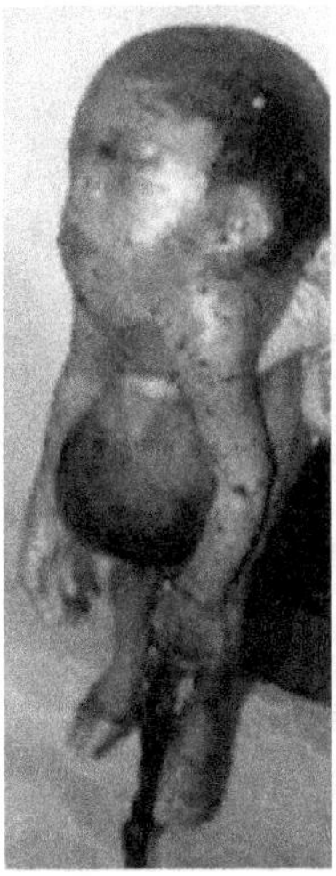

Ungulates are hooved mammals. Two broad types are recognized, those with a pair of weight-bearing hooves on each foot (Order Artiodactyla) and those with one (Order Perissodactyla). The former includes animals such as cows, sheep, goats, deer, camels, and antelope. Examples of the latter are horses, asses, zebras, tapirs, and rhinoceroses. Since they are assigned to different orders, artiodactyls and perissodactyls are usually considered rather distantly related—as different, say, as a chimpanzee and a pig. But in this book, creatures commingling the traits of any ungulate with those of a human being will be termed *homungulates*.

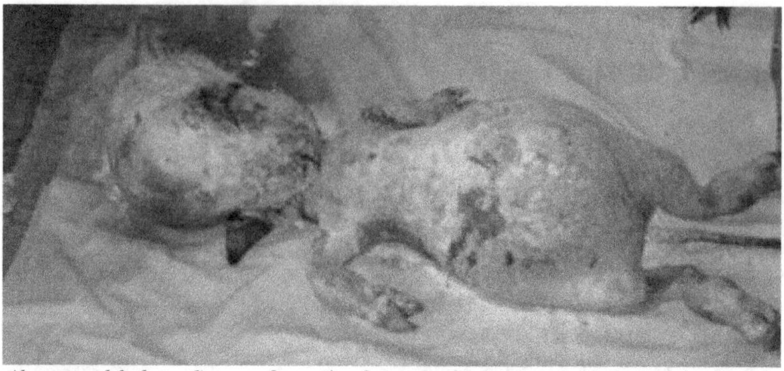

Above and below: Scenes from the funeral of a boshominid that took place in Thailand in 2008. Note the large size of the head, the cloven hooves on both fore and hind limbs, the pointed, low-set ears, the presence of a tail, and apparent lack of eyes (anophthalmia). It can also be seen, in the photo below, that the overall length of this creature is about four feet. Its distended stomach seems to echo the description given in the report published by Wieland (see p. 63), in which he says the boshominid described looked as if it were pregnant. The legs of the Thai boshominid, though short, are far thinner than a human's, and more widely spaced at the hips. The head appears to be larger than those of the spectators, though this may be an illusion due to perspective. If it was the offspring of a human mother, its overall size shows it was not a newborn and that it therefore must have survived birth. Wieland said it took 12 years for a boshominid born to a woman to reach a length of 3.5 feet.

BOSHOMINIDS (COW-HUMANS)

> *Believe that Pasiphae coupled herself with the Cretan bull. We have seen it. The old-time tale receives our faith.*†
>
> —MARTIAL, *Liber Spectaculorum,* 5

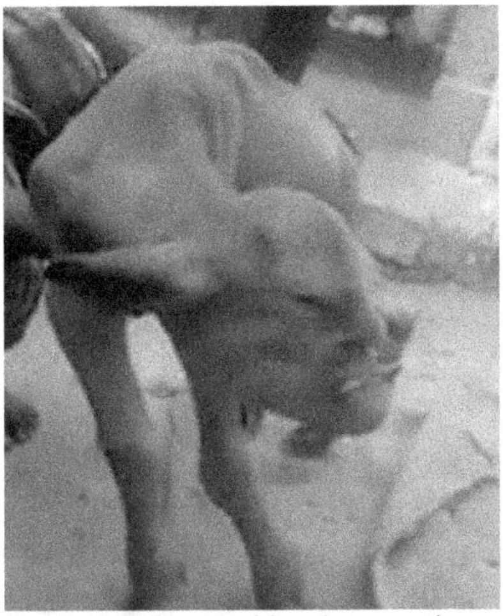

Screenshot from a video documenting a living boshominid calved in 2019 in the Dominican Republic (||2p8hv5t8).

Though they are but rarely seen, creatures commingling the traits of cattle with those of human beings have been reported from ancient times right up to the present. Such composites appear in some of the earliest records of mankind, for example, at Ur, but are still encountered in YouTube videos. Their real existence has been documented by news reports, photos, old chronicles, veterinary records and medical journals. In all, this chapter cites about 180 separate cases.

So, there can be no reasonable doubt that hybridization between humans and cattle does occur. The information collected here makes that clear. There are too many similarities between unconnected sources. Many different witnesses, living at different times, in different places, give similar accounts. Moreover, those who claim that such things cannot occur cannot have witnessed that nonoccurrence. For nonexistence can never be witnessed. They can only say they've never seen it and tell how they've heard, or how they've been taught that it does not happen. They can cite factors that supposedly preclude it: They can say the genetic differences are too great or that the requisite mating events cannot or should not occur. They can claim that even if insemination did occur, fertilization would not follow, and that if it did follow, the resulting embryo would be inviable. They can cite various authorities who say it can't happen.

†. "Iunctam Pasiphaen Dictaeo credite tauro: Vidimus. Accepit fabula prisca fidem."

Such claims crop up again and again when hybrids are discussed. And people who use such arguments clearly believe they are verified and true and not mere hearsay. But anyone who does take such ideas seriously should take the time to examine the reports about hybrids cited and/or quoted in this book, many of which were written by people who had firsthand experience with such creatures. Direct observation has a wonderful way of dispelling the murk of hearsay.

For my own part, experience with hybrids has led me to regard some of my fellow scientists in a different light. I'm now more like Rabelais, who in his book *Pantagruel* described the students of "a little old hunchback, all misshapen and monstrous," named Hearsay, headmaster of a popular school. "Around him," writes Rabelais,

> I saw innumerable men and women listening to him attentively, and among the group I recognized several with very important looks, among them one who held a chart of the world and was explaining it to them succinctly, in little aphorisms. Thus they became clerks and scholars in no time and spoke in choice language—having good memories—about a host of tremendous matters, which a man's whole lifetime would not be enough for him to know a hundredth part of. They spoke about the Pyramids, the Nile, Babylon, the Troglodites, the Himantopodes, the Blemmyae, the Pygmies, the Cannibals, the Hyperborean Mountains, the Aegipans, and all the devils—and all from Hearsay.

One might suppose that the boshominids described in the reports cited in this chapter are not hybrids. But any such notion can be dismissed. For if these animals were not human-cow hybrids, the cases cited would imply that complex human traits could suddenly appear in the offspring of cattle via random mutation. Such an event is excessively unlikely because it would require the conversion of many different bovine genes, by chance, in a single generation, into ones programming the development of human traits. How would the genes needed to produce human body parts arise in the offspring of a cow, if they were not obtained from a human sire? Likewise, in the case of boshominids birthed by women, how would cattle traits—like horns, hoofs or the head of a calf—be produced by a genome that was strictly of human origin? Therefore, the remaining possibility, that these are cow-human hybrids, has to be right. Indeed, the many cases cited represent ironclad evidence, demonstrating not only that such hybridization does occur, but also that it is widespread and ongoing.

In many of the reports cited, the boshominids described resemble Mesopotamian depictions of bull-men (see pp. 72-74). Other accounts describe creatures much like the Greco-Roman Minotaur. These similarities suggest that the forms of these two mythic beings were likely first inculcated into human imagination by the occurrence of such births. However, while bull-men and the Minotaur are each usually represented in a specific form, descriptions of boshominids vary from one report to another. Thus, some of these creatures reportedly look like a cow except in having what appears to be a human arm attached at the shoulder (see pp. 77-79). Composites of this last type seem to have no equivalent in the descriptions of religion or myth. Also aberrant are those cases where a human head supposedly stood atop the head of what otherwise looked like a cow (see pp. 76-77).

Amid this variation, certain rules are apparent. According to reports, most boshominids calved by cattle mothers resemble humans in their foreparts (head and breast) but cattle in their hind parts. Those with human mothers typically have cattle foreparts and human hind parts. The degree and precise nature of the resemblance varies. This conclusion is seemingly corroborated by a statement in the Wikipedia article on one of the Japanese yokai, the kudan, which is widely believed in that country to be the product of hybridization between humans and cattle (||39h8yv5s; accessed 11/19/2023). "The writer Hirokatsu Kihara argues that the kudan and the ushi-onna ought to be distinguished, due to a number of differences: the kudan is a calf born from cattle, the ushi-onna is a daughter of human parents; the kudan is human-faced and bovine-bodied, the ushi-onna the other way around."

The ears of boshominids are usually like small cattle ears but in some cases like human ones. Usually, boshominids have mostly naked skin, as do human beings. But such is by no means always the case. Some have at birth a beard like a man and breasts like a woman. Many are stillborn or die soon after birth. Some reach maturity, especially the offspring of women.

But now—to the reports!

The 2008 Thai boshominid.

RECENT REPORTS

Most recent reports about boshominids appear on the Internet. Such reports are visually rich and verbally poor. Indeed, they often fail to provide even specifics about the exact time and place where the birth in question occurred. Some reports fail to identify the mother.

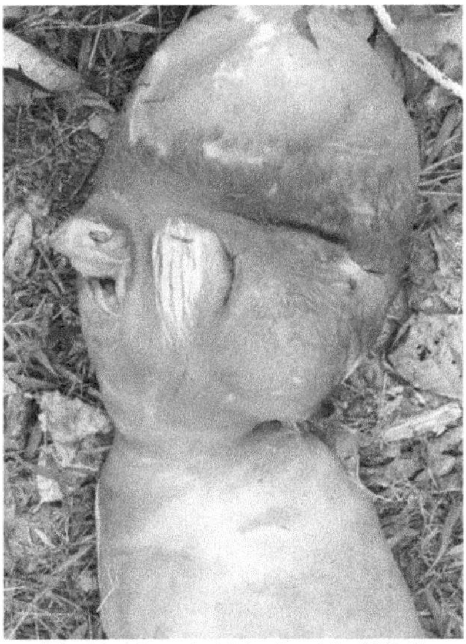

Thai boshominid (calved in 2024) ||2p9k5554

Thailand (2024). At least two recent births have occurred in Thailand. One is the calving of a boshominid somewhere in that country in January 2024. This animal was mostly naked, as are many boshominids. Various pictures are available (e.g., ||2p9k5554) but little specific information.

Thailand (2008). Many pictures of another boshominid appeared on the Internet in 2008. The accompanying reports were scanty, but the photos show a funeral, apparently in Thailand, attended by many people, of a creature dusted with a white powder, seemingly as part of the ceremony. Cloven hooves and a tail are clearly visible, while the general configuration of the body is more human than bovid. The photos are the only real evidence in this case. But it seems unlikely a funeral would have been held if the mother were not human, which would imply this creature was viable since it grew to a size far beyond what a woman could bear.

* With respect to its size and general conformation the 2008 Thai boshominid is generally similar to the one described by Wieland (see pp. 63-66). The main difference is that Wieland's had stump arms, whereas the Thailand animal's arms ended in hooves.

Villa Ana, Santa Fe, Argentina. Internet reports say a cow at Villa Ana, in 2019, calved an animal with a face resembling a human's. Videos show that the rest of its body, including the legs and tail, were similar to those of an ordinary calf. Screenshot from a video (||y593qb2o).

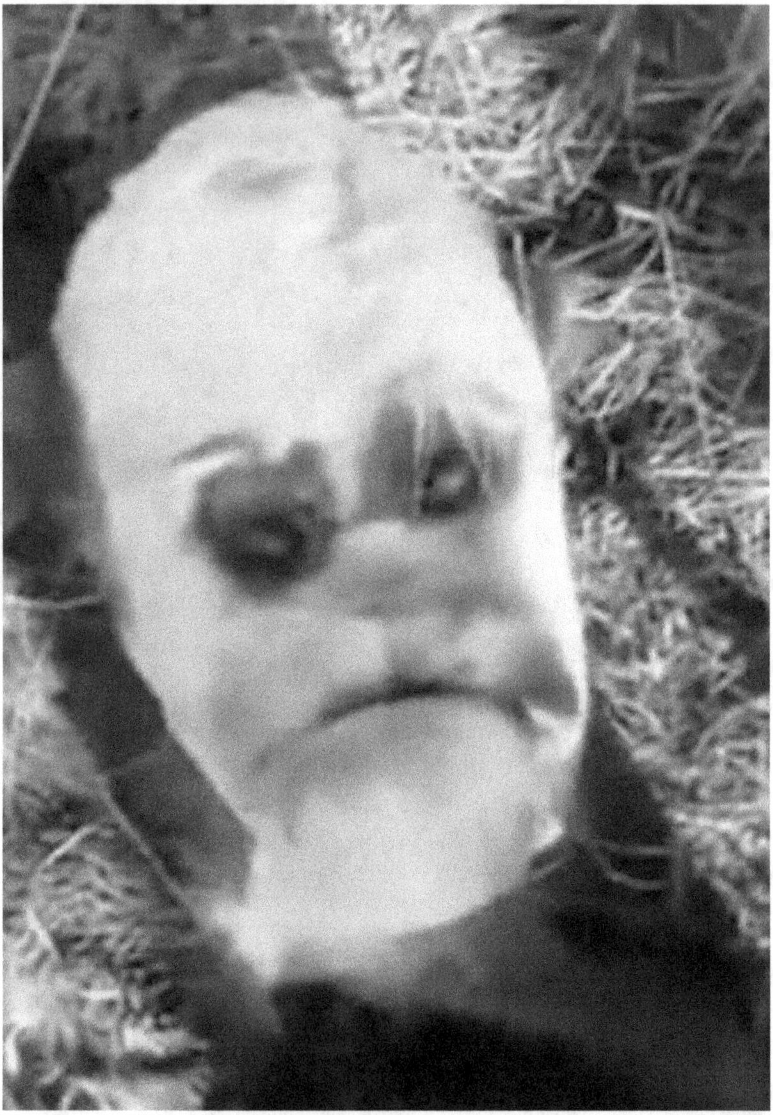

A boshominid that Internet reports say a cow calved at Villa Ana, Santa Fe, Argentina in 2019. Videos show that the rest of its body, including the legs and tail, were similar to those of an ordinary calf. Screenshot from a video (||y593qb2o).

Uttar Pradesh. At least two boshominids have been reported from Uttar Pradesh in India, one in 1899, the other much more recently. The following, about a birth at Agra, is from the *Madras Mail* (quoted in the *Northern Star*, Lismore, NSW, Australia, Aug. 9, 1899, p. 6b; ||ooegnpx). The same report appeared in many British Empire papers.

A CALF WITH A HUMAN FACE

An Agra correspondent of the MADRAS MAIL, writing recently, remarks: "An extraordinary freak of nature was brought to light at Sultanpur, one of the suburban villages of Agra Cantonment, situated not quite a mile from the Cantonment Church, and on the outskirts of what is known as Pensioners' Lines. A cow, it was said, had given birth on the previous day to a wonderful calf, which survived only a few hours, the head of which resembled that of a human being, while the rest of the body was that of any ordinary calf." A correspondent of the paper immediately proceeded to the spot, and says: "When I got there, I found an immense concourse of people gathered inside, outside, and round about a Gwala's hut. With some difficulty I managed to make headway through the crowd, and at length got to the 'sacred' spot. There I found what at first appeared to me to be an ordinary dead calf, but on closer inspection I found that it was one of the most extraordinary calves I had ever seen. The lower portion of the body from the neck downwards was perfectly symmetrical and natural, but the head in every respect was that of an overgrown babe, say about 8 months old. There was the short round face in the place of the long-pointed snout, the ears small, and fixed flatly in the head, instead of protruding out—in fact, a perfect human face. Hindus of all castes, from the sacerdotal Brahmin to the low caste Chamar, were all there to do reverence, as they told me, to this more than sacred cow. Some of the better class of men among the crowd told me that they attach a great deal of importance to this extraordinary phenomenon of nature, as being prophetic of some marvelous development of an abnormal event that will occur in the history of this country at no distant date. So much for idle superstition of uncultured minds. What they mean to do with the carcass I do not know, my offer to purchase it being positively declined."

* The locale implies the cow was a zebu (*Bos indicus*).

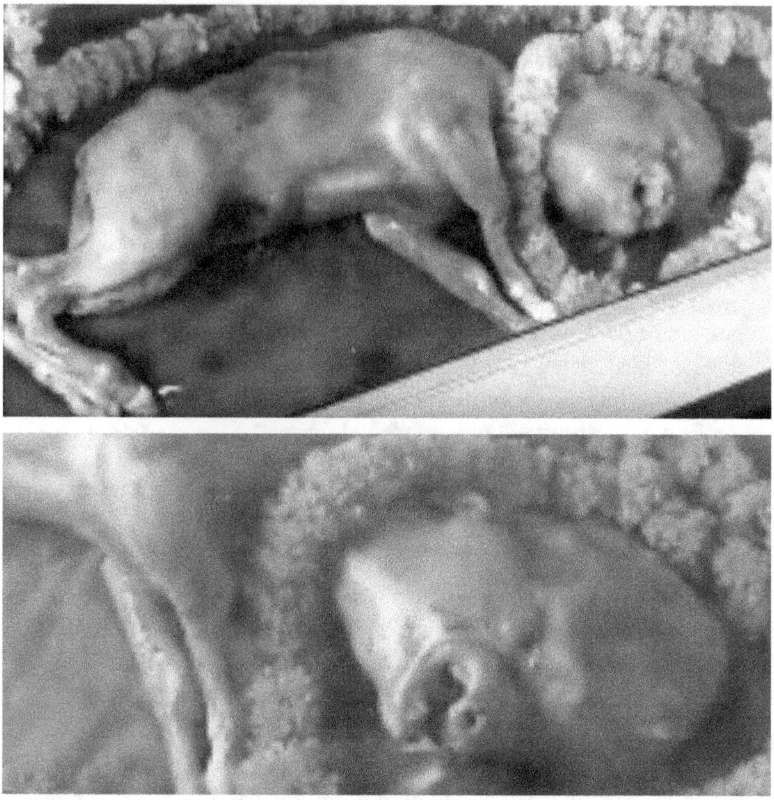

Screenshots from a video (||y4nbgup4) showing the boshominid born in Uttar Pradesh in 2017. As is reportedly the case with many boshominids, it was naked as are humans, but not calves.

The second Uttar Pradesh birth was reported from the city of Muzaffarnagar in June 2017 (see screenshots above). This creature, too, had a head resembling a human being's, and it had a naked body like that reported for most such births. Reports say it survived for about half an hour. Locals worshiped it as a god. Mahesh Kathuria, visiting the shelter where this creature was born, expressed his opinion that:

> God has taken birth from the body of a local cow. We came here to seek his blessings. Religiously, it is an avatar of Vishnu. We believe it's a similar character to that mentioned in the Bhagavata Purana.

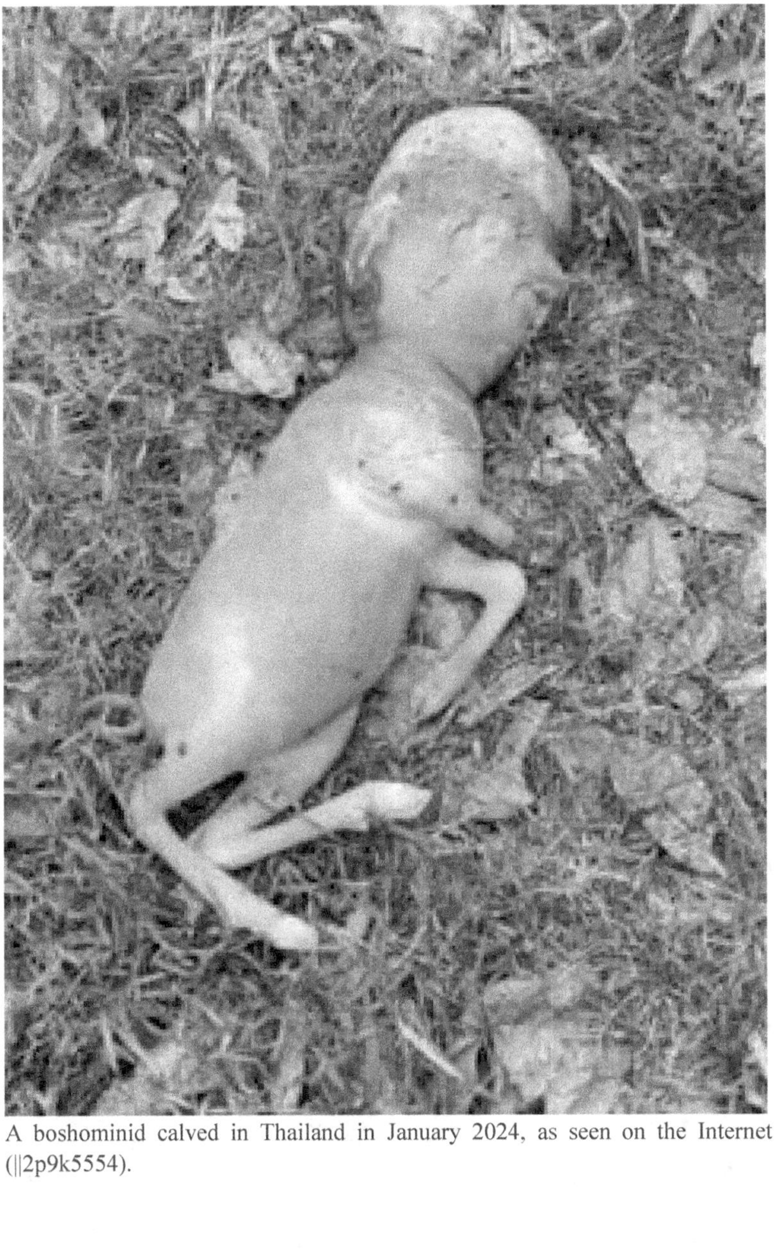

A boshominid calved in Thailand in January 2024, as seen on the Internet (||2p9k5554).

HIGH FERTILIZATION RATE

The multiple births cited below suggest the fertilization rate in this cross is high, as do the findings of Terada et al. (2004), who created early-stage cow-human hybrids by injecting human sperm into cattle eggs (83% of the eggs were activated).†

Fernandina, Florida. The *Arizona Sentinel* (Yuma, Az., Nov. 4, 1876, p. 3c; ||3kdejcsy) said, "Last year a cow delivered in the streets of Fernandina a calf, perfect in every particular, except it had a human head. … But how is it to be accounted for that the same cow has this year delivered another monstrosity of the same kind … ?"

Laurel, Indiana. (case #1) From the *Exponent* (Hagerstown, Sep. 8, 1880, p. 3b): "CONNERSVILLE, IND., Sep. 3.—There is on exhibition at Andres' opera house, this city, the mounted skin of a calf with a human head. This interesting curiosity was born near Laurel, two weeks ago, and belongs to a farmer named [James] Carpenter [1813-1896]. The body is about a foot long, perfectly formed, and is supported by four legs, each about ten inches long. The head, which very much resembles a well-formed human head, has two perfect eyes, ears, and chin, with distinct traces of nose and mouth. The entire body is covered with fine hair." The *Cincinnati Enquirer* (Aug. 20, 1880, p. 2g; ||2s34r48r) said this creature was born Aug. 18 and had a tail on its back. (case #2) The *Daily Examiner* (Connersville, Mar. 21, 1890, p. 2b): "A small animal resembling a calf but having a head like that of a child so that if the skull alone was seen it would be taken for that of an infant, made its advent at the little town of Laurel, a few days ago, but died a few minutes after birth. It is almost completely enveloped in a soft reddish hair, closely akin to that of a calf. Its mother is a young heifer. It is about fourteen inches high and is now in the possession of Dr. Joshua Chitwood [d. 1903, at Connersville] of this city."

Trzcianka, Poland. Rzaczynski (1721, p. 358) said twin calves with human heads were shown in 1709 at Trzcianka.

Würzburg, Bavaria. Fincelius (1566b) recorded the birth of twin boshominids, calves with human faces, at Würzburg in 1550.

Hanover. Twin cow-humans were reported, too, at Hanover (see p. 49).

†. Researchers also produced cow-human hybrids, by inserting human chromosomes into a cattle genome, but terminated development during its early stages (more information: ||y69yuwdm). Sánchez-Calabuig et al. (2015) produced cattle-dolphin embryos via IVF.

The Juan Bautista Alberdi boshominid (||4d4jd3ma).

20TH-CENTURY REPORTS

Juan Bautista Alberdi, Tucumán, Argentina. Caras y Caretas (Buenos Aires, Sep. 21, 1901, 4, no. 155, p. 42; ||4d4jd3ma) said, in translation, "A surprising monster has been sent to the Tucumán Health Department from Alberdi: a calf with a human head. A butcher found the freak in a cow and took it to Police Commissioner D'Angeli, who sent it on to the Health Department. The monster is the size of a four- or five-month-old piglet. Its body is no different from that of other calves, except that the skin is very pink. The head is wondrous and surprising. According to those who have seen it, the cranial region, the eyebrows, eyes and cheekbones, are like a human's. Some of its features are like a calf's, but the rest is more like a human being."

Vertaizon, Puy-de-Dôme, France. L'Écho d'Alger (Algiers, May 30, 1951, p. 8e; ||32kxx2ru) said, in translation, "A freak calf has been born on a farm in the commune of Regnat, Vertaizon (Puy-de-Dôme). The animal has one hind leg that ends in a human foot, the other is a stump, ending with the thigh. It nurses normally and has no trouble keeping itself upright" (cf. the Recow HOMINEQUID, p. 209).

Saint-Priest-sous-Aixe, Haute-Vienne, France. Le Populaire du Centre (Limoges, Aug. 10, 1949, p. 2; ||4z89rbtx) said, in translation,

> A farmer at Vaux, a commune of Saint-Priest-sous-Aixe (Haute-Vienne), Mr. Ménieux Martial, has a freak calf.
>
> Indeed, the head of the animal resembles that of an ape [which resembles that of a human].
>
> Moreover, one of the beast's hind legs is much shorter than the others and ends in a club foot.

Lima, Peru. La Opinión (Los Angeles, Feb. 16, 1940, p. 3b; ||mr2n5rxw) said a Lima newspaper, *La Crónica,* had reported the previous year (1939) the calving of boshominid. In translation, the report reads, "We speak of the calf with the head and arms of a human being, that cried like a baby when it was born. The glory of this tale falls to Colombia and, according to the testimony of witnesses incapable of falsehood, this calf had all the features of an adult human being, bushy, well-delineated eyebrows, big eyes, thick eyelashes, thick lips, a black mustache, a flat nose, and the most horrible thing was that it had the fingers and arms of a human. Almost a centaur, without exaggeration, like the Greek sphinxes, with a star on its forehead. When this animal wanted to walk it did so on its hind legs."

Malang, Dutch East Indies. De Indische Courant (Malang, Java, Mar. 31, 1933, p. 2e; ||yaj7tzgu) said, in translation, "Yesterday a calf with a human head was found in a cow at the municipal slaughterhouse. It had a cleft lip, and the eyes were far apart."[10]

Mazières, Indre-et-Loire, France. L'Avenir de Touraine (Aug. 29, 1931, p. 3c; ||yc7vekd7) said, in translation, "A cow belonging to Mr. Raymault, a farmer in Mazières (Indre-et-Loire), just gave birth to a true freak. The calf, which was stillborn, had the head of a gorilla."

* In this book, when animals are described as having ape-like body parts but were born in locales where apes are absent, it is assumed that the resemblance was to humans instead.

Waycross, Georgia. From the *Birmingham News* (Birmingham, Al., Nov. 19, 1927, p. 3g; ||56rjr6hb): "WAYCROSS. Ga.—A calf with feet shaped like those of a human has been found in lower Charlton County by T. W. French, while investigating a truancy case in that section. The feet are elongated with well-formed toes and a heel. In other respects the calf is normal."

Douzat, Charente, France. La Charente (Angoulême, Jan. 18, 1926, p. 3b; ||4bj29hyp) said, in translation, "Saturday morning at Douzat, a cow belonging to a cattle merchant gave birth to a freak calf that had a body that has much in common with the body of a human being and which died a short time later. The body of the little monster was examined by a veterinarian."

Coney Island, New York. Joseph Lotterman, an *Evening World* (New York City, Aug. 18, 1922, p. 10f; ||pva3uy96) reporter, said he saw at Coney Island "a cow with a woman's face."

Hoste, France. Le Patriote Lorrain (Metz, Nov. 5, 1922, p. 3c; ||bdhjuryd) reported an anophthalmic calf with feet like a human's, calved in the village of Hoste near Puttelange-aux-Lacs, France. Its owner killed and buried it.

Chicago. La Crónica (San Francisco, Nov. 12, 1916, p. 14b; ||mpfr2km) said, in translation, "CALF WITH A HUMAN BODY.—A shocking event has occurred in one of the Chicago slaughterhouses. ... In a slaughtered cow, a fetus was found having the head, back and shoulders in all ways like a human being."

Auburn, Maine. The *Evening Journal* (Lewiston, Me., May 12, 1915, p. 12; ||y9uwua9j) reported on specimens stuffed by Auburn taxidermist, Edmond Boucher [1884-1931], including "a calf with a human head. The eyes were close together, eyebrows were present, the nose was ape-like and the jaw perfectly human."

Milton, Oregon. The *Milton Eagle* (quoted in the *Gate City Journal*, Nyssa, Oct. 2, 1913, p. 5; bd9ttvpn) said, "A freak of nature was prematurely born near here last Friday morning. The monstrosity was a cross between a baby and a calf. Its mother was a cow, and Dr. C. W. Thomas, who has had the specimen put in a chemical preservative, says its father was without doubt a human. The creature died soon after birth. In many respects it resembled a baby although it weighed about 60 pounds. Its lower limbs and shoulders, with the head, were much like a human, while its feet resembled those of a calf. The head was as round as a marble and its mouth was small. It had small ears, also."

Châtenois, Bas-Rhin, France. L'Abeille des Vosges (Neufchâteau, May 4, 1913, p. 3c-3d; ||8xp2part) said, in translation, "A FREAK.—Recently, at Mannecourt, a Châtenois hamlet, a cow belonging to M. Louis Flambeau dropped a stillborn calf that is a true freak. The body is formed like that of any other calf, but the back of the head and the ears are those of a human being. It was hydrocephalic."

Ashland, Kentucky. The *Daily Public Ledger* (Maysville, Ky. Jul. 11, 1912, p. 4c; ||yagzwcjw) said, "What is considered one of the greatest freaks of nature ever seen in Cincinnati was brought from Ashland Ky., Wednesday, by Dr. Edward Fanning. The freak has the body of a calf but the face of a human being. The creature was born several days ago and lived but a few hours."

Lunenburg, Vermont. The *St. Johnsbury Caledonian* (Sep. 3, 1913; ||y7xbcwpk) said, "Frank T. Harris [1857-1923] of Lunenburg has in his possession the body of a calf that has a perfect human face, the only thing of its kind ever known to scientists. He will exhibit this freak at the Caledonia County Fair. The calf is Holstein and has a black and white body, but the head and face are in human form, except the ears. The calf weighed about 30 pounds and had no hair, except a slight beard on the face. It is a wonder that everyone will want to see. An affidavit, signed by many well-known people of Lunenburg, states they saw the freak just after it was born and that it is unquestionably genuine. It has been called to the attention of the medical authorities of Harvard, and they state that such a thing was never known before and will revolutionize medicine. The animal has no marks of sex."

Laclede, Missouri. The *Laclede Blade* (Feb. 17, 1911, p. 1a; ||4s6abntv) briefly reported the calving near Laclede of a strange animal with naked skin, except for hair around its ankles and mouth. This description agrees well with that given in many lengthier descriptions of boshominids.

Orange, Texas. The *Palestine Daily Herald* (Sep. 3, 1910; ||hjeml35) said, "Sheriff Davis of Orange has in his possession a calf with a human head." And the *Brownsville Herald* (Oct. 17, 1910, p. 2; ||ybl5nr6m) said, "ORANGE, TEX., Oct. 9.—Among the freaks of nature to be exhibited here during the fair next week will be a calf with a human head. The calf is owned by E. M. Davis, deputy sheriff and jailer, and was born only a few days ago. It has the human head almost perfect, with the natural expression. It is pronounced a wonder by those who have seen it."

* An Edward McFarlane Davis (1851-1937), a jailer, lived in Orange in 1910 (||y3tz9pnu, ||yyp4ajpl).

Sedalia, Missouri. The *Weekly Times* (Butler, Mo., Jan. 9, 1908, p. 10b; ||j5fgp3z) said, "John Griessen, a farmer living near Sedalia, is the owner of a young calf whose head greatly resembles that of a human. The calf was born August 28, 1906, and was Lilliputian in size, its head and face bearing likeness to a human. Its mother was a half-blood Jersey and his sire a registered shorthorn. When strong enough to stand on its feet the calf only weighed 9 pounds and 10 oz. [The average birth weight of calves is ~64 lbs.] It is now over a year old, healthy and active, but has never grown any taller and will never be any larger than an Angora goat. Its little body is covered with a fine silky hair similar to that of a Persian goat, and its legs are as short as a shoat's and as spindling as a deer's. In height the freak is less than 28 inches and weighs 60 pounds."

Monette, Arkansas. The *Log Cabin Democrat* (Conway, Ark., Feb. 28, 1907, p. 3c; ||5n89p7ep) said J. M. Osborn of Monette had an animal with the "form of a calf, with the exception of the head, which was almost a perfect man's head, with long hair, eyes and thick lips. It was two feet long, and the body was in proportion to its length. It lived only a short time after birth."

Santiago, Chile. Le Courrier de Metz (France, Mar. 25, 1904, p. 1c-1d; ||27jcrxwk) said a cow at a place called Paneros, near Santiago, had recently calved a viable composite that was human from the waist up but like a calf from there down.

Goncelin, Isère, France. L'Ouest-Éclair (Rennes, France, Mar. 9, 1909, p. 2a; ||y3h3nwf4) said, in translation,

> GRENOBLE, March 8.—A very odd freak has just been born at Goncelin [about 32 km northeast of Grenoble].
>
> A cow belonging to M. Henri Tissot, a merchant, just gave birth to a monstrous animal with the body of an ordinary calf, but with a head greatly resembling that of a human being, except that it is about one-third larger than that of a human adult.
>
> M. Ruillier, a veterinarian at Pontcharra, when contacted, said he had never before seen anything like it. Given that the monster had been born dead, he asked M. Tissot, who agreed, to allow him to dissect the animal's head in order to report on the case to the [French] national school [of veterinary medicine] at Lyon. The head of this monster weighed 5 kilograms.

San Juan Guichicovi, Oaxaca, Mexico. The *Mexican Herald* (Mexico City, Nov. 28, 1908, p. 6), reporting that a fetus with human features was found at San Juan Guichicovi in a slaughtered cow and buried in a churchyard, said, "The object had all the appearance of a human being both in the eyes, mouth, face and other details of the anatomy … this freak had four ears, two of which were placed near the eyes and the other two in the neck. The tail and hoofs were exactly similar to that of a calf, but in other respects the head and shoulders were in human likeness."

South Fork, Colorado. El Anunciador (Trinidad, Colo., quoted in *El Independiente,* Las Vegas, N.M., May 17, 1906, p. 4b; ||mvnkdy6k) said, in translation,

> A letter sent to us by our friend Francisco Gomez of Alamosa, Colorado, dated May 6, says that in South Fork, Rio Grande County, a calf has been born (they call it a calf because it was calved by a cow). It has the hindlegs and hoofs of a calf. But from the waist up, it has the body, arms, hands, fingernails and face of a human. The freak was stillborn, and the owner of the cow has stripped off its hide and put it on display for all to see. And it's the strangest combination that has ever been seen around here. People are coming from everywhere to marvel at this amazing thing. The cow was purchased a few weeks ago from some Indians from New Mexico. The letter also says that the owner has been offered a thousand pesos for the hide, but he doesn't want to sell.

Leon, Iowa. The *Leon Reporter* (May 2, 1901, p. 11; ||yxm25urr) said, "Dr. Frank Goodson, the well-known veterinary, showed us a decided curiosity last week in the shape of a calf's head which was indeed a monstrosity. The mother was a Short Horn heifer owned by Samuel Overholtzer, of Franklin township, and the calf was like other calves except its head, the front part of which instead of being straight resembled very much a human skull. The calf was born dead, and Dr. Goodson secured the skull as a trophy."

* Leon is in Decatur Co. A Frank Goodson (1851-1917) lived in Decatur Co. in 1900 (||y6979qks). A Samuel Overholtzer (b. 1839) lived there in 1910 (||y44jo57a).

19TH-CENTURY REPORTS

Brive, Corrèze, France. Le Bonhomme Limousin (Limoges, Apr. 23, 1899, p. 2b; ||22hnt92j) reported that a cow at Brive had, a month before, produced a calf with a human face ("un veau à face humaine"). The report quoted a judicial decision saying a local man had been sentenced after he was determined with certainty to be the animal's father.

Preux-au-Bois, Nord, France. L'Écho du Nord (Lille, Nov. 14, 1898, p. 2e; ||bddbun2a) said, in translation, "A heifer belonging to M. Lebrun, a farmer at Preux-au-Bois, canton of Avesnes, gave birth in the abattoir to a freak calf on exhibit at the Café Leclercq, Rue Desurmont. The animal has the torso of a child and a single eye on its right forehead."

Gadsden, Tennessee. The *Alamo Signal* (quoted in the *Savannah Courier* Savannah, Ga., Mar. 25, 1898, p. 2c; ||olpukn7) said, "Some excitement was created in the neighborhood of Gadsden last Thursday when it was reported that a cow had given birth to a calf with a human head and hand. The hand was where the right forefoot should have been, and the ears were unnaturally far back on the head; otherwise, the head was human in every detail. A large crowd viewed the curiosity, which was born at Esq. John Williams' place, one mile west of Gadsden. It was born dead."

* A John Williams (1839-1901) lived in Gadsden in 1898 (||y2ml3eay).

Laurens, Iowa. The *Pocahontas County Sun* (Laurens, Ia., Feb. 24, 1898, p. 8a) said, "Fred Post is now the proud possessor of a freak in the shape of a calf with a human head. The head, neck, and ears are shaped exactly like a human head and the nose bears some resemblance to that of a human nose. The calf is dead and was the property of Carl Skogstrom. Post will embalm it, and the freak will be worth some money. Already it has drawn large crowds."

* Karl Johannesson Skogstrom (d. 1906) lived in Laurens in 1892 (||y5bq7qak).

Philadelphia. The following is excerpted from a news report about a collection of abnormal specimens in Philadelphia (*Los Angeles Herald,* Jan. 6, 1898, p. 10f; ||jynl2md). "The freaks that have been alcoholically preserved are kept in glass jars of various sizes in the back parlor of the dwelling. In one of the largest of these vessels is a calf with a face with features so distinctively human as to almost suggest that it is a human head upon a calf's body. It is an ugly face and at the same time a startling one. The neck, exceedingly narrow, is set upon the body of an ordinary new-born calf, somewhat bloated by nineteen years' soaking in alcohol. Seventeen or eighteen years ago, when "Bill" Long was the proprietor of the museum at Fourth and German streets, this was one of his chief attractions."

Limaville, Ohio. The *Chillicothe Gazette* (Chillicothe, O., Jan. 26, 1897, p. 1; ||bde5xcf7) reported that Heston B. Paxson (1854-1927), a farmer of Limaville, had a heifer that bore a calf with a human head. Its description, excerpted from a much longer report, reads,

> On Thursday of last week, the Jersey heifer gave birth to the most remarkable freak, says Mr. Paxson, that he ever saw, heard talked of, or saw written about. This was nothing less than a human calf, a foetus weighing ten pounds and having a head formed precisely as an infant's, only with ears shaped like those of a calf.
>
> One of the most remarkable peculiarities of the "what-is-it" lay in the fact that it had no hair on its body or head, save marked eyelashes, a plainly visible brown moustache, and a jet-black goatee. The arms and legs, says the writer, are formed precisely like those of a human being, save where the hands and feet should be, there are hoofs, a very satyr for all the world.
>
> The extraordinary creature has attracted the attention of the entire community, and among them two reputable physicians of Limaville, who made a careful examination of the monstrosity with a view of reporting it to a medical journal. The creature was born dead, but the body has been turned over, temporarily at least, to the museum at Mount Union, with a view of having it put in shape for preservation.

Paris, Texas. From the *Houston Daily Post* (Dec. 4, 1895, p. 11g; ||zzr3a47): "PARIS, TEXAS, December 3.—Last night a cow belonging to Mr. W. N. Ellis of Blossom [Texas] gave birth to a calf with a body of normal size and shape, but its neck and head are human. The strange creature lived only a few hours, but Mr. Ellis skinned and stuffed the hide. A great many people here have looked with astonishment at the wonderful freak of nature, half human and half animal."

* If the middle initial is in error, "W. N. Ellis" is likely William Akin Ellis (1868-1936), a farmer who lived in Lamar County near Paris his entire life (||y5vozt6o).

St. Louis. The *Times-Picayune* (New Orleans, Jun. 18, 1894, p. 4e; ||yc747pjb) said, "St. Louis papers mention a curious monstrosity which is now in that city. It is a 2-year-old bull, with a head exactly resembling that of a monkey. There is no indication of horns, and the mouth presents anything but a pleasing appearance. The head is nearly round, and about the size of an ordinary water pail. The eyes, which are of a sky blue, bulge out of their sockets, presenting quite a ghastly appearance. The mouth bears a striking resemblance to a monkey's. There are only nine lower teeth in its head, and the tongue is not much larger than the tongue of a human being."

Beaumont-sur-Sarthe, France. The *Journal d'Indre-et-Loire* (Tours, Mar. 29, 1894, p. 2e-2f; ||yj27bhsv) said a cow near Beaumont-sur-Sarthe had calved a stillborn creature with naked skin (except near the hooves), hind legs that would have forced it to walk upright, a human-like head, and breasts shaped and placed as in a woman.

Sparr, Florida. From the *Morning News* (Savannah, Ga., July 3, 1894, p. 2d; ||2xechub2): "OCALA, FLA., July 2.—A queer freak of nature is reported by C[arl]. L. Harter [b. ca. 1860 in Indiana] of Sparr. He says there was born near his place a few days ago a half human calf. From head to navel it is a perfect human being, while the other half is that of a calf. It is still alive and will be brought to the Ocala Zoo for exhibition."

* Sparr is near Ocala.

Abbeville, Louisiana. In the *Medical Brief,* John Bartlett Ramsey (1893), a physician practicing in Abbeville (||ypj2f5h7), replied to another doctor who had asked whether there were any authenticated cases of a human impregnating an animal or vice versa. "In reply to Dr. S. J. Smith's inquiry in the June BRIEF, page 704, I desire to say that I have full notes of a most remarkable hybrid I dissected. The whole body and limbs were covered with hair, cow hair; round smooth head; tongue of man and some teeth; round neck; no dewlap; genitals resembling those of woman; no legs below knees. If desired I can give full notes and circumstances."

* In a later issue, Ramsey (1895, p. 48) said he had gone blind and had therefore been unable to supply his notes. He also said two other, similar cases were reported to him.

Eckerty, Indiana. The *Indiana State Sentinel* (Indianapolis, May 3, 1893, p. 8c; ||yysrbuwg) reported a viable boshominid.

> ECKERTY, April 30.—[Special.]—Of the monstrosities which have recently been recorded, Crawford County must stand well into the foreground. One was born last Friday night of a three-year-old heifer—her first production.
>
> The curious story had gone abroad, but unwilling to believe it, the SENTINEL correspondent waited till Mr. Lesson, the owner, invited him to witness the monstrosity. The calf, or what the owner chooses to call it, is almost human from the loins forward, including tolerable shapely hands and fairly interesting countenance, but from the loins backward the creature is a perfect bovine. Though more than two days old now, it will weigh perhaps sixty pounds. Owing to its peculiar shape it has failed to stand, and its awkward attempts to sit are an utter failure.

* The Eckerty boshominid was unusual, as it was described as like a human from the waist forward but like a cow from there back.

Shreveport, Louisiana. From the *Fort Worth Gazette* (Mar. 9, 1892, p. 3d; ||y2vqhlgx): "SHREVEPORT, LA., March 8.—There is on exhibition in this city a monstrosity in the shape of a calf with human characteristics. The "what is it?" came into the world Sunday last [i.e., Mar. 6, 1892] on the Freewater place, this parish, and was dead at birth. Several physicians have examined the freak and pronounce it marvelous. Its head is shaped somewhat like that of a [man], and the breast is of perfect human shape. It is sexless, and the fore feet are shorter than the hind ones, lending them a look of arms."

Jackson, Mississippi. From the *Pascagoula Democrat-Star* (Jan. 15, 1892, p. 4d; ||y4f5ucjs): "JACKSON, MISS., Jan. 3.—On the farm of Steve Britten, situated eleven and a half miles from here, was a few days ago born a most wonderful freak of nature, a calf with a well-developed [human] head. Mr. Britten cut off the head and sent it to a friend in Jackson to be placed on exhibition. The receipt of the head by the gentleman so frightened him that he has been prostrated and he is now in a critical condition. The head was secured by a gentleman here and will be placed on exhibition. There is little to distinguish the head from that of a [man], except the ears are remarkably like those of a calf."

Orange City, Iowa. From the *Cincinnati Enquirer* (Jun. 11, 1892, p. 10; ||bdetysmh):

> ORANGE CITY, IOWA, June 10.—While riding through a heavy forest, about three miles west of this place yesterday, Judge Brown, of the County Court, and Sheriff Marlin discovered in a small clearing near the center of the wood a human monstrosity whose existence has been concealed for nearly twenty years by the family.
>
> In a small log cabin, which looked like it had formerly been used for a crib or stable, the two men found securely fastened in one end, penned like a fattening hog, a nondescript monstrosity, which in some respects resembled a man. It has a human face, head and shoulders, but the rest of its body resembled that of a calf. It was covered with a coarse, red bristle. The frame was small and delicate, and the expression on the face was so strangely human that the visitors were perplexed. After the cessation of the shower which had driven them to shelter, they came to town and related the story of their find.
>
> The history of the monstrosity which they discovered in what is known as Barrett's Wood, is recalled by many of the older inhabitants. Twenty years ago John Barrett and his wife lived alone in a large mansion where the cabin now stands. Barrett was a besotted individual whose mad pranks in his drunken sprees were the talk and wonder of the community.
>
> In one of them he tied his wife, who was at the time in a delicate condition, on the back of an unruly steer, and then, turning the animal into a field watched it race and run until it burst a blood vessel in its efforts to free itself from the burden. Soon after his wife gave birth to a singularly misshapen monstrosity and died. The occurrence sobered Barrett. The monstrosity lived. It is more animal than human. It could not talk, walk or show anything human in its makeup. For nearly twenty years Barrett has had it kept in the cabin where it was found. A man is engaged to care for it while it lives. Its existence had been forgotten by the older inhabitants, and it has been concealed from the younger ones.

East Buffalo, New York. The *Buffalo Times* (Dec. 28, 1891, p. 5; ||jp2vp7ex): "An East Buffalo butcher discovered a horrible monstrosity while dressing a cow. It was an unborn calf with a man's head and calf's body. The features were plainly marked but the tongue was that of a calf. There were but two legs, which were those of a calf and the body and legs are covered with hair, while on the head there is but little. The freak is at the Niagara Medical College."

Binghamton, New York. From the *Sullivan County Record* (Jeffersonville, N.Y., Aug. 2, 1889, p. 4; ||y8gntd3j): "A calf with a human head is on exhibition in Binghamton. The monstrosity was born in Florida, is several months old and is apparently in the enjoyment of excellent health. Its forehead is high and receding, its eyes are round, have heavy lashes and are arched with a fine pair of eyebrows. The nose is flat and turned slightly upwards. The lips are thin and close over a set of even teeth. The chin is square and perfectly formed and the cheek bones are high. The face is black, and the lips and nose are hairless."

Silver Creek, Nebraska. The *Silver Creek Times* (in the *Columbus Journal,* Columbus, Neb., Feb. 22, 1888, p. 2e; ||h53le72) said, "One of the most remarkable freaks of nature recorded comes from the farm of Mat Harry, who resides three and one-half miles from this place. It is a calf with a human head, neck and shoulders, being almost as perfect as those of a person. The cow owned by Mr. Harry gave birth to the wonderful animal last week; but the calf did not live."

Boswell's Depot, Virginia. The *Richmond Dispatch* (quoted in the *Cambria Freeman,* Ebensburg, Pa., Sep. 2, 1887, p. 2f; ||y3q2gwgz) said, "A few days since, a gentleman near Boswell's depot, on the Richmond and Allegheny railroad, missed a cow. He searched for her and when he found her, she had given birth to what should have been a calf, and yet, with all other formations of such, from the shoulder out the neck and head of a woman with all the symmetry thereof, and, furthermore, the hair on this strange head was parted or had gracefully fallen back on each side as ladies usually part theirs. The nose, eyes and other features of its face were [as] clearly and beautifully developed as that of a woman's. In a word, from the shoulder out was human in form and the body that of a beast."

Deep Run, Ohio. The *Spirit of Democracy* (Woodsfield, O., Jan. 13, 1885, p. 3; ||h99u8ht) said, "It is said that a man by the name of Doyle, living on Deep Run, Pease Township [Belmont County, Ohio], is the owner of a young heifer which on Wednesday gave birth to a calf that is half human and half animal. The head and shoulders are exactly like that of a man's, while the remainder of the monstrosity is that of an ordinary calf. The heifer had to be shot, and the monstrosity was skinned and stuffed, and will no doubt be placed in some museum."[2]

Glendale, New York. From the *Commercial Advertiser* (Potsdam Junction, N.Y., Mar. 5, 1885, p. 2b; ||l6s9m4l): "A "freak of nature" in the way of a calf with a head similar to that of a human being, has been exhibited by Elmer A. Tiffany, of Glendale, Lewis County. The calf was dead when born. Mr. Tiffany had the animal photographed and then left it with a taxidermist to be set up."

* An Elmer A. Tiffany lived in Martindale near Glendale (||yy5mqb26).

Indianapolis, Indiana. From the *Bloomington Daily Leader* (Mar. 15, 1884, p. 2a):

> INDIANAPOLIS, INDIANA, March 14.—Quite a freak of nature was found here last evening. It is a horrible looking monstrosity, it having the body of a calf and the head of a human being and is on exhibition at Bergman's grocery store on West Morris Street. It comes from the slaughterhouse of [Charles] Gardner and was dead when found. The features are well developed. It is a wonderful curiosity and was viewed yesterday by hundreds of people.

And the *Cincinnati Enquirer* (Mar. 14, 1884, p. 1b; ||y6nnycjt) said,

> The foreman of the Moody cattle-pens, west of this city, reports the finding of a monstrosity—half human, half bovine—in the yards. It is human to the waist. The head is as large as a man's but the back of the head and the body in color and character are like raw pork. There is no hair save near the tip of the tail, which is of ordinary length, and near the hoofs. All other portions are hairless, flabby and soft of touch. The eyes are not in front, but upon either side; the mouth is human, save that the upper lip looks like a harelip. The nose is almost flat and of human shape, while the chin is distinctly human. [A later report in the *Enquirer* (Mar. 29, 1884, p. 4d; ||6be3yy88) said "the chin is covered with coarse hair, giving the appearance of a full beard"]. The forelegs are bow shaped and could never have been used in walking. The hind legs, instead of setting on the side, were as if the animal, had it lived, would have walked upright. They were shaped like a calf, with no hair save near the hoof. There was nothing to indicate the sex, and no means of tracing the mother. It may be added that the body is about two feet in length. The monstrosity was viewed by scores of people today, among them several physicians, and it is regarded as an exceedingly peculiar formation.

Stonewall, Virginia. From the *Inter Ocean* (Chicago, Jul. 8, 1883, p. 11d; ||y4gstvp4): "A calf with a human head was born recently on the farm of Joseph Hiett, near Winchester, Va. Its body was that of an ordinary calf, while its head, mouth, nose, etc., was the shape of that of a human being. Upon being dissected its head was found to be totally without brains, containing nothing but water, and its body a total malformation throughout. It was pronounced to have been alive up to the time of birth, but to have died instantly upon that event."

* A Joseph Thomas Hiett (1839-1907) lived near Winchester in 1880 (||y2ft3jdo).

Prattsville, New York. The *Prattsville News* (quoted in the *Cobleskill Index*, Cobleskill, N.Y., Mar. 20, 1879, p. 3a; ||mhb5ccx) said, "Mr. Charles Lewis, a farmer residing between this village [i.e., Prattsville] and Gilboa [New York], has a cow which recently gave birth to a calf with a human head."

* A Charles Lewis (1847-1927) lived near Gilboa in Schohair Co. (||y3dqzdhf).

Fort Scott, Kansas. The *Dodge City Times* (Aug. 2, 1879, p. 8b; ||jbhot3g): said a cow at Fort Scott bore a calf with "a head shaped like that of a man with the nose where the chin is usually situated in a person, with a mouth above and below the nose, making two mouths in one head."

Garret Mountain, New Jersey. Paterson Guardian (quoted in the *Tri-states Union,* Port Jervis, N.Y., Sep. 13, 1878, p. 3b; ||2p8e4f3k) said, "Perhaps our readers will not think this heading suggests anything extraordinary, as a human calf, as well as a human ass, is quite a common thing. But the freak of nature to which we now refer is not a usual thing. The other day a farmer, living back of Garret Mountain, had an addition to his lacteal family in the shape of a calf. But such a calf! The body was the same as any other juvenile bovine, but the head, as we are informed, was a perfect human head! The monstrosity died shortly after it was born, fortunately. The facts are given us by a man who alleges that he saw the beast."

Salonica, Greece. Bozner Zeitung (Jul. 13, 1876, p. 5b; ||bdf7du3r) said a human-headed fetus had been found in a butchered cow in Salonica.

Shelby, Indiana. The *Shelby Republican* (quoted in the *Evening News,* Jeffersonville, Ind., Nov. 27, 1875, p. 1b) said, "Captain J. S. Searcey's cow dropped a calf recently that would have been a valuable acquisition to Barnum's museum. The body was well developed, though supported by three legs only. The right foreleg is missing, and a well-developed hand, though boneless, hung pendant from the side. The head bore a striking resemblance to a human head, excepting the mouth was located on the side of the face, from which a low, plaintive wail occasionally issued. A well-developed eye, and only one, was near the center of the head. This monstrosity lived two days, constantly attended by Mr. S. or his family, and was visited by all the neighbors and by many strangers."

Schlierbach, Göppingen, Germany. The *Allgemeiner Anzeiger für Rheinland-Westphalen* (Cologne, Dec. 9, 1871, p. 4b; ||pcxm5shd) said, in translation, "HEIDELBERG, Dec. 5.—Last Sunday, in Schlierbach, a cow dropped a 134-pound calf with a human head. The monstrosity was purchased for the anatomical collection here for four florins."

Würzburg, Bavaria. The *Courrier Franco-italien* (Paris, May 13, 1858; ||yxtohsxf) said, in translation, "FROM MUNICH: I have just received from Würzburg a curious piece of information I thought I should communicate. An amazing freak was presented to the Faculty of Medicine there: A peasant has donated a calf born alive with a human head to the school's anatomical collection. The scholars of our city are holding a large meeting to discuss this freak and are expected to offer their explanation soon."[3]

Kleinprießnitz, Thuringia, Germany. The *Großenhayner Unterhaltungs-und Intelligenz-Blatt* (Großenhain, Mar. 18, 1837, p. 1; ||mvz9sun4) said, in translation, "Near Camburg, when a fallen cow had been opened in the village of Kleinprießnitz, a monstrosity was found. Its head was like a man's with a very pointed chin. The shoulders were wide, but the arms were like calf legs. The hind legs extended directly toward the rear, as in a human. This monstrosity can be seen in Camburg, stuffed."

Egypt. In 1847, *Al-Waqa'i' al-Misriyya* said a cow bore a calf with a human head in a village in Upper Egypt (||yaveptux).

Nálepkovo, Slovakia. One rather well attested birth of this kind took place in Slovakia in 1825. In 1827, a report describing a boshominid, born alive, appeared in *Magazin der ausländischen Literatur der gesammten Heilkunde.* The *Magazin* was published by two Hamburg physicians, Nikolaus Heinrich Julius (1783-1862) and Georg Hartog Gerson (1788-1844). The author of the report, David Schreter (1827), an Austro-Hungarian general practitioner, stated that he had examined the subject himself.

> DESCRIPTION OF A MONSTROUS CREATURE CALVED BY A COW AT WAGENDRÜSSEL IN THE HUNGARIAN COUNTY OF ZIPSER
>
> On March 14, 1825, Benjamin Münich, resident of the mountain town of Wagendrüssel [present-day Nálepkovo], bought—at least so he claims—a pregnant cow from one Johann Krall of Stellbach. On the 25th of the following month, in the afternoon, the beast was having difficulty giving birth and both the owner and his wife were assisting. They were appalled when, instead of a normal calf, they were confronted with a monstrosity, which they at once put to death. A local businessman had this strange animal stuffed, and it was also painted by Johann Müller, an artist from the nearby town of Leutschau. Eight or ten days passed before the authorities there in Zipser launched a legal inquiry into the matter.
>
> This deformed creature has a crown-rump length of three feet [~90 cm] and, when placed upon its feet, is about two feet tall. The head is large and looks quite similar to a human being's. From the upper margin of the frontal bone down the face to the chin, it measures ten inches [~25 cm]. The frontal and parietal bones define a fontanelle like that in the skull of an ordinary human infant [a feature never present in the skulls of ordinary calves]. The sagittal suture is one inch long. On its head, from the fontanelle back, it has one-inch-long golden-brown hair. On both sides the ears are like small human ears, but to their lobes are connected three-inch-long calf ears bearing sparse hair at their tips. The face is smooth and hairless, the eyes

a beautiful blue, and the eyebrows a dark brown. The nose is flat, with the nostrils separated by a septum wider than that seen in human beings. The upper jaw, which lacks teeth [cows lack upper incisors], bears an upper lip like a human's; the lower has ten thin, sharp teeth, and is more like a calf's. On the chest are two breasts with well-formed, prominent areolae 2.5 inches in circumference. These mammae are elevated somewhat (about half an inch) above the surrounding surface, as in a young woman. The torso and buttocks are like those of a human being. But the body is longer in proportion to the extremities. There is a naked eight-inch tail, about half an inch in diameter. The genitalia are female. Between the hind legs is an udder, and some of the umbilical cord remains attached. The torso is naked, as is the upper portion of each of the four extremities. But the lower portions are covered with glossy brown hair. Each leg ends in a cloven hoof like that of a cow.

The fact that the birth of this creature took place is attested by the entire municipal authority of Wagendrüssel and nearly all the inhabitants of the town, as well as by the members of the committee empowered by the County of Zipser to investigate the matter.

David Schreter, M.D., general practitioner at Leutschau.[4]

Carnesville, Georgia. A letter from "a gentleman of unquestionable integrity, in Carnesville, Franklin County, Georgia" published in the *Reflector* (Milledgeville, Ga., Jul. 7, 1818, p. 3; ||yyzftfbg):

Sir,—Amongst the many and curious novelties of nature there has been witnessed in the county of Franklin one of the most astonishing. The fact probably would require ocular demonstration before we could be led to belief of its truth, was it not that so many persons of respectability had actually seen the astonishing phenomenon.—The fact stands briefly thus—in the upper end of this county, was found dead, an animal which in part represented the human species, and in part that of a brute. When it was found, there was a cow (which was the property of Hugh Greenwood[†]) standing over the dead body. The cow had been expected to have calved, and in searching for her she was found, standing in the above position, and from all appearances

> she had been delivered of the dead body. Those parts which represented the human body most strikingly were its arms, which came out in a manner nearly representing that of human, except the part that should have formed the hand, was cloven, its head was round, and the ears represented nearly those of a rabbit. The chin was prominent; the nose was flat ... Its teeth upon examining the mouth were found to be well set above and below, those of the former were about an inch in length. The body was as nearly representing the human as any other part, except that it was covered over with red hair. The legs still more striking except that of the hoofs—the dead body had the appearance of the male kind. The cow remained in the presence of her lost young and it appeared that all exertions to drive her from the place would be fruitless, and in fact the dead body had to be drawn home before she would abandon the place. Those who have seen it have been struck with silence with regard to the cause—many persons were called upon to witness the phenomenon, and as such the truth (though astonishing) can be manifested by a number of affidavits.

* A Hugh Greenwood (1750-1825) was living in Franklin Co. in 1818 (||y4vjcp77).

Chamonix, France. In a letter (dated Geneva, 1806) to a friend in Paris (quoted by Jouard 1807, p. 70), Antoine Joseph Barruel-Beauvert described a boshominid. His account reads, in translation,

> Mr Gosse showed us—and this you'll hardly believe—a true minotaur he claimed to be the male offspring of a heifer and Chamoniard peasant. Mr Gosse was quite blasé, as he showed us this remarkable creature, which he held in his lap. M. Torcia [another person present] seemed quite pleased with this monster. And as one, we both exclaimed, the old tale was true! And in fact, this minotaur in the upper parts of its head was formed just as we are, the forehead, crown, occiput and temples, and the protuberant nose—around seven or eight centimeters long—was both osseous and cartilaginous, stood out from the face and ended in a rounded point. The cheeks and the eyes also were formed like ours, as was the neck—who knows whether, in lacking the brain of a beast, it might have had a cognition like our own? It had teeth only in the lower jaw, as in cattle, and had

only two, so that the palate and upper jaw resembled those of an infant still lacking its teeth. Its ears differed from ours and were placed lower than those of other animals. It had four legs, and the long body of a bull, but on the hind legs had calves somewhat like a human being's. The body from the crown of the head to the end of the back, was about two-and-a-half-feet long (80 cm)—very large for a stillborn infant, and quite small for a calf. The tail of this minotaur had nothing in common with the tails of other beasts, it being little more than a prolongation of the spine. It was just a piece of flesh, as long as your little finger, and which must have been limp. It was hairless, as was the rest of the body, covered with smooth, dry skin. This animal male and the scrotum, quite voluminous. The feet are cloven and end in hooves like those of his mother, who in this respect dominated over the sire's traits.

Albany, New York. Another account (Yates 1808) appeared in a journal edited by John Redman Coxe (1773-1864), a professor at the University of Pennsylvania School of Medicine. Christopher Yates, the report's author, was a physician practicing in Albany, New York. His report was passed on to Coxe by another professor at the university, James Woodhouse (1770-1809).

ACCOUNT OF A MONSTER BY DR. YATES
COMMUNICATED IN A LETTER TO PROFESSOR WOODHOUSE

Albany, June 15, 1807.

Dear Sir,

I have it in my power to communicate something new and extraordinary to you.

A very extraordinary animal was produced by a three-year-old heifer, near the Unadilla River, in the western district of this state. On the 26 of last April, in the afternoon, the heifer exhibited symptoms of great distress, which increased, and urged an almost constant bellowing; this, together, with the great agony she appeared to be in from her conduct, induced the owner to shut her up in the stable under the impression that "she was getting mad." Her bellowing continued till near daybreak; the owner supposing her dead, got out of bed and went with a light to the stable, when to his surprise he found her licking what he supposed was her calf; on a nearer approach he discovered its form different from what he expected, and attempted to take it up by the legs, when the heifer darted at him with violence, struck her hoof against the monster's head and broke it; he however took it home, but is not positive whether it shewed signs of life, he thinks it did; he says he did not suspect the heifer to be with calf.

The owner of the heifer sold it to two countrymen, who brought it to this city and immediately procured a five-gallon jar, put it in and filled it with spirits. I prevailed on the man to take it out and permit me to examine it for a few moments; he gratified me, and I shall endeavour to give you an outline of this monster.

From the forehead over to the back of the neck, it very much resembles a child; indeed, the head from a back view would be mistaken for that of a child. The nose resembles about as much that of a [man] as of a calf, I think rather more. The eyes (or rather sockets without eye-balls) are situated about three inches from each side of the upper part of the nose, in the side of the head, on a line with the os nasi.

It has no distinct upper jaw; a few loose bones are felt immediately on the introduction of the finger into the mouth, under the nostrils.

The under jaw extends about one inch beyond a line with the tip of the nose, and measures in all about two and a-half or three inches and is shaped like a harpoon (thus >). It has no tongue. The mouth inside, apparently rough, but feels smooth. Ears much like those of a calf, though small in proportion and placed unnaturally back, near, or on the neck; they are about two inches long, and one and a half broad.

The arms or forelegs resemble the human; there is one joint more; it measures from the shoulder to the elbow about four inches; from the elbow to the next joint (which is half way between the elbow and hoofs) three inches; from thence three inches to the wrist; in this division are two bones as in the human frame; from here the hoofs or webbed fingers begin to extend themselves, dividing into five parts, distinguishing thumb and fingers, having at the extremity small marks or spots where the nails should have been, the substance of the hand or hoof the same as of the calf. From below the under jaw or chin down to the pubes, human; breast broad, abdomen full.

Immediately below the pubes, are two appendages like the teats of a cow (and about as far apart from each other) or, from their flaccidity I might more properly compare each to a separate infant scrotum; the one is as large again as the other.

No organs of generation—has an anus situated properly; and a small tail about three or four inches above the anus up the back, about the thickness of a crow's quill, and an inch and a-half or two inches long.

From the haunches down to the knees apparently human, no knee pan, in other respects, down to the hoofs or toes, similar to the arms and hands.

Its form evinces that its natural posture in walking would have been erect![5] The skin appears exactly like the human, and the colour a darkish yellow. No hairs, excepting a little on the arms and legs, of a reddish cast and hardly perceptible. At the first glance everyone is impressed with its resemblance to a … child, and many suppose it the issue of a man with the heifer. I hope it will go to your city.

Yours &c,
CHRIS. C. YATES.

Dr. John Redman Coxe [Editor of the journal]

Coshocton, Ohio. The *Gettysburg Compiler* (Gettysburg, Pa., Oct. 19, 1825, p. 2e; ||2p87t573) said, "A letter from a gentleman at Coshocton, O. to his friend in Washington, dated September 15th, 1825, furnishes the following extraordinary account: "On Tuesday last, a monstrous calf was suspended on a tree about ten miles from this place, on the great road. Its head, shoulders and part of its arms or forelegs resembled the human species, except that its ears were rather high upon its head, the nose flat, and the eyes rather in the side of the head than the front; some parts hairy, others bare like a newborn infant; its fore feet had hoofs like a calf. The cow that had this calf belonged to a person who lives near the place where it was suspended."

Bull-man (Tell Brak, Akkadian, ca. 2350-2150 B.C.). This depiction shows an animal with the typical form of cow-calved boshominids.

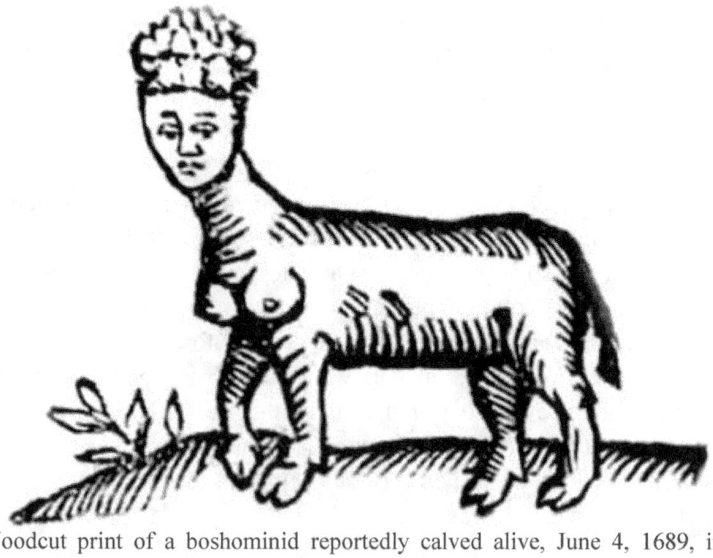

Woodcut print of a boshominid reportedly calved alive, June 4, 1689, in Goselitz, a village near Hanau, Germany (Anonymous 1689). The growth atop the head may represent exencephaly.

The Nemyriv boshominid

18TH-CENTURY REPORTS

Nemyriv, Ukraine. Rozier and Mongez (1784) pictured a boshominid (their illustration is reproduced above). Their report, in translation, reads, "An important event for the study of natural history has taken place in Nemyriv. There, on September 23rd of this year [1783], a Jew slaughtered a cow in her seventh month of pregnancy. A male fetus of singular form was extracted alive but died minutes later. Its torso (Plate 2, July 1784 issue) is that of a naked calf with dark red skin. The hooves are unusually large, and the skin above the hooves, covered with short, light brown hair. The upper part of the head, completely bald, is just like a human's. But instead of a nose there is a kind of snout on the forehead [i.e., a frontal proboscis[14]]. Without bone or tendon, it is three inches long by 1.5 inches thick. Short hair rings a one-third-inch opening at its end. The eyes lie below the base of the snout opening in a one-inch-wide opening. They are rather deeply embedded and have lashes like a human's. On either side, above the eyes, are five long hairs serving as eyebrows. The rest of the face is just like a human's. The cheeks are round and covered with a smooth skin. The large upper lip bears a thick, curly mustache. The tongue tip is split to a depth of one-quarter inch. The upper jaw is cartilaginous; the lower has four incisors. The chin is rather large and covered with a bushy, light brown beard. The creature is 26 1/4 inches long; shoulder height 12 1/2 inches; 15 inches from the ends of the hind legs to the top of the spine. It weighs 23 1/2 pounds. Count Vincent de Potoski ordered it preserved in alcohol. It is now in the court apothecary.

* The brief description of an animal calved at Deutelhausen, Bavaria (*Tiroler Grenzbote*, May 13, 1914, p. 4a; ||yw77y9hy) seems to connect it with the Nemyriv boshominid. It had a frontal proboscis and a naked face. The *Montreal Star* (Aug. 24, 1900, p. 6c; ||5n83uxy7) reported a cyclops calf with a frontal proboscis, so, perhaps a boshominid.

Clayworth, Nottinghamshire. In the *Gentleman's Magazine*, an account (Anonymous 1752) entitled "Extract of a Letter from a Clergyman at Clayworth in Nottinghamshire" reads, "I should have wrote sooner but that I wanted to be satisfy'd in the truth of a report of a monstrous production in a neighbouring village. The animal in question is the offspring of a cow, is about the size of a child of 10 years old, and formed in all respects like a human creature, except the ears and hoofs, the latter of which are cloven, and the ears resembling a calf's, are covered with a kind of down. But no hair appears in any part, except for about three inches above the hoof on each foot or hand (which you will please to call it) and on the upper lip, like a Spanish mustacho. The face is much like that of an old man, the chest perfectly resembles that of a woman, to which sex 'tis said that it has also a very distinguishing analogy; but I was not very curious in that part of the scrutiny; the skin is soft, smooth and of a complexion at least equal to a French foot soldier after a summer's campaign."

* The author was likely Rev. James Carrington (1718-1794), then the Rector of Clayworth (||pavk74zu, ||y5xpz8l4).

Grimma, Germany. Quellmaltz (1736) reported a boshominid calved at Nimbschen Monastery near Grimma, April 26, 1736, a premature fetus, "The forehead, eyes, nose, cheeks, lips and mouth were like a human in all ways except that the skin was hairy" (translation mine).

Samogitia, Lithuania. Michalowski (1857, p. 188), mentions that in 1788, "in Samogitia, a calf with a human head was born."

Ufton Nervet, Berkshire. From the *Derby Mercury* (Mar. 12, 1740, p. 1; ||39hvu67x): "We hear from Round Oak in the Parish of Ufton, that a cow had lately calv'd there, which besides its Calf brought forth a Monster that had a Face like an human Creature, with an Hair [*sic*] Lip, a Tooth in the Mouth, one Ear, and one Arm, but no Legs; the lower Part of the body being like a Calf, and hairy: Which is now in the possession of Thomas Bushnell, at Sullamstead near Theale."

Boston. Cotton Mather (1663-1728) reported a human-faced calf born near Boston in 1716 (Kittredge 1916, p. 37).

Lviv, Ukraine. The *Breslauische Sammlungen* (1724, p. 566; ||yyx7zozx) said a cow near Lviv calved a half man and half calf in May 1723.

* *Breslauische Sammlungen* is the short title of what was perhaps the most important German-language scientific journal of the early eighteenth century (||ya4h3769).

One of the Hanover boshominids (Schmuck 1683).

EARLY REPORTS

> *There are cases of men—if they may be called men—uniting with female beasts, with cows, among others, to produce not human progeny, but vile monsters beyond human imagination.*
>
> —JOACHIM BECHER, *Physicae subterraneae,* 1669†

Hanover, Germany. Friedrich Wilhelm Schmuck (1683) pictured and described two boshominids that he said were calved June 27, 1682, by the same cow, in the principality of Hanover. He said one looked as if it had a turban atop its head (exencephaly?) and a mark on its side like a saber. The other's head reportedly looked like a woman's (Schmuck depiction of the latter nondescript appears above).

Bohemia. A description (Anonymous 1696, p. 4) of a boshominid reportedly calved in Bohemia on April 7, 1689, recalls the boshominid calved in June of that same year at Goselitz (which is pictured on p. 46).

†. Volume II, p. 242 (||me7myhv). Original Latin: "E contra exempla sunt, viros (si homines vocandi sint) cum brutis fœmellis, et inter reliquas cum vacca coiisse, quae non hominem, sed turpe monstrum in lucem edidit, quod phantasia humana defuerit."

Lanz, Germany. Bekmann and Bekmann (1751, p. 884) said, in translation, "In 1691 at Lanz in Prignitz, a dead calf was taken out of a cow. It had a huge head that in front was more like a man's than a calf's. The eyes were also big and at least a handspan apart. There were no ears except for some long hair that grew in their place, longer than elsewhere on the head, which had no neck and was wholly deformed."

Drebkau, Germany. Bekmann and Bekmann (1751, p. 884) said a calf "with many human parts" was born January 24, 1673, at Drebkau.[6]

Atzpach, Germany. In translation, Franci (1673, p. 91) said that on June 27, 1673, in the village of Atzpach near Giessen, a cow was slaughtered "in which a gruesome fetus half human and half calf was found. The lower part of the mouth was right for a calf but pointed and hairy. Its tongue, long, like that of a calf, hung down. The upper jaw was wide, and a harelip took the place of a nose. The eyes stood well back on the face. The ears were a calf's, red and hairy. But the upper part of the head was rounded, as in a human. And the neck could scarcely have been more like a man's, and the shoulders and back, as well, and really, all its body, except that it had four teats like a heifer's. The legs in their upper parts were like a human, but the lower, like a calf, hairy and hoofed. It lived for two and a half hours and shrieked all the while like a child. It was so well developed that the doctors, who said it had been conceived out of coitus with a man, said it would have been born in a few days."

Kalkaboda, Sweden. In his book, *On Meteors*,[7] Joan Petri Klint (translated in Rosen 1994) told of a boshominid born in March 1588. "Near Örebro in the parish of Svensta, in a croft called Kalkaboda just outside Stenkulla, a cow gave birth to a calf with the appearance of a human head and human feet, but without forelegs. The calf seemed to be a half-man, half-calf. He had big eyes, stretched out his tongue and wanted to get at the people who were watching him; he rose bellowing and off he went, but was slain."

Madagascar. Urbain Souchu de Rennefort (1688, p. 78) claimed that while he was in Madagascar (ca. 1665) a cow produced "a monster half-human and half-ox." He went on to say that "In that country such creatures are not very extraordinary but are regarded as evil omens for anyone living where they happen to be born" (my translation).

Ox-eyed Girls. Bierling (1679, p. 40 obs. 19) reported seeing in 1659 at Halle, Saxony, a deaf and blind two-year-old girl whose left eye "in truth looked like the eye of a calf" ("vero oculum vitulinum repraesentabat"). Christianus Helwich (1700), Stadtphysicus at Wrocław, said he saw a girl with two calf-like eyes.

* In ancient Greece, did ocular boshominids, resembling those reported by Bierling and Helwich, perhaps inspire Hera's frequent epithet *βοῶπις* ("ox-eyed")?

Horned Humans. In his *De gula, ventriculo, intestinis tractatus,* published first in 1618, Hieronymus Fabricius (1738, p. 137b) reported dissecting a Paduan noble who chewed cud and whose father had a "horn on his forehead as thick his little finger, and as long as a Spanish olive and made of a substance much like horn" (my translation). He concluded such individuals might have some hereditary connection with horned animals.[8] Bartholin (1661, p. 44), too, reported seeing a 60-year-old nun in Padua who had a small horn on her forehead, "the same as cow horn in hardness, flexibility and color" (my translation). Amatus Lusitanus (1567, 1: 51) reported seeing a boy with a congenital horn (no locus in quo). Ferrara physician Giuseppe Lanzoni (1697) said a woman gave birth, on July 29, 1693, to a "boy" with a cleft lip and a horn in the middle of his forehead. Zacutus Lusitanus (1637, 3: 93), another physician, reported another such, who supposedly reached maturity and became well read in philosophy. He lived in Portugal's Alentejo region, on an estate called Villa Amoena.

HAIRLESS CALVES

Since hairlessness is a human trait, hairless offspring of cattle are boshominids. Those listed here were all reportedly viable.

1885	Pawnee City, Nebraska	‖yn2d2df8
1902	Preble (now defunct), Brown County, Wisconsin	‖v8xkyfsk
1906	Hartford, Wisconsin	‖phjrw2pc
1908	Seaford, Delaware	‖59x8s7h9
1916	Sparta, Tennessee	‖4x6mph37)
1922	Brebotte, Territoire de Belfort, France	‖rmr4unhp
1924	Green Bay, Wisconsin	‖57hn5rhr
1926	Neuville-sous-Montreuil, Pas-de-Calais, France	‖yebyfxr4
1932	Nedde, Haute-Vienne, France	‖yzhee8y7
1936	La Bérangerie, Loire-Atlantique, France	‖5n7hbck5
1940	East Lampeter Township, Pennsylvania	‖2n9h98ju
1953	Amelia, Holt County, Nebraska	‖35mwkmxc

See http://www.macroevolution.net/permalink7878.html for the sources corresponding to the TinyURL keys in this table.

A handbill depicting the Bamberg boshominid, published in the reported year of its birth, 1556. In translation the text read, "An appalling birth, that of a calf, took place July 24th at Kleisdorf, on the Itz, half of a mile from Ebern and three miles from Bamberg. It had four calf's feet, a large human head, a black beard, two small human ears, as well as the breasts and back of a human being. Its tail was like a dog's, but naked. It did not live long. The cow belonged to a lady residing there at Kleisdorf. [speculation omitted.] Printed at Nuremberg by Endres Obermayer." Nuremberg is about 60 km from Bamberg (Wickiana Collection; ||r7pmvrsc).

Bamberg, Bavaria. Jobus Fincelius, the sixteenth-century German humanist and physician, was a professor of philosophy and medicine at the University of Jena. In his book *Wunderzeichen,* he listed events he interpreted as miraculous signs. Among them was a creature with human features he said had been calved by a cow in the village of Kleisdorf, near Bamberg (Fincelius 1566a).

> In the same year of 1556, on the 24th of July, a calf was born in the village of Kleisdorf on the River Itz about three miles [north of] Bamberg. This dreadful creature was fat and had the hooves of an ordinary calf but a large human head and a black beard, as well as ears and breasts like a human being's. The torso, too, was like a human's, but longer, and the naked tail was like a dog's. This monstrosity, which did not live long, was calved on the farm of a noblewoman there at Kleisdorf.[9]

The Bamberg boshominid
(Schenck 1609, p. 104)

Although Fincelius said this creature did not survive, if he was mistaken, or if some other like it was calved near Bamberg at about the same time, it might then explain Philipp Camerarius' account describing what is known in English as the Bamberger cattle-boy.† In translation, Camerarius (1624, vol. 1, ch. 75, p. 343) said,

> In the court of the Prince-Bishop of Bamberg, we wondered at a man who, as he himself would say, was raised among the cattle of the neighboring mountains. He was so swift and agile that those who beheld him were filled with awe. There were many therefore who thought he deceived the eyes with magic. But it seemed to me that this was in no way certain, nor did I think he had the power to do such a thing even if he had so desired. But the most surprising thing about him was that he accomplished these displays of agility, not standing erect, but on all fours.
>
> There in his court, the Prince-Bishop kept a dwarf named Marinette, who mounted this agile being as if he were a horse, turned him in circles and rode him about, drilling him in various ways, though, in truth if he wanted, he could buck off his rider with a single leap, no matter how hard he might try to hold on.

†. Camerarius' report prompted Linnaeus to list "*Iuvenis bovinus bambergensis*" in the *Systema naturae* (1758) as a specimen of *Homo sapiens ferus* ("wild man"). Later authors construed the cattle-boy as an example of a feral child.

Afterward this quadruped would pick a fight with purebred mastiffs—some of which were extremely ferocious—and then, with his doglike barking and growling, and his dangling hair, drive them from the room. Sometimes they would chase after him and try to seize him with their teeth. But this quadruped, with amazing leaps, would hop all around the room, as even a monkey could scarcely have done, as he was from the country and in such good physical condition. Seemingly he did this without the least difficulty.

Once, while I was lunching with the Prince-Bishop, after the quadruped had shaken off his dwarf rider and shagged the dogs from the room, I saw him leap from behind over the head of one of the guests onto the table, without disturbing the cups and dishes, and then leap onward to the upper parts of the room, with such speed that … he seemed to fly. He would run around high up in the roof beams as if he were a cat, and it were the easiest and most ordinary thing to do. Moreover, he performed various other tricks with his agility, so that now he is spoken of everywhere as an unprecedented phenomenon.

Bitterfeld, Germany. Peucer (1553, pp. 326-327) said, in translation, "In 1547, a calf found in a field near Bitterfeld, had the eyes, nose and ears of a human. The crown of its skull appeared shorn and was bright red.† It had a calf's mouth and chest. The forelegs were a calf's, the hind legs, like a human's and rather short. But each hind leg ended in hooves, over which human toes hung down."[10]

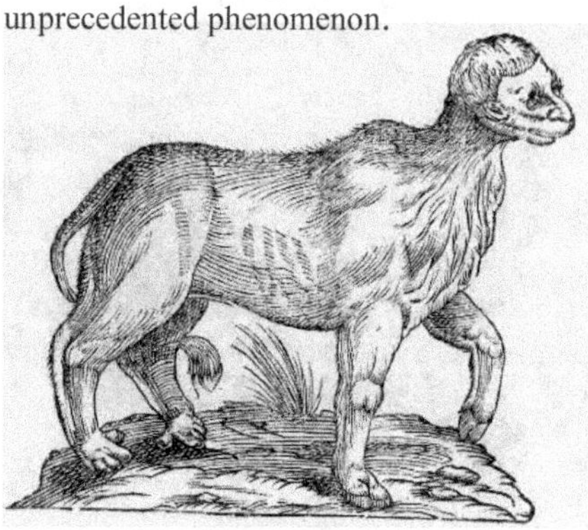

The Bitterfeld boshominid (Aldrovandi 1642)

Berge, Osnabrück, Germany. A Wickiana report (||4awb447u) says a cow at Berge calved a creature with human foreparts in 1578.

* Also see Franck et al. (1585, p. 1038) and Irenäus (1584; *sub anno* 1578).

Muri, Switzerland. A Wickiana report (||3bcmva54) says a cow at Muri calved a boshominid in December 1571.

Brunnen, Switzerland. A Wickiana report (||2yw2k522) says a butcher at Brunnen found a fetus with a human head in a cow, January 11, 1554.

Abbey of Corvey. A chronicle records a cow calving an infant with the legs of a calf at Corvey in 1462 (Leibniz 1710, p. 318).

†. Bloody? Peucer says "vertice tonso & miniato."

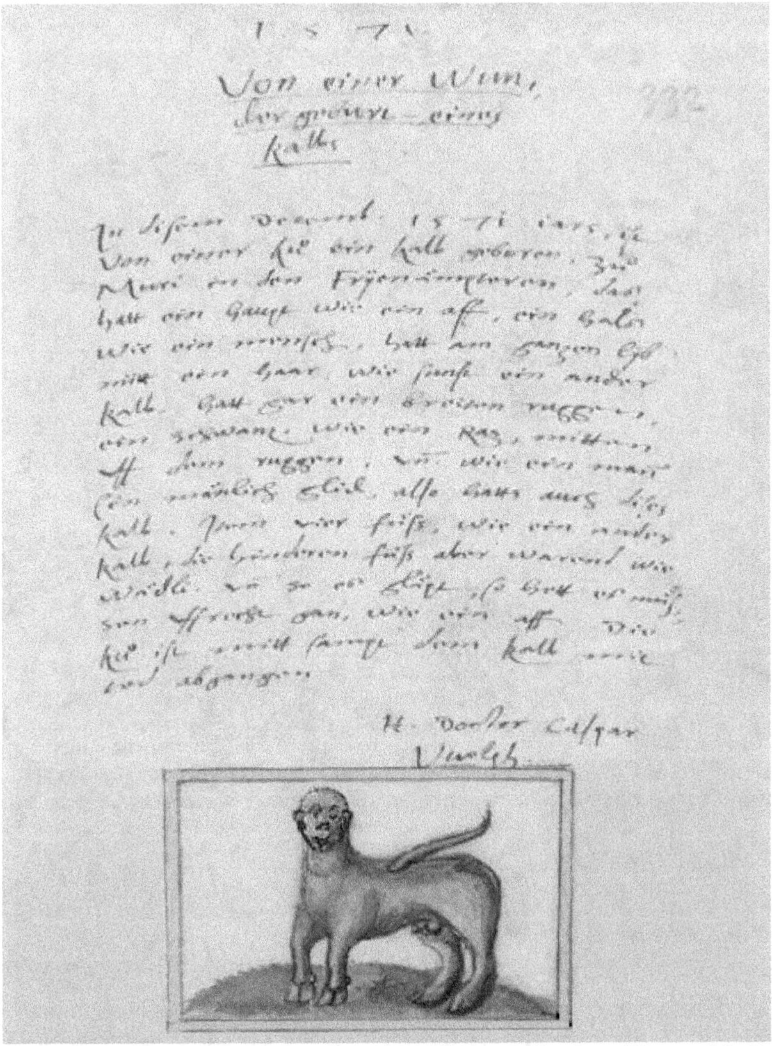

A letter about the birth of a boshominid (Wickiana Collection; ||3bcmva54). In translation, it reads “In this December of 1571 a cow at Muri in Freiamt [in the Swiss Canton of Aaagau] had a calf that had a head like an ape, a neck like a human being’s, calf’s hair elsewhere on its body, a broad back, and a tail like a rat’s in the middle of its back. It also had genitals like a man’s. … Both cow and calf died.” In this book, when animals are described as having ape-like body parts but were born in locales where apes are absent, it is assumed that the resemblance was to humans instead.

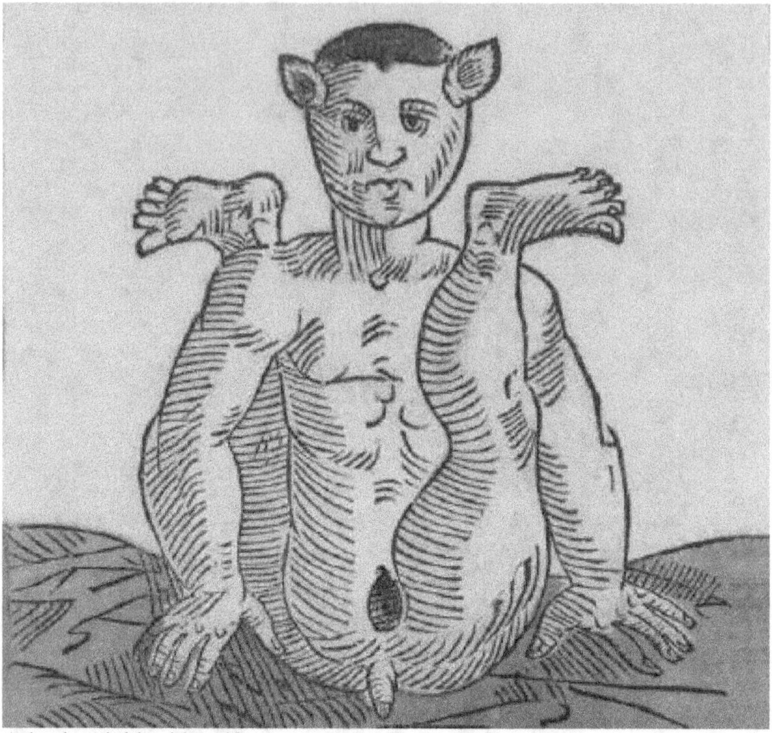

A boshominid with calf's ears reportedly borne by a woman near Augsburg, Dec. 18, 1561 (Wickiana Collection; ||ms9tk5ub). The illustration suggests it was intersex. Bovine features were limited to the upper parts of the head.

Boshominids Borne by Women

> *Some born of a cow have the foreparts of a man; others, on the contrary, spring up begotten of a woman but with the head of a cow.*
>
> —Empedocles, *Fragments*

Though less frequent than reports about cows birthing boshominids, reports do exist about such creatures being borne by women. Such accounts seem to describe offspring that differ from those calved by cattle, in that they supposedly have more human-like bodies, and more cow-like faces, than those born to bovine mothers. Some reports also describe these woman-mothered boshominids as walking erect. Such accounts, then, tend to agree with the ancient reports about Queen Pasiphae of Crete giving birth to the Minotaur. The next report, describing a boshominid with the face of a calf, borne by a woman, recalls the ancient case—but it was published in 1905.

Louisville, Kentucky. Excerpted from the *Fairmont West Virginian* (Nov. 22, 1905, p. 4e; ||ybuefq5a): "There is the skin, stuffed so to speak, in the museum of the Louisville Medical College, of a monstrosity born in Kentucky. It has the face of a calf, with hoof-like hands, small horns and a tail. It was dead when born."

New York City. Robert B. Talbot (1888, p. 459), a physician practicing in Manhattan, read a paper in which he briefly referred to a woman whose "child was born with a head closely resembling a bull's."

Tavistock, Devon. In *The Lancet,* John T. Hislop (1890), of Tavistock, reported going in early October 1890 to attend a woman who bore a stillborn boshominid. "The child's head was the extraordinary part," he said, "and exactly resembled a miniature cow's head. The occipital bone was entirely absent, the parietal bones only slightly developed; the eyes were placed on the top of the frontal bone, which was quite flat, and each of the superior angles of which was twisted into a rudimentary horn. The striking resemblance to, or, more correctly speaking, almost exact imitation of, a cow's head was unmistakable."

Le Nayrac, Aveyron, France. L'Express du Midi (Toulouse, Jan. 15, 1899, p. 3b; ||5n7e6n7p) said a woman in the village of Le Nayrac had recently given birth to a child with the muzzle and ears of a calf.

Perpignan, France. L'Indépendant des Basses-Pyrénées (Pau, Sep. 7, 1892, p. 3c; ||3v2hektn) said, in translation, "It is reported from Perpignan that a triple and curious freak of nature has occurred in the Saint-Mathieu quarter. A woman there just gave birth to three children, of which one has the head of an ape, another, the head of a calf, and the third, two mouths. Despite these deformities the three children are doing well."

Pebble Township, Ohio. The following is a report about a fully viable minotaur that survived for several decades in Ohio (*Chillicothe Leader* quoted in the *Cairo Bulletin,* May 28, 1884, p. 24b; ||y62lwlgf).

> Marvelous as are the "freaks" gathered together by the show people, they haven't anything in the way of a human curiosity to compare with the remarkable being confined in the county infirmary of Pike County. The creature is a man—but his right to that title rests solely upon the fact that he was born of a woman. In every other respect he differs from his kind and possesses no attribute in common with ordinary representatives of his sex. He bears upon his person, in his habits, and vocal organs all the characteristics of a bull in as nearly a perfect form as it is possible for a two-legged creature to possess them.

This remarkable being is named John Haines and is 40 years old. He is the son of very respectable parents, who at the time of the monster's birth lived in Pebble township [in Pike County]. Twelve years ago, the mother who gave birth to the unnatural object died, but his father is still living.[11] Prior to the mother's death John was watched and cared for at home but since her demise he has been confined at the infirmary. The existence of the monstrosity is known to comparatively few people, and many of those who do know of his being have never seen him, being deterred from visiting his cell because of the terrible sight that would meet their eyes.

Haines, or "John," as he is called by everybody who has occasion to speak about him, has been confined at the infirmary twelve years. He is of medium height, with a very large head, and the forehead broad and bold, with a strongly marked ridge running down the center of it. The sides of the head are flat, running back almost at right angles with a forehead, while the top of the head slopes backward and downward at a sharp incline, leaving the cranium without brain room. His face is dark skinned, heavy and brutish in expression, and very repulsive. The eyes are like those of a bovine and roll about in his head in an animalish sort of way. A heavy mouth, in which a restless tongue is almost constantly rolling a quid [of tobacco], strengthens the creature's likeness to a bull. His large ears stand out from his head like those of an alarmed beast.

The long, thick, and bushy head is covered by a close growth of short, coarse, stubby hair. His shoulders and breast are remarkable features of the monster, they being extraordinarily thick and heavy. He has a remarkable depth of chest, the formation of which bears a strong resemblance to that of his shaggy counterpart. From the breast downward his body gradually tapers to the thighs. His lower limbs are slender and joined together like those of the noble steer. The feet and hands are those of a man.

John is a remarkably strong and vigorous combination of flesh and bone, and prior to his confinement in the infirmary he was noted for his wonderful speed of foot. He would dart away from his home into the woods and run like a hound for miles, making the air melodious as he went, bellowing like a bull. His actions are governed by instinct. His reason is an infinitesimal quantity.

He lacks the power of speech, and the only word that he can say that possesses meaning to his hearers is "bacca." He is intensely fond of tobacco, and the first thing a visitor hears when nearing his apartment is his cry "Bacca! Bacca!"

When he is given a piece of the palatable leaf, he tears it into little pieces, and puts them into his mouth, one at a time. He rolls the tobacco around with his tongue, like a cow treats a quid of grass, and finally swallows the mass—tobacco and all. Tobacco is a luxury that fills the poor devil with keen delight, and when he sees a pouch produced his eyes roll in pleasurable anticipation. John is kept in confinement in a little eight by ten cell, entrance to which is had through a barred door of heavy hickory strips. Just in the rear of this is an enclosure, twenty- or thirty-feet square, and unroofed. The fence surrounding it is about twenty feet in height. This is John's exercising yard. He possesses the instincts of a bull and is strongly affected by changes in the weather. He becomes greatly excited just before a rain, thunder, snow, or windstorm, and will plunge out into his exercising-pen, tear about at a fearful rate, paw up the earth, and bellowing most frightfully. He has a voice of wonderful power, and the bullish noise he makes can be heard for a mile around. They cannot be distinguished by a stranger from those of a gorgeous bull. So unerring is John in his demonstrations, that the residents in that locality rely upon him as their barometer, and he never fails to acquaint them with pending meteorological changes.

He wears men's coarse clothing, after a manner peculiar to himself. He cannot be induced to wear suspenders. He is frequently presented with those useful appendages to the male's toilet, but it matters not how bright and gaudy the colors may be, John will tear the suspenders up and ornament his neck, arms and legs with the pieces. He keeps his pantaloons in place by constantly holding them up by the band in front. When one hand is tired, he catches a fresh grip with the other and struts around his dismal quarters with all the pomp and circumstance of a prize bull at an agricultural fair. He sleeps on a small cot in one corner of his cell, having for his bedfellows a number of old tin cans, bones, and brickbats. He never removes his pantaloons, and nothing can induce him to wear shoes. He is a rugged, healthy creature, with a voice like a Kansas cyclone.

The man's history is a sad one. His mother, a very pleasant and intelligent woman, was one day crossing a field not far from their home a few months prior to John's birth. A vicious bull was pastured in the field and when he caught sight of Mrs. Haines, he came plunging toward her at full tilt, snorting angrily. The terribly affrighted woman ran with all speed to the nearest fence and succeeded in getting on the other side of it before the bull, which was close in her wake, could harm her. The shock was a terrible one, and Mrs. Haines suffered for days from nervous prostration. The terrible effects of that fright were impressed upon her unborn child, and when he was ushered into the world, the poor mother found that her offspring partook more of the nature of a bull than he did that of a human being. [for an explanation of this odd claim, see MATERNAL IMPRESSIONS] Her grief over this misfortune was pitiable, and for almost thirty years her life was embittered by a contemplation of the hideous child. When the wretched woman died, John was removed to the infirmary and has been confined there ever since.

* A U.S. census record lists a John Haynes, not Haines, aged 33, as living in Pebble Township, Pike Co., Ohio in 1880 (||yy3d63d2). It also states that he was unmarried, idiotic, insane, and that he could not read or write. This individual, then, would quite likely have been the *tertium quid* described in the report. The *Daily Globe* (St. Paul, Minn., Mar. 21, 1880, p. 5d) mentioned another boshominid as residing in the Suffolk County Almshouse at Yaphank, Long Island: "The Yaphank (L.I.) monster … is now at the Poor-house there and has hoofs like an ox."

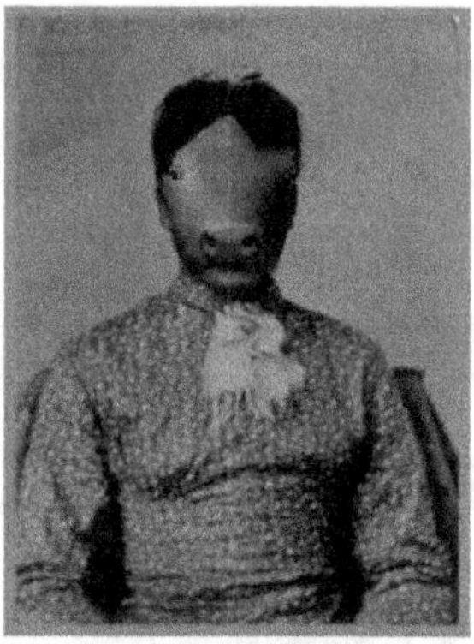

Frances McClellan
(reconstruction)

Jackson, Tennessee. Reports describe a viable minotaur named Frances McClellan. That she was named and dressed as a woman suggests her mother was human. The *Weekly Chronicle* (Clarksville, May 20, 1876, p. 1i; ||kyotcbe) said, "Last Wednesday a mulatto girl named Frances McClellan, living on the farm of Mrs. McClellan, about ten miles from Jackson, visited that city, and created quite a sensation. She was raised in Madison County, and is now about seventeen or

> eighteen years old [ergo, b. ca. 1858]. Her face appears to be that of a human, with a mask-like covering on the upper portion, resembling very forcibly the front and nose of a calf. From the top of her forehead down to her lower lip, the face is a calf's face. The nostrils perform the duties of the human nose and extend down on the lower lip like the nose and upper lip of a calf. Her eyes are several inches apart, large and peculiar in appearance. The skin of her calf-face is smooth like other parts of her person. The lower lip, chin and under jaws are natural human features, and of common size. The large eyes, the broad, flat, calf-like face, the wide nostrils, and the thick, heavy upper lip, joined to the human features below, gave to the face at once a horrible, revolting and hideous appearance. All who saw the girl were excited with curiosity and eyed her very closely. She soon noticed this and hid herself from public gaze.

* If honestly reported, McClellan's case is remarkable in that she supposedly reached maturity, was raised in a human family, and could communicate (perhaps talk?) well enough to go shopping in town, which suggests boshominids, at least in some instances, attain a human level of intelligence.

Yuma, Arizona. The *Los Angeles Herald* (Jul. 21, 1875, p. 1; ||y8u6kjvf): "YUMA, JULY 15: … A male child was born here a day or two since which was a most singular monstrosity. The head, from the lower jaw up, was that of a calf, with indescribable appendages. It lived one day. Dr. De Courcey, the attending physician, and Dr. Loring, Medical Director of Fort Yuma, examined this and are satisfied as to the cause."

Lozère, France. Le Journal du Midi (Avignon, Aug. 5, 1877, p. 2c; ||3a74hd5j) reported that a woman in Lozère had given birth to a stillborn offspring with three heads, one of them like a calf's.

Paris. L'Echo de France (Paris, Feb. 7, 1892, p. 3d; ||5n6jtmwx; also see: ||mr3jpkw5) reported that a 27-year-old Polish woman named Katjouja living in a caravan at Pantin, a northern suburb of Paris, had given birth to a child, in good health, with the head of a calf.

Vernon, Florida. The *Troy Messenger* (Troy, Ala., Jun. 2, 1909, p. 2; ||y3emzrgd) said,

> The report published in THE MESSENGER several days since [May 26, 1909, p. 4b; ||yy3953xs] of the birth of a strange-looking child near the line of Alabama and Florida did not impress ... as being strictly truthful. Not that they doubted the source from whence it came to Troy, but it seemed impossible that such a thing could happen in these days.
>
> But THE MESSENGER is in position to verify the report and add a few interesting facts to it.
>
> The child (if it can be called a child) was born and lived several days. It was covered from head to foot with hair; it had two short horns, one over each ear, and its feet were like the hoofs of a cow. After living several days, the child died, some think it was murdered, and it is now in Dothan where its remains are being embalmed for preservation.
>
> The child was born in Vernon, Fla., ... and its mother when she saw it asked the doctor to kill it. Being refused, some think it was murdered.
>
> The story sounded pretty fishy all along, but today the story is in the press dispatches and there can be no doubt of it.[12]

Okpo, Burma. The *Friend of Burma* (Rangoon, quoted in the *Times*, Wellington, N.Z., Sep. 20, 1902, p. 15b; ||y68w4nuq) said, "in the Tharrawaddy district, at a place called Okpo, may now be witnessed a strange freak of nature. A baby has been born there with a kind of cow's head covering its own. The parents are delighted, and they intend to make money out of it by exhibiting the monstrosity. They exhibited it in Okpo without charge, and ... they intend going on tour. The first place they will visit will be Rangoon. It is said they were offered 300 rupees for their offspring, but they refused to part with it."

Brazil. Le Resumen (Mayagüez, P.R., quoted in the *La Correspondencia de Puerto Rico,* San Juan, P.R., Jul. 4, 1892, p. 3a; ||wdvh5f2n) said, in translation, "A woman in Brazil just gave birth to a true freak, nothing less than a child with the head of a cow. ... One thing is certain: the mother has rejoiced since the first moment that she was told about the monstrosity she had birthed." When her doctor pressed her to donate her offspring, who was in good health, to science, she exclaimed "What? Don't you understand that this is going to make us a fortune?"

Florence, Italy. Matteo Villani (1846, I, book 4, lx, p. 361) recorded a woman at Florence giving birth, in February 1355, to a boshominid with a face like that of a calf and bovine eyes but with a human body.

Paris. Nicolas de Blégny (1680, obs. 5, p. 61), personal physician to Louis XIV, reported delivering a woman of a male offspring with a calf's head, which died two hours after birth.

Osnabrück, Germany. Physician August Droste (1845), practicing in Osnabrück, reported a girl born with a circle of calf hair on her elbow.

Winnsboro, Texas. Winnsboro physician M. B. Pollard (1912, p. 1137b) reported seeing a "child born with a head resembling a cow."

France (no locus in quo). Duchesne (1772, p. 392) said a woman in France bore an intersex nondescript like a human in its upper parts but with a leg bearing a calf's hoof attached to its stomach.

Lyon. Physician Jean-Ferapie Dufieu (1763, p. 228) said, in translation, that "January 20, 1759, the vicar of Saint-Nizier, Lyon, baptized an infant [implying a human mother] with features like a calf's. The lower portion of the face was human and the upper, a mingling of human and calf. He had the nose of a calf above another nose like a human being's, and his skin was covered with hair from the upper part of the small of the back upward over the top of his head, as far as his calf-like nose. To baptize the child, the priest had to move this shaggy hair aside. There were no eyes, just large, empty sockets. The arms were extremely short and had only two digits [cloven hooves?]. The child died the next day, the 21st, and was buried in the cemetery there at St. Nizier."

Champagne, France. Lugeris (1691) said a woman in Champagne bore a stillborn boshominid. In translation, he said it lacked external sexual organs and had "a malformed stomach, mouth and nose. The eyes were above the forehead and very large ... Above, one saw two horns emerging like those of a two-month-old lamb. The rear of the skull was flat and black, like that of a cow; the back and shoulders were of that same color down to the loins."

Volhynia. The *Breslauische Sammlungen* (1726, p. 197; ||2p8ehzc4; also see: ||2p8c5ewh) said a woman in Volhynia in 1724 had borne a creature that was half human and half calf, which lived eight days.

* *Breslauische Sammlungen* is the short title of what was perhaps the most important German-language scientific journal of the early eighteenth century (||ya4h3769).

Freiberg, Germany. Old records say a boshominid was born December 12, 1522, at Freiberg in Saxony. Aldrovandi (1642, p. 372) says it was born to a woman, and Schenck (1609, p. 89), that she was the wife of a butcher. Martin Luther claimed its birth was a sign from God that the Catholic Church was corrupt (Melanchthon and Luther 1523). Catholic partisans (e.g., Aldrovandi, ibid., Cochlaeus 1523) said it foretold Luther's defection. Illustrations by Lucas Cranach the Elder and in Schenck depict it with a misshapen head, protruding tongue, low-set ears, two nub horns, upright posture, and hooves instead of hands and feet. It was known as the Münchkalb (Monk Calf) because Luther likened the flap of skin on its shoulders to a monk's cowl. Its black skin, too, supposedly recalled a monk's habit. Schenck (ibid.) said Henry IV, Duke of Saxony, had it preserved.[13]

The Münchkalb
(Lucas Cranach the Elder)

* Some authors suggest the mother of the Freiburg boshominid was a cow (see Ratzeberger 1850, p. 200). But primary sources suggest this is not the case.

* Other telenothians played a propaganda role in the Reformation, e.g., the Sea Monk (p. 943), the Papstesel (pp. 1162-1164), and the Priesterschwein (note, p. 194).

Wieland. Christoph Martin Wieland (1784, pp. 253-263) published a report about a boshominid. It tells of three sisters who kept, in a room at the back of their house, a 32-year-old, viable cow-human. Though no mother is named, their caring for this creature for so many years, suggests it was born to a human, not a cow. As in most other cases involving a human mother, this boshominid was described as having a cow-like head. In translation, the report reads,

The Münchkalb
(Schenck 1609, p. 89)

People seek to pursue Nature down her most secret paths, with the hope of engaging themselves with the infinite peculiarities of her productions, a truly fruitful occupation for the reflective mind. And here in this letter, my friend, you have a contribution to the natural sciences whose subject you will find most remarkably strange.

In one of the local suburbs lives a family of three sisters. But there is also a fourth member of that household, a creature that could just as well be classified as either human or animal. It has a human-like form but eats and acts like a beast. And, had I not been hardened by the study of natural history, I would have been shaken with horror at the very sight of it.

The three sisters are fully aware of the dread this creature inspires, which is why they make it so hard to see. They keep it hidden away in a back room, and for years have been covering up its existence by saying it was dead. But I learned it was not, and I went to the utmost trouble to see this nondescript with my own eyes. Public opinion indeed made a great monster of it. But what I in the end saw myself was even more terrible than anything anyone had described. I was taken to a small back room and, as we went in, an appalling stench struck my nostrils, which became even worse as I approached a misshapen form, asleep upon a mattress. As it lay there, I could see the thing had an unnaturally thick head, upon which, here and there, were scattered wisps of blond hair. The forehead was broad and flat as the brow of an ox, but smooth and without the slightest wrinkle. Of eyebrows there was no trace. The nose was a full two inches wide and seemed to be pressed out wide. It had only one nostril, the other being closed off with cartilage. The upper lip was missing. In its place was a close-set row of white incisors, which were unusually wide. The lower lip was round and very thick, blue and repulsive. Just below the chin, a fleshy growth stood out, larger than an ordinary goiter, but hanging and limp. Close beneath it rose a pair of prominent breasts. A pair of small round teats, like a fourteen-year-old girl's (the creature is female) waved about, and beneath them was a belly swollen like that of a pregnant woman. The feet were crooked and turned inward. There were no toes. They were fused, as were her fingers. Her arms were thin and undeveloped. The bone seemed to be covered with only a thin yellow skin. Indeed, the entire creature was bright yellow, which made her inexpressibly repulsive.

The creature still slept, and I expressed my wish to see it awake. The maid, my guide, told me I must not scare her, and gave her a shake. Suddenly she opened her eyes and uttered a cry that pierced me to my core. It was the natural voice of an ordinary calf. She began beating the air with her hands and feet, and to calm her, the girl placed a flat bowl of porridge, a mixture of milk and grated bread, near her mouth. She sniffed and at once lowered her whole head into the bowl and, with her eyes closed, lapped up the milk like a dog. When bread got stuck between her teeth, she chewed and lifted her head slightly from the bowl. It was a horrible thing to see.

The creature is so greedy that one cannot take away the dish without raising a terrible mooing and causing her to thrash about with her arms and legs. I had a small dog with me who was lured by the scent of the milk. Before venturing to the bowl, he paced round the creature three times, and when she failed to make way, he began to bark. The thing seemed to be listening but never stopped licking until the whole bowl was empty. Afterward, she sniffed round the bowl and licked the floor.

And then she began once again with the lowing, not so loud as before, but rather more like the weak voice of a drunkard whose tongue is too heavy, and stomach, too full. Now she began to throw herself about and to roll from side to side, thrashing the air with her arms and legs. The girl said that she always does this after she's eaten, and that often she seems to smile. But I myself saw no smile. Presumably, such expressions are elicited by a feeling of well-being from satisfying a basic animal need.

At first, I thought she would go to sleep again. But she didn't. Instead, she thrashed and rolled about, harder than ever, and finally rolled right off the mattress onto the hard floor but seemed not to notice. Rather, she continued the same thrashings even more frenetically, and in the end rolled right out of the cloth in which she was wrapped and so lay completely naked before me, for she had no undergarments. Really, I have no idea how they dress her at all, given that she is always rolling and dragging herself about from one place to another on her belly, or on her hands and feet.

She is always hungry. And if you place a full bowl before her, she will fall upon it at once and never stop eating before it is emptied to the bottom. She usually eats six times a day, and her forceful cries leave no doubt as to when the feedings should occur.

She refuses to eat bread unless it has been crumbled. Nor will she eat meat or legumes. If one offers her such food, she may eat for a while but seems to taste at once that it is not her usual fare. And then she raises a great cry and keeps on until she gets her customary porridge of milk and crumbled bread. And then, when she has finished, she always behaves as I have already described. This dish must therefore always be at hand because often this creature will cry out in the middle of the night or early in the morning and wake the entire house and never cease until someone has seen to her desires.

Unable to speak, she could only bawl like a calf. Nor, seemingly, does she hear properly either. For I stood next to her and stamped loudly against the floor, and I also struck a nearby leather-covered chair with a metal pipe. It brought no reaction. But she must be able to hear the door, for the girl said she was always drooling when it opened, which is not surprising, since she only does so when food is being delivered. The sisters themselves never trouble themselves directly in the care of this creature—and I must say, I myself am not eager to view her again.

Nevertheless, I did inquire about many things, so that I might determine whether this monstrosity[14] showed any human desires and impulses, other than the most basic animal cravings. But I failed to get any information, partly because the girl sees this thing only a few times a day—and that for a short time—and partly because she has grown too accustomed to her to feel any curiosity. I would have preferred to have spent a week or two on my observations. But it simply wasn't possible. I had enough trouble even getting in there once. In fact, my sister had to promise the maid that if she were caught by her employers and dismissed, that she would take her into her own service. But I can certainly understand why these sisters take the attitude they do.

In the summer they often put the creature out in a high-walled garden. The neighbors became aware of its presence and began peering over the wall. But the sisters complained to the authorities, who expressly prohibited this intrusive behavior. For the neighbors were extremely curious and, at one point, a throng standing on ladders rang the walls.

And then the sisters built a roofed garden house, where the creature could spend summer days, and planted a lawn around it. For her room stank so much that opening the door made the whole house reek. At first, however, she refused to stay in her new house, until they finally realized that she loved the sun. The house was turned to admit the sunlight and she now will spend the entire day there or lolling on the lawn, where she is often heard mooing, not in a distressed way, but more in joy. They often leave her outdoors at night and find her in the morning out there in the open, all fresh and frisky.

And now a bit of a detour. The chambermaid told me that sometimes, especially in the spring, the creature lies on the lawn in the sun and rolls about, so much that she ends up entirely naked, and then she moans, thrusts, pants and seems to smile. "Thrusts?" I said in all innocence. At this the girl blushed a deep red, and I suddenly realized what she had meant. These movements were nothing more or less than the expression of that same natural drive seen in many young girls.

While the creature is out in the garden, they fumigate her room. But they cannot do so as often as they would like, as it distresses her and she doesn't sleep. When they bring her in, she begins sneezing over and over. She then starts up a loud bawling and keeps it up throughout the night. She must be very irritable. So the sisters are put to a lot of bother in the winter, given that the beast is easily chilled and already in September must be kept in a heated room.

What is most surprising is that this unfortunate creature is 32 years old. Ordinarily such beings die soon after birth. But this one is an exception. But it seems she was formerly friskier and less thin, and that she smiled and ate more. She has only been ill a few times in her life. And even on those rare occasions she recovered her appetite and bounced back in just a few days. She does not at all like to drink but gets what liquid she needs from her milk porridge, which she laps up, allowing no one to feed her. But three and a half feet long, she reached her mature size at the age of twelve. I wanted to include a drawing of this creature. But I myself cannot draw. And it wasn't possible to bring someone along who could.[15]

Ghent, Belgium. Gerlin (1624, XII) said a woman at Ghent bore a three-headed infant with a cow's tail in 1160 A.D. but gave no details.

Buchdorf, Bavaria. Irenäus (1584) said a woman at Buchdorf bore a child with a calf's head in 1567, but he failed to elaborate.

Cracow, Poland. Maciej Miechowita (1521, p. 360) recorded the birth of a two-headed, seven-legged calf in 1499, in Cracow's Kazimierz district. He was living in Cracow at the time. But certain later writers (e.g. Bielski 1564; Stryjkowski 1584, pp. 117-118) claimed, apparently without basis, that a Jewish woman had given it birth, so that this birth would represent a boshominid. Given that no human characteristics were mentioned, the mother could have been a cow, in which case the birth in question would have been a conjoined twin, not a boshominid.

A boshominid, as pictured in a Swiss manuscript dating to 1578. In translation, the text reads, "In early January 1578, a cow at Embrach, gave birth to a calf with a human face and head. It was put to death at birth and secretly buried." Embrach is a municipality in the Canton of Zürich. (Wickiana Collection; ||yc3wf8s2).

The Swiss surgeon Johannes de Muralto (1683) reported that Elizabeth Linggin (b. ca. 1639, d. May 7, 1678), pictured above, was born at Wädenswil, near Zürich ("Vadis-Villa, ditionis Tigurinæ"), with an udder, which grew over the years until, at her death, it weighed 73 pounds. In a related report, George Ashe (1698), Lord Bishop of Cloyne, said an Irish girl named Elizabeth Dooly, living in County Kilkenny and the daughter of a milkmaid, had beside her left eye "a piece of Flesh growing exactly like a Cow's Teat in Bigness, Shape, &c." When Ashe saw her in 1687, she was 30 (ergo, born ca. 1657).

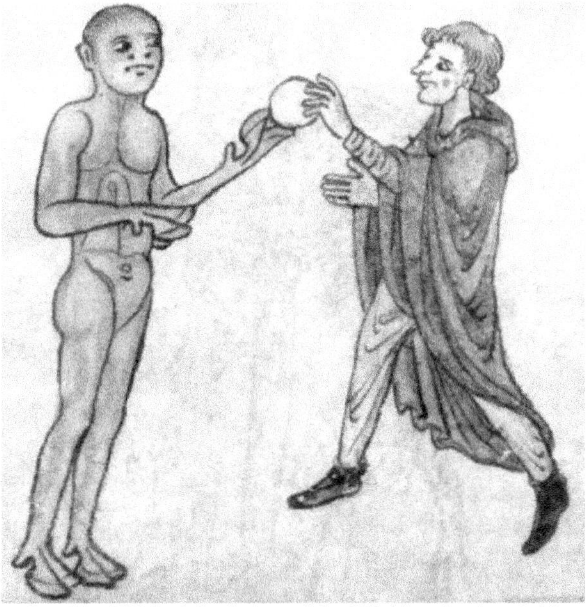

Gerald of Wales' depiction of the Wicklow boshominid.

Wicklow, Ireland. In the *Topographia Hibernica,* Gerald of Wales (b. ca. 1147, d. ca. 1223) says, In translation, "Near Wicklow, when Maurice FitzGerald first gained lordship, there was seen a human prodigy, if indeed it is correct to say "human." For while the creature had a human body, his extremities were bovine. To the wrists and ankles, instead of hands and feet, were attached cow hooves. He was bald except for a few patches of down. The eyes were large, cow-like in their roundness and color. The face below was flat, just two nostrils, with no protruding nose. He spoke no words but could only low. Long an attendant of Maurice's court, he came every day to meals, and what he was given to eat, he gripped in the cleft of his hooves, which served him as hands, and so conveyed it to his mouth. But because the nobles of the castle often despised the Irish for getting such things on cattle, they at last, in their malice and spite, waylaid and killed him, which he little deserved. For shortly before the English arrived on that island a cow had borne this man-calf in the mountains of Glendalough … Thus, you might well suppose that 'a man half-bull and a bull half-man' had been created once again. For almost a year he remained with the cattle, following and nursing his mother. But at last, as he was more human than cow, he was brought over into human society."[16]

MYTH

Daedalus, Pasiphae and wooden cow (Casa dei Vettii, Pompeii, before 79 A.D.).

When Gerald of Wales wrote of "a man half-bull and a bull half-man," he meant the Minotaur, scion of the Cretan queen Pasiphae, Queen of Crete. In the story her husband King Minos prays for a white bull to sacrifice to the sea god Poseidon. But then, as Apollodorus relates, when such an animal did emerge from the sea, Minos sent the bull to the herds and sacrificed another. But angry at him for not sacrificing the bull, Poseidon made the animal savage, and contrived that Pasiphae

> should conceive a passion for it. In her love for the bull, she found an accomplice in Daedalus, an architect, who had been banished from Athens for murder. He constructed a wooden cow on wheels, took it, hollowed it out, sewed it up in the hide of a cow which he had skinned, and set it in the meadow where the bull was kept. Then he introduced Pasiphae into it; and the bull came and coupled with it, as if it were a real cow. And she gave birth to Asterius, who was called the Minotaur. He had the face of a bull, but the rest of him was human; and Minos, in compliance with certain oracles, shut him up and guarded him in the Labyrinth.[17]

In ancient times, not only pagans believed the tale of Pasiphae, but also Jews. Philo Judaeus (*Special Laws,* 8:46) said Jewish law had banned bestiality to prevent the birth of such commingled beings.

> But some persons, imitating the sensual indulgences of the Sybarites and of other nations more licentious still, have in the first place devoted themselves to gluttony and wine-bibbing, and other pleasures affecting the belly and the parts adjacent to the belly, and then when fully sated have behaved with such extraordinary insolence … that in their insanity of passion they have gone frantic and been so maddened as to desire no longer human beings, whether male or female, but even brute beasts, as they say that in ancient times in Crete, the wife of Minos the king, by

name Pasiphae, fell in love with a bull, and became very violent in her passion from her despair of being able to gratify it (for love which fails in its object is usually increased in no ordinary degree), so that at last she reported to Daedalus the affliction by which she was overwhelmed, and he was the most skillful of all workmen of his time.

And he, being very ingenious, so as by his contrivances to discover things undiscoverable to anyone else, made a cow of wood, and put Pasiphae into it at one of the sides, and the bull rushed at the wooden cow as if it had been an animal of its own kind. And Pasiphae, becoming pregnant at a certain period, brought forth an animal half man and half beast, called the Minotaur.

And it is very likely that there may be other Pasiphaes also, with passions equally unbridled, and that not women only, but men likewise may fall madly in love with animals, from whom, perhaps, indescribable monsters may be born, being memorials of the excessive pollution of men; owing to which, perhaps, those unnatural creations of unprecedented and fabulous monsters will exist, such as hippocentaurs and chimaeras, and other similar animals. [translated by Yonge (1855)]

Among the Greeks and Romans, images of boshominids were common. In the murals and mosaics of Pompeii, minotaurs are innumerable. Similar images decorated cups and dinner plates. Mints struck coins that bore representations of boshominids. The city of Gela, in Sicily, issued silver tetradrachms stamped with a human-headed bull, honoring the river god Gelas, for whom the city had been named.

Tetradrachm
(Gela, 5th cen. B.C.)

In Hellenic culture, river gods were often represented as man-headed bulls, as they had been by the Etruscans from as early as the eighth century B.C. (Molinari and Sisci 2016). They were also described as man-bodied and bull-faced. In Sophocles' *Women of Trachis* (Torrance trans., 1966, 1.9), a woman describes being wooed by a river god, who in one of his forms was a bull-man.

My suitor was the river Achelous,
who took three forms to ask me of my father:
a rambling bull once, then a writhing snake
of gleaming colors, then again, a man
with ox-like face.

And thousands of years before the Greeks, there was Mesopotamia, where depictions of boshominids are known from the Uruk period (ca. 4000 to 3100 B.C.). The plaque at right, from Mesopotamia (Old Babylonian Period, ca. 2000-1600 B.C.), shows two bull-men, and the image below is from a lyre found in the Royal Cemetery of Ur (Sumerian, Early Dynastic III Period, 2550-2450 B.C.).

Man embracing two boshominids. From a lyre found at Ur (2550-2450 B.C.).

Bas-relief at Pasargadae (Iran, 6th cen. B.C.). *Left:* Legs of a bull-man. *Right:* Lower parts of OANNES, the Mesopotamian fish-human deity.

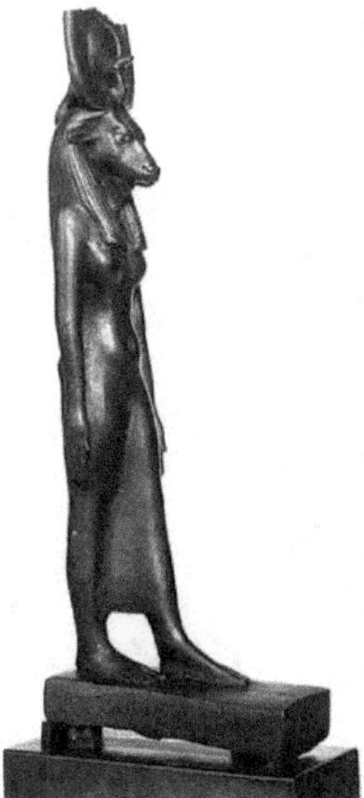

Left: The goddess Hathor (Egypt, 7th cen. B.C.; ||26h6hwk5). *Right:* Bull-man, incised shell (Sumerian, ca. 2600 B.C., Metropolitan Museum of Art).

Sumerian bull-man (2500 B.C.)

Bull-men and HOMINAVIANS (cylinder seal impression, Neo-Assyrian, ca. 700 B.C.; ||yyuxaw35).

Fragment of a terracotta frieze plaque showing a minotaur and felines (Regia, Rome, ca. 600–550 BC, Antiquarium Museo del Foro Romano, Wikimedia, Sailko; ||y9zc2jwo).

Two boshominids (Wat Pho, Bangkok, 16th century; Image: Wikimedia, Photo Dharma from Sadao, Thailand; ||ptp2kryy).

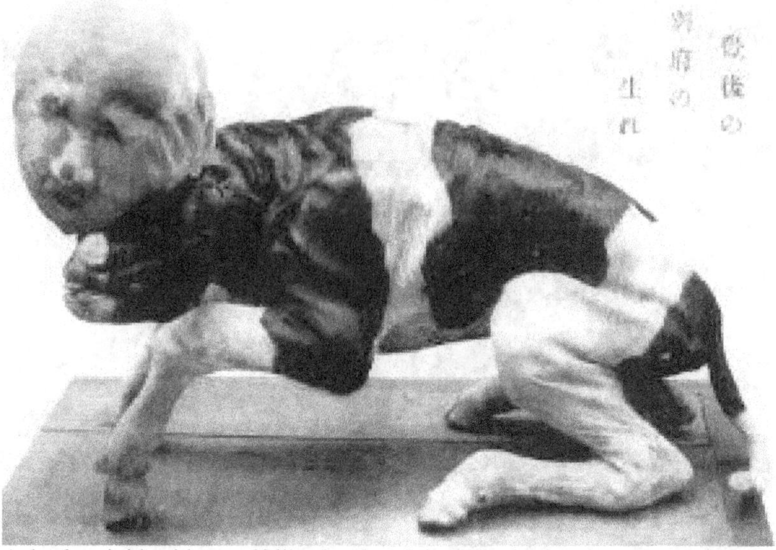

A boshominid with a calf-like head and body and a supernumerary human head. The caption, translated, reads: "I was born in Beppu, Bungo." Bungo is an old name of Oita Prefecture.[18]

XENOMELIA

Rather often, a telenothian will look like one kind of animal but have what appears to be part of some other kind of animal attached, a condition that may be termed *xenomelia* (/ZEE-nō-MAIL-ee-yə/).

Beppu, Oita, Japan. One boshominid seems to be known only from a Japanese postcard widely circulated on the Internet. The card, pictured above, is a reproduction of a photo originally taken in Kobe. The animal looks like a nearly complete calf but with an attached growth resembling a human head sheathed in cow hair. A website says this creature was born in 1926 in Oita Prefecture, Kyushu.

Durant, Mississippi. The *Intelligencer* (Anderson, S.C., May 18, 1916, p. 4; ||f6cje4cx) said a calf born near Durant "has a body and feet like any other calf but it has a calf head with a human head between its eyes and ears. Its nose and mouth are like a calf. Its eyes are on the beginning of the large head and this head has ears like a human with calf ears behind it."

Jean-François-Clément Morand
1726-1784

Paris. The Parisian physician Jean-François-Clément Morand (1812) described a boshominid he saw at the St. Germain Fair in 1748. He said, in translation, that it had "atop its true head, a growth of the same size and form as a human head."[19] So its conformation was the same as reported for the Beppu, Durant, and Shelby animals. Morand also notes that this same BOSHOMINID had calved 12 offspring. So, if it was a hybrid, it was fertile, nevertheless.

Shelby, Ohio. From the *Ohio Democrat* (Canal Dover, O., Mar. 18, 1886, p. 2c): "SHELBY, O., March 12—A cow belonging to W. H. Morris of this city gave birth this forenoon to one of the most remarkable monstrosities on record. It was a well-formed calf, with a large-sized lump, a perfect representation of a human head, between the ears. The head of the calf was otherwise well formed. It will probably be given into the hands of an expert taxidermist and properly prepared for exhibition."

Rubiera, Italy. Antonio Vallisneri (1715, p. 197, no. 11) stated that he was present when a woman in Rubiera gave birth to what seems to have been the reverse of the bovine-human composites just described, that is, an offspring that appeared to be a little girl with the head of a calf attached to her back. The supernumerary head was the size of a fist, he said, and contained bone and nerve tissue. After its surgical removal, the girl survived.

Brooklyn, New York. The *Evening Gazette* (Port Jervis, N.Y., Nov. 26, 1874, p. 2b; ||y53uprpz; also see: ||4uamfdsn) said, "The British ship *Scindia,* Captain George Harrison, arrived at New York a few days since from Calcutta, having on board one of the most remarkable natural curiosities the world has probably ever seen. It consists of a short horned, sacred Brahmin bull, born in Nepal, northern India, in 1871, and imported here by Captain William Denison Folger, formerly of this city, but latterly a resident of Calcutta. The bull itself is a remarkable curiosity and is one of that description of animals held in great reverence by the natives of India on account of its traditional sacred character. But in this case a novel *lusus naturae* renders it a most singular creature. The brute portion is symmetrically formed with a glossy coat of fawn colored hair and well-shaped body and limbs. But protruding as it were from the left side of the hump on the back of the neck of the animal is a wonderful and regularly shaped, in nearly all respects, human arm. The deltoid and triceps and biceps muscles are well developed, particularly the latter; the joint at the elbow is flexible, the forearm rather attenuated, the wrist as fully flexible as in the human arm proper, while the hand is composed of four distinctly marked fingers, two of them connected together. This extraordinary appendage to an otherwise well-formed body does not appear to give the creature any inconvenience or pain, generally hanging listlessly by its side except when it is angered or annoyed, when the arm rises as if its possessor felt inclined to 'strike from the shoulder.' The animal is quite healthy."

Picardy, France. In a report about freaks on display in Paris, *Le Petit Gessien* (Gex, Apr. 24, 1930, p. 2a; ||y27k72z8) said a calf from Picardy, by the name of "Nicholas," had a fifth limb attached between his shoulders. "This is no ordinary leg, but a true arm, with a shoulder, elbow, wrist, and three fingers, two with finger joints, and a thumb that 'Nicholas' sucks and gnaws in a very human fashion" (my translation).

Owensville, Missouri. The *Gasconade County Republican* (Owensville, May 27, 1904, p. 3b; ||y4rzybw8) reported another such, "a cow with a fifth limb greatly resembling a human arm, which came out near the backbone and just behind the shoulder."

Ora, Mississippi. The *Brunswick News* (Brunswick, Ga., July 23, 1908, p. 3; ||2bbv5ntb) reported another, an 18-month-old "calf" with a human arm attached to its withers. Its owner J. W. Holloway lived in Ora. The report said doctors detected a pulse of 72 in the arm but of 44 in the forelegs, which suggests the arm was fed by a separate heart with a pulse like that of a human. The pulse rate is, on average, lower in cattle than in human beings. According to the report, "The arm, as it is termed, hangs by the side of the calf. At the juncture on the back there is a ball and socket similar to that of a human arm, an elbow joint, knuckle joints similar to those of a hand, while five claws form the so-called fingers."

Bessay, Vendée, France. Le Réveil Populaire (Luçon, Apr. 20, 1907, p. 3e; ||5dka9tkd) reported a viable boshominid, born on the farm of a Mr. Sacre at Bessay, that looked like a calf with the left arm of an adult human being attached to its right shoulder.

Montauban-de-Bretagne, France. La Dépêche Bretonne (Rennes, Jun. 29, 1901, p. 3d; ||7yyvzc4x) said a two-month-old calf had on its shoulders a humanlike arm and hand, with five fingers bearing nails like hooves.

Nîmes, France. Le Républicain du Gard (Nîmes, Sep. 26, 1910, p. 3c; ||nhdt7tby) said a cow at a local fair had a human arm.

Paris. The *Joker* (Paris, Feb. 24, 1928, p. 3b; ||4b7nm3sk) said, in translation, "Among the freaks once shown at the fair in Boulevard Richard-Lenoir, was an excellent man, born without arms, who exhibited himself next to another freak, a calf born with two human arms attached to the lower part of its neck. By what caprice did Nature grace this calf with two arms it never uses, though they would have been most useful indeed to his neighbor?"

Marcillat-en-Combraille, Allier, France. The *Journal de Cette* (Cette, Jul. 22, 1914, p. 3c; ||3c2b4jfn) said, in translation, "MONTLUÇON.—M. Auguste Cayet, a farmer at Marcillat, has a calf with an appendage perfectly like a human hand on its withers. The animal is three months old and in good health."

Bligny-sur-Ouche, Côte-d'Or, France. Le Bien Public (Dijon, Aug. 25, 1895, p. 2f; ||4weu4sha) said that a cow belonging to a Mr. Thévenot, a resident of Bligny-sur-Ouche, bore a calf that had an extra "tail," which was 55 cm long and attached to the lumbar region of its back, and that the tail ended in a human hand with five fingers. Per the report, it could grip things with the hand.

Rushville, Indiana. The *Cincinnati Enquirer* (Jun. 2, 1879, p. 4e; ||ya3r84yf) said a calf at Rushville had a single human finger on its tail.

An early representation of a kudan. The Japanese in the picture translates as "In December 1836, in Tango province, a monster with the face of a human and the body of a cow called 'kudan' appeared on Kurahashi Mountain." In addition it says that "a kudan was produced also in December of 1705, and a good harvest followed. Any who wore an amulet depicting the kudan experienced great family prosperity even in times of sickness, and large harvests in times of famine. Indeed, it is an auspicious beast! Because the kudan is an honest beast, it is customary to write 'like the Kudan' at the end of every act and deed." (Translation taken from Wikipedia.)

A kudan on display at a yokai exhibition in Osaka.

KUDANS

Is it possible you never heard of the Kudan? The Kudan has the face of a man and the body of a bull. Sometimes it is born of a cow, and that is a Sign-of-things-going-to-happen. And the Kudan always tells the truth. Therefore in Japanese letters and documents it is customary to use the phrase, Kudan no gotoshi,—"like the Kudan," or "on the truth of the Kudan."

—LAFCADIO HEARN, *Glimpses of Unfamiliar Japan* (II, p. 567)

In Japan, there is a centuries-old belief that rare hybridization between human beings and cows occurs. There, such hybrids, known as kudans, are widely believed to be short-lived and infallible prophets—they are born, speak their prophecies and promptly die.

Such trust is placed in these seers that sayings have entered the language, such as "honest as a kudan." And people widely wear, or at least have worn in the past, pictures of kudans as amulets, with the idea that they serve a tutelary function.

A kudan, reportedly calved in Kagawa Prefecture in 1930, appears at right. Others are listed below.

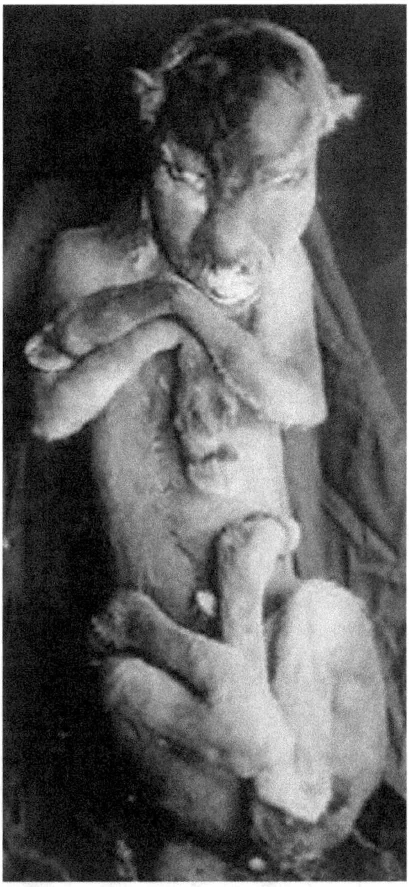

The Kagawa kudan.

REPORTED KUDAN BIRTHS (JAPAN)

1819	Kaminoseki, Suō Province
1836	Tango Province
1892	Mionoseki (stuffed specimen)
1899	Gotō Islands
1930	Kagawa Prefecture
1943	Iwakuni
1945	Kobe
2004	Gunma Province (stuffed specimen)

* The cases in the table are from a Wikipedia article on kudans (||39h8yv5s). Chinese Wikipedia (||5n77aapf) lists two other boshominid births in Taiwan in 1862 and 1913.

INCERTAE SEDIS

Radstadt, Austria. Lycosthenes (1557) said a woman at Radstadt bore a homungulate ("a child with horns") in 1233. Mezger (1692, p. 432) also mentioned this birth.

The Radstadt homungulate (Lycosthenes 1557)

Molova, Turkey. Der Bothe von und für Tirol und Borarberg (Sep. 6, 1832, No. 72, p. 2; ||bddjer4w) said the kadi of Molova filed an official report about a woman there having borne a child with two eight-inch horns.

Rabun County, Georgia. According to numerous news reports (e.g., *Gwinnett Herald,* Lawrenceville, Ga., Jul. 9, 1879, p. 1; ||2p8usr4a), a certain Sarah Cawkins, of Rabun County, gave birth in 1876 to a devil-like homungulate with bovine traits, such as small horns and cloven hooves.

Farnroda, Germany. Irenäus (1584) records the birth of what might be interpreted as either a boshominid or a CANIBOVID in 1562 at Farnroda, on or near All Saints Day (early November).

Insquinatoyac, Guerrero, Mexico. The *Star and Herald* (Panama, Nov. 1, 1895, p. 1) said a woman at Insquinatoyac bore "a well-formed male child with horns on its head."

Rossow, Pomerania. Lycosthenes (1557, p. 649) said a woman bore, in December 1555, a boshominid at Rossow. But the case is uncertain because his description is typical of those given of cow-calved boshominids. He said, in translation, "In December [1555] in the village of Rossow, which is only a mile from Pasewalk, a woman gave birth to a horrible monster. The body was that of a calf, the head bald and round, similar to that of an ape, with a hairy chin and protruding tongue. The forefeet, without hoofs, were round balls, but the hind ones like a true cow. The tail was stiff, without joints."

* Palfijn and Mauriceau (1708, p. 203; ||p8pnsrz) and Liceti (1665, p. 188; ||nuhnb2k) reiterate Lycosthenes' claim. The mother may have been a cow (as such a birth in a rural village might well have been blamed on a "witch").

SOME OF THE BOSHOMINIDS CITED IN THIS BOOK

1355	Florence, Italy	1748	Paris	1892	Orange City, Iowa
1522	Freiberg, Germany	1752	Clayworth, England	1892	Shreveport, La.
1547	Bitterfeld, Germany	1759	Lyon	1893	Abbeville, La.
1550	Würzburg, Bavaria	1772	France	1893	Eckerty, Indiana
1554	Brunnen, Switz.	1784	Nemyriv, Ukraine	1894	St. Louis
1555	Rossow, Germany	1784	Wieland	1895	Paris, Texas
1556	Bamberg, Bavaria	1788	Samogitia, Lithuania	1895	Shallotte, Twp., N.C.
1561	Augsburg, Germany	1807	Albany, New York	1898	Gadsden, Tennessee
1562	Farnroda, Germany	1818	Franklin County, Ga.	1898	Laurens, Iowa
1567	Buchdorf, Bavaria	1825	Coshocton, Ohio	1898	Los Angeles Herald
1571	Muri, Switzerland	1825	Nálepkovo, Slovakia	1899	Uttar Pradesh, India
1578	Berge, Germany	1845	Upper Egypt	1901	Leon, Iowa
1578	Embrach, Switz.	1858	Würzburg, Bavaria	1902	Okpo, Burma
1586	Passau, Bavaria	1874	Brooklyn, New York	1904	Owensville, Mo.
1588	Kalkaboda, Sweden	1875	Fernandina, Florida	1905	Louisville, Kentucky
1597	Burgos, Spain	1875	Shelby, Indiana	1908	Ora, Mississippi
1639	Wädenswil, Switz.	1875	Yuma, Arizona	1908	San Juan Guichicovi
1659	Halle, Germany	1876	Fernandina, Florida	1908	Sedalia, Missouri
1661	Padua, Italy	1876	Jackson, Tennessee	1909	Goncelin, France
1665	Madagascar	1878	Garret Mountain	1909	Tepljuh, Croatia
1673	Atzpach, Germany	1879	Fort Scott, Kansas	1909	Vernon, Florida
1673	Drebkau, Germany	1879	Prattsville, N.Y.	1910	Moravská Třebová
1678	France (Blégny)	1879	Rushville, Indiana	1910	Orange, Texas
1681	Zeitz, Germany	1880	Laurel, Indiana	1912	Ashland, Kentucky
1682	Hanover, Germany	1883	Stonewall, Virginia	1913	Buckeye, Arizona
1689	Goselitz, Germany	1884	Indianapolis, Indiana	1913	Lunenburg, Vermont
1691	Champagne, France	1884	Pebble, Ohio	1915	Auburn, Maine
1691	Lanz, Germany	1885	Deep Run, Ohio	1916	Durant, Mississippi
1700	Wrocław, Silesia	1885	Friendship, N.Y.	1926	Beppu, Oita, Japan
1709	Trzcianka, Poland	1885	Glendale, New York	1933	Malang, Indonesia
1716	Boston	1886	Shelby, Ohio	1949	St.-Priest-sous-Aixe
1723	Lviv, Ukraine	1887	Boswell's Depot	1951	Vertaizon, France
1724	Volhynia	1888	Silver Creek, Ne.	2008	Thailand
1736	Grimma, Germany	1889	Binghamton, N.Y.	2017	Uttar Pradesh, India
1737	Italy	1890	Laurel, Indiana	2019	Villa Ana, Argentina
1741	Ufton Nervet, UK	1892	Jackson, Mississippi	2024	Thailand

NOTES

1. In general, Javan farmers do not keep European domestic cattle (*Bos taurus*), so the mother cow in this case would likely have been a banteng (*B. javanicus*).
2. This report originally appeared in the Bellaire, Ohio, *Independent*.
3. Though the material quoted is from a Paris newspaper, Bavarian newspapers also attest this event (||p4htep9, ||puw4njm). Twin boshominids had reportedly been calved at Würzburg three centuries earlier, in 1550 (see p. 24).
4. Schreter (1827, pp. 487-489). Translation mine. French translation: ||nwh2y7o. Spanish translation: ||ky757yy.
5. The subject, with its cow-like face and allegedly erect posture, seems to fit better with the various cases in which boshominids have reportedly been borne by women, so it may be that the mother is here incorrectly reported as a cow. Note that the report says the farmer "did not suspect the heifer to be with calf."
6. Bekmann and Bekmann (ibid., pp. 885-886) mention a three-volume work on reported prodigies by Martin Friedrich Seidel (1621-1693), a manuscript, now lost, and from which it seems they cited the Drebkau case.
7. Its title page has been lost, but Rosen (1994, p. 151) says Klint's "book is usually referred to by the name *On Meteors*. It seems originally to have been called *Om the tekn och widunder som föregingo thet liturgiske owäsendet*. The book is now in the Stifts- och Landsbiblioteket (Box 3085, S-580 03 Linköping, Sweden) and has a new title page crediting the work to Jonas, not Joan, Petri Klint.
8. Wikipedia (||ybgrczns, accessed Apr. 2, 2019) states that "Rumination syndrome, or merycism, is an under-diagnosed chronic motility disorder characterized by effortless regurgitation of most meals following consumption, due to the involuntary contraction of the muscles around the abdomen."
9. Translation mine. Lammert (1861, p. 236) lists sources attesting the Kleisdorf birth. Scheuchzer (1716, p. 69) gives the date as July 28 ("28. Heum. 1556. zu Kleissdorff an der Itz ½ Miel von Eber, 3. Meil von Bamberg").
10. Fincelius (1566b) also describes this case.
11. John's parents were likely Joshua Haynes (b. 1815, Patrick, Virginia, d. Jan. 12, 1887, Pebble Township, Pike Co., Ohio) and Lydia Jennie Cooper (b. 1816, Ohio, d. Mar. 1, 1869, Pike Co.), who had a son John in 1842, not 1847 (||y38y85e2 and ||y5veobso). The date of her death, 1869, approximately corresponds to the date given in the news report (12 years before 1884). John is listed as 8 years old in an 1850 census record for Pebble (||yyxoztmb). The same record states that Joshua was a farmer. An 1860 census record for Pebble lists John, the son of Joshua and Lydia, as a 16-year-old idiot (||y4asabuc). Joshua and a 28-year-old John are listed in an 1870 census record (||y2ppyfu7) for Pebble, but no Lydia. Pebble Township is about 20 miles south of Chillicothe in Pike County. In 1866, Pike County purchased a farm near the village of Idaho, Pebble Township, and built there an infirmary, also known as the poorhouse (for more information, see: ||y3x5sxjz).

12. A freely accessible copy of the *Troy Messenger* report appeared in the *Journal* (Pensacola, Fla., Jul. 25, 1909, section 2, p. 9; ||yynarkze). See, too, the June 2 edition of the *Messenger* (p. 4c; ||mum8pd2j).
13. Aldrovandi cited Ambroise Paré and Jakob Ruf. Ruf, a Swiss surgeon and obstetrician, pictured and described the Monk Calf in books on human obstetrics (Ruf 1554, LXXII; ||4h95err7; 1587, Book 5, ch. 3, p. 44; ||y35wc3o2), which suggests he thought it was the offspring of a woman. He said it had four hooves like a cow, that its eyes, mouth and nose recalled a calf, that it had red flesh atop its head in the shape of a tonsure, that a flap of skin like a monk's hood hung down its shoulders and that the skin was everywhere "fissured." This last trait, as well as the presence of the supposed monk's hood, may indicate that the skin was too large for the underlying body with extensive, deep wrinkles and loose flaps of skin being the result. The red, fleshy circle on its scalp may refer to an incomplete closure of the cranium as seen in a variety of other telenothian births.
14. The German word used was *Mondkalb*, equivalent to the obsolete English word *mooncalf*, meaning monstrosity or abortion. The old idea was that exposure to celestial influences, such as moonbeams, could cause such births.
15. The final two paragraphs are omitted here, as they amount to mere speculation about why the mother had given birth to such a creature.
16. See, too, the *Expugnatio Hiberniae* for the year 1176.
17. *Library and Epitome,* 3.1.4; ||le6nco6.
18. A Japanese correspondent, who wished to be anonymous, translated the characters on the postcard: 私しや (I was) 豊後の別府の (Bungo, Beppu) 生れ (born in).
19. Both Geoffroy Saint-Hilaire (1836b, pp. 270-271) and de la Fond (1802, I, p. 376) mention what was likely the same animal. They say it had an extra leg attached to its withers and was exhibited in Paris in 1745.
20. *Colophon:* A boshominid, as depicted in Liceti (1665, p. 190).

A hominocaprinid taken from a white Saanen goat The pink region is a protruding tongue, not lips. This individual appears to be anophthalmic. *Image: screenshot of a video taken by automotive mechanic Osmam de Jesus in Brazil, Aug. 4, 2021 (||63yx5edv, ||ndjv76kx).*

HOMINOCAPRINIDS (GOAT- AND SHEEP-HUMANS)

> *Herodotus in Euterpe says that in his time a goat had intercourse openly with a woman.*
>
> —JEAN TIXIER DE RAVISI, *Officina*

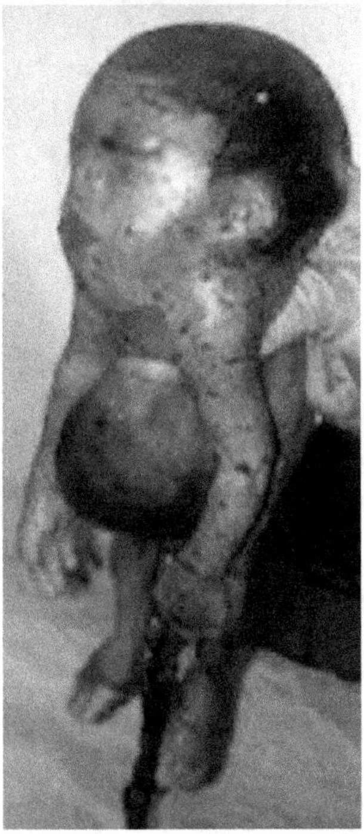

Though in ancient times the existence of such creatures was widely accepted, they are now generally considered legendary. But is such really the case? In 2012, a photo of what looked like a half goat and half human, born in Jahun, Nigeria, went viral on the Internet (see image at right). Few details are available for the Jahun birth, but from the account two facts are clear: 1) its mother was a human being; and 2) its body looked human above the waist but goatlike below—the same structure described for the satyrs and fauns of old. In the photo, note the presence of hooves on the hind extremities and what appears to be hair. The *Daily Post* (Yaba, Nigeria; ||yyk2n62r) reported the following about this composite.

> A general hospital located in Jahun, Jigawa state became a theatre of the absurd and bizarre when a woman who had labored for 9 months gave birth to what looks like a combination of human head and goat legs. The woman, whose identity is yet to be revealed, had delivered peacefully only to have had a child with human head with no neck, and goat legs handed to her by the doctor.

A variety of other hominocaprinids are pictured on the next page.

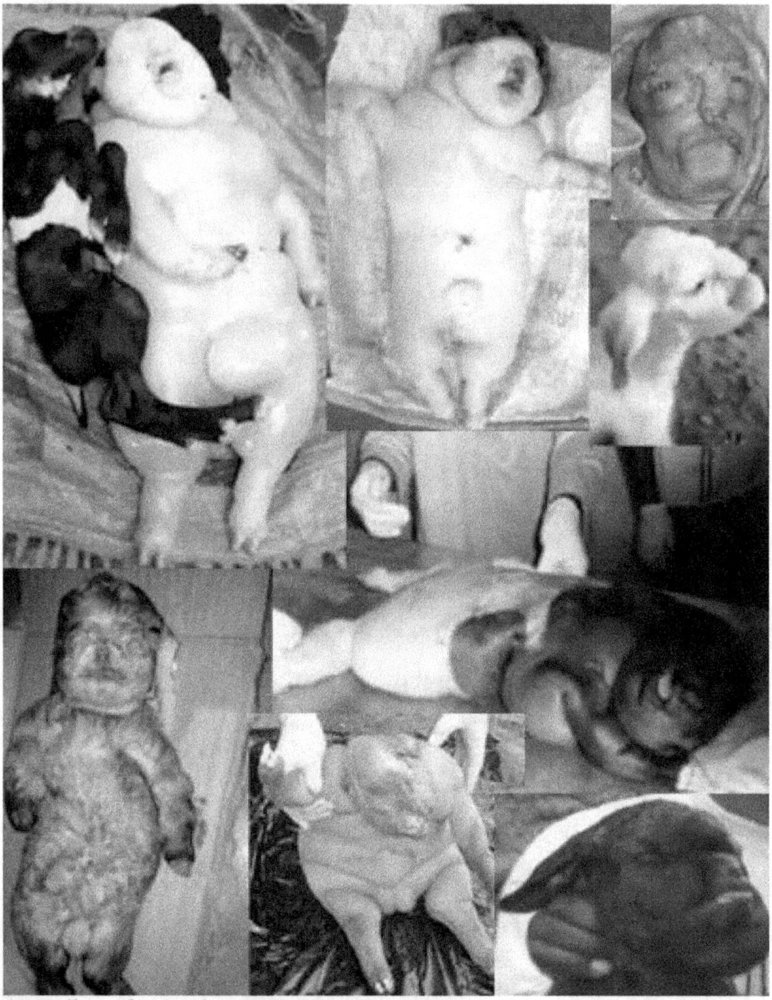

A medley of recently reported hominocaprinids. From upper left clockwise: Yeaned by a goat, Sultan Kudarat, Philippines (2018; ||to65ut3); by a goat, Solapur hamlet, Karnataka, India (2014; ||uzwlsxt); by a goat in the village of Saltipada, Gujarat, India (2021; ||nnrpjjxb); by a ewe, in the village of Chirkei (Чиркей) in the Buinaksky district of Dagestan (2015; ||s3ugf9p); by a ewe, Nigeria (2012; ||slwgcd4); by a goat, Mombasa, Kenya (2009; ||v2ja2l9, ||rrlvj5s); by a ewe, Lady Frere, South Africa (2017; ||wkzhcfc); by a goat, Kota Tinggi, Malaysia (2016; ||sbvy6sl). As can be seen, such births vary widely, some hairy, others not. All had hooves instead of hands and feet. Another recent case (not pictured) was borne by a goat at Maboleni, in Zimbabwe, in 2009 (||2p9fypzt).

A Sheep-Human Hospital Birth

In 2018, one of my British correspondents told me his midwife sister was asked at a hospital in 2004 to help care for an infant sheep-human. Shaken, she changed careers. Her description of the birth follows, as it was sent to me by her brother, except information tending to identify those involved has been omitted. "I just spoke to my sister," he wrote, "and got some more information from her. Here's what she said:

- Witness [his sister] was a midwife on night shift at a hospital in an English city.
- Witness had worked at the hospital for three years and had witnessed over 100 deliveries.
- She had seen birth defects before, but never like this.
- She did not witness the delivery.
- Witness was told to look after the infant as part of her normal duties during a night shift on the maternity ward.
- Witness was afraid of the infant and refused to attend to it, so somebody else fed it during that shift.
- Another midwife told her "we get these all the time."
- Says the mother rejected it.
- Hospital could do no more, so infant was to be sent home.
- Infant was to be "discharged to the care of the parents" (to "let nature take its course," which meant that it was expected to die of neglect; death would be attributed to "cot death").
- Infant was bottle-fed, as the mother refused to have anything to do with it.
- Spina bifida [often seen in telenothians] was present (there was a hole in the base of the spine, near the coccyx; hairs protruding from the hole).

LIMBS

- Cloven hooves in place of hands and feet.
- Skin covering the hooves was as normal human skin.
- No tail was present.

HEAD AND VOCALIZATIONS

- A "proper head of hair" but further back on the scalp than normal.
- Hair was black and normal for a human.
- Body was hairless [commonly reported in homungulates].
- Mouth was a muzzle-shaped snout like a dog or sheep.
- Long tongue like a sheep or goat that reached the end of the muzzle and could protrude outside the mouth.
- Bleated exactly like a lamb.
- Mouth shaped like a sheep's, extending halfway up the snout.
- Ears abnormally low [also common in homungulates].

* A doctor told his sister that "care homes and adoption are common in such cases."

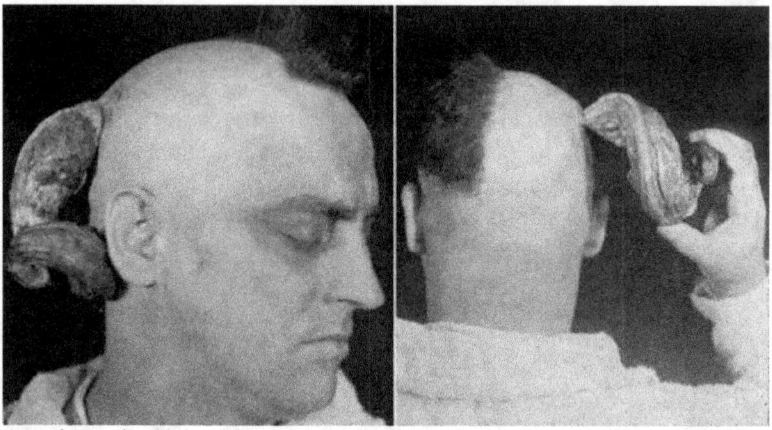

A patient who had a 20-cm-long, ramlike, spiral horn excised at a Brooklyn hospital (*American Journal of Surgery,* Charache 1935).

20TH-CENTURY REPORTS

Nulvi, Sardinia. From the *Star-Independent* (Harrisburg, Pa., Dec. 10, 1914, p. 1g; ||h6vpvx7): "CAGLIARI, ITALY, Dec. 9.—From Nulvi comes a strange story of a sheep giving birth to a monstrosity that in all respects resembles a human being. It created great consternation among the populace and the attention of noted men of science has been attracted to an event that is almost beyond belief. ... Giovanni Fiori is the name of the shepherd who was in charge of the flock when one of its number is said to have given birth to the monstrosity which weighed twelve pounds. The head is round, and the rear legs are much longer than those of the front part of the body. All who have seen the phenomenon pronounce it a close resemblance to a human being and Fiori has donated the 'lamb boy' to the University of Sassari."

Salins, Jura, France. Le Petit Comtois (Besançon, Aug. 30, 1909, p. 2f; ||ycvj7vua) said a fetus taken from a slaughtered sheep had a nose and mouth like a human's. Its ears, ovine in form, were placed low against the chest.

La Mare Carel, Normandy. La Liberté (Paris, May 3, 1923, p. 1f; ||y9ezb2wc) said a goat belonging to a Mr. Prétat, at La Mare Carel, bore two kids, one of which had nearly the head of a newborn infant with a rounded forehead and turned-up nose. The ears were at the rear near the base of the head. It lived a few minutes. In its bloated belly, autopsy revealed two large kidneys like those of a pig. Pig and human kidneys are much alike.

Gómez Palacio, Durango, Mexico. La Opinión (Los Angeles, Aug. 29, 1931, p. 7; ||3bbb2dsm) said, in translation, "GÓMEZ PALACIO, DGO., Aug. 28.—The Director of Public Health, Dr. Francisco Hernández Barrena, reported a freak discovered today at the Office of Public Revenue, in the city of Gómez Palacio, where a goat bore a kid that had various human members. Juan Arenas, the Foreign Tax Agent, brought this animal, which died during birth, to the sanitation authorities, who will carefully investigate this animal to determine its origins. The kid has certain human features, the lips, nose, forehead and the hair on the head being the most notable. The genitals resemble a man's. It has a double jaw, and the feet have fingers, albeit small ones."

La Monselie, Cantal, France. A goat at La Monselie reportedly bore a stillborn, naked kid with a human head in each of two years in a row (*Le Réveil du Cantal,* Mauriac, Apr. 25, 1908, p. 3c; ||36w2c8vw).

Ontario, Oregon. The *Butte Miner* (Mar. 17, 1903, p. 8a; ||2d8f273w) said a lamb yeaned near Ontario had "two eyes in the center of the forehead, which are protected by a single diamond-shaped eyelid, the eyeballs being connected. The upper portion of the skull and lower jawbone resemble that of a human being. This monstrosity was brought to the city yesterday by G. A. Darr, a farmer living four miles west of the town, who is feeding about three thousand head of sheep on his place."

* A George Alexander Darr (1870-1955), a farmer, was living near Ontario (||2p9n6ts5).

Warialda, New South Wales. The *Tumut and Adelong Times* (Oct. 18, 1932, p. 4; ||kgp4h6n; also see: ||h9y8lys, p. 6d).

> A freak of nature, a lamb with the characteristics of a human face, is being shown by the Moree Ambulance Queen Committee.
>
> The lamb, which was born six weeks ago, but lived only three days, has eyes with pupils and skin between the eyes resembling that of a human. It has woolly eyebrows, short fluffy eyelashes, even teeth and jaws like a human being.
>
> Its body is that of a normal lamb. It was born at 'Fishpond,' Warialda, and Mr. Mick Byrnes is the owner.
>
> He tried to raise it by spoon-feeding and offered £10 to anyone who could rear it.
>
> Its nose is three inches long, a piece of protruding flesh with one nostril [likely, a frontal proboscis], quite different from that of an ordinary lamb.

The *Moree Gwydir Examiner* (Oct. 6, 1932, p. 2; ||m78d75h) ran a report about the same case:

> On Friday there will be on exhibition in Moree of what is perhaps the greatest freak of nature in the world. This is a lamb with the features of a human being. There are the high forehead, two eyes with an indication of a nose, the mouth, teeth, bottom lip and chin. The features on the whole are very distinct.
>
> The lamb, which was born six weeks ago, but lived only three days, has eyes with pupils and skin between the eyes resembling that of a human. It has woolly eyebrows, short fluffy eyelashes, even teeth and jaws like a human being.
>
> The lamb was dropped on a property between Warialda and Bingara[1] and lived for three days. It then came into the possession of Mr. Eli Brown, a prominent resident of Bingara and well-known north-western bookmaker, who had the lamb preserved in a big jar.

Mexico City. El Mensajero (Phoenix, Az., Mar. 9, 1936, p. 1b; ||57amsuva see too: ||257haczn) said, in translation, "In the slaughterhouse in Tacuba [a northwestern section of Mexico City], a goat produced a fetus that has the face of a human being and—except for its hoof-tipped extremities—could be taken for an infant."

Sevastopol, Crimea. The *Mährisch-Schlesische Presse* (Jeseník, Austria-Hungary, Dec. 7, 1910, p. 4g; ||y6kv243f) said, in translation, "A piece of almost unbelievable news from Sevastopol: A sheep yeaned twins, one of which has the complete head of a human being."

Breaffy, Ireland. The *Aberdeen Journal* (Aberdeen, Scotland, Apr. 28, 1919, p. 2g) said, "A ewe, the property of John Murphy, Breaffy, Ballina, recently gave birth to a lamb normal in body but with a human head. This freak has caused no small commotion in the district, and many are the opinions expressed regarding it."

Torremaggiore, Italy. The *Dolomiten* (Bolzano, May 7, 1930, p. 2c; ||d385wrhh) said, in translation, "A MONSTROSITY WITH THREE LEGS AND A GOAT'S HEAD.—FOGGIA, May 6.—In Torremaggiore, a woman gave birth yesterday to a child that had three legs and a head that looks very similar to a goat's. Two incisors were very well developed. This monstrosity died soon after birth."

Balatonberény, Hungary. The *Innsbrucker Nachrichten* (Dec. 27, 1909, p. 7b; ||49sbfdnz) said a woman at Balatonberény bore an infant with the head of a sheep and that both "were doing well" ("befinden sich woll").

Shaw, Mississippi. The *Democrat-Star* (Pascagoula, Apr. 28, 1893, p. 1c; ||h9vldh9) said a "woman in Sunflower County, Miss., has given birth to a freak of nature which is half human and half billy goat." The *Sea Coast Echo* (Bay Saint Louis, Ms., Apr. 8, 1893, p. 4b; ||y569rqxx) said it had "the peculiar cry of the kid." The *Grenada Sentinel* (Grenada, Ms., Apr. 15, 1893, p. 4a; ||y4rlph47) said it was born in Shaw.

Liverpool, Ontario. The *Ottawa Citizen* (May 7, 1900, p. 1; ||4u9ad853) said that in the (now-defunct) city of Liverpool. "Thousands are coming from far and near to see the child which has three eyes, rudimentary horns set in its forehead and club feet, which with only a slight stretch of the imagination can be considered cloven."

Avenza, Italy. L'Italia (San Francisco, Oct. 20, 1904, p. 3a; ||yj8k2xae) said a woman at Avenza gave birth to a hairy, goatlike nondescript with large ears and legs like an ape. It was born dead.

19TH-CENTURY REPORTS

Newark, Ohio. The *Star* (Marion, O., May 18, 1882, p. 4a; ||mr2ty5ua) said, "A ewe on the farm of William Freese, Licking County, gave birth to a dead lamb, the head of which was exactly like that of a human being. It had the perfect body of a sheep, but without wool, the head and neck had the perfect shape and expression of a human being, and with a slight appearance of wool on the top part of the head."

* A William A. Freese (1849-1899) lived near Newark in 1882 (||y3ujohln).

Lawrenceville, Illinois. From the *Weekly Gazette* (Fort Worth, Tex., Mar. 9, 1888, p. 7c; ||h6vpvx7): "LAWRENCEVILLE, ILL., Feb. 26—A remarkable freak of nature was discovered last Wednesday afternoon on the farm of Mr. J. E. Roberts, a few miles from this place. It is a young lamb with a human head and body. Its weight is about seven pounds. Its skull and face are almost perfectly human except the ears and a little wool on the forehead. Its body is human and, although the feet are cloven, the formation and attachments of the shoulders and hips indicate that the most natural position of the animal would have been that of a biped. When found the monstrosity was dead, but there were signs that it had been born alive. Many physicians and eminent persons have examined it and pronounce it as more human than animal. It is causing great excitement. Mr. Roberts has entrusted the curiosity to Mr. S. D. Dills, who has taken steps for its preservation."

* James Ellsworth Roberts (1853-1903) lived near Lawrenceville in 1880 (||y4em2mct).

Eden Township, Wyandot County, Ohio. The *Cincinnati Enquirer* (May 7, 1894, p. 7; ||yszdbesv) said, "Isaac Corfman, of Eden Township, has a monstrosity that is attracting considerable attention. It is a lamb with a head as near human as could be imagined. The face, however, is a fright. It has two lower jaws, a nose with a long hook and an eye with no eyeball."

* An Isaac Corfman (1838-1919), a farmer, lived in Eden Twp. in 1894 (||h7kbtc4u).

Horse Creek, Wyoming. The *Democratic Leader* (Cheyenne, Wyo., Apr. 21, 1885, p. 3a; ||4vpau4ec) said, "A monstrous lamb was recently born at Mortimer's sheep ranch on Horse Creek. The lamb had a body of its kind, but its head was that of a human being, and was covered, however, with wool. The monstrosity lived but a few moments."

Butte County, California. The *Tribune* (*San Luis Obisbo,* Nov. 19, 1881, p. 2; ||yjvsbj9e) reported a dead sheep with human traits found in Butte Co.

Villejuif, Val-de-Marne, France. Le Journal (Paris, May 6, 1899, p. 3d; ||3wpnyscv) reported a goat yeaning a stillborn hominocaprinid in the Parisian suburb of Villejuif. In translation, the description read, "The stillborn kid, born this morning, had a baby's head, crowned with an attractive crop of blond hair. The forehead is very well developed, the eyes, open, the nose, flat, the teeth, white, the goatee, prominent. The chest is that of a child. The navel visible. The legs, well shaped, end in hooves. In a word, it's a freak. In the afternoon, the same goat yeaned a completely normal kid."

Tehama, California. The *Marysville Daily Appeal* (Mar. 1, 1874, p. 3c; ||kqukzch) said, "The people of Tehama[2] were much excited on the morning of Wednesday last, by someone coming to town and reporting that a lamb had been born on the opposite side of the river with a human head. On this news coming to the ears of Mr. Charles Harvey,[3] he immediately repaired to the locality, captured the what-is-it, and brought it to town, where it was viewed and examined by nearly every person in the place and in almost every instance was declared to be a sheep body and a baby head, a hybrid."

Tulare, California. The *Tulare Times* (quoted in the *San Francisco Chronicle,* Mar. 2, 1874, p. 2e; ||mrenxn6b) reported another hominocaprinid yeaned in 1874, *"Editors Times:* I have the pleasure of recording a freak in nature which never met my eyes before. I had a ewe that gave birth to a lamb of perfect form except the head which was as that of a babe. Its head was about four inches across and five inches long; its eyes were about one inch apart, and about one inch below its eyes extended its nose in a raised form, as that of a human being. Its mouth was hardly an inch from one corner to the other. Its chin extended about one inch below the mouth. Its cheeks were full and long, rather than broad. Its ears lopped close down by the side of the head, and its face was covered with a short, fine wool, and the under and back part of the chin and neck, and the top of the head, with long coarse hair. When I discovered it, I showed it to G. H. Perry, who turned away and said he could not bear to look at it. Unfortunately, before I thought it would be of any value, it was destroyed."

Wallula, Washington. The *Yakima Herald* (Yakima, Wa., Jun. 20, 1895, p. 4a; ||jl4887z) said, "A singular freak of nature was born recently at a sheep ranch near Wallula.[4] It was a perfect lamb, so far as the body was concerned, but had the head and face of an infant. The little thing only lived about a week when it died, as it could not nurse from the mother ewe. Reliable parties who saw the monstrosity, while living, said its life could have been indefinitely prolonged, had proper sustenance been given. It is stated that the oddity has been preserved in alcohol and will be sent to a prominent museum in the east."

Banyuwangi, Java. The *Soerabaiasch-Handelsblad* (Surabaya, Java, Aug. 7, 1896, p. 1e; ||ybht7d9b) said (translated by M. Butter), "From the city of Banyuwangi there is a report that a kid has been born with a head that is clearly human. … This miraculous product has only been alive for a few hours."

Los Angeles. The *Anaheim Gazette* (quoted in the *Santa Barbara Weekly Press*, Santa Barbara, Ca., Feb. 28, 1880, p. 7e; ||45r7s5kd) said, "An employee of Mr. H. M. Johnston brought to the LOS ANGELES HERALD office yesterday evening the body of a dead lamb that had been born with one eye, which was located in the center of its forehead and surmounted by a brow resembling that of a human being. The upper jaw was also found to be like that of a man. The monstrosity was one of twins and died immediately after birth, its mate living."

* A Los Angeles rancher Hancock McClung Johnston (1845-1945) did exist (||yhssp2nc).

Arrondissement de Baume, Doubs, France. Le Courrier Franc-Comtois (quoted in the *La Presse,* Paris, Jul. 4, 1876, p. 4d; ||3ppf5fj6) said, in translation, that a ewe in the Arrondissement de Baume had lambed "a hybrid being, a human in the upper parts of its body, a lamb in the lower. The monster would have lived but the owner … killed it at birth. In doing so, he made a big mistake. For, not mentioning the creature's scientific interest, it would certainly have made him a fortune. He would no doubt have made a considerable sum exhibiting it."

Souk Ahras, Algeria. In a list of reports from Souk Ahras, *Le Républicain de Constantine* (Feb. 13, 1889, p. 3b; ||2x7phf7k) said, in translation, "A freak … was found last Thursday at the abattoir during the slaughter of a pregnant ewe. The fetus having been extracted, those present found themselves in the presence of a monster more like a human being than an animal. The head, the arms, the chest, completely hairless, were those of an infant, while the lower members were identical to a lamb's. Hundreds of curiosity seekers demanded to see the freak when it was brought into town and witnessed its resemblance to a newborn infant. This little monster was breathing when it came into the world. So, if it had been swaddled and cared for by a midwife, instead of being left exposed in the sharp morning air, our town would have had a faun or, if you prefer, a satyr, which could have been adopted and raised by the members of our municipal council."

Armidale, New South Wales. The *Armidale Express* (quoted in the *Logan Witness,* Beenleigh, Qld., Mar. 19, 1892, p. 2; ||zh39aom) said, "There are at present on view in Armidale some most singular freaks of nature, in the shape of sheep with several tails and extra feet, but one sheep in particular is a most extraordinary specimen. This woolly individual, which rejoices in the name of Jimmy Ah Poo, while being like other sheep in formation, wool, &c., bears a face of human resemblance startlingly like that of a Chinaman. There is the distinct flat nose, thick lips and jaw that bears the unmistakable hallmark of the Mongol."

Stockton, Wisconsin. The *Oshkosh Northwestern* (Oshkosh Wis., Jun. 10, 1878, p. 4c; ||4a47ff3v) said, "A description of a very curious monstrosity comes to us from the southern part of the town of Stockton, which we would look upon with incredulity did it not appear to be so well authenticated. One man who lives near the farm where it was given birth, tells us that it was seen by several persons, all of whom agree in their description of it. We allude to a lamb with a human face. Not only that, but it had no wool on it, and is ears, nose, mouth and shoulders were unmistakably human in shape, while what should have been its forelegs resembled the arms of a child, only they were supplied with hoofs where the hands ought to have been."

Fontenay-Saint-Père, France. L'Industriel de Saint-Germain-en-Laye (Mar. 30, 1878, p. 1d; ||jd2dtuzb) reported a conjoined twin with eight legs, two sheeplike bodies, and a head like an ape, yeaned on a farm at

Fontenay-Saint-Père. It died during delivery. In this book, when animals are described as having ape-like body parts but were born in locales where apes are absent, it is assumed that the resemblance was to humans instead.

Buire, Aisne, France. Le Guetteur (Saint-Quentin, Jan. 31, 1895, p. 2e; ||3m7m8dzd) said, in translation, "BUIRE.—A FREAK.—On the farm of Mr. Albert Drubigny, … the mayor, a ewe just yeaned a freak lamb. It has the very head of an ape, surmounted by a trunk [i.e., a frontal proboscis] that serves as a nasal apparatus by which it breathes. The eyes, the boundaries of which are edged in horrible naked flesh, and the pointed chin complete the bizarre face. A cleft upper lip keeps it from nursing, and it has to be fed with a spoon. It bleats like its siblings."

Paris. L'Est Républicain (Nancy, Feb. 15, 1899, p. 2a; ||2dtjyawk) said, in translation, "An amazing freak was found in the womb of a pregnant sheep recently in a slaughterhouse in La Villette [a Paris neighborhood]. Its face resembles both that of a newborn infant and that of a lamb. It had a flat nose, huge jaws, almond-shaped eyes, narrow cheeks and devil ears. From the front the face looked almost like a baby's, but in profile almost like a lamb's. The arms and legs were like the human members, except that they ended in hooves instead of hands and feet. The chest and stomach were those of an infant."

Houston, Texas. The *Houston Post* (Feb. 8, 1897, p. 8; ||jcygurc) said, "Last night Mr. J. Drenkle, Jr.,[5] blacksmith at the Houston and Texas Central shops, exhibited at the office of the POST a very remarkable freak, in the form of a young goat, which was taken yesterday from its mother, stillborn, and which was remarkable in that nature had placed the face of a human being upon the body of a goat. The mouth, nose and eyes, and, in fact, the entire face was that of a human, while the back of the head and ears were those of a goat. The front shoulders also had the semblance of a human formation, but nature, as though revolting at its repugnant work, failed to provide a backbone and thus deprived the monstrosity of the means of existence even had it come into the world alive."

Doubravník, Czechia. Der Humorist (Vienna, Mar. 13, 1852; ||y28r4chb) said, in translation, "On Feb. 18, in Doubravník, near our city of Brünn, a mother sheep in a merchant's flock gave birth to an abominable monstrosity, with a well-formed human head, but otherwise like a sheep."

London. Gazette Nationale ou le Moniteur Universel (Paris, Oct. 8, 1825, p. 3a; ||55jsh2zd) reported a creature captured on the Gulf of Guinea and exhibited in London, a living adult with the body of a woman but the head of a goat, topped by two great horns. The body was, according to the report, covered with soft, white hair "like that of a Tibetan goat."†

Calma, Serbia. The *Drau* (Serbia, quoted in the *Prager Tagblatt,* Prague, Mar. 30, 1883; ||yxdrrgm8) said, in translation, that in Calma, a nearby village, "there was born today a creature half human and half goat. The head, body and legs are normal for a human, but the mouth pokes out like a goat's, the ears are soft, drooping folds of skin, and the legs have hooves instead of feet. The nose is that of a human being, but the arms are mere stumps. On its rear, this wondrous little goat-man—fortunately born dead—bears a prominent tail. The mother of this monstrosity is said to be a goat that had not previously given birth."

Valparaiso, Indiana. The *Semi-weekly Interior Journal* (Stanford, Ky., Apr. 14, 1899, p. 4b; ||hmclnas) said, "A lamb with a human head was born near Valparaiso, Ind."

Anklam, Mecklenburg-Vorpommern. Jurende's Vaterländischer Pilger (1828, p. 373, #8; ||yxq9x6k4) reported a sheep-human born in 1822. In translation, it said "Its face was "round with light blue eyes, eyebrows, and plump, hairless cheeks. The scalp was covered in shaggy hair. The nose, chin, mouth and tongue were like a human's. The body, too, was human, large and round. The feet were cloven hooves, but the front pair faced sideways, while the rear pair was directed forward. In addition, at birth, this un-beast ["Unthier"] was twice as big as a normal lamb. Born dead, it was sent to a museum in Berlin."

* Also see *Mittheilungen der K.K. Mährisch-Schlesischen Gesellschaft zur Beförderung des Ackerbaues, der Natur- und Landeskunde in Brunn,* 1822, p. 144; ||ntyjkmum.

Goshen, Indiana. From the *Comet* (Johnson City, Tn., Apr. 23, 1891, p. 1h; ||jjwto37): "GOSHEN, IND., April 22.—William Clark, a farmer living east of here,[6] brought to town a lamb with a perfect human face. The lamb was born only a few days ago, but the features are well developed and look more like the features of a mature woman than a child. The lamb is alive and apparently perfectly happy."

†. Also see *Jurende's Vaterländischer Pilger* (1828, p. 373, #13; ||yxq9x6k4).

HOMINOCAPRINIDS BORNE BY WOMEN

Mexico City. Cevallos (1820) described a Mexican packer named Paul Rodriguez with three large horns like those of a ram. Rodriguez had always kept his head covered, but due to an accident at work, in which a barrel partially broke off one of his horns, his secret was exposed. They not only looked like ram horns but were also made of the same material. They were not, however, directly connected to the bone of the skull.

Paul Rodriguez

Marlton, New Jersey. The *Daily Argus* (Rock Island, Il., Dec. 8, 1891, p. 1b; ||klnxq2j): "MARLTON, N.J., Dec. 8.—The young and pretty wife of Mr. Schaeffner gave birth Monday week to a child who is one of the strangest freaks of nature. The child is without eyes and, just where they should be, are two horns. Otherwise, the child is perfectly formed and healthy."

Henderson, Kentucky. Dr. P. J. Thompson (in Furman 1880, p. 465) said he attended a woman who bore a child with "the peculiar white looking eyes of the sheep, and as he grew up, his hair, though kinky, was perfectly white like wool. In addition to this, and more remarkable than all, he always gave forth the peculiar scent of the sheep, it being specially marked when he was heated and perspiring freely."

* A Dr. P. J. Thompson (b. circa 1842) lived in Henderson in 1880 (||qtvom64).

Mosiertown, Pennsylvania. Leroy Lambert Doty (1881), a physician residing in Mosiertown, reported his experience with a hominocaprinid. "I was called, about three o'clock on the morning of March 27th, 1878, by Mr. H. He said his wife had been delivered of a child about half an hour before, and that the child was not right. On examination, I found the head drawn out in the form of a sheep's head and covered with black wool, which also extended down the neck, across the shoulders and back and down the outside of the arms. The two central incisor teeth on each jaw were large and presented the same appearance that a lamb does when first born. The eyes were large and circular and looked just like a sheep's eyes. A stump of a tail, about four inches long extended out from the coccyx. Every few moments it would give a low, bleating cry, which sounded just like a young lamb. The body and limbs were those of a well-formed female child. ... The child died with convulsions on the third day."

* A Leroy Lambert Doty (1848-1917), a physician, lived in Mosiertown (||nwbmctup).

McKenzie, Tennessee. The *Memphis Commercial* (Mar. 10, 1893, p. 2b; ||3rb7vkrc) reported a McKenzie child whose "entire lower extremities, from the hips down, were that of a 'nanny' goat, the body, shoulders, head and arms being well-developed human. The freak lived about a month."

Vieux-Condé, Nord, France. Le Pays (Paris, Mar. 14, 1865, p. 3c; ||2wsd5hcb) reported a woman at Vieux-Condé giving birth to a goat-headed offspring that survived only a few hours.

Temuco, Chile. The *Star and Herald* (Panama City, Panama, Sep. 21, 1894, p. 2c) said, "The freak of nature reported from Tebuco, [*sic*] and a Vichuquen newspaper is responsible for this story. This journal states that a woman of the former place lately gave birth to a creature of the feminine gender whose body with the exception of the head and arms, resembled any ordinary infant, while the head was that of a sheep, and the arms two masses of unshapely flesh without bones. The cries of the creature are also exactly the same as the bleatings of a lamb."

Union County, North Carolina. The *Weekly Raleigh Register* (Raleigh, N.C., Apr. 21 1852, p. 4e; ||5n86t7kj), describing a hominocaprinid then living in Union County, stated that he is a negro boy named 'Squire,' ... who is 14 years old [ergo, born ca. 1838], 38 inches high, and weighs just 27 pounds. His face and head are about the size of an ordinary child one month old, and his head is shaped like the head of a sheep. His eyes are quick and restless, somewhat like those of a monkey. His feet are 6½ inches long, and only 2 inches wide, and quite thin. He has not grown any for two years and cannot stand erect without assistance. He appears to hear distinctly but cannot talk or make any intelligible sign whatever. ... His feet and hands are placed flat on the ground; and he trots off very much like a dog in an ordinary gait. His hands are shaped like the hands of a monkey. ... He evidently possesses no mind above the brute creation, but seems to be as susceptible of teaching, as a dog or a horse.

Sabana Grande, Puerto Rico. The *La Correspondencia de Puerto Rico* (San Juan, P.R., Dec. 29, 1899, p. 3f; ||pt4k3azb) said, in translation, "SABANA GRANDE.—Today a notable freak was brought to the public eye. A woman living near here has given birth to a child with a head shaped like the head of a sheep. The bones overlying the brain are absent and its mouth has three tongues. It's still alive."

EARLY REPORTS

Some assert, too, that in that region the pleasant, shady hills belong to the goat-pans and satyrs.
—PLINY, *Natural History,* 6.35

One report seems to bridge the gap between modern and ancient times. The *Kimball Graphic* (Kimball, S.D., May 11, 1888, p. 4e; ||y3355buo) said, "The residents of Jeffersonville, Ind., were startled a few days ago, it is reported, at sight of a strange being passing through the town. The freak was a negro, who walked on all fours, and whose body in many respects resembled that of some animal. His hands and feet looked like hoofs. He could not talk so anyone could understand him and was supposed to be bound for Cincinnati."

This tale of an incoherent satyr shares much with a story told in Plutarch's biography of the Roman general Sulla (ca. 138 - 78 B.C.):

> Sulla, having marched through Thessaly and Macedon to the seacoast, prepared with twelve hundred vessels to cross over from Dyrrhachium to Brundusium. Not far from thence is Apollonia, and near it the Nymphaeum ... there they say, a satyr, such as statuaries and painters represent, was caught asleep, and brought before Sulla, where he was asked by several interpreters who he was, and, after much trouble, at last uttered nothing intelligible, but a harsh noise, something between the neighing of a horse and the crying of a goat.

Pan and Daphnis
(Pompeii, ca. 79 A.D.)

The ancients took the existence of such creatures for granted. Plutarch likely chose to mention the Nymphaeum because it was sacred to the god Pan and to the nymphs. Pan, as usually depicted, had the form of a satyr[7] (see figure, right) and had the same lustful habits. And the nymphs, too, were often described as the objects of satyrs' lechery.

Classical literature is rife with accounts like Plutarch's. And, as Plutarch seemingly accepted his own story about Sulla, the authors of other such accounts accepted their own, or even expressed their belief outright. As Pliny (12.1) observed, "we also believe that the Silvani and

Fauns and various kinds of goddesses are as it were assigned to the forests from heaven and as their own special divinities." Pausanius (*Description of Greece,* 1.23.5-6) states that, "Euphemus the Carian said that on a voyage to Italy he was driven out of his course by winds and was carried into the outer sea, beyond the course of seamen. He affirmed that there were many uninhabited islands, while in others lived wild men. The sailors did not wish to put in at the latter, because, having put in before, they had some experience of the inhabitants, but on this occasion, they had no choice in the matter. The islands were called Satyrides by the sailors, and the inhabitants were red haired, and had upon their rumps tails not much smaller than those of horses. As soon as they caught sight of their visitors, they ran down to the ship without uttering a cry and assaulted the women in the ship. At last, the sailors in fear cast a foreign woman onto the island. Her the Satyrs outraged not only in the usual way, but also in a most shocking manner."

In his geographical description of the Roman Empire, Pliny (5.1.7), while giving an otherwise factual account of the Atlas Mountains, comments that at night its highest peak "flashes with frequent fires and swarms with the wanton gambols of Goat-Pans and Satyrs, and echoes with the music of flutes and pipes and the sound of drums and cymbals." And regarding the satyr Marsyas who, after losing a piping competition with Apollo, was flayed alive by the god, he says (16.89, Rackham translation), "We have mentioned the region of Aulocrene, traversed by the route leading from Apamea into Phrygia; in it travellers are shown the plane-tree from which Marsyas was hanged after losing his match with Apollo."

The Romans accepted that humans could hybridize with animals. In his *De Natura Animalium,* Aelian (ca. 175 - ca. 235 A.D.) tells a story that reflects a belief in, and even an approval of, events that most people today would regard as myth, and that many would even reject as obscene and impossible. The Scholfield translation (1958, 6.42) reads,

> An Italian story, which records an event that occurred when affairs were at their prime in the city of Sybaris, has reached my ears and is worth relating. A mere boy, a goatherd by occupation, whose name was Crathis, under an erotic impulse lay with the prettiest of his goats, and took pleasure in the union, and whenever he wanted sexual pleasure, he would go to her; and he kept her as his darling. Moreover, the amorous goatherd would bring to his loved one aforesaid such gifts as he could procure, offering her sometimes the loveliest twigs of

tree-medick, and often bindweed and mastic to eat, so making her mouth fragrant for him if he should want to kiss her. And he even prepared for her, as for a bride, a leafy bed ever so luxurious and soft to sleep in. But the he-goat, the leader of the flock, did not observe these proceedings with indifference, but was filled with jealousy. For a time however he dissembled his anger and watched for the boy to be seated and asleep; and there he was, his face dropped forward on his chest. So, with all the force at his command, the he-goat dashed his head against him and smashed the fore-part of his skull. The event reached the ears of the inhabitants, and it was no mean tomb that they erected for the boy; and they called their river "the Crathis" after him. From his union with the she-goat a baby was born with the legs of a goat and the face of a man. The story goes that he was deified and was worshipped as a god of the woods and vales.

Aldrovandi's depiction of the goat-human of Sybaris.[8]

In Egypt and Mesopotamia, such hybrids were not only accepted but even viewed as sacred. It seems that any births commingling the traits of humans and animals were interpreted as prodigious avatars of the gods—even as they still are among the Hindus (see pp. 21-22). When Alexander the Great entered Egypt in 332 B.C., he was deified as the son of the god Ammon, who was often represented either as a ram or, at least, as ram headed. Thereafter, coins depicted Alexander with the horns of Ammon, effectively making him into a hominocaprinid.

Indeed, it seems that there were even attempts to produce such deities, which the Christians later chose to forget. Thus, Budge (*Gods of the Egyptians*, 1904, p. 353) writes: "At several places in the Delta, e.g., Hermopolis, Lycopolis and Mendes, the god Pan and a goat were worshipped; Strabo quoting (xxii. 1, 19) Pindar says that, in these places, goats had intercourse with women. Herodotus (II. 46) instances a case which was said to have taken place in the open day." The passage in Herodotus to which Budge refers (quoted here from Macaulay 1890) reads, "Now the reason why those of the Egyptians whom I have mentioned do not sacrifice goats, female or male, is this: "the Mendesians count Pan to be one of the eight gods … and the painters and image-makers represent in painting and in sculpture the figure of Pan, just as the Hellenes do, with goat's face and legs … The Mendesians then reverence all goats and the males more than the females (and the goatherds too have greater honour than other herdsmen), but of the goats one especially is reverenced, and when he dies there is great mourning in all the Mendesian district: and both the goat and Pan are called in the Egyptian tongue Mendes. Moreover, in my lifetime there happened in that district this marvel, that is to say a he-goat had intercourse with a woman publicly, and this was so done that all men might have evidence of it."

Herodotus traveled extensively in Egypt and may have witnessed such rituals himself. Indeed, it seems the Egyptians were far more intimate with animals than would be considered normal in Western culture today, or even in certain other ancient cultures. Herodotus, an eyewitness of their customs, listed many ways in which the people of the Nile differed from those of other nations. He saw one such difference in their relationships with animals: "All other men pass their lives separate from animals; the Egyptians have animals always living with them" (*Persian Wars,* 2.36).

Left: The Egyptian god Khnum, a hominocaprinid.[9] *Right:* Silver tetradrachm depicting Alexander the Great with rams horns (Arados, 242/241 B.C.).

A hominocaprinid with two proboscides reportedly yeaned in Germany near Frankfurt in 1557 (Wickiana Collection; ||mheej772).

Mary Davis (b. ca. 1594, aged 74; British Museum ||ym37hnyu). On the painting, an inscription (not shown) reads, "Portraiture of Mary Davis, an inhabitant of Great Saughall near Chester ... was taken Ano. Dom. 1668, Aetatis 74. When she was 28 years old, an excrescence rose upon her head which continued thirty years then grew into two hornes after 5 years. She cast them, then grew 2 more. After 5 years she cast them. These upon her head have grown 4 years and are to be seen ..." (ellipses illegible). Also see Morgan (1732, pp. 248-250). For other pictures of Davis, see Leigh (1700, p. 192), as well as ||2wactr8w and ||52k8rhpp.

A headless fetus yeaned by a ewe, Sep. 10, 1571, reportedly at Zürich. It was said to have the feet and genitals of a man but was otherwise lamb-like. (Wickiana Collection; ||bdfkh7rb; also see ||2ph5mv3c).

Even the Jews, who had stern laws against sexual interactions with animals, must have had transgressors who did engage in such acts. Else, such laws would not have been required. And the Old Testament does refer to such behavior. It is widely said that it was from jealousy that the brothers of Joseph sold him into slavery. But it seems English translations have been bowdlerized. The Vulgate (Genesis 37:2) states that after tending the flocks with his brothers, Joseph went to his father and "accused them of the worst of crimes" ("accusavitque fratres suos apud patrem crimine pessimo"). Aquinas confirms that Joseph accused his brothers of bestiality. In the *Summa Theologica* (II-II, 154, 12; ||ykwl6lg), he said the most grievous sin "is the sin of bestiality, because use of the due species is not observed. Hence a gloss on Gn. 37:2, 'He accused his brethren of a most wicked crime,' says that 'they copulated with cattle.'" So, Joseph's brother's reason for selling him would have been more than jealousy. The penalty for the crime of which he had accused them, under Jewish law, was death.

In Christian times, reports about satyrs continued to crop up. In *The Life of Paul the First Hermit* (§8), Jerome tells how, during the reign of Constantine (306-337 A.D.), a satyr "was brought alive to Alexandria and shown as a wonderful sight to the people. Afterwards his lifeless body, to prevent its decay through the summer heat, was preserved in salt and brought to Antioch that the Emperor might see it." Jerome adds that "it was a matter of which the whole world was witness."[10] Augustine (*De Civitate Dei,* 15.23) says, in translation, "Inasmuch as it is very widely reported, and many people who have either experienced it or who have heard from those who have, and whose trustworthiness cannot be denied, do confirm that sylvans, pans and fauns, which are commonly called "incubi," have often wantonly approached women, and have sought and procured intercourse with them … it seems impudent to deny it." Writing in the seventh century, Isidore of Seville (*Etymologiae,* 6) asserted that satyrs lived in the interior of Africa and "had no characteristics of human beings except their faces" ("Satyri, hominum nihil habent nisi facie").

In *De animalibus* (2.1.4; ||bdcs4cxz), Albertus Magnus tells of reports, he says he believed, of a wild man, recently captured in the forests of Saxony. This creature, hairy all over but human in form, he says, was taught to walk erect and to speak a few words with the voice of a goat. John Zonaras (*Annali,* XVIII, 289) recorded a woman at Constantinople, under the reign of Michael VII Doukas, bearing a cyclops with the feet of a goat.[†] Lanfranc of Milan (*Chirurgia Magna,* 1295, 3.2.3) says he was consulted by a man with multiple "eminences" on his head, one as large and sharp as a kid's horn, rooted in the cranium and as long as a thumb (see Schenck 1644, p. 13, obs. 2). And Matteo Villani (1846, I, Bk. 5, liii, p. 442) tells how, in 1355, "a seven-year-old girl with wool like a sheep," was presented to Holy Roman Emperor Charles IV at Pietrasanta.

During medieval times, Christian jurists came to believe that Satan seduced witches in animal form, often that of a goat. Any seeming half-human birth could lead to an accusation of sexual congress with the Devil. Until about 1700, women bearing such progeny were often executed. Stephen Batman (1581, in Bates 2005) wrote,

> At Birdham near Chichester in Sussex, about twenty-three years ago [ergo, ca. 1558], there was a monster found upon the common, having the form and figure of a man in the fore-part, having two arms and hands,[‡] and a human visage, with only one eye in the middle of his forehead: the hinder part was like a lamb … This young monster was nailed up in the church porch of the said parish and exposed to public view a long time.

Batman's tone reflects the marked shift in societal attitudes that had taken place since ancient times. What to the pagan had been mere foible, to the Christian had become a grave sin against nature.

†. An event Irenäus (1584) dated to 1072. Carion (1588, p. 614) mentions this birth.

‡. Hands are atypical in ewe-yeaned hominocaprinids. But the *Bend Bulletin* (Bend, Ore., May 17, 1907, p. 1f; ||yck2vsks) did say that there was "born to one of J. M. Reeder's ewes in the lambing camp recently a lamb with perfectly formed hands."

But not everyone shared Batman's views. Pierre Bayle (1735, vol. III, p. 20) said the Duke of Nevers, coming to the succor of Charles IX in 1562, brought 3,000 Italian auxiliaries who kept 2,000 she-goats, caparisoned in green velvet trimmed with gold, which "were so many mistresses for himself and all his officers."

A hominocaprinid yeaned by a ewe at Postell, Poland, Jan. 18, 1630 (Schmuck 1680). The structure atop the head may represent either hydrocephaly or exencephaly, both common in telenothians.

Reports of hominocaprinids are rife. On January 17, 1578, a woman at Chieri, now in Italy, reportedly† bore a child with five ram's horns, backward knees like hocks, and fleshy growths hanging from the back of its head (a depiction appears on p. 118). Pedro de Peramato (1576, p. 92), body physician to the Duke of Medina Sidonia, describes a stillborn *tertium quid* with a human mother, hircine features and nascent horns, born on April 6, 1573, in Mexico at San Lorenzo Sayula. ⸸

Lehmann (1699, p. 713) reported a woman bearing a male offspring with a complete woolly sheep's head in 1630 at Schletta, a village in Germany.‡ A report in the Wickiana collection says a woman bore a nondescript with ram's horns and a fleecy torso at Buchs, near Zürich, in 1581 (||44rkh2p7). And Praetorius (1678a, ch. 2, iv) says that on March 28, 1670, a ewe at Plößte, Germany, yeaned a lamb with a human head, a naked body, lamb-like ears, and one forefoot bearing four digits, the other, three. The Danish anatomist Thomas Bartholin (1657b, §86, p. 172) wrote, in translation,

> In the town of Skånevik, Diocese of Bergen [Norway], during Lent, 1640, a goat kidded a creature in which the face, arms, eyes, nose, mouth, cheeks, forehead, and a blonde head of hair, were

†. Paré (1641, p. 648); Aldrovandi (1642, p. 457); Belleforest (1578, p. 511); Schenck (1644, p. 13, obs. 1).

⸸. A place near Tulancingo, Hidalgo. Peramato said it was taken to the duke, who asked Peramato to publish a report. Medina Sidonia was later admiral of the Spanish Armada.

‡. Lehmann also mentions a girl born with a sheep tail, in Germany, on May 31, 1624.

entirely like a human's, though the fingers had hooves instead of nails. In short, from the waist up it was just like a human, but the lower part was like a formless bladder. And yet one could distinguish the hind feet and hooves, though they did not turn under the belly as in beasts, but directly outward, or even bent somewhat toward the back. And in this part of its body alone did it lack human form. The face was not hairy, but rather rosy and white, so that people could not agree as to whether it was a thing of beauty or not. It was stillborn, so the goat may have given birth before her time. The pastor there, Elias Andersen, wrote Bishop Ludvig [Hanssøn] Munthe [1593-1649] at Bergen, to relate the story just as it occurred.

Years later, Bartholin (1677, p. 99) received a letter, dated May 27, 1675, from Stevelin Adamsen Reutz (1640-1687; ||y8bkbf9a), parish priest at Luster, Norway, reporting a recently lambed hominocaprinid, which Reutz said he had seen himself. In translation, it read,

> I'm writing to let you know that here in our parish of Luster, in the Sogn district of the Diocese of Bergen, in a village called Offre Staar, a prodigious monster was brought forth on March 13 of this year, to wit, a lamb with a rounded head, not obviously different from that of a human being, but having four ears and two faces, of which the one in front had no eyes, nor even their vestiges, and which was wondrously provided with a cloven mouth and two tongues. The left one, longer than the right, lolled down out of the mouth. The hind face, which was in fact placed to the right, had a single great, red eye without a pupil, and again a split mouth and double tongues, of which, here, the right was longer and so, again, more prominent. The rest of the body was frightful, with eight legs, a single thorax, and a divided body from the navel down. Except in being a bit thinner than usual, the bodily members, which were clad in black wool, were otherwise perfect, even to the extent of female sexual organs being present on both sides.

Moritz Hoffmann (1678), a German physician now remembered for his discovery of the pancreatic duct, described and pictured a pair of aborted conjoined twins yeaned by a ewe at Altorf, Oct. 8, 1672. Altorf now lies in Bas-Rhin, France. The Altorf nondescript is treated here as a hominocaprinid because it was hairless, as are so many reported human-animal composites. Hoffmann said it had a single head and two bodies. The testes were large, but the genital members were small.

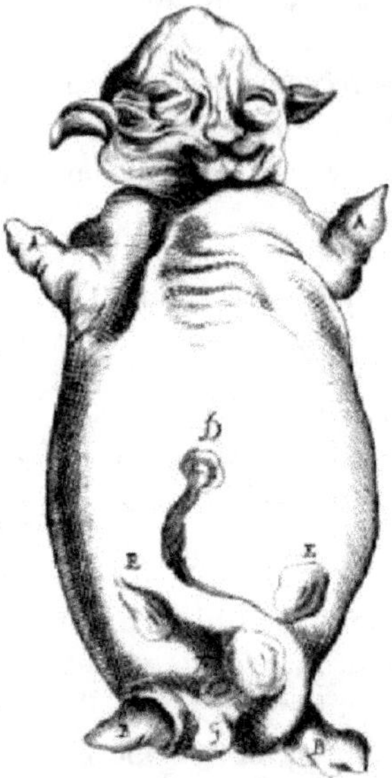

The Copenhagen hominocaprinid

The Lviv homungulate

Bartholin (1654, I, obs. 77, p. 128) recorded his examination, in 1646, of a certain Margarite Mayners, a 70-year-old woman (ergo, b. ca 1576) in the Netherlands at Purmerend, who had a large horn on her head. He said in thickness, hardness and appearance it was just like a goat's horn[†] and arose from the soft tissue over her right temporal muscle. He also reported (Bartholin 1677, vol. 5, obs. 6, p. 21), dissecting at Copenhagen an ovine hominocaprinid (shown at left) yeaned by a ewe. According to his account, it had naked skin and a nose like a human being's; the anus was imperforate; the kidneys, absent; the sex, indefinite, and the limbs, stumps.

Hacquet (1789) described a female horned homungulate, born to a woman at Lviv, Ukraine, Apr. 21, 1788 (see his depiction at left).

Leigh (1700, p. 3) referred to the "very remarkable Phœnomenon of one Alice Green whose Picture I have seen in Whalley-Abby, in Lancashire. This Woman had Two Horns which grew out at the back part of her Head. They grew backwards like those of Rams and were about three Inches long. These she cast once in three years and always had intolerable pains before the Horns broke out."

†. Dauxais (1820, p. 30) said it was 16 cm long.

Christian Lehmann (1699, p. 725) described an infant born in 1678 to the wife of one Hans Bäßer, a resident of Weipert, in Bohemia (now Vejprty in Czechia). He said this “horrific monstrosity” (“heßliche Mißgeburt”) had a goat’s head, with two horns on its forehead and large goat’s eyes on its cheeks, and that its sex could not be determined.

In connection with the picture reproduced above, Topsell (1607) informed his readers that “there is a kind of Ape called *Aegopithecus,* an Ape like a Goat.”

In a book describing medical cases, surgeon Job van Meekeren (1682, p. 353) reported that in the public market at Amsterdam in 1655 a “girl” had been put on exhibit whose whole body was covered in goat hair, especially on the chin.

And in a medical journal, van Meekeren’s contemporary, the German physician Gabriel Clauder (1688a, pp. 328-330) reported an event he said took place in the village of Schleenhain, near where he lived (Leipzig). There, according to Clauder, a goat gave birth to a fetus that was “in many ways similar to a human being.” It had a completely naked white skin, he said, and head and ears like a human, except that there were merely two pits in place of eyes. The legs and body, however, were “nearly like those of a goat.” It had been dead in the womb for several days. The same nanny gave birth to an ordinary, living kid about an hour later. Thus, per Clauder’s report, this birth was like one that occurred in the Philippines in 2018.

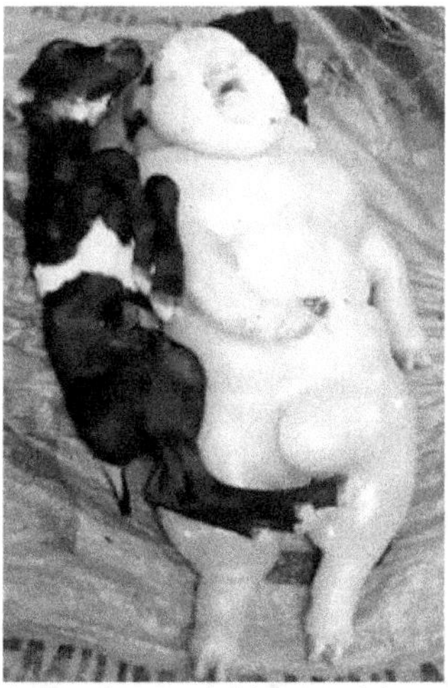

A hominocaprinid yeaned by a Philippine goat in 2018 recalls the 1688 birth at Schleenhain (described at left).

The Dutch surgeon Nicolaes Tulp (1672, IV, ch. 10, pp. 296-298) gave an account of a "bleating youth" (*iuvenis balans*), in which he describes a hominocaprinid who grew up among sheep in Ireland. "A young man from Ireland, about sixteen years old, was brought to Amsterdam and seen there by all. Having apparently been separated from his parents during infancy, and living all his life among wild sheep, he had become much like a sheep himself. Nimble of body and agile of foot, his expressions

A portrait in da Vinci's visi monstruosi series. Supposedly drawn from life, it recalls Tulp's *Iuvenis balans*.

were wild. His body was rough, his skin scorched by the sun. His limbs were drawn together, his forehead so flat and sloping that his head bulged at the rear. In temper he was wild, fearless, and devoid of all human feeling, but in other respects healthy and extremely vigorous. Lacking a human voice, he bleated like a sheep and rejected human food and drink. Rather, he ate only grass, hay, and the other foods of sheep. In eating, he continually turned everything over, choosing each single morsel by its odor and flavor. He had lived in the rugged hills and wild places, and was no less fierce and wild himself, haunting caves and places remote and uncultivated. Used to living in the open air, he endured equally hot and cold weather, avoiding the snares of hunters. And yet, though he kept to the cliffs and rocks, and cast himself recklessly into thorny thickets, he at last was caught. Living as he did, he had taken on more the character of a wild beast than a man. And even restrained and living long among humans, he set his wild temper aside only unwillingly and with the passage of much time. [translation mine]

But how would simply being raised by sheep change the shape of someone's head? And how would a human survive solely on sheep fodder? Or was this a viable, mature hominocaprinid brought to Amsterdam and seen by many witnesses? At any rate, it's interesting that Tulp says it lacked a human voice and could only bleat.

* The *Systema naturae* (1766) treats Tulp's bleating youth as a species separate from *Homo sapiens*, under the name *Juvenis Ovinus hibernus* (Irish sheep-man).

Benito Jerónimo Feijóo (1751, 30, §12, p. 364), a leader of the Enlightenment in Spain, described a hominocaprinid a butcher had recently taken from a goat in the town of Fernán Caballero, Ciudad Real (ibid. p. 358). "The head was round," he said, "like a human being's, the eyes, open and in their normal position. The eyebrows and eyelashes were red and so fine they could scarcely be seen.

Benito Jerónimo Feijóo
1676-1764

> The blunt nose was human, as was the mouth, and the tongue as well, except in having two tips. The ears were those of a goat, but in their concavity resembled a human's. The lower jaw protruded somewhat beyond the upper. The lips and gums were human, as were the neck and shoulders. The arms began as in a human being but ended in circular hands, one of which bore along its edges five fingers, the other six. These, instead of nails, ended in little hooves. Beneath was a palm, as in a human hand. Tendons could be observed extending from the arms and wrists to the fingers ... The stomach and adjacent hind parts were similarly human in form. The testicles were divided into two little bags [goats have a split scrotum] ... The legs and feet were similar in appearance to what has already been described for the arms and hands. ... Other than the brows and lashes, the skin was white, smooth and hairless, as in infants. It was a little over a foot long. [translation mine]

Ulm physician Eberhard Gockel (1700, p. 151) said, in translation, "a few years ago, at Esslingen am Neckar, a butcher's wife, daughter of a tradesmen known to me, bore a child with a sheep's head." Rotari (1762, p. 113) said he saw in Verona a female infant, with four goat horns and an extra ear, borne by a woman, March 20, 1701.

In 1746, Léonard Chabelard (1751, pp. 45-46), a surgeon at Tours, described a hominocaprinid borne by a woman. In translation, he said, "The head was a flat oval. A muscle from the base of the occipital covered it like a monk's hood. Two horns, hard, nearly osseous, 1.2-1.3 inches [14 ou 15 lignes] long, stood on the supero-anterior portion of the parietals. The hands, forearms, feet and legs were hairy. It was a male, and the genitals, like a dog's, were enclosed in a sheath affixed to the belly" (as in goats and sheep).

Philippe Pinel (1809, pp. 182-184) described an ovine inmate of the Salpêtrière asylum in Paris. In translation, he said, "One of the most unusual cases ever seen is that of an 11-year-old imbecile, of whose skull I have had made an engraving, and who by the form of her head, her tastes, and mode of living, seemed to recall a sheep. While here at the asylum, she showed a strong aversion to meat but avidly ate vegetarian food such as pears, apples, raw vegetables or bread, likewise a kind of cake her mother

Philippe Pinel
1745-1826

> sometimes brought. She only drank water and showed an appreciation for the efforts of the young woman who was her caretaker, which led her to utter the only words she could, *bé* [French for English *baa*] and *ma tante* [*my aunt*]. Otherwise, she seemed entirely mute, due to a lack of ideas, though her tongue appeared fully mobile. She would also rub her head up and down on the belly of her caretaker, in the way that sheep do, to express her gratitude. In quarrels with other children, she assumed the same pose, trying to butt them with her head. When angry, having only the blind instinct of an animal, she could not control her movements, and her reactions, at slightest provocation—or even for no reason—could be convulsive. She never sat in a chair, to rest or eat, and slept rolled up like a sheep. Her whole back, flanks and shoulders were covered with revolting black hair, one and a half or two inches long, fine as wool. ... This young imbecile, homesick, grew ever more depressed and died after two and a half months at Salpêtrière. I preserved her skull, which is remarkable in size and form.

* Writing during the same era, Mirabeau (1783, p. 152) claimed that bestiality occurred more commonly in France than was generally believed and that it was known through the confessional that every shepherd of the Pyrenees had his favorite goat.

Hominocaprinids continue to be reported right up to the present. In July 2017, a goat with a human face was born in the Spanish province of La Rioja. Late in 2023, another goat, in Nuav, Uttar Pradesh, India, yeaned a viable hominocaprinid, seen on the Internet.

Athena Driving the Vices from the Garden of Virtue (Mantegna, detail). Note that the vices being expelled are depicted as homungulates.

Deformity as Evil

> *He is deformed, crooked, old and sere,*
> *Ill-faced, worse bodied, shapeless everywhere;*
> *Vicious, ungentle, foolish, blunt, unkind;*
> *Stigmatical in making, worse in mind.*
> —*A Comedy of Errors*, IV, 2

The medieval mind linked deformity with evil and witchcraft (Huizinga 1922, p. 35). Children of abnormal structure were deemed the result of a mother's having had intercourse with the Devil, who, it was said, would come to them in the form of an animal (Barnes 1991). This notion of deformity as evil was ingrained in the mentality of the era. Albertus Magnus argued that monstrosity of form was often associated with monstrosity in morals (Borgnet 1890, p. 300b). Martin Luther said he saw the Devil in a profoundly disabled child. If these children lived, Luther recommended killing them (Barnes 1991, p. 12). An alternative explanation for any composite births would be to suppose the human parent had mated, not with the Devil, but with a nonhuman animal. Such ideas find their roots, in part, in the Bible. Leviticus (21:16-21) forbids anyone with a physical defect from approaching the altar. This connection between physical and moral defect survives even today in metaphor when we use such terms as "crook" or "crooked dealing." Such is the reason we call an honest dealer "straight."

A woman with ram's horns (da Vinci, *visi mostruosi* series, allegedly drawn from life). Her chin is far more protrusive than that of any normal human being. Some accounts claim the horns are "tied" to the woman's head. But the band, as depicted, does not appear to be appropriately positioned, or large enough to hold the horns in position. Also, when these drawings were created (ca. 1500) women often wore a style of headband, known as a ferronnière, circling the wearer's forehead, usually with a jewel suspended at the center. A contemporary example of a woman wearing a ferronnière appears in the inset (from *La Belle Ferronnière*, school of da Vinci, 1490-96). Enlarged views of the centered object visible in the drawing show it is a jewel, not a knot.

HORNED HOMUNGULATES

Les Mézières, Sarthe, France. In 1599, Jean de Beaumanoir, Marquis de Lavardin, brought to Paris from Sarthe a certain François Trouillu (b. ca. 1564), who, according to Dauxais (1820, p. 31, vii) and Thou (1733, p. 825), bore a horn as large and hard as a ram's or a goat's. Trouillu, born in the hamlet of Les Mézières, was exhibited widely. He said he was seven when his horn first budded and that it had kept on growing throughout his life. It must have arisen from the cranium, not from the overlying soft tissue, as it was so hard and stiff that it had to be trimmed periodically to keep it from piercing his skull.

François Trouillu
(Wellcome Collection; ||28y5kh5k)

* Also see Eloffe (1866).

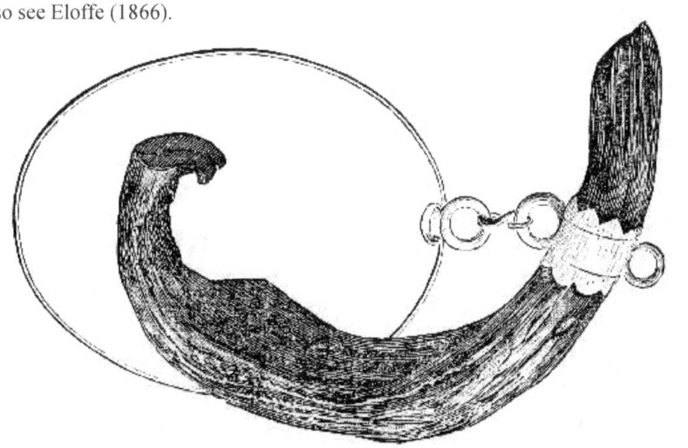

A horn, as depicted in an old medical text (Bennett 1874, p. 216, fig. 241). Its caption reads, "From an old preparation in the Edinburgh University Museum. The medal attached bears the following quaint inscription: 'This horn was cut by Arthur Semple, Chirurgeon, out of the head of Elizabeth Low, being three inches above the right ear, before these witnesses, Andrew Temple, Thomas Burne, George Smith, John Smytone, James Twedie, the 14 of May, 1671.—It was agrowing seven yeares; her age 50 yeares'" (ergo, she was born ca. 1621). Damon (1869, p. 213) said this specimen was about five inches long. The museum website has a photo of this object (||2tdussxf).

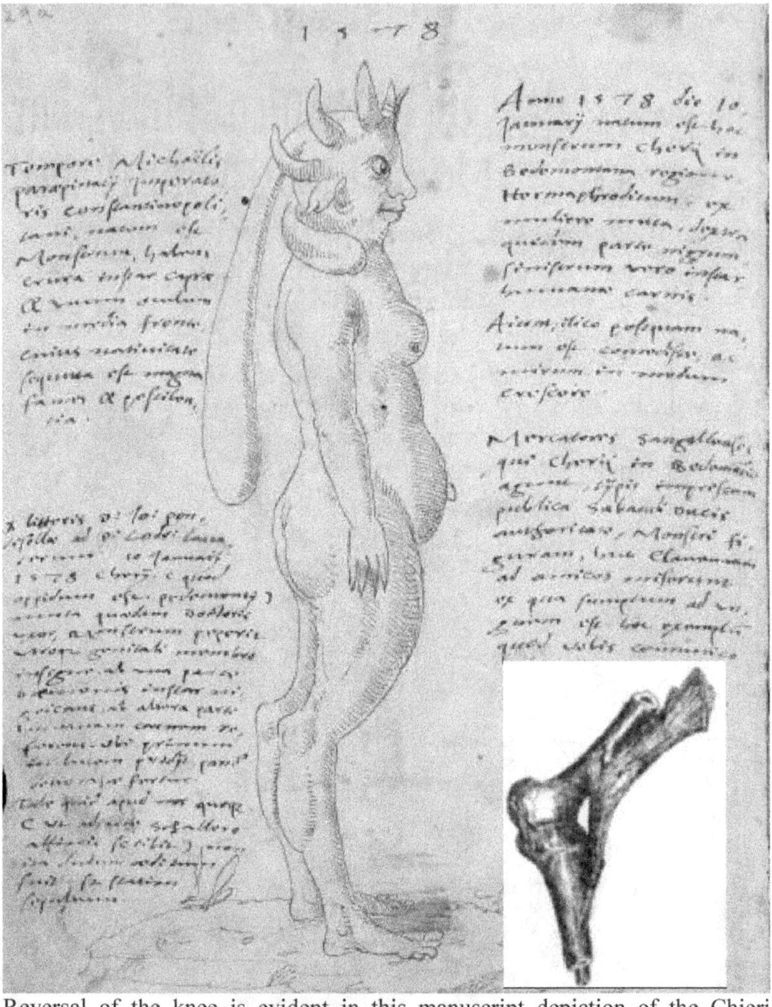

Reversal of the knee is evident in this manuscript depiction of the Chieri hominocaprinid in a letter, dating to 1578, from grammarian and historian Scipio Lentulus (ca. 1526-1599) to Zürich clergyman Johann Jakob Wick (Wickiana Collection; ||yc3wf8s2). In translation, the two paragraphs at upper right read, "On January 10, 1578, this monster was born at Chieri, in the Piedmont region, a hermaphrodite borne by a woman, the righthand part of which was in fact black, truly half beast and half human. They say he could immediately walk at birth and that he grew at an amazing rate." In medicine, knee reversal has been termed called *genu recurvatum* (Owen 1891, Sheild 1898). The inset is an illustration from Sheild (ibid., Fig. 2) showing the structure of the underlying joint in a infant affected by this condition. It is shown here in the same orientation as is the knee of the Chieri nondescript.

Ella Harper (1870-1921). Like an ungulate, she lacked large muscles in her calves (as did Huddleston). *Lower inset:* Bovine hindleg. The rear-bending joint equates with our ankle, the long bone below it, with our metatarsus. *Upper inset:* Her shoes were as short as her hands. Did they conceal hooves?

BACKWARD KNEE

Those with "backward knees," such as sideshow stars Ella Harper[11] and Robert Huddleston,[12] have knees bending in the same direction as in ungulates. This leg structure is like that of the Chieri hominocaprinid (opposite page). Harper was billed as the Camel Girl and Huddleston as the Pony Boy. Both ordinarily went on all fours. A video shows one such in Lyon (||4v6xdtab). John Pelter (pers. comm.) says he saw another in Barcelona in 2024.

Robert Huddleston
1895-1970

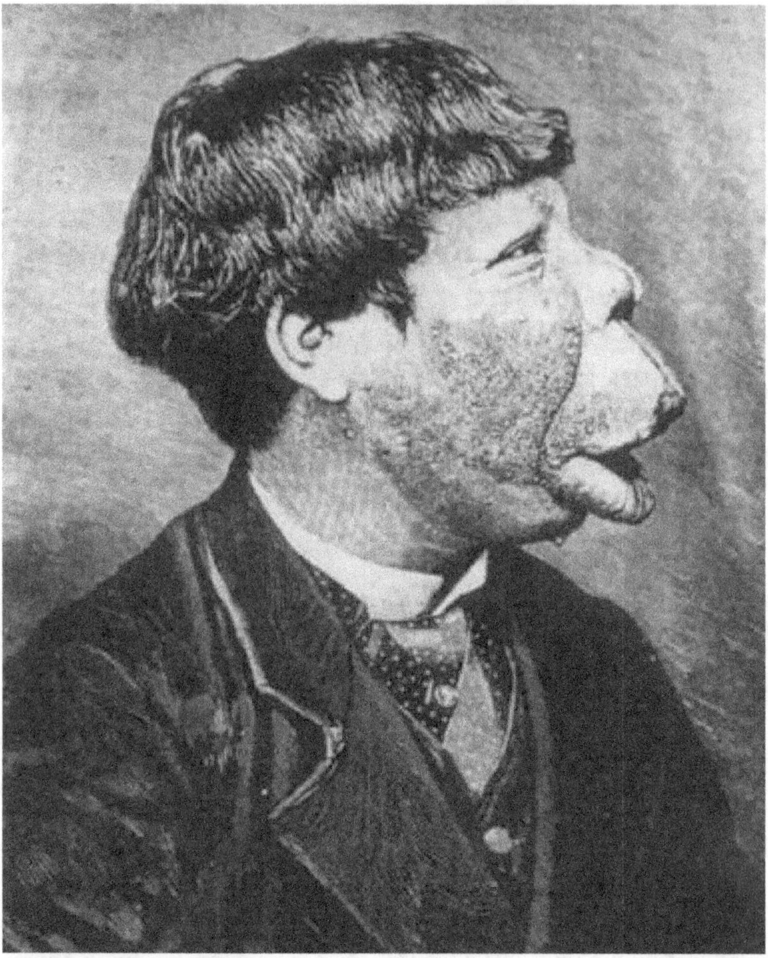

Eugène Boudou, known as L'Homme à la Tête de Veau (The Calf-headed Man). He performed in Paris cabarets in 1884. Various reports (e.g., *Le Gaulois* Paris, Mar. 4, 1884, p. 2f; ||46n3nccy) say he was born the son of a butcher near Cherbourg, circa 1860. Though described as having a calf head, photos suggest he differed from ordinary humans only with respect to the structure of his mouth. Yet, from available images, he does not seem to have specifically resembled a calf, even in that region of his face (which was reportedly blood red, ibid.). Despite his stage name, Boudou's mouth connects him about as well with sheep as with cattle. In 1891, news reports said Boudou ran off with a lover, his manager's wife (||mv4xfwmb).

SHEEP MOUTH

In some individuals, ovine and caprine traits are reportedly limited to the oral region. Johann Georg Schenck (1609, p. 11, no. 11), town physician at Haguenau, Alsace, reported delivering a hominocaprinid in all ways like a boy, except for his sheeplike mouth. In translation, his report reads,

> A MONSTROUS BOY BORN AT HAGUENAU, beneath whose nose was a growth like the mouth and tongue of a sheep and whose upper lip was cleft. At Haguenau in Alsace, September 13, 1608, at 5 a.m., after a difficult labor, a boy was born, whose body was perfect in every way, except that his upper lip and the adjacent region of the nose formed a monstrous growth or proboscis that was lamblike in appearance. This formation recalled the mouth of a sheep, and the tongue protruded, as if bleating, through the lips. Only that region was affected. So, without this lamentable mouth and gaping upper lip, he would have had a lovely face. I found the nose was inwardly affected, with the nostrils blocked, which caused marked problems when the child was fed … This boy survived and was baptized by Johann Jacob.[5] I, Johann Georg Schenck, town physician, witnessed this case during my practice in Haguenau.

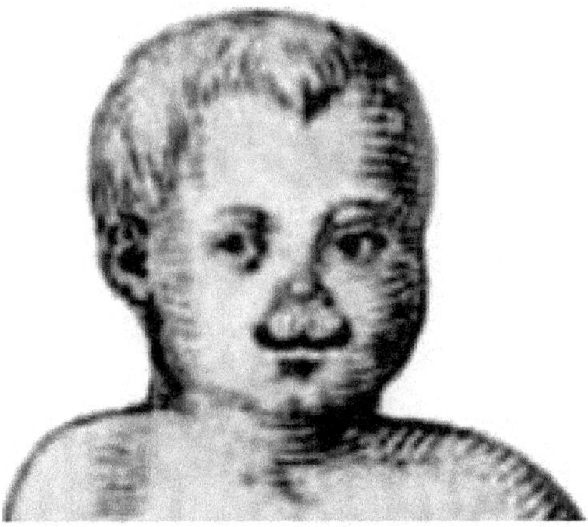

Left: The Haguenau hominocaprinid (Schenck 1609). *Right:* Woman with a sheeplike mouth (Hollar 1665, after one of da Vinci's *visi mostruosi*).

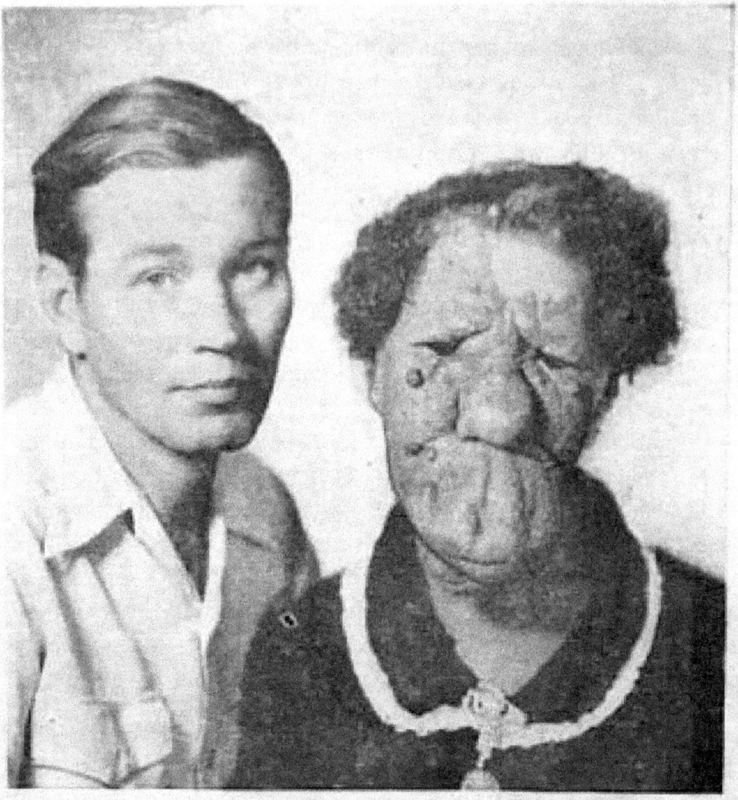

I was born on a farm near Numa, Iowa, in 1888. I am 5 feet tall and weigh 115 lbs.

The best medical talent of the United States and Canada have had me before them repeatedly, for close examination and scrutiny.

My remarkable facial condition is a birth-mark, due to a confusion in the capillary, arterial and nervous systems previous to my birth. My father, mother, sisters and brothers were normal folks in every way — just good healthy people. I lead a normal life — prefer housekeeping, although the show life is an enjoyable one and healthful.

I have one son, Elmer, 32 years of age, perfectly normal.

MRS. GRACE McDANIEL

Grace McDaniel and her son Elmer. Born near Numa, Iowa, March 17, 1888, she had a facial structure that was similar to that of Eugene Boudou (see p. 120). In the present context, that she was born on a farm is significant. Though placed here with hominocaprinids, the condition of McDaniel's skin is similar to that seen in HOMOGALLOPAVIDS.

Other educated men have reported the like. In discussing the theory of MATERNAL IMPRESSIONS, Hermann Conring (1662, §V), a University of Helmstedt professor, said, in translation,

> A grievous but, indeed, notable example of these impressions can be seen in the son of a pious and respectable man living a few miles from here. His mother, while carrying him in her womb, looked too intently upon the mouth and dental structure of a sheep, which caused the teeth of the embryonic infant to be formed in a similar way and, in the boy, later, the development of the teeth of a sheep, which made it difficult for him either to eat or to speak, and created a gross facial deformity that cannot be corrected. The unhappy child lives even now.

Letzner (1596, 51b) claimed another such, born at Dassel in Germany in 1576, had a mouth like a sheep, a single ear, and an intestine-like mass on the back of its head. Schurig (1731, p. 590) stated that some forty years before (ergo, ca. 1691) there was a boy with a sheep's mouth and teeth living in Leipzig. He said the condition was common enough to have a name: *Schaafmaul* ("sheep mouth").

Study for the Head of a Sheep (Artist: Charles Émile Jacque).

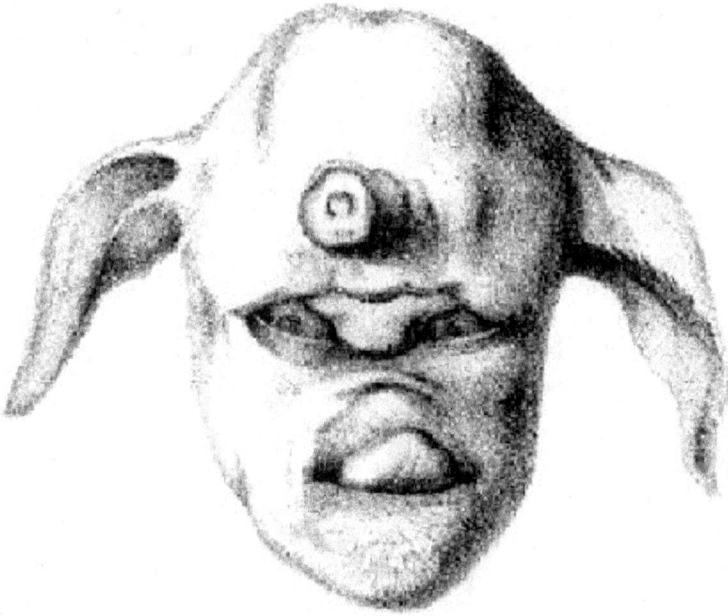

A hominocaprinid with a frontal proboscis, yeaned by a ewe in 1845 at Natura Artis Magistra, a zoo in Amsterdam (Vrolik 1849, plate 57, fig. 7). Gerard Westerman (1807-1890), one of the zoo's founders, donated the specimen to Willem Vrolik, who published an anatomical description (ibid.). During dissection, he found its eyes were both connected to a single optic nerve, a form of partial cyclopia (see figure, p. 174).

HOMINOCAPRINIDS WITH FRONTAL PROBOSCIDES

1557	Frankfurt, Germany	Wickiana (‖mheej772)	EO
1573	Sézanne, France	Paré (1641, p. 666)	EO
1740	Bayreuth, Bavaria	Superville (1740, p. 302)	S, EO
1839	Catania, Sicily	Galvagni (1839)	CY, ECP
	Catania, Sicily	ibid.	S, EO
1845	Amsterdam	Vrolik (1849, plate 57, fig. 7)	S, EO
1872	London	Lowne (1872, p. 44, 160 & 161)	CY, EO
1850	Rayton, Nottinghamshire	‖bde6x834	CY, EO
1882	Marton, New Zealand	‖4dptxxxc	CY, EO
	West Middleburg, Ohio	‖4s6hfae9	CY, EO
1891	Cathlamet, Washington	‖4yc4633a	CY, EO
1895	Westphalia	Landois (1895)	CY, ECP
2013	No locus in quo (video)	‖2p85ktj2	S, EO

Key: CY: cyclops; S: synophthalmic cyclops; EO: from a ewe; ECP: from a goat. For URL sources see: http://www.macroevolution.net/permalink7878.html

A contemporary depiction of three hoofed homungulates reportedly borne by a woman at Mesocco, Switzerland, in 1578 (Wickiana Collection; ||mpphz82h).

INCERTAE SEDIS

Damascus, Syria. Reports tell of a boy captured among gazelles in the Syrian Desert in 1946 (||mq7satl, ||kw52956). Emaciated and unable to speak, he could supposedly run faster than any human.

Waterford, Ireland. Ash (1685) described a girl born in Waterford, ca. 1772, who had horns growing on her body and out of the tips of her fingers and toes. From each of her elbows, he said, grew a horn, curled like a ram's.

Chauny, Aisne, France. Le Réveil de l'Aisne (Chauny, Jun. 24, 1906, p. 2d; ||33u3p5cm) told of an ape-faced offspring of a goat, sired by a chimpanzee at Chauny.

Campos dos Goytacazes, Rio de Janeiro, Brazil. A video (||4d4wcwxw) shows a boy with hoof-like feet living in Campos dos Goytacazes.

Bolzano, Italy. The *Bozner Zeitung* (May 10, 1876, p. 3b; ||2s3njr4c) said, in translation, "A MONSTROSITY.—A woman in our suburbs bore a child that had four horns on his head. Two were excised. And it is hoped the other two can be removed in the same way. The child is completely viable and healthy."

Goats with breasts located as in a woman. *Left:* Jonston (1650, Tab. XXV); *Right:* An animal Paullini (1687b, fig. 8) said his father saw at Eisenach.

NOTES

1. Warialda and Bingara are small towns in northeastern New South Wales.
2. Tehama is on the west bank of the Sacramento River.
3. A Charles Harvey (ca. 1831 - 1884) was living in Tehama at the time in question.
4. Wallula is a census-designated place in Walla Walla County, Washington.
5. A Jacob Drenkle, Jr. (1872-1948) lived in Houston in 1897 (||2z3sms23).
6. Some reports say the Goshen hominocaprinid was born on a farm near Lagrange, a village in north central Ohio near the western end of Lake Erie (||y2pfjkez, ||y65o5msn).
7. For the Greeks, a satyr was like a human with horse ears and a permanent erection.
8. Aldrovandi, who cited Caelius Rhodigenius, thought this event occurred in 1547.
9. Khnum was often depicted, as here, with the horns of *Ovis longipes palaeoaegypticus,* an extinct form of sheep once common in Egypt.
10. Fremantle et al. (1893).
11. Ella Harper (b. Jan. 5, 1870, Gallatin, Tn., d. Nashville, Dec. 19, 1921, Spring Hill Cemetery) had a twin brother Everett, who lived three months, and two female offspring (one perhaps adopted), who both died when a few months old (||yxfaxf3u, ||y3qug6bd).
12. Robert Lee Huddleston (b. Feb. 25, 1895, Excelsior Springs, Mo., d. Sep. 3, 1970, Fremont, Mo.; Pine Lawn Cemetery, Winona, Mo.; ||2vxdtnes). A WWI draft card says Huddleston had "to walk on hands & feet." His death certificate lists his mother as Clarcia E. Huddleston, his father as unknown (||2vxdtnes).
13. *Colophon:* Lycosthenes (1557, p. 669).

A hominocervid (Gundestrup cauldron, ca. 150 B.C.).

HOMINOCERVIDS (DEER-HUMANS)

Hercules' son Telephus, having been exposed by the orders of his grandfather, was reared in the woods by a doe.

—JEAN TIXIER DE RAVISI, *Officina*

All reported hominocervids, it would seem, are the offspring of human mothers.

Christoph Wilhelm Hufeland 1762-1846

France. The noted German physician Christoph Wilhelm Hufeland (1791) stated, in translation, that a certain French patient, one "Marie Marjorat, 70 years old [ergo, born ca. 1720], and 16 years a widow, had on her forehead, at the hairline, an ash-gray outgrowth divided into three extremely hard branches, which in color, form and strength very much resembled the antlers of a deer. Two of the tines were short, but the middle one was four inches long, bent forward, and covered her right eye. The other tines had grown quite long but had broken off, being only as thick as a little finger. These horns move when she wrinkles her brow, so there is no firm bond to the underlying bone. The core material is softer, and lighter in color, than the cortex. In other respects, she is strong and healthy and may live a long while. In her 56th year, her widowhood's second, she began having severe headaches with twitching, and it was then this thing appeared. And even now, if it gets bumped, she suffers from headaches and twitching.

Dijon, France. Mannagetta (1670) and Dauxais (1820, p. 45, §34) describe the case of another hominocervid, that of a certain M. Delaine, President of the Parliament of Dijon, sixty years of age, who, after recovering from a severe fever, developed over his last thoracic vertebra a hard walnut-sized tumor that remained unchanged for ten years. But five years before Mannagetta wrote his report, it sprouted into what looked like the antler of a young buck. "It grew so much," he said, in translation, "that, had it not been cut off, at the point where the pain began to be exquisite (a finger's width from the skin), it would have been more than half a foot long." The dates given imply Delaine was born circa 1595.

Caddo Parish, Louisiana. A letter, signed Thomas Adams, published in the *Caddo*[1] *Gazette* (quoted in the *Washington Sentinel,* Washington, D.C., Aug. 7, 1855, p. 3; ||59c57szt) said, "I have been overseer[†] of what is known as the Jones Farm, in this parish, near Ferry Lake. There is on the farm a negro boy about eighteen months old, who about a month and a half since had a pair of horns to make their appearance on the front of the head, precisely in appearance to those of a young deer; they are now about one and a half months old and are full six inches long, well covered with velvet, and on one of them, two spikes making their appearance, and are still growing remarkably fast. I have succeeded in purchasing said negro from the former owner and will exhibit him as above on the 4th of July at Shreveport free of charge and shall start on from Shreveport to Vicksburg and other western cities, also to New York, to make an arrangement with Barnum to exhibit this as one of Nature's greatest freaks, to all the world and the rest of mankind." A sequel appeared in the *Highland Weekly News* (Hillsboro, O., Feb. 23, 1860, p. 1f; ||y2lboosr): "The HINDS COUNTY (Miss.) GAZETTE gives an account of a negro boy in that region, eight years old, who has horns on his head like a "young devil." The horns are said to be three inches long, to project from the head above each ear, and to interfere considerably with wearing a hat. They are stiff and hard, but not flinty, as the horns of quadrupeds."

†. A Thomas S. Adams, an overseer, lived in Caddo Parish, in 1850 (||y2and86b).

Göttingen, Germany. Johann Jakob Huber (1748, pp. 6-8, sec. 6), a Göttingen professor, described a patient who, in translation, was "abnormal only with respect to his skin. He is a healthy and vigorous four-year-old [ergo, born ca. 1744] of ordinary size, with all the usual parts except that on his back, from nape to rump, as well as his belly, especially, and even some of his chest, the skin is not smooth and white but a light brown, perfused with a reddish tint, and entirely covered with hair. Along the spine and lower back, this pelt is quite dense and tufted, and somewhat whitish. But elsewhere it is reddish and hardly less hairy. It is not the down seen in infants but true hair, scarcely differing from a deer's, except perhaps softer, shorter and a bit less dense. This coat of hair, then, is genuinely like that of a deer, and the skin, too, in the affected regions, is colored like a deer's. It has already been noted that this child can scarcely speak, being able to say only a few babbled words. ... But the most important thing that I wish to emphasize is that this coat of hair is truly deerlike, even in color."

Dachstahl, Lorraine. Buffon (1831, p. 199) wrote, "A child was exhibited in Paris in 1774, whose name was Anna Maria Herig. She was born at Dachstahl, in Lorraine, [Nov. 11] 1770. Neither her father nor mother had any spots whatsoever on their skin. But the child was spotted all over the body, and even on the face, with marks, some large and some small, the greater part of which were raised above the skin, ... in some other parts they were covered with hair, like that of a deer. Both the skin and the hair were of a brown colour; there were also spots quite destitute of hair, and the naked skin in those spots resembled tanned leather. In other respects the child was of a pleasant countenance; had handsome eyes, but very large eyebrows. She appeared uneasy in summer from the heat; for besides the hair spots in other parts of her body, her stomach and belly were covered with longish hair, of a brown colour on one side, and lighter on the other; and her back seemed to be covered with a tunic of hairy skin, ... The lower part of her loins, and the tops of her shoulders were covered with hair more than two inches in length." And Johann Caspar Lavater (1804, pp. 270-271), who said he saw Herig himself, wrote, "I can only say that the colour and growth of the hair were like that of a stag. The hair, also, of the forehead, arms, and legs differed from the hair of the head; and the former likewise had some resemblance to the hair of a stag, which was a very extraordinary circumstance."

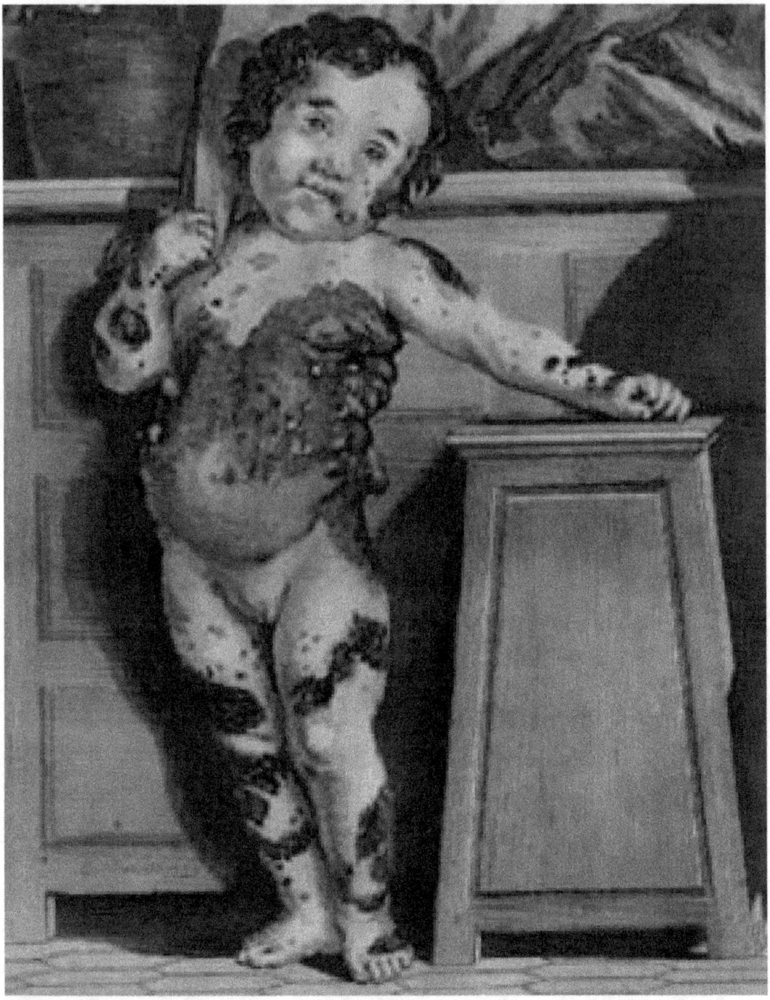

Above, and opposite: Anna Maria Herig (Buffon 1777, plates 3 & 4). Other pictures of Herig appear both in Lavater (1804, plates 59 and 60) and in Wünsch (1796, Tab. XIII). Wünsch's color depiction appears to be based on Buffon's depictions, reproduced here. Similar cases of hypertrichosis are shown in the chapter on HOMOSIMIANS (p. 333). The resemblance is with respect to the patchy distribution of the hairy nevi. In the case of the HOMOSIMIANS, however, the hair on the nevi was described as being like an ape's, not a deer's.

NOTES

1. Caddo Parish is named after the Caddo Nation. A mask found at a Caddo site, Craig Mound in Spiro, Oklahoma, was in the form of a human face with deer antlers.
2. *Colophon*: Sketch of an antlered figure depicted in the Cave of the Trois-Frères, Ariège, France (Breuil 1954, p. 166). It dates to circa 13,000 B.C.

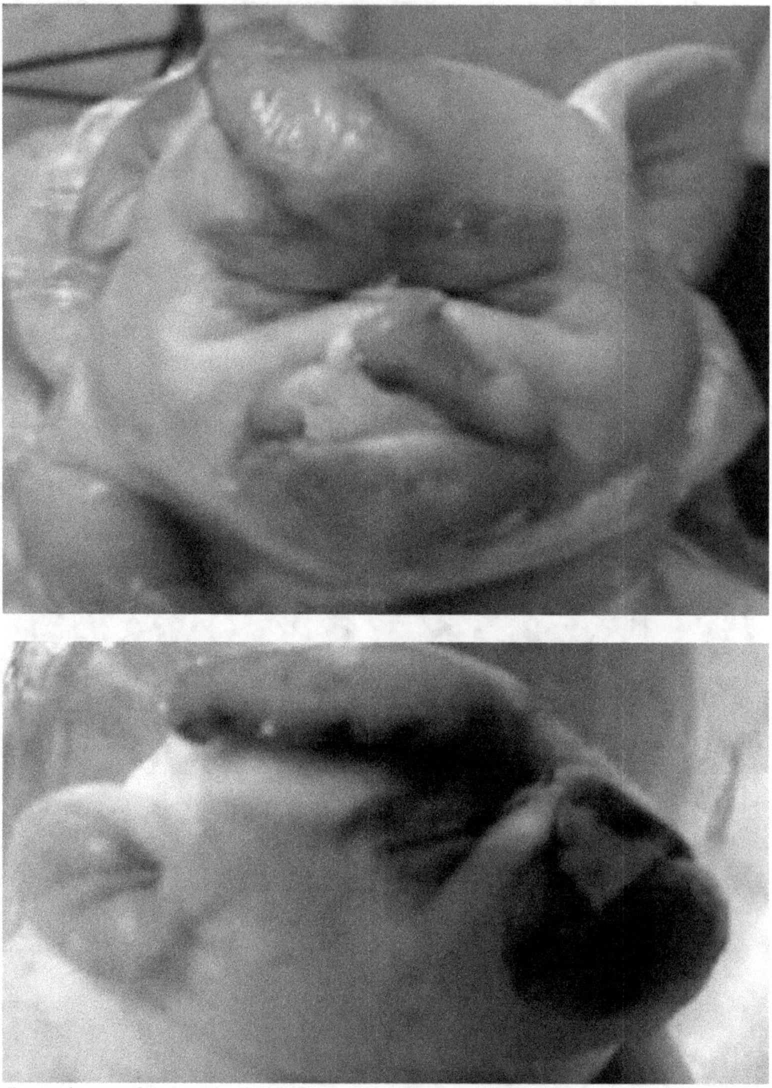

Two views of a hominosuid farrowed, along with three normal piglets, by a sow on a Brazilian farm in 2012 (from a YouTube video; ||wj98wdc). As is often the case with hominosuids, the tongue protrudes and the ears are directed toward the rear of the head, as in humans, instead of upright, as in pigs. A frontal proboscis (the fleshy tube attached to the forehead) is present. This specimen, about 10 inches long, was preserved in alcohol.

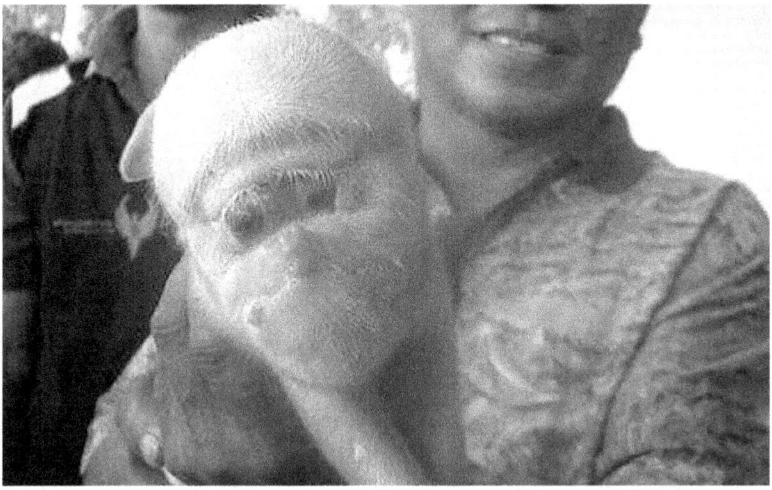

The Chinese hominosuid farrowed in 2008.

HOMINOSUIDS (PIG-HUMANS)

Sows can give birth to piglets with human heads.
—ALBERTUS MAGNUS[1]

Pig-humans are more frequently reported than any other telenothian. In this book alone, some 290 separate cases have been compiled. There are many newspaper reports and videos about such creatures, and many additional accounts in the medical and historical literature, as well.

Hominosuids are more often birthed by sows than by women. Those of the former type are often described as having humanlike faces, with rounded skulls, and bodies like those of a pig, except for the presence of hands and, often, arms like those of a human being. In other words, the foreparts tend more to resemble humans, the rear parts, pigs. In the case of those borne by women, the reverse is usually the case, as the foreparts tend to be more porcine. These two rules can be, to some extent generalized to all kinds of human-animal composites, in that there is a strong, but not invariate, tendency in all such composites for the hindparts to resemble the mother's kind.

The descriptions differ with respect to the claimed degree of human resemblance. Some, however, do allege an exact similarity with respect to the facial features and/or the hands (these reports are among those quoted below). Available photos of hominosuids depict creatures with faces reminiscent of, but certainly not identical to, those of human beings.

Varying degrees of eye fusion are often reported. The degree of fusion ranges from eyes that are simply close set, to ones that have two irises and pupils in a single eye, to individuals that are fully cyclopean. In many cases, eyes are absent. The reported degree of similarity to a human physiognomy varies, but hairless facial skin and the presence of a chin are often mentioned. The skin of the whole body is often said to be like that of a newborn human infant. The ears are usually described as like those of a pig but as lying close to the head and turned toward the rear, as in humans, instead of being upright, as in pigs (some reports mention animals with one pig, and one human, ear). The mouth often gapes. The hooves of these creatures often turn up at their ends like ski tips. Several reports mention individuals lacking mouths, and quite a few say the subject's sex was indeterminate. Often, the chin is described as prominent, rounded and like that of a human being. Many hominosuids have, instead of a snout, a frontal proboscis, which is a fleshy structure like an elephant trunk, albeit a small one. It is nearly always attached to the center of the forehead, just above the eyes or (in cyclopses) the eye. The combination of a proboscis with large rear-turned ears and a small mouth, often seen in these creatures, has led many a reporter to describe such a birth as a piglet with the head of an elephant. The presence of a frontal proboscis is sometimes termed "holoprosencephaly." Merely naming a condition does not explain it.

Human Cells Can Combine with Pig Cells. When piglets were injected *in utero* with human stem cells, researchers investigating the resulting animals detected the presence of cells containing the DNA of both humans and pigs (Ogle et al. 2004). In other words, pig and human cells had fused to become composite, living cells that were pig-human hybrids. Moreover, experimentation has shown that human sperm can bind and penetrate pig egg cells. The acrosome reaction (AR) is the process by which a spermatozoon penetrates an ovum. The zona pellucida (ZP) is the outer layer of a mammalian oocyte (female germ cell). Canovas et al. (2007) state that

> The results shown in this work indicate that human spermatozoa can bind tightly to the ZP from in vitro matured pig oocytes. Besides, the induction of the human AR when human sperm were incubated with pig oocytes for 2.5 hours indicated that the binding was active, and it led to AR.

RECENT REPORTS

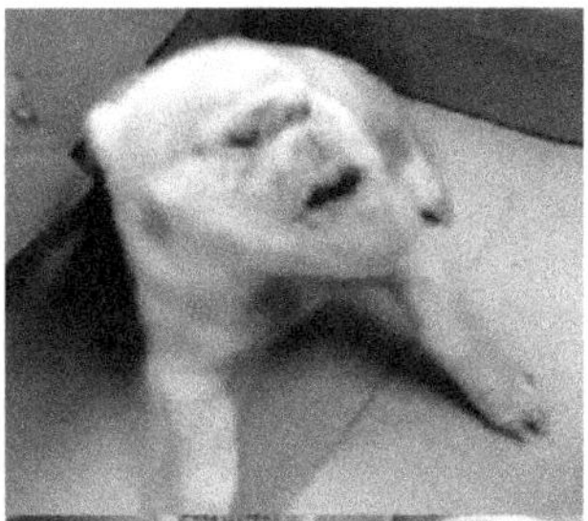

Hungary. In 2018 a woman in Hungary communicated that a hominosuid had recently been farrowed there. She provided the uppermost picture at right and tried to obtain the subject for genetic testing, but the owner refused to cooperate.

Cuba. A case first reported from Ciego de Ávila, Cuba on January 5, 2016, is shown alive in the middle picture at right, which is a screenshot taken from a YouTube video (||y3m7584x), seemingly the only source of information about this case. This individual died for unknown reasons four days after birth. Many hominosuids, however, are reportedly unable to nurse.

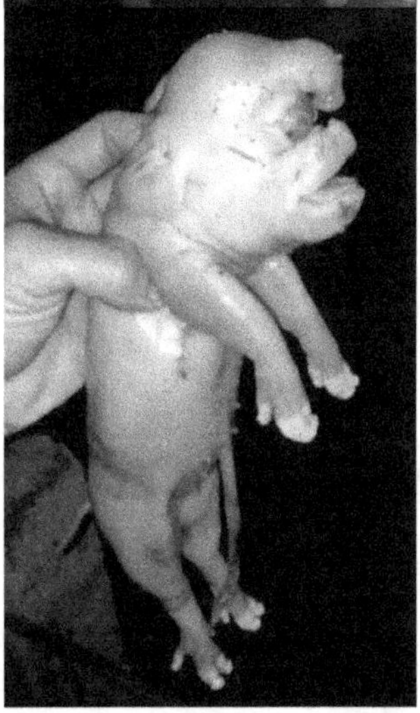

Nanning, China. The birth pictured in the lowermost image at right reportedly took place February 9, 2015, in China. Farrowed by a sow belonging to Tao Lu, a resident of Yanan township in the city of Nanning, it was described as a "piglet" with a human-like face. Pictures accompanied the news stories. The animal was born alive but died of starvation soon after birth. Nearly the last of a litter of 19, it was crowded out by siblings and couldn't nurse. Newborn ordinary piglets, deprived of milk, usually live only a few hours.

In the Nanning animal, note the fleshy tube extending outward from the forehead and hanging down in front of the eyes like a small elephant's trunk. This structure is known as a *frontal proboscis.* A process of this sort is present in many hominosuids. They have also been reported in BOSHOMINIDS and HOMINOCAPRINIDS but are not so common as in hominosuids.

* That only one in a litter of 19 resembled a human is not inconsistent with the possibility of it being a hybrid. When a litter-bearing mother, such as a sow, is multiply inseminated by different males, members of the same litter can have different fathers. Within the context of hybrid crosses, this means that a mother inseminated by two males—one of her own kind, and one not—can produce litters composed of both pure and hybrid offspring. The former will usually make up most of the litter since the fertilization rate will generally be higher. It is, however, well established that coition between human and pig, while usually kept secret, is widespread (Evans 1906; Laws and O'Donohue 2008, p. 391; Miletski 2009; Valcuende del Río and Cáceres-Feria 2020).

Scotland. In 2015, the *Mirror* (||qetuwns) said a Scottish case, much resembling the 2015 Chinese hominosuid died at birth.

Guizhou, China. Internet reports published pictures of a viable hominosuid farrowed in Guizhou Province in 2016. A permanent record of one of these images is available (||yyrfbvrn).

No locus in quo. A video (||y6kh236r) taken in 2015 in an unknown locale, shows a dead specimen with a frontal proboscis and an eye with two irises ("synophthalmic cyclops"), a middle stage along the continuum between a pair of normal eyes and a single, fully cyclopean eye (see p. 189). The protruding tongue is also common in hominosuids, as well as in CANIBOVIDS. Such cases may result from the tongue growing to the size normal in a pig while the oral cavity grows only to the size it does in humans.

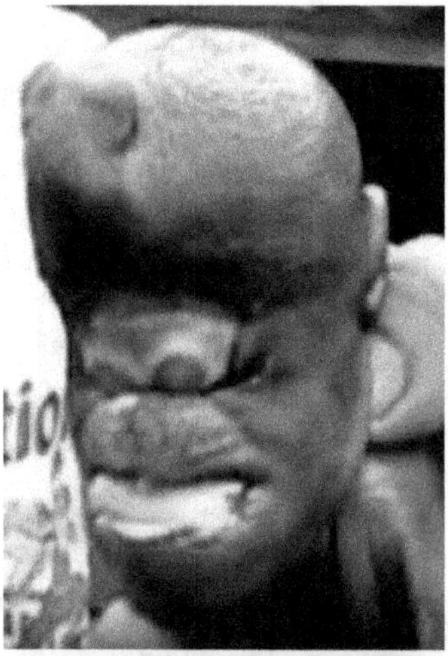

Myanmar. Another, viable individual was born in Myanmar in 2014 and was recorded in a video (||y2eokhdn), from which the screenshot at right is taken. No background information is available about this birth. But the video is among the most convincing of any of the videos documenting the existence of hominosuids. The sleeping creature is alive and very clearly shown.[2]

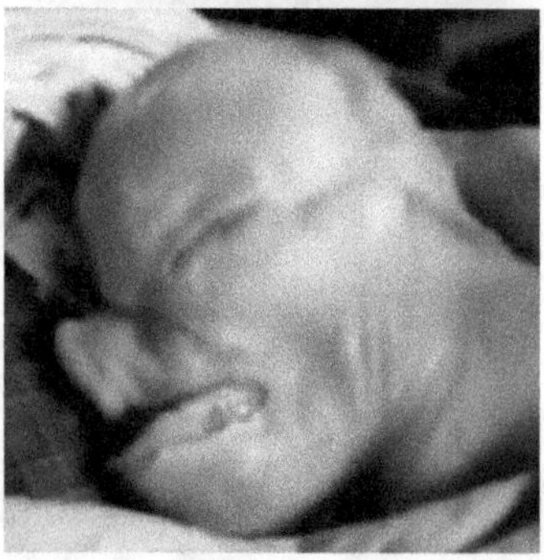

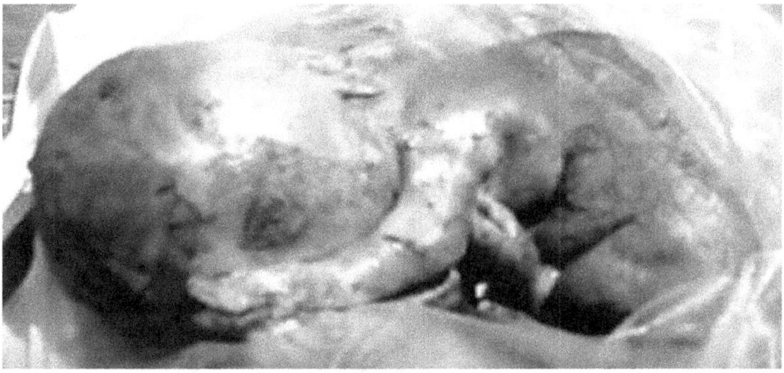

A hominosuid born in 2012 (||y32ccezk).

No locus in quo. Another hominosuid is known only from a 2012 video (||y32ccezk), from which a screenshot appears above. In this picture it can be seen that the ears turn toward the back instead of standing upright as do those of ordinary swine. This orientation of the ears, which is humanlike, seems to be typical in hominosuids. Note that human ears, too, are directed backward, not upward. The subject appears to be anophthalmic.

* Anophthalmia and cyclopia occur at high rates in telenothians.

Guatemala. In 2011, a sow farrowed in Guatemala a hominosuid much like the 2012 birth pictured above. The video (||y4y6xbk7) is of very low quality, but it can be seen from the screenshot at right that, in comparison with an ordinary piglet, the subject had a greatly enlarged cranium with a generally rounded appearance, as seen in humans. Moreover, it appears to be anophthalmic, as was the 2012 specimen.

The Guatemala hominosuid.

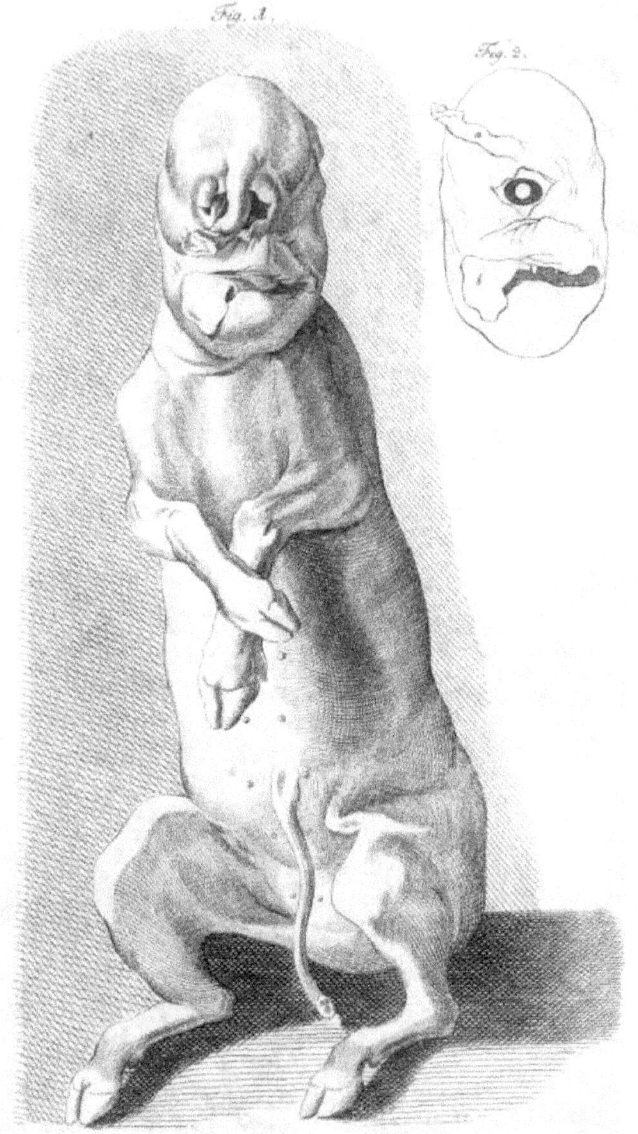

A hominosuid in the Kassel anatomical collection in Germany, a cyclopean female with a frontal proboscis, farrowed by a sow (Sömmerring, 1791, Tab. XII; ||yykgzb87). The collection had another such specimen (ibid., § 80; ||yc3kjsvy). Lowne (1872, pp. 44-45; nos. 162-172) briefly described 11 such specimens in the collection of the Royal College of Surgeons, London.

A sow farrowed this cyclopean hominosuid, with a frontal proboscis, Feb. 25, 1923, at a mental hospital in Rydalmere, New South Wales (Prior et al. 1924). Stillborn, it weighed one kg (shoulder height: 15 cm; girth: 25 cm; length: 32 cm). Its body was hairless. A second cyclopean hominosuid, stillborn at the same facility (Aug. 6, 1923), weighed 650 g (shoulder height: 16 cm; length: 29 cm). Prior et al. reported numerous additional anatomical details.

20TH-CENTURY REPORTS

> *Each of these creatures, despite its human form, its rag of clothing, and the rough humanity of its bodily form, had woven into it—into its movements, into the expression of its countenance, into its whole presence—some now irresistible suggestion of a hog, a swinish taint, the unmistakeable mark of the beast.*
>
> —H. G. WELLS, *The Island of Dr. Moreau*

Ignacio de la Llave, Veracruz, Mexico. The *Alger Républicain* (Algiers, Jan. 28, 1947, p. 4; ||bdec9hcm) reported the farrowing of a viable, human-faced piglet at Ignacio de la Llave.

Galax, North Carolina. The *Skyland Post* (West Jefferson, N.C., Oct. 8, 1936, p. 3; ||3zw8y32w) said a sow farrowed a creature that had a human hand, with nails, but that looked otherwise like a pig.

Sparta, North Carolina. The *Alleghany Times* (Sparta, N.C.; June 25, 1936, p. 1; ||4ums5rxs) reported a female synophthalmic cyclops, with a frontal proboscis, farrowed in Sparta.

Contestee, Greenville County, South Carolina. The *Index-Journal* (Greenwood, S.C., Nov. 23, 1934, p. 4e; ||yy7eg49u) said an upcoming show included "two Half-Human Pigs, which were born recently near Contestee, and which have been exhibited in several cities in this neighborhood. These pigs have been seen by over ten thousand people and have been examined by doctors from far and near … One of these half-human pigs has a perfect human head, has fingernails instead of hoofs, and before it died at the age of six days, cried like a baby. This pig has apparently two human eyes that grew together into one, and other human features that are deeply amazing. The other pig, from the same litter, has a perfect human skull, large blue eyes, and three human teeth. It also died after living for a week."

Saint-André, Réunion. Le Peuple (Saint-Denis, Oct. 24, 1931, p. 2b; ||dput6shu) said, in translation, "On October 14 at Saint-André, Ravine Creuse, a sow farrowed a monster amidst a litter. It had the body of a pig, the head of a human and hands. It only lived an instant. Instead of sending it to a museum, the owner, frightened, buried it at once."

A Coruña, Galicia, Spain. The *Paris-midi* (Paris, Jun. 19, 1926, p. 4a; ||2hwz3exs) said, in translation, "Correspondents of the Madrid press at A Coruña have reported telegraphically the birth of a robust little piglet that, rather unusually, has arms and hands. The newspaper *El Orzan* provides some details. The piglet, farrowed on the estate of Mrs. Elvire Salgado, widow of the commandant, has very short forelegs like two arms, of which the elbows are exactly the same as those of an infant. At their ends are two perfect hands. One of the hands has five digits, the other, only has four. Each finger has a nail, and the palms each bear the characteristic "M" of the human palm."

Taixing, China. The crude image at right accompanied a news report about a hominosuid farrowed at Taixing in 1923.

The Taixing hominosuid (||y4k7f65o).

Coney Island, New York. Reporter Joseph Lotterman said he saw at Coney Island "a pig with a human face" (*Evening World,* New York City, Aug. 18, 1922, p. 10f; ||pva3uy96).

Concordia, Sinaloa, Mexico. El Tucsonense (Tucson, Az., Aug. 17, 1922, p. 2a; ||4dajf4d5) said, in translation, "PIG BORN WITH HUMAN FEATURES.—MAZÁTLAN, SINALOA, Aug. 15—A weird event, unique in the history of Concordia, has shocked inhabitants. This shock, having resulted from the presence of one of those strange freaks that nature produces from time to time, has lasted several days. Indeed, a week ago a pig was farrowed in that town with a face very much like that of a man. All the details of a human face can be seen in that animal, whose body, unlike other pigs, is hairless, and covered with an extremely white skin. A veritable horde of pilgrims are filing through the neighborhood to that house, with curiosity seekers snapping photos of the freak. The pig died a few hours after birth."

West Goshen, Indiana. The *Ketchikan Miner* (Dec. 24, 1915, p. 1; ||2xkk4cp9) described two pigs born in the same litter in West Goshen: "One has two bodies, but only one head; also has eight legs, two tails, four ears, one having a horn growing from it. The other pig has a trunk similar to that of a baby elephant; has one large eye in the center of its forehead, with a three-inch tail [i.e., frontal proboscis] just above the eye."

Dayton, Kentucky. The *Cincinnati* Enquirer (May 28, 1915, p. 9c; ||yaktw895) said one hominosuid farrowed at Dayton had "half of its face similar to that of a baby," "four fingers and a thumb on the right side of its body in place of the front and rear legs and feet" and "a human arm and shoulder."

South Norwalk, Connecticut. From the *Guthrie Daily Leader* (Guthrie, Ok., Apr. 23, 1915, p. 6f; ||goq2sk3): "SOUTH NORWALK, CONN., April 23—A pig with ears, nose and mouth almost human in character, yet otherwise a perfect pig, was born here and is causing a tumult among scientists. The pig was born in the piggery of Albert Kunze, and while a sow usually gives birth to eight or ten pigs, this porcine monstrosity carried out its human resemblance by being born alone."

* An Albert Kunze lived on North Taylor Ave., Norwalk in 1910 (||y3kx7cbs).

Sikeston, Missouri. The *Sikeston Herald* (quoted in the *Dunklin Democrat,* Kennett, Mo., May 7, 1915, p. 7c; ||34zhyc32) said "Last Sunday a sow of James Klein's gave birth to [a] pig, which was undeveloped and born dead, had a head slightly resembling that of a human and hands, with nails on the fingers like that of a person."

Decatur, Indiana. The *Decatur Democrat* (quoted in the *News-Times*, South Bend, Ind., Apr. 2, 1915, p. 6e; ||h9l4u42): "D. M. Hensley, the well-known taxidermist in this city [i.e., Decatur], is busily at work stuffing the little pig which was born on the Gephart farm some days ago, the feet of which resembled the hands of a human being. The pig is a freak, and Mr. Hensley will have it on display in a few days. It will no doubt attract attention."

* In 1910, a David Martin Hensley (1866-1927) lived in Decatur (||y5opmdny). In 1915, a Phillip Gephart (1854-1951) lived just outside Decatur (||y4jqatm3).

Rock Port, Missouri. From the *San Bernardino News* (San Bernardino, Ca., Jun. 2, 1914, p. 1; ||yyvx9cq2): "ROCK PORT, MO., June 2.—Mrs. Lillie L. Breauzeal of this town has a freak hog which is different from anything ever heard of here. The front of the head of the animal resembles a human face and it has no hoofs. It is without hair except small patches over its eyes."

* A Lillie Braazeal (*sic*) was living in Tarkio, near Rock Port, in 1920 (||y349os49).

Knox, Indiana. The *Bristol Banner* (Bristol, Ind., May 26, 1911, p. 2b; ||24d8xxkp) reported a likely hominosuid. "Henry Cole, at Knox, Ind., discovered a freak among a litter of pigs that was farrowed, Monday. The eyes of the monstrosity were located between snout and the mouth and were united. Other deformations occurred about the head, and the body was hairless, making the pig a fearful sight. The creature lived but a few hours."

Sturgis, Kentucky. From the *Crittenden Record* (Marion, Ky., May 21, 1908, p. 1f; ||kx8y7g2): "STURGIS, KY., May 16th.—Perhaps one of the most peculiar freaks ever known is the human-faced pig now three days old, at the home of J[osiah] S. Hancock, on the Jones farm. Mrs. Hancock feeds the pig with a spoon as she would feed a baby. The mouth, nose, eyes, forehead and chin are identical with that of a human. It is one of ten pigs having a body as perfect as any of them."

* A Josiah S. Hancock, aged 46, a laborer, and his wife Ida, lived near Sturgis on the farm of Elijah Banner Jones (1872-1926) in 1908 (||yxkdxa33, ||y4sjh355).

La Crosse, Wisconsin. "One of the most extraordinary freaks of nature ever seen in this city is a small pig about the size of a cat, with small human hands, the property of John Desmond, St. Paul and Caledonia streets" (*La Crosse Tribune*, Jan. 11, 1911, p. 4d; ||4w2drzu6).

Umatilla Reservation, Oregon. The *Athena Press* (quoted in the *East Oregonian*, Pendleton, Ore., Dec. 3, 1904, p. 2b; ||hh25x2c) said, "A freak is to be seen at the farm of Frank Gagnon, five miles south of town on the [Umatilla] reservation. It is a well-developed pig, born with a litter of several others, but differing from them in this point: For one leg and foot it has a perfectly formed arm and hand, and on the remaining three legs, instead of feet, are hands. The mother sow refuses to accept her freak offspring, and Mrs. Gagnon is raising it 'on the bottle.' She has refused two offers, one of $50 and one of $100 for the animal."

* In 1900, a Frank Gagnon, or Gunyon, Jr. (b. 1854), a member of the Cree Nation, lived on the Umatilla Reservation between Athena and Pendleton, Oregon (||ysksmupf). Athena, where the report originated, is 18 miles from Pendleton.

Kinston, North Carolina. The *Kinston Free Press* (Apr. 29, 1904, p. 4a; ||4c7nhc4h) said a pig farrowed near Kinston had "a left foot in the shape of a human hand. The hind feet had two hoofs each, but these were as long as those of a full-grown hog and were turned up."

Thionville, Moselle, France. L'Est Républicain (Nancy, Apr. 30, 1902, p. 2f; ||e4pfz4pt) said, in translation, "In the stable of Mr. Rische, a miller at Gandren near Thionville, a sow farrowed a piglet that had, instead of two forelegs, the arms and hands of an infant. Each hand had five perfectly formed fingers with very long nails. It couldn't nurse, so Mr. Rische tried to bottle feed it. But it died after three days. So he preserved it in alcohol."

A hominosuid farrowed at Easton, Illinois. In its mouth, it had a single, double-pupiled eye (*Daily News*, Champaign, Ill., Nov. 7, 1896, p. 7d; ||ycdep8vq).

The Schwarzpaul hominosuid (Landois 1874).

"APE-HEADED" PIGS

Several accounts describe "ape-headed" piglets. However, in this book, when animals are described as having ape-like body parts but were born in locales where apes are absent, it is assumed that the resemblance was to humans instead. Here are a few examples:

- *Schwarzpaul, Germany.* Hermann Johann Theodor Landois (1874), a zoology professor, described a hominosuid farrowed in December 1873 at Schwarzpaul. He called it a "piglet with a human, or at least ape-like, head" (Ferkelchen mit einem menschlichen oder wenigstens affenartigen Kopfe"). He said that it lived several hours, and seemed viable, but that the farmer killed it at the insistence of his superstitious wife.
- *Le Guetteur* (Saint-Quentin, Jul. 26, 1885, p. 2e; ||4v9kb44s) reported another case of a sow farrowing a viable animal with an ape head, at Festieux (Aisne, France). The report said it had "the head of an ape, with the ears of a pig and the tongue of a human ("la tête d'un singe, avec the les oreilles de porc et une langue conforme à la langue humaine").
- *La Semaine Illustrée* (Paris, Jan. 1, 1899, p. 9d; ||msx6mtwn) reported an ape-headed animal farrowed in the hamlet of Le Corbet, Isère, France.
- *Le Télégramme* (Toulouse, Apr. 23, 1908, p. 2; ||2s4hvas7) said a sow farrowed a viable ape-headed animal, farrowed near Voiron, Isère.
- *Le Mémorial* (Saint-Étienne, Jun. 17, 1933, p. 2g; ||5xy9ek5h) reported one with a frontal proboscis, farrowed at Chalon-sur-Saône, Saône-et-Loire, France. Its owner killed it.
- The *Osborne County Farmer* (Oct. 12, 1882, p. 3b; ||28cysbvy) said the head of a pig farrowed on the farm of Thomas Edmund Fultz in Osborne, Kansas, "resembles a baboon exactly, and of course approximates the human. It has only one eye located in the middle of its forehead."
- The *Ithaca Journal* (Ithaca, N.Y., Mar. 31, 1882, p. 4; ||bdda7hk7) said a P. Robertson, of Ulysses, New York, had "a pig with "the face of a monkey and with but one eye, the organ of vision being in the center of the forehead. The monstrosity … lived only a short time."

19TH-CENTURY REPORTS

Saint-Marc-sur-Seine, Côte-d'Or, France. Le Bien Public (Dijon, Feb. 27, 1899, p. 3; ||2dy8wcwf) said a sow belonging to Paul Mongin, a farmer at Saint-Marc, farrowed a cyclops with a frontal proboscis.

Chrisman, Illinois. The *Chrisman Courier* (quoted in the *Evansville Journal,* Evansville, Ill., Sep. 11, 1898, p. 11b; ||2hvntrdx) said "Quite a freak of nature is owned by James Connor near Logan. ... Last Friday Connor had a sow that gave birth to a litter of pigs ... one of these said pigs is undoubtedly a valuable curiosity. It has four legs the same as any other pig, but on the front legs there projects on each one a perfect human hand which consists of four fingers and one thumb. The only difference between the hands on the pig's legs and a human is that the thumb is on the under side of the hand instead of the inside. His pigship is in good health."

* Logan is an unincorporated community 6 miles southeast of Chrisman.

Wilmington, North Carolina. The *Goldsboro Headlight* (Goldsboro, N.C., Dec. 24, 1896, p. 1d; ||zj4alp4) said, "Laura Galloway, a colored woman who resides on Fourth and Dawson streets, in Wilmington, is the owner of a sow with a litter of six pigs. ... one of the pigs has the perfect ears of a child, has a human hand on the right side, and sits erect like a child. ... It also has a full set of teeth and long tusks like a grown hog. The pig's face is also partly human, and it has to be fed from a bottle."

* A Laura Galloway lived at 410 Dawson St., Wilmington in 1900 (||3y7ah9sd).

Blackwater, Virginia. The *Fisherman & Farmer* (Edenton, N.C., May 29, 1896, p. 2d) said, "A curious freak of nature can be seen at the farm of R. C. Ausell, living near Blackwater, Va. It being no less than a young pig that is part human. It has a human head, also jaws, nose, eyebrows and feet—there being no eyes at all, and no hair on the body."

Winchester, Kentucky. From the *Public Ledger* (Maysville, Ky, Nov. 3, 1896, p. 1f; ||yy79u8us): "LEXINGTON, KY., Nov. 3.—William McAndrews, of this city, came home from Winchester Monday night with a strange freak of nature in the shape of a pig with a human head. It has but one eye,[8] that just above the nose. Otherwise, the formation is perfect. The pig is but three days old and was one of a litter of eight."

* An 1898 Lexington directory lists a William B. McAndrew (||y29y8pfl).

Brooklyn, New York. A report about a pig with an old man's face, from the *St. Louis Post-Dispatch* (Aug. 26, 1896, p. 29; ||y5a5dwgw):

The Brooklyn hominosuid.

> A pig with a human face is in the possession of D. A. Sammis of No. 64 Vanderbilt Avenue, Brooklyn. It is one of the most remarkable monsters ever produced by an aberration of nature.
>
> The pig is dead and is preserved in a spirit jar, which enables its strangely human features to be inspected.
>
> The pig's head differs from a human being's only in having the long, pointed ears belonging to normal members of its family. This gives it somewhat the appearance of a goblin or other creature of the imagination, with a partly human shape.
>
> The head, apart from the ears, is like that of an old man. It is free from all hair except eyebrows. These are well grown and are a remarkable abnormality in a pig. The forehead is high, and the skull is rounded at the top of human shape. It suggests considerable intelligence and a well-balanced character.
>
> The whole face is oval in shape and is similar to that of an old man of dignified appearance. There is rather too little nose for a very good-looking man, but still, this member is distinctly human in shape. It is small and snub, and is utterly different from the sharp snout of a pig
>
> The chin is heavy and well rounded. It is perhaps the most human of all the features. The eyes are much larger than those of an ordinary pig. The skin is as white and smooth as that of a delicate woman. Although it must be called a monster, on account of its strange physical abnormality, the little pig is not repulsive. On the contrary, it is quite amusing in appearance. If it were alive and well, it would be a decidedly interesting animal to have about the house.
>
> How it came to die is a curious story. It was born on May 12 last on a rancho near Arecibo, on the island of Porto Rico, in the West Indies. The mother produced only this one animal at the time, itself a very unusual circumstance.
>
> In spite of its grotesque appearance the little pig seemed to enjoy good physical health. But his mother had no affection for him. She was disgusted and alarmed at his unnatural features. She gave him little nourishment, and in consequence he died on May 18. The fact that he remained alive for six days in a half-starved condition shows that he must have entered life with fair health.

Flatrock, Ohio. The *Democratic Northwest and Henry County News* (Napoleon, O., Jul. 4, 1895, p. 1; ||kc7anrj) said,

> Mr. Albert C. Knipp, who resides five miles south of Napoleon on the farm of the late Peter Knipp [in Flatrock], reported the birth of a monstrosity on Saturday, in the shape of a pig, which instead of having natural legs and feet had three appendages which resembled the human arm upon which are three shapes like the human hand. The other leg and foot are natural.
>
> This monstrosity is one of a litter of pigs numbering fifteen, twelve of which are living, and in all respects, it is healthy with good prospects of maturing. Mr. Knipp intends to care for it and take all the necessary pains to have it live.

Marquand, Missouri. The *St. Louis Republic* (quoted in the *Philadelphia Times,* Aug. 7, 1892, p. 17d; ||y2p8huf8) said, "A strange malformation is attracting considerable attention here [in Marquand]. A sow on the premises of Mr. [Thomas Pitkin] Russell [1864-1952], a resident of this place, gave birth to five pigs, four of which were ordinary and are at present thriving. The fifth, however, is a wonder to behold. It is a perfectly formed pig with a human head. Its resemblance to a baby being perfect. Many persons have been to see it, and every resident of this place will vouch for the truth of this statement.

Indianapolis, Indiana. The *Indianapolis Journal* (Jun. 8, 1892, p. 8d; ||y9589mqz) said, "Mrs. Mary Mason, who is a grocery keeper, has a freak of nature in the form of a pig which was born with a human face. It lived only eleven hours and has been preserved in alcohol. It was one of a litter of eleven, the rest of which did not offend nature."

* Mary Mason, a widowed grocer, lived in Indianapolis in 1892 (||y479okuj).

Lake Providence, Louisiana. David Llewellyn Morgan (1820-1893), editor of the *Carroll Democrat* (Lake Providence, quoted in the *Meridional,* Abbeville, La., Oct. 24, 1891, p. 1e; ||dx3prrny), said that "a sow in that parish recently brought forth … a regular nondescript, a monstrosity. On the right side of the head was a regular flopping over hog's ear, but on the left side it had a well-defined human ear … The body was white, but the breast was flat, and instead of forelegs, it had two human arms and hands or paws; instead of the cloven hoof, four well defined fingers with a thumb on each paw; the elbow joints were well defined."

Martin, Tennessee. From the *Tennessean* (Nashville, Feb. 7, 1891, p. 6b; ||yfvc9jfk): "MARTIN, Feb. 6.—News has just reached here of an animal being born a few miles north of here that is a very curious freak of nature. It is a hog with a body very much like that of a human; it has no eyes; its skin and muscles are the same as that of a human. It lived only a short while after its birth, but while it was alive it would cry like a baby. Its toes resemble the fingers of a person. It is now in the possession of Dr. Sims, about eight miles north of this place. He has it preserved in alcohol."

New Albany, Indiana. In news from New Albany, the *Courier-Journal* (Louisville, Ky., Jun. 28, 1889, p. 6; ||3kczszyb) said, "A freak of nature in the shape of a pig with a human face, and two feet formed like hands is attracting considerable attention in Chartres Street."

Tenterfield, New South Wales. A report from Tenterfield in the *Armidale Express* (Armidale, NSW, Jan. 29, 1886, p. 4d; ||y2k9wdv6) said, "An extraordinary specimen of the "freaks of nature" is in the possession of Mr. C. I. Cavanough, chemist, in the shape or form of an infant pig with a head of human appearance. The 'monstrosity' comes from Mr. [Charles] Brandt's farm, Six-Mile"

* A Charles Innes Declair Cavanough (1847-1921) lived in Tenterfield (||y56yggu7). Charles Brandt (1838-1911) had a farm at Six Mile Swamp in northern New South Wales near Kippenduff, about 100 miles east of Tenterfield (||y632law7, ||y632law7).

Urbana, Ohio. The *Cincinnati Enquirer* said one hominosuid, at Urbana, was like a pig with one human hand (1886, p. 4h; ||bddfs682).

Norristown, Pennsylvania. The *Hickman Courier* (Hickman, Ky., Nov. 20, 1885, p. 4f; ||yxz394lq) said, "Henry Sclater [ca. 1810-1892], of Norristown, has a litter of seven young pigs, one of which is a monstrosity of an unusual type, combining with its porcine construction a portion of the human anatomy. The left forelimb, instead of being a leg, is like the arm of an infant. At the extremity is a hand, containing a thumb and four taper fingers, which are regular in form, even to the nails."

Concordia, Kansas. The *Concordia Blade* (Apr. 3, 1885, p. 2a; ||3pppvwwr) said a pig had instead of forelegs, two arms with human hands, each bearing four fingers and a thumb, each of which bore a human-like nail.

Groton, New York. Reports say Professor Simon Henry Gage (1851-1944) purchased for the natural history collection of Cornell University a *tertium quid* with a pig's body and a human face. The owner was John M. Grettenberger, a farmer in Groton. Gage's purchase received widespread coverage. In translation, one article, in *Le Voleur* (Paris, Aug. 28, 1884, no. 1417, pp. 557-558; ||yypskge9), said the natural history collection of Cornell University had recently acquired a piglet with the face of a woman, "and indeed of a pretty woman." According to the report, the eyes, ears, mouth and dimpled chin were all those of a human being. The report below, from the *New York Times* (Jul. 29, 1884, p. 5d; ||y4xkobnv), mentions no woman, but otherwise tells the same story: "ELMIRA, N.Y., July 28.—Prof. Gage, of Cornell University, has added to the museum of that institution a curious monstrosity in the shape of a pig with a head closely resembling a human being's. It was one of a litter of 10 pigs born on the Gretenberger farm, near Peruville, Tompkins County. When the creature is placed on its feet the line from the chin to the forehead is perpendicular. The top and back of the skull have an almost perfect human development. The ears are strikingly like a man's in shape and are attached to the head at a similar angle. The mouth is large, but almost perfect in its human resemblance. The chin is well defined and perfectly formed. The eyes are imperfect. Prof. Gage paid $25 for the curiosity."

* A John M. Grettenberger (b. ca. 1827), a farmer, lived in Groton (||u7l4ubf).

Crawford, Kentucky. The RICHMOND REGISTER (quoted in the *Semi-Weekly Interior Journal,* Stanford, Ky, Feb. 12, 1884, p. 4c; ||m9u98w8n) said, "Some weeks ago, the REGISTER contained a news item to the effect that Mr. Miles Crawford, of Breathitt County [Kentucky], found on his doorstep one morning an infant, half human and half dog. Last Monday a messenger from Mr. Crawford came into our office to inform us that the item was partially incorrect. A horrible monstrosity—a mixture of hog and human—was found in Mr. Crawford's corn field one cold morning last January frozen stiff. It was in a little brown sack together with some corn. Some farm-hands came close to the sack, and it was accidentally knocked over, when the hideous lump of deformity rolled out. It had three legs, with split hoofs. Its head was of human formation well covered with hair. The mouth was large; the upper lip was hard and gristly, like a hog's "rooter." The nose and eyes were considerably more human than hog. The body resembled the human body and was perfectly smooth."

* A Miles C. Crawford, a farmer, was living at Crawford in 1900 (||y339xlnm).

Washington, D.C. Billings (1884, p. 24, No. 1275) listed an anophthalmic piglet with a rounded cranium and frontal proboscis.

Occoneechee Neck, North Carolina. The *Weldon News* (quoted in the *Charlotte Home and Democrat*, Jun. 8, 1883, p. 2b; ||kpwlfvf) said, "A correspondent whose truthfulness we will vouch for, sends us the following account of a 'child-pig' … born in Northampton a few days ago: On Thursday, 24th, a sow, the property of Jacob Benjamin, a tenant on the Maratock plantation in Occoneechee Neck, Northampton County, farrowed and among the pigs there appeared a monstrosity, astonishing and almost incredible. The thing was a female, assuming the form of a child and pig. The body was that of a natural pig to the shoulders, while from the latter arms, instead of legs, protruded, with elbows, wrists, hands and requisite number of fingers with nails, and inclined towards the head, as a child in the act of clasping its hands together. The head was human in form possessing the expression and features of a pig. It lived eight or ten hours, its cries being those of an infant in distress. Over a hundred people visited the premises, to witness this curious freak of nature—'half child, half pig,' and will corroborate the facts as given."

* Jacob Benjamin, a farmer, lived in Occoneechee, N.C., in 1880 (||yxk6wrql).

Hawaii. The *Pacific Commercial Advertiser* (Honolulu, Apr. 7, 1883, p. 5c; ||bddm3pct) said, "A correspondent on Hawaii writes us that in a litter of twenty-eight little pigs there was one with a human head, chest like a man, teeth an inch long, forelegs like a man, no tail."

Canton, Massachusetts. The *Boston Globe* (Feb. 15, 1883, p. 1f; ||yckv94nr) said the editor of the *Canton Journal* had a pig's foot with a "perfect human thumb" preserved in alcohol.

Constantine, Algeria. The *Akhbar* (Algiers, Nov. 8, 1882, p. 3; ||4kh6r3f5) said a sow in the Sidi-Mabrouk quarter of Constantine gave birth to a viable nondescript with a human head.

Kansas City, Kansas. From the *Cowley County Courant* (Winfield, Ks., Dec. 22, 1881, p. 2d; ||2p8hzpx5): "Kansas City is excited over a horrible monstrosity taken from a sow in Plankington & Armour's packing house. The terrible thing is about the size of a two-months-old pig, has the body and head of a pig, but where the forelegs should be, are human arms perfectly formed below the elbow, showing fingers and fingernails in perfect shape. This is enough to shake up the average citizen. The sow was one of thousands that are daily shipped in, and there is no telling where she came from."

Sturgeon, Missouri. The *Columbia Herald-Statesman* (Columbia, Mo., May 20, 1881, p. 2f; ||24rpuu6z) said a W. H. Harris had a sow farrow a pig with forelegs like arms and "5 perfect fingers on each hand."

Sacramento, California. The *Record-Union* (Sacramento, Jul. 14, 1881, p. 3; ||ycky9u2n) said, "a strange freak … has just occurred in the eastern part of the city, at the ranch of M. M. Odell, Thirty-first and Y streets. It consists of a young pig, which was dead when found, and the head of which strongly resembles that of a human being."

South Point, North Carolina. The *Charlotte Observer* (quoted in the *Anderson Intelligencer,* Anderson, S.C., May 26, 1881, p. 2f; ||yd62byse) said, "On the place of a gentleman named McKee in South Point Township, Gaston County, about 15 miles from this city, is a pig, born like other pigs with the exception that where one of the forelegs ought to be is in place a perfectly shaped human hand with four fingers and a thumb, with well-developed nails upon them. There is another exception. On the other foreleg is a toe like those of a human being. The pig is six weeks old. It carries its hand in front and parallel with the body as if in a sling and runs on its three legs much faster, even, than the other pigs with four. … Dr. J. C. Bauman, a practicing physician of Gaston County, made a scientific examination of it and says the bones and ligaments are those of a perfectly formed hand."

* Farmer George Logan McKee (1853-1909) lived in South Point in 1880 (||5du28ya9).

Ingeldorf, Luxembourg. The *Neuigkeits-Welt-Blatt* (Vienna, May 27, 1880; ||y4wsuqm3) said, in translation, "A sow has farrowed an appalling monstrosity at Ingeldorf. Among nine piglets, one had a well-developed human head. It lived but half an hour."

* Ingeldorf, or Angelduerf, is a town in the commune of Erpeldange in Luxembourg.

Franconia, Pennsylvania. The *Lansdale Reporter* (quoted in the *Cambria Freeman,* Ebensburg, Pa., Mar. 26, 1880, p. 2; ||y3d46rj9) said, "A short time ago a sow belonging to Mr. Wimerlick, of Franconia Township, near Souderton, gave birth to a litter of pigs, and one of the number, instead of having its front legs like a pig, had human arms, elbows, hands, four fingers and a thumb, and even the nails on the fingers were perfect like those of a human being. It did not live."

* In 1880, farmer John Wunnerlich (b. 1808) lived in Franconia (||3sah9663).

Friendship, Tennessee. The next, rather well-attested report is from the *Memphis Daily Appeal* (Memphis, Tn., Nov. 31, 1877, p. 4c; ||pa25tuh).

> FRIENDSHIP, CROCKETT CO., August 28.—Having seen a statement going the rounds of the papers that a pig with a human face was found near Dyersburg on the seventh instant [i.e., Aug. 7, 1877], we propose to make some corrections and give the case alluded to more in detail: Said pig was delivered of a fine young Berkshire sow in this village and county on the fourth instant and was one of a litter of ten pigs. The sow had been bred to a Jersey red boar previously and, as stated, brought a litter of ten pigs on the fourth instant. The entire litter was beautifully spotted, red and black, exhibiting a genuine cross between the black and Jersey red. The pig in question was the fourth farrowed and was dead when expelled. It was brought to our office a few minutes afterward by Mr. Wyatt [A.] Lunsford[4] [1825 - May 15, 1887] of this place. It was a perfect, well-developed pig, from its head back. Its head was shaped after that of a rather thin-visaged child, with a perfect, though rather narrow, human face. The nose resembled a human nose, somewhat shrunken from loss of the nasal bones. The right ear was a well-formed hog's ear, while the left, in rather a rudimentary state, resembled a human ear. The hair on the head was uniformly red, very soft and downy, making quite a contrast with the coarser spotted hair covering the body. The face,

including the forehead, which was very high and free from hair, was perfectly smooth and as tender as an infant's. That portion of the head formed by the frontal and temporal bones was as perfect human in resemblance as could be. The eyes, when the lids were drawn apart, presented a beautiful blue color, with a freshness unusual in a dead infant. Its tongue was wide and thin, just like an infant's tongue. It had but one tooth, and that was a perfectly formed incisor. This monster pig was put by us in alcohol, and intended to be preserved for future study, but unfortunately it began to decay in a few days, and our best efforts to preserve it were unavailing. It was visited, however, by hundreds of people (and among the number several intelligent physicians), by whom the above facts can be corroborated.

Respectfully,
J. D. Smith, M.D.
M. M. Smith, M.D.

* A John Dinkins Smith (1804-1881), a physician, lived at Henderson, Tenn., in 1877 (||yxvs4stp). A physician, Millard Smith, lived in Henderson Co. in 1880 (||y2z9xqga). M. M. Smith at Cedar Chapel, Henderson Co. in 1876 (||y3vrdwcr). A Milliand M. Smith (b. 1851), a physician, lived in nearby Whiteville in 1900 (||y4fc9s85).

The *Clarksville Weekly Chronicle* (Clarksville, Tn., Sep. 1, 1877, p. 1; ||px4mwfk) said,

They had strange doings up at Friendship the other day. Mr. Adams, the blacksmith at that place, has a sow, which on the 7th instant dropped a litter of pigs, one of which has a perfect human face and head, the rest of its formation being that of any other pig. It has red hair on its head; its mouth, teeth, and eyes are strangely human, while one ear is that of a hog, the other is that of a little child. This new and strange production is the wonder of the neighborhood. Dr. Duffie, the druggist at Friendship, has preserved it in alcohol and hundreds are flocking to see it daily. We got our information from Capt. Tom Jones and Mr. John McFee, both of whom we have known for years to be truthful and honorable gentlemen. Both of them have seen the pig and we are satisfied that the facts are as stated above.

* Records (||y3pyxlhe) document the existence of a John Henry Adams (1851-1944). A Joseph W. Duffey was a druggist in Friendship in 1880 (||y6lv82wd; ||y3mphcu4).

Berryville, Virginia. From the *Alexandria Gazette* (Aug. 18, 1876, p. 2; ||2hw9wpe6): "Major J. D. Richardson left at Dr. Page's drug store, in Berryville, a pig having a head and face like that of a child. The forehead, chin and mouth are unmistakably human in appearance, but the nose is represented by a fleshy, beak-like shape about a half inch in length. The eyes are wide apart and set somewhat up in the forehead. The ears, body, legs and feet are those of a pig. This singular deformity lived 24 hours."

* Dr. Robert Powell Page (1838-1914) died at Berryville (||y6dypdxa). John D. Richardson, a farmer, lived in Berryville (then called Battletown) in 1870 (||y625xhas).

Pietermaritzburg, Natal. From the *Natal Witness* (Nov. 21, 1876, p. 2): "A remarkable freak of nature transpired in this city a few days ago. A sow owned by a resident, gave birth to a litter of pigs. All were perfect specimens of the porcine family except one. This one was brought forth in the middle of the litter and had the head of a monkey while its body was that of a pig. The jaws were exactly those of a monkey, as was also the forehead and mouth. Indeed, its cries were so plaintively like those of a child or a young monkey that it must have taken the semblance of an evil spirit to the superstitious owner, for he turned it away and it died for want of sustenance. From the character of its noises the vocal organs must have been like a monkey's also. But the strangest feature of the monstrosity was its eyes. There was but one eyeball, and that was in the center of the head over the bridge of the nose, the eyelids and eyelashes sloping evenly to the corner, on each side. The eye was large, and there were two separate and perfect pupils in the same iris."

Holly Springs, Mississippi. The *Holly Springs Reporter* (quoted in the *Times-Picayune* Aug. 1, 1869, p. 10d; ||y652k628) said, "On Tuesday evening last we saw one of the greatest curiosities we ever beheld. It is a perfectly white pig, Chester White breed, with a human head. Looking at the head alone, one sees what he believes to be a small, yet tolerably well shaped human head. Its chin, jaws, mouth and nose are not at all of the pig kind, but look exactly like the chin, jaws, mouth and nose of a human being. Its head is shaped like the human head. Although only ten hours old when we saw it, and fatigued and fretted, having been handled and examined by over a hundred people, yet it was as active and as strong as a pig usually is at the age of two weeks.

This curious thing has no eyes,[3] and its nostrils are inside the mouth, rendering respiration difficult. It has rows of teeth as regular and as perfect as the rows in the human mouth and uses its tongue like a human being. All in all, it is a wonderful freak of nature, and if raised will be a fortune to its owner."

Elizabeth City, North Carolina. The Elizabeth City *North Carolinian* (quoted in the Wilmington Journal, Jul. 16, 1869, p. 1d; ||yctnvnqh) said, "One of the most singular freaks of nature we ever saw has been on exhibition at the post office, at this place, during the past week. It is a pig with a human face! Its shape and structure are natural up to the ears, but from the ears to the top of head and down to the chin the shape and features are those of the human, excepting that it is covered with hair like the rest of the body and has no nostrils, being obliged to breathe through the mouth. The pig is about two weeks old."

Carmi, Illinois. The *Evansville Journal* (Oct. 16, 1867, p. 8; ||2hu63adr) said a viable piglet farrowed at Carmi "had fingers like a newborn infant, instead of hoofs, with five nails on each foot."

Louisville, Kentucky. The *Louisville Journal* (quoted in the *Public Ledger,* Memphis, Tn., Aug. 8, 1867, p. 4a; ||y2obxfxt) said, "At the drugstore of G. W. Anderson [ergo, born ca. 1831], corner of Fifteenth and Walnut streets, yesterday afternoon, we saw a most extraordinary monstrosity—shocking, enormous, and horrible. It was a pig with a clearly defined human face. The chin, mouth and eyes are particularly noticeable for their similarity to those of a man, though the nose is not so much so; it turns up, very unlike a pig's, and has no nostrils. The animal was therefore necessitated to eat and breathe through the mouth at the same time. This strange creature lived about thirty hours, and during that time cried like a child, made a noise like an ape, and sometimes grunted like a pig. Its body is like a well-shaped pig, and rather larger than those animals generally are at such an age. The owner of the sow, its mother, was afraid of it and gave it to the druggist above named, who had it preserved in alcohol. The curious can see it by calling at the drugstore."

* A George W. Anderson, a druggist, lived in Louisville in 1870 (||6b4ek32w).

Consett, Durham, England. The *Newcastle Guardian and Tyne Mercury* (Newcastle-upon-Tyne, England, Oct. 15, 1864, p. 5f) said, "A great sensation has been created in Consett since Tuesday by a current rumour that a pig had been littered, bearing the resemblance to the arms of the human race. On examination and enquiry it was found that a sow, the property of Mr Harrison, had had a litter of young pigs, ten in number, one of which was a monstrosity. The forelegs resembled the human arms and hands; what ought to have been the hoof was split into four fingers and had a side claw, resembling the thumb." This nondescript seems to have survived.

Crawley, Gloucestershire, England. From the *Merthyr Telegraph* (Merthyr Tydfil, Wales, Aug. 18, 1860, p. 4; ||y6lf6751): "A short time since Charles Mill, living at [the hamlet of] Crawley, near [the village of] Uley [in Gloucestershire], had a pig which brought forth young to the number of ten, several of which were more or less deformed, one in a most extraordinary and wonderful manner. It had the perfect form of a human face (with eyes, nose, mouth, teeth and tongue) which was perfectly upright. There were no hind legs, but two marks where legs ought to have been.[5] There were two forelegs, on each of which were five fingers, and the hand represented a human being. The ears, which were not upright, but were situated along the side of the face, were half pig and half human in appearance. It was born alive. The face and forehead were quite bare of hair, but the hinder part was covered with hair, as of a pig."

Bratislava, Slovakia. Kornhuber (1860) said a conjoined cyclops hominosuid had been farrowed near Bratislava ("Pressburg"). It lacked a lower jaw and had large rear-directed ears and a frontal proboscis.

Milton, Pennsylvania. The *Penn Miltonian* (quoted in the *Southern Patriot,* Charleston, S.C., Jun. 24, 1843, p. 2) said "On Thursday last, we saw a pig 3 or 4 days old at the farm of Jacob Schitz … which had an arm with fingers, something of the human kind, projecting out above one of the forelegs."

Armley, Yorkshire. The *Leeds Intelligencer* (quoted in the *Manchester Courier and Lancashire General Advertiser,* Sept. 30, 1837, p. 3) said a sow at Armley had farrowed a pig "with the face of a child."

A jawless, sow-farrowed cyclops (Geoffroy Saint-Hilaire (1836a, p. 426; 1837, pl. vii). Jean Méry (1709) described a similar subject borne by a woman in Denmark.

Offenbach, Germany. The *Regensburger Zeitung* (Regensburg, Bavaria, Sep. 8, 1837; ||y2lm7l26) said, in translation, "A few days ago, allegedly, a sow belonging to a jailer in Offenbach bore six offspring, among which one had the face, and indeed the fully formed head, of a human being. A pig with a human head is as great an oddity as a human with a pig's head! One wonders if this curious pig could be schooled and brought up to be a useful and peaceable citizen of human society devoted to salvation and righteousness."

Doncaster, Yorkshire, England. The *Hull Packet* (Hull, England, Dec. 30, 1836, p. 3f) said "A pig was slaughtered in Doncaster, last week, which had on one of its feet, just below the natural toes, a joint protruding something like a man's thumb."

New York City. The *Sun* (New York, quoted in the *Rhode-Island Republican,* Newport, Jun. 11, 1834, p. 2; ||2s4d3aex) said, "We saw a pig night before last at 420 Broadway, the face of which resembled that of a human being. The forehead was round and well turned; the teeth on the lower jaw are very regular and perfect, and there is a small hair mole on one of its cheeks. It also has five toes on its left fore foot."

Leibnitz, Austria. Schallgruber (1823, p. 139) described a hominosuid with a frontal proboscis farrowed at Leibnitz in 1818.

Geneva, Switzerland. Jouard (1807, p. 77) quoted a letter, dated 1806, saying that a piglet with a human head was then on exhibit in Geneva.

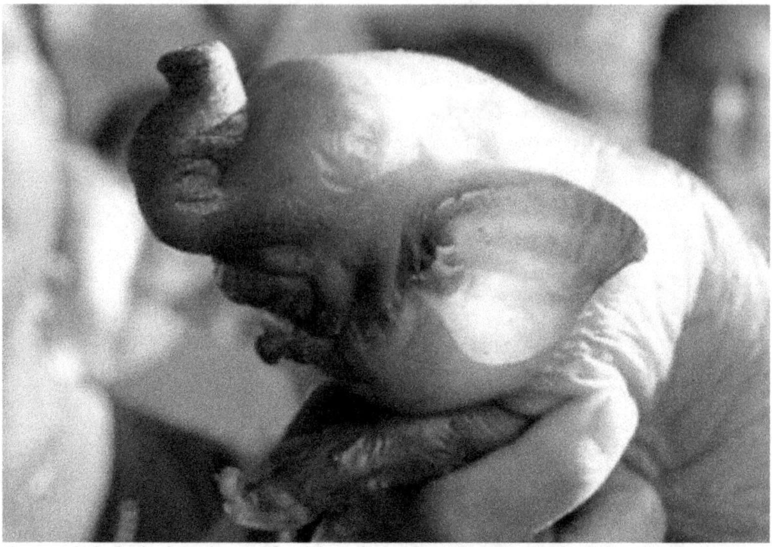

A anophthalmic hominosuid with a frontal proboscis (||y4w95mv2). It had upturned hooves and rear-directed ears, both characteristic of hominosuids. Farrowed in the Chinese province of Jilin in 2014, it had seven ordinary siblings. Its hairless skin contrasts with that of the Cambodian subject.

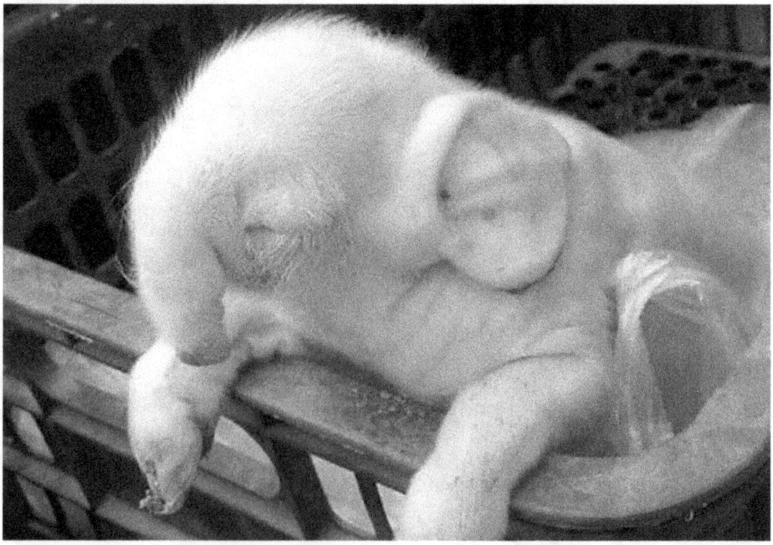

A hominosuid with a frontal proboscis born in Cambodia in 2016 (||y8kyargf). It could not open its eyes and was likely anophthalmic. Ears are rear-directed, hooves, upturned. Mouthless, it could not eat and died soon after birth.

“ELEPHANT-HEADED” PIGS

> *The people of Rome were disturbed by prodigies; for a wolf was seen in the city, and a pig was born that resembled an elephant.*
>
> —DIO CASSIUS, *Roman History,* 43.2.1

Sow-farrowed animals that old reports say had elephant heads likely resembled the creatures pictured on the opposite page. But note that though the pictured animals lack human heads and hands, they do have traits often seen in hominosuids (frontal proboscides, upturned hooves, rear-directed ears). So, such animals are classed as hominosuids.

REPORTS ABOUT “ELEPHANT-HEADED” PIGS

<table>
<tr><td>1858</td><td>Cuxton, England</td><td>||34f5ahwn</td><td>“lived some time,” CY</td></tr>
<tr><td>1860</td><td>Lincoln, England</td><td colspan="2">Worcestershire Chronicle (Jul. 4, 1860, p. 4f)</td></tr>
<tr><td>1861</td><td>Eureka, California</td><td>||3wcxjaz6</td><td>S</td></tr>
<tr><td>1873</td><td>Mumbai</td><td>||48xh6cr8</td><td></td></tr>
<tr><td>1876</td><td>Augusta, Georgia</td><td>||2tvyv9ej</td><td>stillborn, CY</td></tr>
<tr><td></td><td>Falmouth, Cornwall</td><td>||yt7u4y9r</td><td></td></tr>
<tr><td></td><td>Newville, Pennsylvania</td><td>||426vxzr2</td><td>CY</td></tr>
<tr><td></td><td>Palmerston North, N. Z.</td><td>||y4gqzrgm</td><td></td></tr>
<tr><td>1877</td><td>Angleur, Liège, Belgium</td><td>||38n2ec4r</td><td>2 “ape-headed” littermates</td></tr>
<tr><td></td><td>Lathrop, Missouri</td><td>||4mxcdkxj</td><td></td></tr>
<tr><td>1878</td><td>Waverly, Nebraska</td><td>||bddzsrxx</td><td>misplaced eyes, no mouth</td></tr>
<tr><td>1880</td><td>Quakertown, Pennsylvania</td><td>||wcnr7cc3</td><td>died at birth, CY, CJ</td></tr>
<tr><td>1881</td><td>Holt County, Missouri</td><td>||4a9833ay</td><td>CY, found dead</td></tr>
<tr><td>1883</td><td>New Orleans</td><td>||yx4uczcu</td><td>round ears</td></tr>
<tr><td>1887</td><td>Mound Valley, Kansas</td><td>||5fk5jane</td><td></td></tr>
<tr><td>1888</td><td>Blair, Nebraska</td><td>||28tnfk9e</td><td></td></tr>
<tr><td></td><td>Colwich, Kansas</td><td>||j8efukab</td><td>CY</td></tr>
<tr><td></td><td>Ireton, Iowa</td><td>||45etfzhp</td><td>CY</td></tr>
<tr><td></td><td>West Pennard, Somerset</td><td>||47fk7bew</td><td>2 in litter</td></tr>
<tr><td>1891</td><td>Texas (no locus in quo)</td><td>||bdz4m65d</td><td>CY, viable</td></tr>
<tr><td>1892</td><td>Topeka, Kansas</td><td>||v39uu4nd</td><td>S</td></tr>
<tr><td></td><td>Montpelier, Indiana</td><td>||uxs28shv</td><td></td></tr>
<tr><td>1893</td><td>Gannon, Texas</td><td>||mrcw25r9</td><td>no mouth CY, CJ (3 ears)</td></tr>
<tr><td>1905</td><td>Fort Dodge, Iowa</td><td>||yjxds2uf</td><td></td></tr>
<tr><td></td><td>Mattituck, New York</td><td>||spfwyxf8</td><td></td></tr>
<tr><td>1906</td><td>Paray-sous-Briailles, France</td><td>||3uyhb3ww</td><td></td></tr>
<tr><td>1907</td><td>Butler, Oklahoma</td><td>||4e46drfe</td><td></td></tr>
<tr><td></td><td>Elgin, Kansas</td><td>||munc5j6u</td><td></td></tr>
<tr><td>1909</td><td>Pont-l’Eveque, France</td><td>||n8pvbt8n</td><td></td></tr>
<tr><td>1912</td><td>Bruailles, France</td><td>||55mc8d7d</td><td>lived a few hours</td></tr>
<tr><td>1913</td><td>Medford, Oklahoma</td><td>||y2afj4ya</td><td>viable</td></tr>
<tr><td>1914</td><td>Henry, Illinois</td><td>||5n8em68s</td><td></td></tr>
<tr><td>1915</td><td>Ossian, Iowa</td><td>||4a8c3n5w</td><td>CY</td></tr>
<tr><td></td><td>Willows, California</td><td>||2stwzvhy</td><td>died at birth, CY</td></tr>
<tr><td>1926</td><td>Seville, Spain</td><td>||ykb5b98n</td><td>lived a few days</td></tr>
<tr><td>1929</td><td>Pink Hill, North Carolina</td><td>||3e77wf99</td><td></td></tr>
<tr><td></td><td>Saint-Pierre, Martinique</td><td>||27ufsjyv</td><td></td></tr>
<tr><td>1932</td><td>Arras, Pas-de-Calais, France</td><td>||mr2payf9</td><td>large ears, S</td></tr>
<tr><td>1938</td><td>Blanot, Côte-d'Or, France</td><td>||yaezfbc9</td><td>upturned hooves</td></tr>
<tr><td>1949</td><td>Montbéliard, France</td><td>||yk6544j3</td><td>lived a few hours</td></tr>
<tr><td>1953</td><td>Bailleul, Nord, France</td><td>||ycavvrfc</td><td>CY</td></tr>
</table>

Key: CY: cyclops; S: Synophthalmic cyclops (two eyes in a single socket); CJ: conjoined

A hominosuid with a frontal proboscis reportedly farrowed near Weissenfels, Germany; one ear was human, the other, porcine; the expanded cranium, upturned hooves and protruding tongue are typical in such births (Schreber 1751).

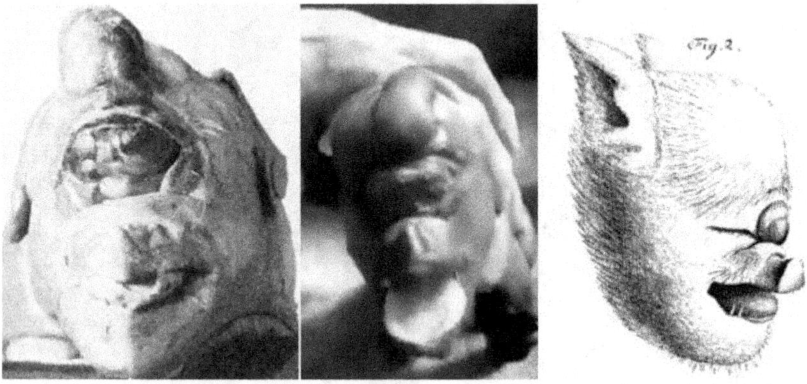

Left: A synophthalmic cyclops hominosuid with a rear-directed frontal proboscis (Schwalbe 1906-1960, p. 208, fig. 133). *Center:* A similar, seemingly viable individual born at Ourolândia, Bahia, Brazil in 2021 (screenshot from a video; see ||2p7zjak). *Right:* A sow-farrowed hominosuid born in the German village of Raboldshausen (Huber 1748, fig. 2). The frontal proboscis was small.

A hominosuid (Cleyer 1690, fig. 5; communicated by Christian Mentzelius).

HOMINOSUIDS WITH FRONTAL PROBOSCIDES

The greater frequency of a monstrous kind of proboscis in the pig than in any other mammal, considering the position of the pig in the mammalian series, has likewise been attributed, perhaps truly, to reversion.

—CHARLES DARWIN

The previous section listed animals described as elephant-headed, in most cases solely due to the creature in question having a frontal proboscis. But in this section, reports are collected that either do not mention elephants, or, if they do, mention not only a proboscis, but also other traits that place them in the hominosuid spectrum.

Oswego, Kansas. From the *Oswego Independent* (Jul. 27, 1906, p. 3d; ||2p87tmhx): "The greatest freak now along the hog line is a male pig at the Deming Ranch, Oswego. This freak of nature is one of a litter of six pigs. It has human hands, three fingers and a thumb; the nails are as perfect human nails as can be and are of the pretty pink color, the hands are laying on the breast. The pig is a jet black in color, except the breast, arms and hands, they being as white as can be. The spot on the breast is hairless and resembles the skin of a child. The eyes are those of a frog, the nose belongs to the bat family, the ears are large and lay back on the shoulder like those of an elephant. The freak has no hind legs, the body is long and round, and tapers on the end like a nice hoe handle. The tail has a natural curl which will not uncurl. It lived some time after birth. ... The pig was one of the fattest and plumpest young animals the writer ever saw, and he has been raising hogs for over forty years. There was another greater freak in this litter. It had four hands, but as it had been dead for some time before birth, it was not a good specimen.—FRED MANNING"

* Richard Orin Deming (1860-1928) owned the Deming Ranch, south of Oswego (||y4psmyst). Frederick Manning lived near Oswego in Montana Twp. (||yy9gvtro).

Bazile Mills, Nebraska. The *Nebraska State Journal* (Lincoln, May 18, 1884, p. 5; ||mr346zat) said a sow, belonging to a Courtland Pendleton at Bazile Mills, dropped a stillborn cyclops with "a protuberance, two inches or more in length, which projected from the forehead."

Austin, Nevada. The Austin Reveille (quoted in the *Gold Hill Daily News,* Gold Hill, Nev., Apr. 24, 1871, p. 2c; ||3bdedsyy) said, "At Barovich's saloon may be seen ... two black pigs joined together from the umbilical cord upward. The lower portions are separated and perfectly formed. But from the navel upward the consolidation into one body is complete. It has four forelegs and terminates in the ugliest … head. This head has two sets of ears … On one side of the head an eye … projects as if it had been gouged out, and on the corresponding side of the head, instead of another eye, a trunk sticks out about an inch in length, which resembles in form an elephant's trunk. It has not got any mouth. It was dropped by a sow belonging to Ed Dunham and died immediately after birth."

Godmanchester, England. From the *Cambridge Independent Press* (Cambridge, UK, Aug. 9, 1873, p. 7e; ||33zz72kx): "GODMANCHESTER.—REMARKABLE FREAK OF NATURE.—On Monday evening last a sow belonging to Mr. F. Markham, of the Swan Inn, in this town, brought forth 13 fine pigs, two of which had heads half monkey and half elephant, apparently. The lower portion of the head being exactly like a monkey, and on the forehead a tubular piece of flesh, exactly like an elephant's trunk, was a very prominent feature. These only lived about ten minutes. The third monster had no trunk, but the same countenance as the other two, only that it had but one eye, and that in the centre of the forehead."

Weatherford, Texas. The *Fort Worth Daily Gazette* (Mar. 19, 1888, p. 2d-e; ||sxt9uts) said, "There was born in Parker County a pig that seemed part elephant, part human and part pig. The monstrosity was born alive and only lived a few minutes. It had the tusks, ears and snout of an elephant, the stout being about two inches long. The eyes and forehead, as well as the forefeet, which had well defined thumbs with nails on them, represented what seemed human, the balance of the animal being a pig. The above-mentioned animal is preserved and can be seen in Weatherford."

Hubbard, Minnesota. The *Courier Democrat* (Langdon, N.D., May 12, 1892, p. 4f; ||yceqdz66) said, "Warren Watkins, a farmer living near Hubbard, Minn., has a pig with an almost human face, the body being formed like the ordinary piglet. Its proboscis is very much like that of an elephant. Hundreds of persons have viewed the freak."

Ottawa, Kansas. The *Ottawa Weekly Republic* (May 16, 1878, p. 3c; ||3kkxx8h3) said "Friend" McQuesten (Ben Chase McQuesten, 1861-1943) brought to town a sow-farrowed cyclops with a frontal proboscis.

New Orleans. The *Daily Crescent* (Jan. 26, 1854, p. 2c; ||y6wq6mcn) said a New Orleans butcher found in a slaughtered sow "a pig with a correctly developed human face, and its front legs growing out from the sides, with terminations to each resembling the four fingers of a child."

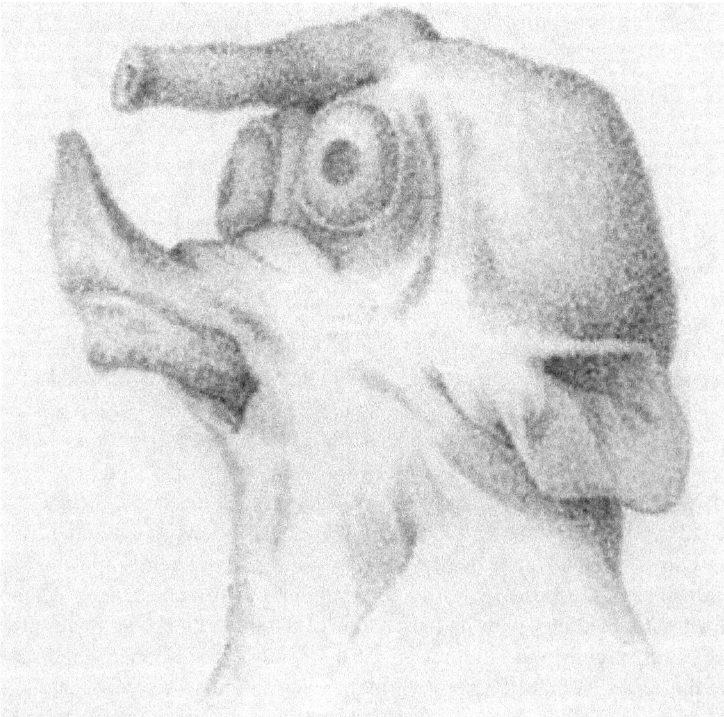

A hominosuid (Museum Vrolik, Amsterdam; Vrolik 1849, pl. 58, fig. 6).

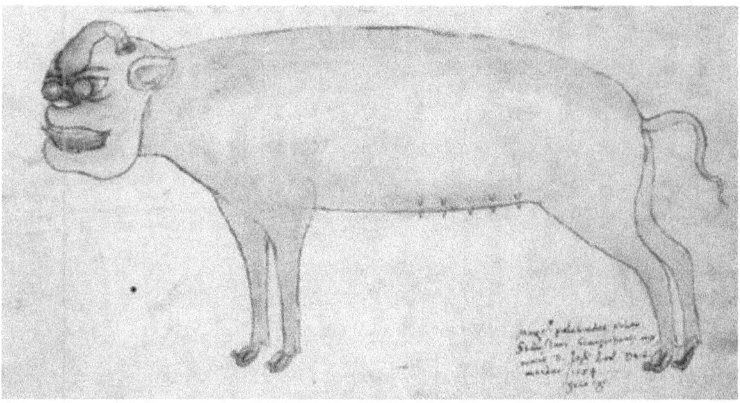

A hominosuid farrowed at Vilnius in 1562 (Wickiana Collection; ||y2eazrv6).

A cyclops hominosuid farrowed in Andhra Pradesh, India (Alexander 1828).

Burlington, Iowa. The *Burlington Hawkeye* (quoted in the *Daily Wabash Express*, Terre Haute, Ind., Feb. 1, 1871, p. 2a; ||y9jyzapf) said, "We were shown, Tuesday, at the office of Dr. G. R. Henry, of this city, a nondescript animal which is without a parallel in our observations of natural history. The body is six or eight inches in length and was evidently that of a pig before birth. The head is a strange combination in some respects almost startling the beholder with its indications of a human face. The lower jaw is like that of a child, and the outline of the head is more human than brute. The upper jaw is that of a hog, while the eyes and nose are unlike anything we ever saw before. The eyes are located side by side, closely together, and below the nostrils. The latter are comprised in a long, fleshy protuberance from the upper part of the forehead, projecting forward and downward, like the trunk of an elephant, to the mouth."

* A Dr. Greenburg Henry lived in Burlington in 1871 (||y5q7vtpm, ||y3rkwy2s).

Johnson, Nebraska. From the *Tri-Weekly Tribune* (Fremont, Ne., Apr. 18, 1889, p. 4; ||mputxjuk): "Enoch Moses, a farmer living near Johnson, has a pig that … has two tails and only one eye. The eye is in the center of its head and the extra tail is just above the eye."

Utrecht, Netherlands. Among the specimens in a Utrecht University collection, Bleuland (1826, p. 401) listed three cyclopean sow-farrowed hominosuids with frontal proboscides (nos. 2515, 2516 and 2517).

Coloma, California. From the *Columbia Gazette* (Columbia, Ca., Dec. 10, 1853, p. 3c; ||bde6u6wm): "A pig was littered near the town of Coloma … a few days since, that has a miniature elephant's trunk projecting from the forehead. The trunk is about three inches in length."

A hominosuid with a frontal proboscis.
(Mutter Museum, Philadelphia)

Red Mills, Arkansas. The *San Antonio Daily Light* (Jul. 30, 1895, p. 3c; ||yv37f9uk) said, "D. K. Persons, a farmer of Red Mills, Ark., is the owner of a pig that … is perfectly devoid of hair, and has a double set of eyes, feet like human hands (even to the nails), one very large ear and a face which looks more like that of a young baby than it does like that of a regulation pig. From between the upper set of eyes projects a proboscis like that of a young elephant. This proboscis has two holes through its entire length."

Nuevo Laredo, Tamaulipas, Mexico. La Opinión (Los Angeles, Nov. 6, 1932, 37e; ||bdm94tz9) reported the birth of a seven-legged "pig" with human hind parts but failed to say whether the mother was a human or a sow. It died soon after birth.

A cyclops farrowed in the Graz cattle market, August 11, 1742 (||y6854ypx). Note the up-turned hooves, rear-directed ears, humanlike chin, and frontal proboscis. The Jacksonville hominosuid seems to have been rather similar.

Sacramento, California. The *Sacramento Union* (quoted in the *Jeffersonian,* Stroudsburg, Pa., May 11, 1854, p. 1e; ||knu8lho) said, "Dropping into the K Street drugstore yesterday, we observed Dr. [Thomas Muldrop] Logan [1808-1876] busily engaged in his office, making a drawing of the most monstrous *lusus naturae* we ever saw. It was no less than an abortive attempt on the part of Dame Nature to manufacture a human being out of a hog. The animal had attained its complete fetal growth and was one of a litter of well and naturally formed pigs. The particular one in question, however, is entirely destitute of hair, while the skin is white and smooth. The forehead and head are full and round … In lieu of the nose, a proboscis, in the exact miniature shape of an elephant's, proceeds from above and between the eyes, and rests on the upper lip. The eyes are large, round, and full as a man's. The lower jaw and chin project beyond the upper, and the tongue protrudes a little beyond the lips. The ears are flattened and laid back against the side of the head. Indeed, the whole contour and figure of the head and face are those of a human being while the rest of the strange animal partakes of the characteristic formation of the porcine species."

Jacksonville, Florida. The *Saturday Evening Mail* (Terre Haute, Ind., Oct. 4, 1890, p. 3d; ||y7nhytz2) said,

> Charles A. Clark, an undertaker and embalmer, has in his establishment the remains of a young pig, which is one of the greatest freaks of nature ever seen in Jacksonville, Fla. It is the property of E. A. Lindsey … and is one of a litter of eleven pigs.

Ten of these pigs are perfectly formed and are thriving, but the one in question was born dead.

It is about twice the size of the other pigs of the same litter, and is perfectly hairless, except for a small patch of whiskers on the chin. The head of this freak is human shaped. The ears are set on the side of the head like those of a human being but are larger and shaped like an elephant's ear. The chin, mouth and cheeks are human shaped, while the teeth are like a shark's. It has no nose, but instead a perfectly shaped trunk extends from the base of the forehead just where the root of the nasal organ should be.

Immediately under the root of the trunk is situated one large egg-shaped eye, with two large, round pupils (one at each end of the eye), which peeps out from under and on either side of the trunk. Each pupil or eye is shaded by white eyebrows, which very much resemble the velvety brows of an infant. The trunk is nearly four inches in length and is perfectly pliable. The hoofs … turn up at the points like those of a mountain goat.

* A Charles A. Clark, an undertaker, lived in Jacksonville in 1910 (||y4x9w9o3).

Chattanooga, Tennessee. The *Chattanooga Times* (quoted in the *Clarksville Chronicle,* Aug. 22, 1874, p. 2g; ||oda8uan) described an exhibited animal: "The curiosity in question is a pig, with the skin of a human infant, a human face in everything except a trunk with the exact conformation of the elephant's. The skin is velvety, entirely devoid of hair. The face is a curious mixture of animal and human. The skin is all human, and so are the eyes and forehead; the eyes having well-defined brows. … The ears, body, legs, feet and tail are pig and nothing else. The trunk projects in the same place as the trunk of an elephant, and beneath it is a mouth so human in expression that it raises a doubt whether it can belong to an infant swine. The pig was born on a farm in the eleventh district and lived fifteen minutes after birth."

Venteuil, Marne, France. Le Bien Public (Dijon, Jun. 26, 1884, p. 3; ||4ftrpzdp) said a sow belonging to Mr. Martin-Noël, a farmer in Venteuil, gave birth to a hairless cyclops with a frontal proboscis.

Saint-Rambert, Loire, France. Le Messin (Metz, Mar. 13, 1887, p. 3b; ||22ad5u7s) reported a sow at Saint-Rambert farrowing a viable cyclops with a double proboscis and said it was kicked to death by its owners soon after birth.

Atlanta. The *Atlanta Journal* (Feb. 10, 1887, p. 4) said Mr. A. W. Crawford, of Edwardsville (now part of Atlanta), brought a sow-farrowed nondescript, preserved in alcohol, to the *Journal* offices. It had two eyes in a single socket and a double proboscis.

* In 1887 William A. Crawford was the Edwardsville postmaster (||yc7fy3yy).

Riom, Auvergne, France. The Riom-Journal (in *Le Facteur,* Ussel, Jun. 19, 1880, p. 2c; ||2s64f833) said a sow in Riom had farrowed a viable nondescript with a human head that had four eyes and a frontal proboscis. The report said it lived and nursed ("a vécu et même teté").

Madison, Indiana. The *Madison Courier* (quoted in the *Fort Wayne Daily Gazette,* Oct. 19, 1869, p. 1; ||mryf5ju7) described another hominosuid with a double proboscis. "A pig owned near that city that is perfectly developed, with the exception of the head, which was most singularly deformed. The head was provided with two trunks or proboscides, that resembled that of an elephant. Both trunks, however, were turned upward. The eyes were situated between the trunks, and were seemingly in the mouth, where the tongue should have been."

Catania, Sicily. Galvagni (1839, p. 87) described a cyclops pig with frontal proboscis in the collection of the Accademia Gioenia di Catania.

Columbus, Ohio. The *Free Press* (Ravenna, O., Sept. 10, 1874, p. 4a; ||y6fkasgz) said, "A pig was born recently in Columbus with a half human face and head, perfect chin and mouth, signs of a large tusk on one side of the mouth, and a perfect elephant's trunk extended from the forehead, with ears similarly shaped to those of an elephant. It will be an interesting fact to physiologists to know that a circus had passed through Columbus some months before this pig was born, and that there was an elephant with it. The maternal sow may have seen the elephant—hence the above monstrosity."

Wellsville, Ohio. The *Cincinnati Commercial Gazette* (quoted in the *Belmont Chronicle,* St. Clairsville, O., Jul. 23, 1891, p. 1; ||y5g3c337) said, "Wellsville leads in the procession of monstrosities. It has a pig eight inches in length, which has four eyes,[6] feet like the hands of a human, distinct fingernails and large ears. The upper part of its face is like that of an elephant, and the lower part like a human. There is a trunk projecting from the middle of the head, which turns backward and lies between the ears. In the end of the trunk are two large holes through which the freak breathes, there being no holes in the nose."

Dothan, Alabama. The *Dothan Eagle* (via the *Guntersville Advertiser*, Jun. 9, 1914, p. 1; ||mv35bz27) said, "A freak pig, having an elephant snout and forehead and chin that resembled that of a human being was brought to town last night by a farmer who lives eight miles from Dothan."

Opelousas, Louisiana. From the Opelousas *Journal* (Mar. 13, 1874, p. 3a; ||mm2j8yc): "Professor Rudolph [Adelbert Sebastian] Mayer [1836-1903], of this town, has a pig whose head and features, with the exception of the ears and snout, are those of a human being. The ears are those of a pig. The snout is neither pig nor human but is a genuine snout like that of an elephant, hanging from the forehead. The skull is evidently human, having apparently the front and back brain, which no ordinary pig has. The face, eyes and mouth are all alike the human. The pig is about the size of a rat, and … was taken from a sow that was slaughtered for pork."

Hampden, New Zealand. The *Oamaru Mail* (quoted in the *Inangahua Times* (Reefton, N.Z., Jan. 16, 1882, p. 2; ||yy95ufnu) published a letter from a Mr. McKay, a resident of Hampden, a township in North Otago. "I have a spotted sow that produced eight young ones, and among them is a most curious nondescript. It is the same size as the other pigs but has not a hair on it beyond a few bristles on the eyelids. It has a perfectly formed head of exactly the same shape as that of an elephant, with a trunk like that of an elephant, over three inches long, and turned up over the forehead. It has no mouth. It has two large eyes under the head, joined together, each being quite as big as a shilling. The ears are extraordinarily large and are placed on each side of the breast, and they are at least four times as large as the ears of the other pigs. The nostrils are placed between the ears and quite prominent. The feet are all alike, and of the most extraordinary shape, all the fore toes being at least 1 in. long, and turned up with nobs on them."

* A sow farrowed a nondescript reminiscent of the Hampden hominosuid at Cortevaix, Saône-et-Loire, France (*Le Progrès,* Lyon, Apr. 20, 1888, p. 3d; ||3mmvwt7y). It had eyes under its lower jaw, ears on its neck and a frontal proboscis. It soon died.

Blenheim, New Zealand. The *Herald* (Wanganui, N.Z., Jun. 16, 1868, p. 2; ||y6j5lpen) said, "A singular lusus naturae (says a Marlborough paper) was shown to us by Mr. John Ewart the other day. It was one of a litter of pigs newly farrowed. It had but one eye,[3] and that of undue proportion, fixed in the middle of the forehead; and in lieu of snout, an appendage to the forehead like the proboscis of an elephant. With the exception of the feet, which were pointed and hooked, the rest of the animal was perfect."

* A John Ewart (1828-1891) lived at Blenheim in 1868 (||y2a49uwq).

Kitzbichl, Austria. The *Tiroler Volksblatt* (Bolzano, Italy, Mar. 17, 1875, p. 4a; ||y9wubafx) said, in translation, "An interesting monstrosity is reported from Kitzbichl. There, a sow farrowed a pig with a human chin, a single large eye (in the middle of its forehead) with a hornlike process [i.e., a frontal proboscis], 1.5 inches long, projecting from the forehead above it. Sadly, it was born dead, but it has been preserved."

St. Wendel, Minnesota. The *Fargo Forum* (Fargo, N.D., Jan. 26, 1915, p. 10; ||bdxmfr7z) said, "ST. CLOUD, MINN. Jan. 26.—A pig, six hours old, with a jaw like a human being, a short trunk about two and a half inches long over one eye and a single tusk in his mouth was being exhibited in the city today. The freak was born yesterday on the farm of Jack Meghean of St. Wendel town."

Sedalia, Missouri. The *Kansas City Times* (Apr. 5, 1889, p. 8e; ||maua77nu) said, "A tiny pig destitute of hair, one abnormally large eye in the center of the forehead, nose shaped like that of a sheep and proboscis resembling an elephant's is the latest freak at Sedalia. The little animal only lived a short time."

Paris. At the French Academy of Sciences, de la Faye (1755) exhibited what seems to have been a synophthalmic hominosuid with a frontal proboscis. He refers to a like specimen shown in 1718 by Cardinal Polignac, and to others in Antoine Ferchault de Réaumur's cabinet.

Wilcox, Missouri. The *Hopkins Journal* (Hopkins, Mo., Mar. 11, 1897, p. 3c; ||25zvauvu) said, "Dr. Hall has in his possession … a hairless little pig of perfect form except the head. The snout is natural, but it has a human face and eyes, and out of the forehead springs an appendage resembling an elephant's trunk. It is now dead. It came from the farm of F. M. Hughes near Wilcox and lived with the litter three days."

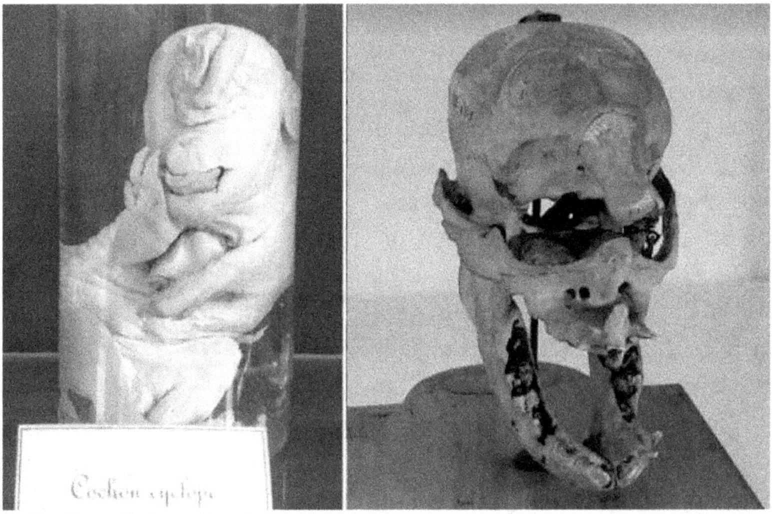

Left: A cyclopean hominosuid (Muséum national d'Histoire naturelle, no. 1902-896). The photographer, John Pelter, stated that though the subject "appeared to present a rather human-shaped ear, I was unable to photograph it at a more oblique angle." A frontal proboscis and upturned hoof tips are visible. *Right:* A synophthalmic hominosuid (Musée Fragonard, no. ter.16.201; image: John Pelter). A hominosuid similar to that shown at left, farrowed at Granite Hill, N.C., is described in the *Stateville Mascot* (Statesville, N.C., Feb. 1, 1900, p. 1c; ||f3pktexx): "Its head was round as a ball, and it had only one eye without a lid and as large as a horse's. It had a trunk four or five inches long on top of its head, and no mouth or nostrils." Another such was exhibited, also in North Carolina, at Salisbury (*Messenger and Intelligencer,* Wadesboro, N.C., Oct. 12, 1925, p. 5a; ||wzy8mkjd).

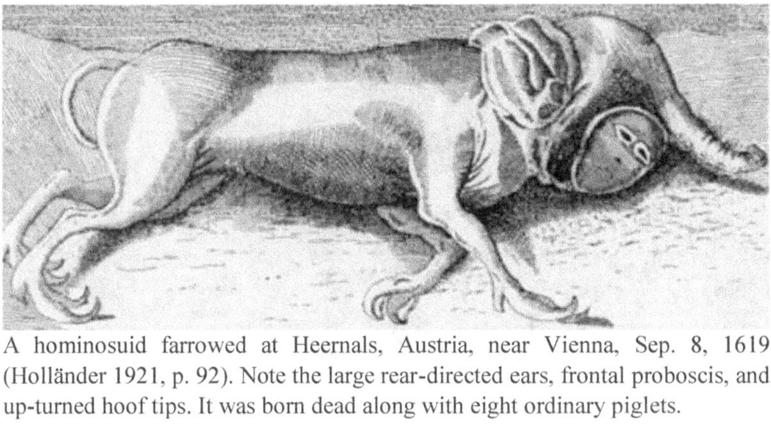

A hominosuid farrowed at Heernals, Austria, near Vienna, Sep. 8, 1619 (Holländer 1921, p. 92). Note the large rear-directed ears, frontal proboscis, and up-turned hoof tips. It was born dead along with eight ordinary piglets.

The Guásimas hominosuid.

Guásimas, Sinaloa, Mexico. In a broadside, José Guadalupe Posada (||rebx5va8) pictured and described a hominosuid farrowed at Guásimas in early June 1899. The skin, he said, was the same as human skin ("la piel de la misma calidad y color del pellejo humano"), the head, with the exception of the frontal proboscis, very much like that of a human being and covered with a thick pate of hair ("la cabeza es muy perfecta y está cubierta de cabello grueso"), the ears, exactly those of a human and placed exactly as in humans ("las orejas son exactísimas á las de gente y colocadas en perfecta posición humana").

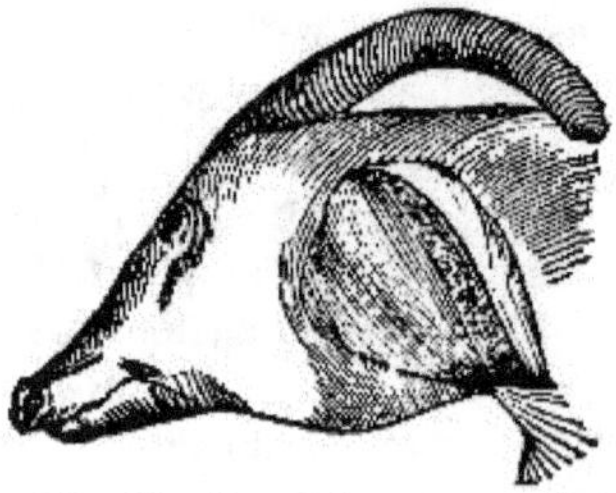

The Cincinnati hominosuid.

Cincinnati, Ohio. The *Cincinnati Commercial Tribune* (May 29, 1847, p. 2) reported a pig with a frontal proboscis and caudally-directed ears, farrowed in that city.

The Münster hominosuid.

Münster, Germany. Landois (1874, 1887) described two cyclopean hominosuids with frontal proboscides farrowed by sows in or near Münster. Shown here is that reported in 1874, his specimen III.

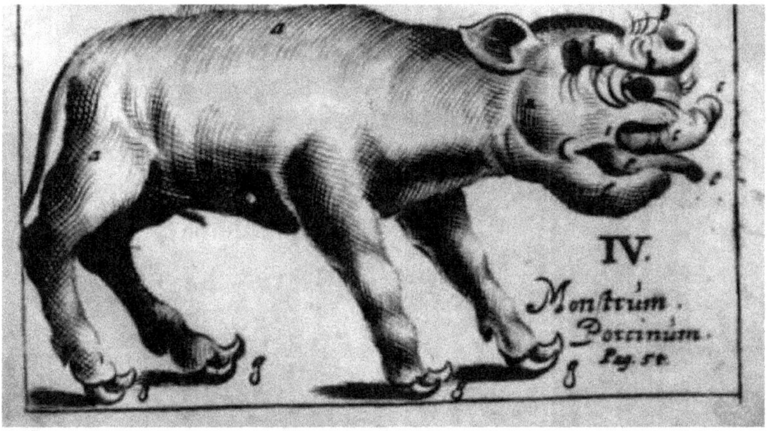

A hominosuid described by Bartholin (1654, p. 54). Farrowed at Helsingør, Denmark, it was born hairless from head to heel, he said, after seven normal piglets. Note the upturned hooves, rear-directed ears, and unusual double proboscides. Both the tongue and the huge eyes protruded. The chin was prominent and rounded.

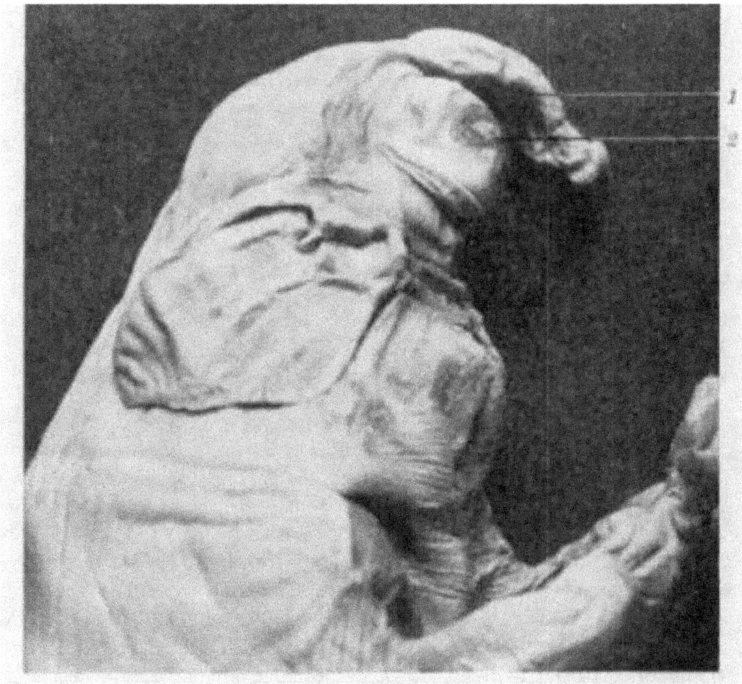

Profile of a cyclopean hominosuid 1: Frontal proboscis; 2: Cyclopean eye. Mouthless, its ears are rear directed. (Schwalbe 1906-1960, fig. 186).

A cyclopean hominosuid, farrowed on the island of Martinique, and at one time (possibly still?) in the collection of the Jardin des Plantes, Paris (Regnault 1775; image: Moreau de la Sarthe 1808). Again, a frontal proboscis is present. Note that the chin is prominent, as in a human, and that the ears are directed toward the rear, as in humans, not upward, as in pigs.[7]

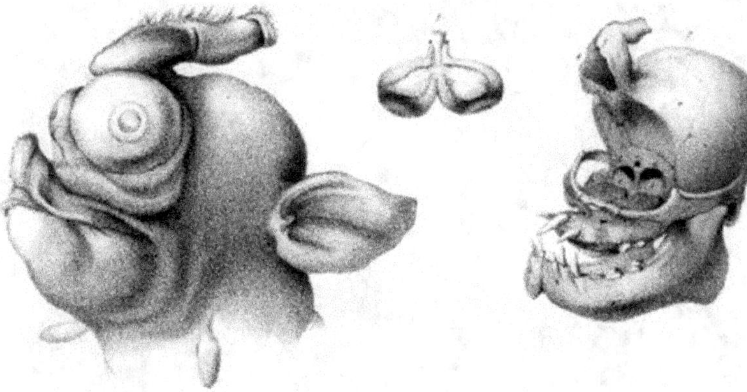

Hominosuid (Vrolik 1849, plate 57). *Left:* Profile before dissection; as seen here, the eyes in these creatures can sometimes be extremely prominent (also see illustrations on pp. 140 & 158). As usual, the ears are rear directed. *Right:* Cranium of the same individual. It can be seen that the eyes lay in a single orbit and that the frontal proboscis was partially supported by an osseous tube (also see the photo of Fragonard specimen reproduced on p. 171). The occiput is more like a human's than a pig's, while the mandibles are more porcine. *Center:* The eyes from a separate specimen (a HOMINOCAPRINID pictured by Vrolik 1849, plate 57, fig. 7), which unite in a common optic nerve, a form of synophthalmic cyclopia. Schwarz (1927) described a similar case, the head of a synophthalmic cyclopean JUMART in which two initially separate optic nerves merged before reaching the brain.

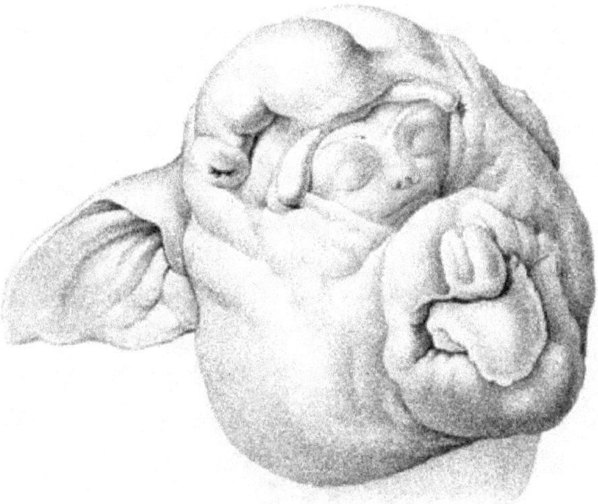

The head of a hominosuid farrowed by a sow (Carus 1841, Tab. 76). In addition to the extraordinary presence of a humanlike face in the cyclopean eye socket, note the presence of typical hominosuid traits such as a frontal proboscis, protrusive tongue, humanlike chin, and rear-directed ears.

Carl Gustav Carus
(self-portrait)

Hildburghausen, Germany. The German physician, anatomist and painter Carl Gustav Carus (1789-1869) was credited by Freud with originating one of the major concepts of psychology, the unconscious. Carus is also said to have devised the idea of a vertebrate archetype, a fundamental component of the theory of evolution as later developed by Charles Darwin. But in connection with telenothology, his most interesting contribution was his description of a specimen.

That specimen was the head of a bizarre hominosuid sent to him from Hildburghausen by a certain Dr. Hartenstein.[8] Carus published a detailed account of the head of this hominosuid, farrowed by a sow (Carus 1841). He not only gave a detailed description of his dissection of this head but also employed his artistic talents to create a fine quality copper plate engraving of the subject. And he explicitly stated that its appearance, as shown in this illustration, was "very truly

Two telenothians reportedly born in Germany in 1578. *Left:* An intersex hominosuid that reports say a woman bore in Praest, near Emmerich am Rhein, on New Year's Eve. Per reports, it had six digits on its right hand, six on each foot, and the left hand was merged into the belly. *Right:* A boshominid reportedly calved by a cow at Berge in Osnabrück. Image: Wickiana Collection (||4awb447u). Also see Scheuchzer (1716, p. 78) and Wickersheimer (1925, p. 256, #93).

represented" ("sehr treu wiedergegeben"). This hominosuid was unusual because its cyclopean eye socket was filled by what appears to be an infant's face. This individual, then, farrowed by a sow, had an indisputably human trait—a human face. Therefore, if the report is true, this creature must certainly have had mixed porcine and human parentage. It was born alive. Remarkably, a second such birth was reported. The *Perrysburg Journal* (Perrysburg, O., Jul. 30, 1914, p. 14; ||22pfvstx) said, "HAWKINSVILLE, GA.—Joseph Fleischman,[9] a merchant of this town, is the owner of one of the greatest freaks of nature ever seen in this part of the south. It is a pig which has a human face in its eye and two noses, one a normal pig's snout, and the other a diminutive elephant's trunk, the latter growing from its forehead. It has no hair on its body. Fleischman bought the pig from a farmer living half a mile from here."

HOMINOSUIDS BORNE BY WOMEN

If a woman gives birth to a child with a swine's head, offspring and wealth will increase in that house.

—A MESOPOTAMIAN BIRTH OMEN[10]

Hominosuid reports alleging human mothers usually mention pig-like foreparts but human hind parts. In other human-animal composites, too, the hind parts tend to resemble the mother. Thus, in the picture on the opposite page, the Praest hominosuid, was reportedly borne by a woman, but the Berge boshominid, by a cow.

Carl Heinrich Rösch, director of an asylum at Mariaberg.

Mariaberg, Germany. Carl Heinrich Rösch (1844, obs. 53) described a 41-year-old patient, who was barely three feet tall and had pea-sized testicles (hybrids often have tiny testes). His head was "more like a pig's than a man's." He could not speak, said Rösch, but "when he wants something, grumbles, or really, much rather, grunts like a pig. Sometimes this grunting is extremely loud and lasts for hours." … "The jaws jut out; the mouth is large and snout-like," … "the nose is blunt and broad, the nostrils gape wide, and the fore part of the nose is extremely wide and flat." … "The ears are large and stand out from the head. The forehead, as it extends backward, is completely flat; and toward the front, the skull tapers like a sugarloaf, and is fully flat behind, so that the bones of the rear of the skull form a broad, flat base."

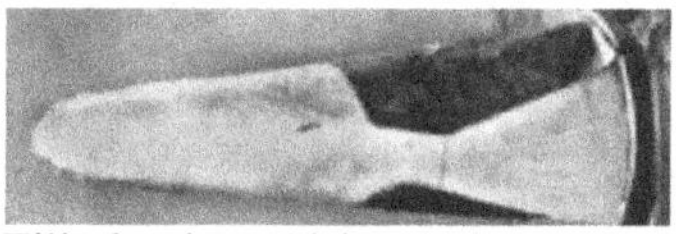

Till the late 19th century, sugar was sold in tapering loaves like the one shown here.

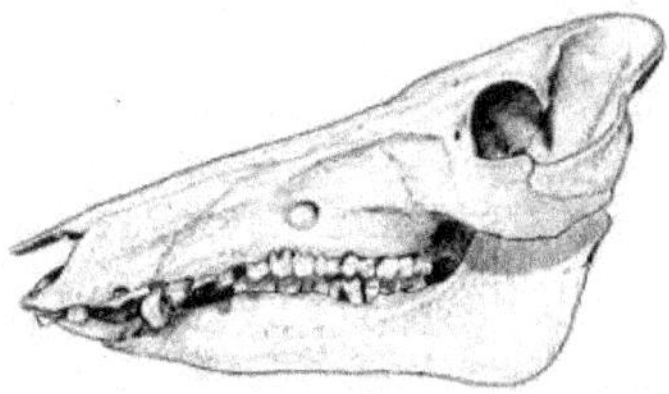

Sugarloaf-shaped skull of a pig.

* The quotations above are all translations of Roche's German.

Amsterdam. Gottfridi (1710, p. 1133) stated that at Amsterdam, on April 17, 1611, a woman bore a "child" with a pig's head with one pig's ear and one human ear. And Theodor Kerckring (1670, obs. 60, p. 122), an Amsterdam anatomist, later described a "human" skull (right), which may have been the same case. This infant had been born normal, he said, except for its piglike head, and except for the missing lower jaw (often absent in hominosuids).

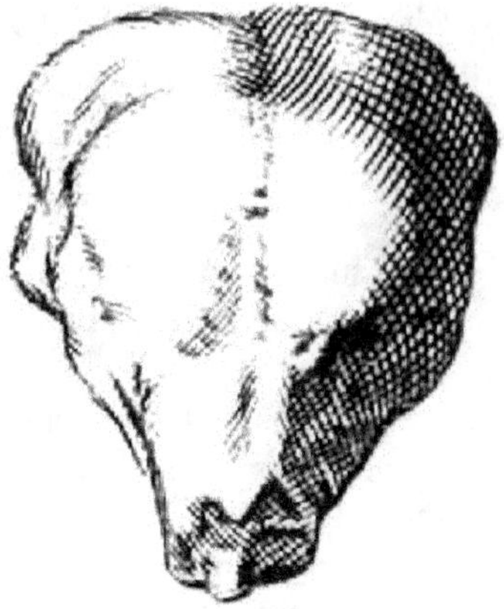

The Amsterdam hominosuid (Kerckring 1670, fig. 19).

Russellville, Kentucky. The *Logan County Enterprise* (Russellville, quoted in the *Hartford Herald,* Hartford, Ky., Dec. 18, 1878, p. 1; ||ydc8kmtp) said, "a colored woman in this county a few days since gave birth to twins, which were from general appearance half pig and half human—the head, eyes and ears of the little ones were those of a pig, while the remainder was that of a human.

St. Albans, Vermont. The *Rutland Daily Herald* (Rutland, Vt., Oct. 30, 1879, p. 4c; ||44wrensh) said, "The ST. ALBANS ADVERTISER tells of a child that was recently born to a family living some two or three miles from St. Albans on the shore of Lake Champlain, that in many respects bears a close resemblance to a pig. Its head is shaped somewhat like that of a pig, while the eyes are very large and also strongly resemble those of swine. Its fingers are webbed and curled together somewhat, and the feet have no toes, but are stubbed and to some extent like a hog's hoofs. It is strong and healthy and gives promise of long life. It is said that the mother of the child, some months ago, took into the house and cared for a young pig that was sick and to which she became attached; after a time the animal died, and the woman mourned greatly over the death of her novel pet."

London. During a lecture at Guy's Hospital, Blundell (1828, p. 231b) exhibited a child with a pig's head, perhaps the same specimen later mentioned by Physician John Charles Atkinson (1833), who stated that "there is in the museum of the London University a preparation of a human fetus, with a face the perfect resemblance of a pig."

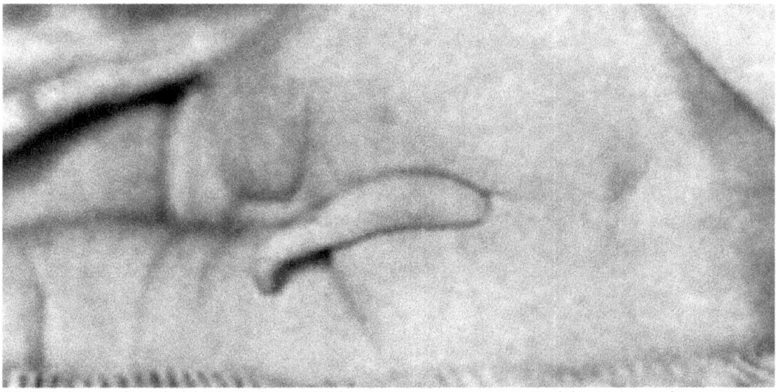

Rear view of a viable hominosuid borne by a woman. Baltimore professor Ross Granville Harrison (1900) said it had a piglike tail, and that the right foot bore four toes (pig trotters have four toes) with "heavy" upturned nails. Hoofs are, in essence, heavy nails, and hominosuid hoofs often turn upward.

Estill County, Kentucky. The *Hickman Courier* (Hickman, Ky., Jun. 27, 1868, p. 2c; ||m3b33pc) said,

> A child was born in Estill County last week, near the Clark line, which from the waist up, was a perfect pig, and in its lower extremities was developed like a human. This is no cock and bull story gotten up to furnish an item, but is vouched for by Dr. Aleck Stewart, who saw the singular creature. It lived eight hours.

Bipemba, D. R. Congo. A report about a hominosuid appeared in 2016 on a now defunct webpage on a French-language news site. In translation, it read,

> Last Tuesday, Feb. 4, at Mbuji-Mayi in Bipemba, Eastern Kasai, D. R. Congo, a 22-year-old bore a child of uncertain kind. Witnesses say the monstrous child had upper parts of a human but lower parts of a pig. The two short arms ended in small hands with eight fingers, and two short legs, in little feet with four toes each. The birth took place in the Meda maternity hospital.[11]

Parras, Coahuila, Mexico. La Gaceta (Parras, quoted in *La Voz del Pueblo,* Santa Fe, N.M., May 8, 1897, p. 2c; ||2x65cz3j) said a woman in Parras had borne a nondescript with a nose "exactly like the snout of a pig" ("exactamente como el hocico de un cerdo") and with arms like the forelegs of a pig, which ended in hooves.

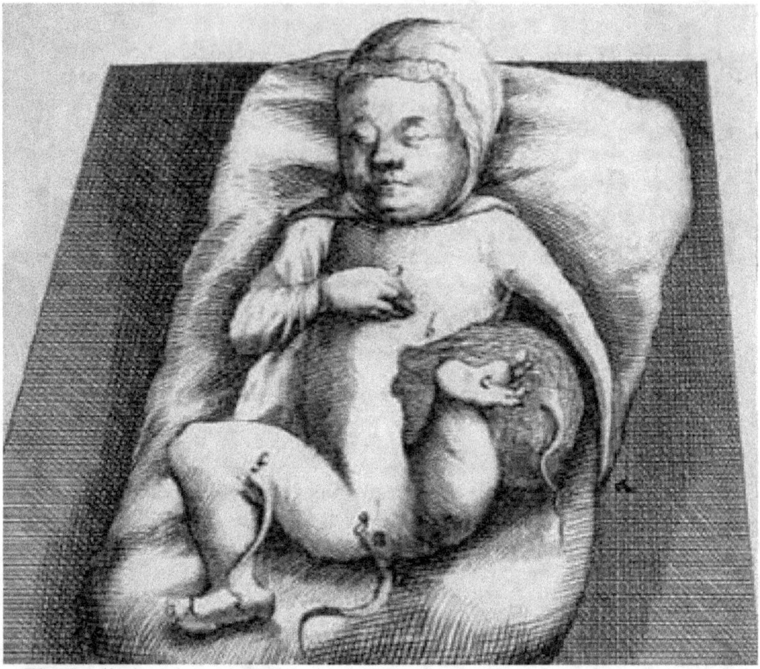

The Berlin hominosuid (Elsholtz 1673). The feet had four toes, as do pigs.

Berlin. German physician Johann Sigismund Elsholtz (1673, p. 80) reported an atypical hominosuid born to a woman in Berlin. He said it had three pig-like tails. In translation, the passage in question reads,

> And indeed, I would like to add something about a monstrous girl recently born in our city, of which I include a picture [shown above]. In Berlin, on October 13, 1669, a female child of the most repulsive appearance was delivered stillborn. The mother was a young woman who, the year previous, had married a soldier of the city guard. When I examined the fetus, three days after its birth, the head at least was normal, the face, charming, even pretty. Likewise, the right arm and the whole of the back were free of deformity.
>
> But the left arm tapered to a single digit tipped by a nail (a). The left leg, as it emerged from the abdomen, began at the knee. Its foot had but four digits, which were noticeably cloven in the middle (c), so much so that from a distance there seemed to be two separate soles on a single foot. From this portion of the abdomen, moreover, there protruded what one might describe as

an umbilical sac, containing a liver, mesentery and, it would seem, other digestive organs (b).

The right foot, too, had four toes, and the right knee lacked a cap (d). Truly, it was amazing to see a naked tail like a pig's attached to the inner thigh (f), and another to that bulging bag of organs. Another tail, shorter and hairy, stood at the usual place by the anus. We could scarcely look upon these things without wonder.

Köpenick, Germany. Muhrre (1588) said Agneten Marten, a woman at Köpenick, bore a child with a hog snout to the left and above its mouth, hands with six fingers, and feet turned rearward, on August 13, 1588. See, too, Angelus (1598, p. 401).

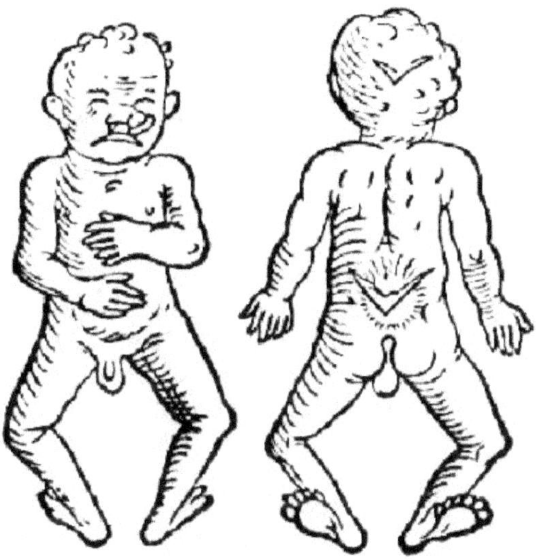

The Köpenick hominosuid (Muhrre 1588).

Přerov, Czech Republic. The *Tiroler Tagblatt* (Aug. 8, 1867, p. 1596; ||yubbd38n) reported a case at Přerov, resembling the Köpenick case, in which the wife of a swineherd bore a hominosuid with a pig snout and six digits on its hands and feet.

Kcynia, Poland. Der Bote für Tirol (Oct. 11, 1836, p. 236b; ||5n7a2jnu) said a woman near what is now Kcynia bore a nondescript with a pig snout and six digits on its hands and feet. It was still alive after 14 days.

Langenau, Baden-Württemberg. Gockel (1700, p. 629) described a fully viable hominosuid born to a woman at Langenau, March 4, 1697. In translation, he said, "He has a snout with a sharp tip, very much like a pig's, and a second, normal nose that lacks nostrils. The palate, as well as the lip on each side of the snout, is so cleft that the brainpan and eye sockets show through the roof of his mouth. He is elsewhere normal. This child eats, drinks and sleeps, is never ill, and is yet alive."

Magara, Romania. The *Innsbrucker Nachrichten* (Sep. 18, 1909, p. 19a; ||37h432ry) said a farmer's wife in Magara bore a "boy" with a fully formed pig snout.

A "woman" with a pig snout.†

Market Deeping, Lincolnshire. Market Deeping physician James Clapperton (1875, p. 169a) exhibited a female hominosuid borne by a woman.

Selma, Alabama. The *Agramer Zeitung* (Zagreb, Feb. 15, 1896, p. 7a; ||yyra8cym) reported a viable hominosuid.

> From New Orleans comes the following: "The faculty of the local school of medicine has brought for investigation from Selma, Alabama, a natural wonder that has no equal on the face of the earth—a monstrosity of male sex, half human and half pig. Their investigation has shown that the creature has a human heart, but lungs unlike those of any human being. Its face has no cheekbones, the mouth more than 100 teeth. The knees and ankles are double jointed. The head is fully three times as long as a human being's. The ears, which are placed far back on the neck, are those of a swine. The feet and hands are solid bone covered with skin, and bear hooves. The structure of the fingers, muscles and tendons was not described. This being, which is four years old, has never spoken a word and only grunts. He only eats food that his mother has previously chewed for him but walks upright like a man. The parents of this nondescript are white and from the interior of the United States. [translation mine]

* The Selma case may be the same as that reported in the *Clarion-Ledger* (Jackson, Miss., Jan. 17, 1895, p. 2; ||yypvldzc), which said a New Orleans woman bore "a monstrosity in the way of a child. It is perfectly formed in every respect except the head and face, which is that of a hog with long snout and no under lip."

Danville, Indiana. The Courier-Journal (Louisville, Ky, Jul. 24, 1890, p. 4; ||uh8m9ab6) said Alexander Johnson, Secretary of the Indiana State Board of Charities, reported that a female inmate of a Hendricks County asylum (at Danville) had a large, piglike head.

†. Leonardo da Vinci (*visi mostruosi,* ca. 1500, from life (||mwh7npbv). I once saw, at a local swimming pool, a "boy" with a pig-like snout like that depicted by Da Vinci.

"CHILDREN" WITH FRONTAL PROBOSCIDES

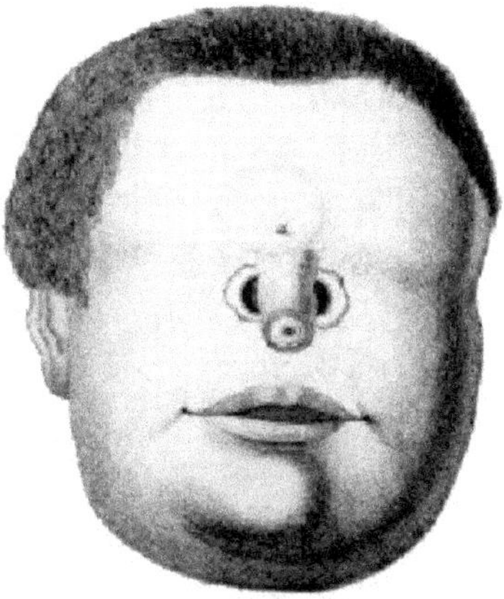

The Esslingen hominosuid.

A frontal proboscis is a frequent but not ubiquitous trait of hominosuids. Indeed, it is so prevalent among sow-farrowed pig-human composites that it can virtually be taken as diagnostic. As has already been noted, this feature has often resulted in hominosuids being described as elephant-headed pigs. These are often cyclopses. So, it makes sense, whenever a nondescript with a frontal proboscis is borne by a woman, to classify it, tentatively, as a hominosuid, especially when the report describes it as elephant headed or cyclopean.

Esslingen am Neckar, Germany. Wilhelm Gottfried Ploucquet (1791), Rector of the University of Tübingen, described and pictured a synophthalmic cyclops with a frontal proboscis borne by a woman near Esslingen am Neckar in 1784.

Erlangen, Germany. Ploucquet (ibid.) appended an account, taken from Erlangen professor Jacob Friedrich Isenflamm (1726-1793), of a specimen resembling the Esslingen hominosuid, in that it had a frontal proboscis and lacked a nose but was not a cyclops. It had six digits on each of its hands and feet.

Pittsfield, Massachusetts. The *Berkshire County Eagle* (Pittsfield, Mass., Jan. 5, 1871, p. 2c; ||yepm5wxb) said "A child was recently born in this town with the perfect body of a boy, but with a head shaped like an elephant's, with flopping ears, and nose elongated into a trunk … The child was born dead."

Bridgeport, Connecticut. The *Gazette* (Alexandria, Va., Mar. 23, 1887, p. 3c; ||59786xxn) said, "Mrs. Krettschuer, a German woman in Bridgeport, Conn., recently gave birth to a child having an elephant's head and a small trunk instead of a nose. The mouth and lips protrude like an elephant's. The child weighs nine pounds and can be fed only with a spoon."

Columbia, South Carolina. The *Daily Record* (Columbia, quoted in the *Lexington Dispatch,* Lexington, S.C., Apr. 19, 1899, p. 1c; ||lqyqqya) said, "A monstrosity is reported to have been born to colored parents who live on Laurel Street recently and is creating much talk among the colored people, who are all in excitement over what they call the 'elephant baby." They say the child has no legs and its nose extends from its face like a snout, while its ears are abnormally large, giving it, as reported, the appearance of an elephant."

Quitman, Georgia. From the *Wyoming County Times* (Warsaw, N.Y., Dec. 26, 1889, p. 8c ||kbsy3xg): "A GEORGIA MONSTROSITY.—ATLANTA, GA. Dec. 19.—About seven miles from Quitman there lives a young, healthy married negro woman who recently gave birth to a child which represents a remarkable appearance. Instead of a nose of the regulation pattern it has a well-developed proboscis of an elephant. On either side of this pendant snout there projects two teeth of abnormal dimensions that look like the tusks of an elephant. [Pigs have tusks, too.] The eyes are small and far apart and altogether the facial appearance is that of an elephant … It is still living.

Paris. Laroche (1823), a Parisian medical student, described and pictured a cyclops with a frontal proboscis borne by a woman.

Aisne, France. In 1810, a woman in the French department of Aisne gave birth to a noseless, mouthless cyclops with a frontal proboscis (Anonymous 1812, p. 83, §25).

Reading, England. John Hooper (1789, p. 332), a Reading physician, said a woman bore, on November 21, 1782, a synophthalmic cyclops after a seven-month gestation. "It was living," said Hooper,

> and struggled for about half a minute, when it died. It was well formed except the head, the appearance of whose forehead was similar to that of the occiput, with scarcely any vestige of a face, no eyebrows, or eyes (where they should have been) or nose; no mouth, but in its place an orbit enclosing two eyes; at first I took them only for one, but on a narrower inspection, I could discover two joined together; they stood very prominent from the orbit. Above this was suspended a substance, whose form was not much unlike a penis of that age; but I could not perceive either glans or prepuce. From the lower part of the orbit, hung by a slender filament of about one-third of an inch long, another fleshy substance, the bigness of a small pea. No chin nor ears in their natural situation, but something like ears inverted, &c. joined together in the neck, close under the maxilla inferior. The sex very obscure, having a kind of wart on the superior part of the pudenda.

Berlin. Eller (1757) described another such, a cyclops with a frontal proboscis born to a woman in Berlin, Feb. 19, 1755.

Copenhagen. In the Nyboder section of Copenhagen, in 1754, Heuermann (1765, p. 313) delivered a stillborn, female synophthalmic cyclops with a frontal proboscis.

Valognes, Normandy. De la Motte (1721, p. 677, obs. 353) described a noseless synophthalmic cyclops with a frontal proboscis born at Valognes, Apr. 13, 1712.

Padua, Italy. Morgagni (1761, p. 246a) described another noseless synophthalmic cyclops, which had a transparent membrane in place of eyelids, born at Padua in 1735.

Dresden, Germany. Bartholin (1671, p. 258) said that in 1648, while visiting a Dresden collection, he had seen a preserved fetus with elephant-like characteristics ("elephantem figura repræsentatem"), a human abortus half a handspan long. A woman had given it birth a few years previously.

Hranice, Czech Republic. Welsch (1668, Episagmata Centuria I, p. 50, LXXVI) reported that in April 1651 a woman at Weißkirchen (now Hranice) had given birth to an "elephant-like" child, which the midwife scalded to death.

Bologna, Italy. Aldrovandi (1642, p. 449, II) says a woman in Bologna bore a cyclops in April 1630, normal otherwise except for a fleshy, horn-shaped ("instar cornu") mass on its forehead. Unable to nurse, it had to be fed with a tube.

England (no locus in quo). Gross (1875) described a synophthalmic cyclops, with a frontal proboscis, borne by a woman in England.

Rennes, France. La Rue (1757), a surgeon and anatomist at Rennes, dissected another such synophthalmic cyclops, a female with a frontal proboscis. Beside each of its little fingers it had a supernumerary finger without osseous connection.

Brussels. A woman bore at Brussels, December 21, 1563, a nondescript that was anophthalmic but otherwise resembled the Rennes birth, also with a supernumerary finger, on the right hand (Wick 1564). Reports said it had an elephant's trunk.

A certaine Relation of the Hog-faced Gentlewoman called Miſtris *Tannakin Skinker*, who was borne at *Wirkham* a Neuter Towne betweene the Emperour and the Hollander, ſcituate on the river *Rhyne*.

Who was bewitched in her mothers wombe in the yeare 1618. and hath lived ever ſince unknowne in this kind to any, but her Parents and a few other neighbours. And can never recover her true ſhape, tell ſhe be married, &c.

Alſo relating the cauſe, as it is ſince conceived, how her mother came ſo bewitched.

London Printed by *J.O.* and are to be ſold by *F. Grove*, at his ſhop on *Snow-hil* neare *St. Sepulchers Church*. 1640.

Title page of a seventeenth-century pamphlet describing Tannakin Skinker, a "hog-faced gentlewoman" supposedly borne by a bewitched human mother in the Netherlands in 1618. A princely sum would be paid, it was said, to any man who would marry her. Sadly, willing suitors were lacking. Was Tannakin fact or fancy?

One of the many depictions of the "Pig-faced Lady," a hominosuid who supposedly lived in one of the upper-class residences of Manchester Square in Marylebone, London, during the early nineteenth century. Beginning in late 1814, rumors of her presence swept the city, and her existence was soon reported as fact in London newspapers. It was said that she spoke in grunting tones and that a maid served her meals in a silver trough. Her story parallels that of Tannakin Skinker, in that a large dowry was supposedly offered to any gentleman willing to take her hand in marriage. Satire or reality? Image: *Illustrated Police News,* Jan. 7, 1882; ||22w43hsa.

A hand-footed hominosuid, with nailed fingers, farrowed by a sow at Önnestad, Sweden, in October 1699 (Wiedemeyer 1699, p. 300 & Tab. 10).

EARLY CASES

Walda, Germany. Schurig (1731, p. 628) said, in translation, "In 1682, in a village near my native Hayna [a village in Herxheim, Rhineland-Palatinate], called Walda, at the home of the prefect of that place, a sow farrowed seven piglets, three of which were normal. But four had true elephant trunks hanging down past their mouths from their foreheads."

SOME GERMAN HOMINOSUIDS

Date	Locale	Comment	Mother
1558	Zembschen[†]	Anophthalmic, frontal proboscis	sow[‡]
1576	Mühlhausen[f]	Stillborn; red lumps replaced eyes	human[b,d,e]
1576	Diocese of Speyer	Human head	sow[b,d]
1578	Praest[h]	6 digits on one hand and both feet	human[b,d,g]
1587	Prenzlau	Human head	sow[q]
1589	Rehna[i]	Stillborn; male; pig snout	human[c]
1638	Český Wiesenthal[j]	Daughter with pig trotters	human[k]
1642	Nuremberg	Human ears	sow[s]
1647	Berlin	Human head	sow[q]
1652	Golnau, Pomerania[u]	Human head	sow[q]
1678	Eisleben	Pig snout	human[e]
1719	Havelberg	Human head	sow[q]
1725	Lüben[m]	Male; pig snout (viable)	human[n]
1733	Bremen	Male; earless; pig snout	human[a]
1865	Trebitsch, Austria-Hungary	Human face	sow[v]

†. Zembschen is now part of Hohenmölsen, Sachsen-Anhalt; ‡. Fincelius (1566b); a. Büchner (1733, p. 773); b. Franck et al. (1585, p. 1020); c. Gregorii (1712, p. 513); d. Irenäus (1584); e. Praetorius (1678a); f. Oct. 13, 1576; g. b. Dec. 31, d. Jan. 1; h. A village near Cleves; i. In Mecklenburg; j. Now in Czechia; k. Lehmann (1699, p. 713); m. A village near Steinau; n. *Breslauische Sammlungen* (1727, p. 682; ||5n76264p); q. Bekmann and Bekmann (1751, p. 883); s. Helwig (1680, p. 20); u. A village near what is now Szczecin, Poland; v. *Innzeitung* (Innsbruck, Oct. 11, 1865, p. 3c; ||2d6ejcjp).

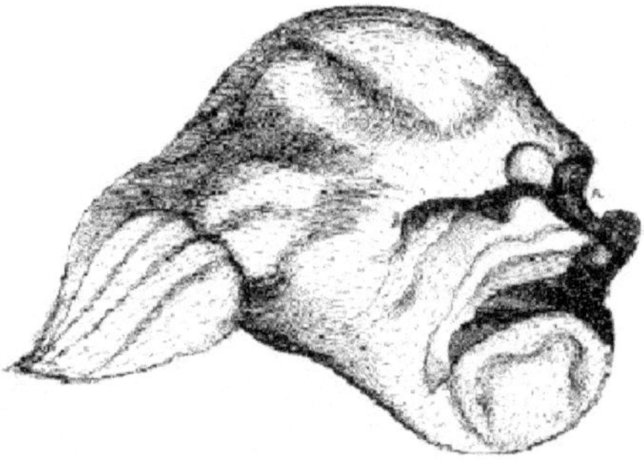

Head of an hominosuid (Floyer 1699, fig. 11). Note the rear-directed ears, anophthalmia, as well as the humanlike chin and cranium. A frontal proboscis may also be shown.

Weeford, Staffordshire. Floyer (1699) said that "In May 1699 there was shewed to me a Pig, at Weeford in Staffordshire, with a Face something representing an Human Fœtus, and the roundness of the Head, and the flatness of the Ears surprised all Persons, and they did usually apprehend it to be a Human Face, produced by the Copulation of two Species."

The head of the animal, as seen in Floyer's illustration, corresponds in many respects with what is often described in other reports about hominosuids: a rounded brow and cranium, a prominent chin, anophthalmia, and rear-directed ears. The Weeford specimen was a male, one of a litter of eight otherwise normal piglets (many other reports refer to a single hominosuid in a litter, likely a result of superfecundation). Floyer notes that it was hairy, unlike another subject sent from Derbyshire. In this latter individual, the sex could not be determined (telenothians are often intersex). He also described the latter as having a "depression betwixt the Eyes."

A continuum. It seems that the eyes of many hominosuids are not separated by the bridge of a nose. Instead, in different individuals the eyes stand at differing distances apart. Specimens are seen in which the eyes are well separated, as in a human; others, in which they are close set; still others, in which the eyes have partially merged to form a wide ocular structure with two separate irises ("synophthalmia"); and others, that are fully cyclopean (as shown, left to right, below). Others have no eyes at all ("anophthalmia").

London. The *Newcastle Weekly Courant* (Newcastle upon Tyne, England, Oct. 30, 1736, p. 2b) noted in news from London that "On Tuesday last one Clowseley of Oxford Road had a Sow pigged a Monster; its Head, Neck, and Skin resembling a Child, and all the rest of the Body, Legs and Tail like a Pig."

London. An anonymous 1676 pamphlet (||hbr6grr) announced the birth in London of a hominosuid. Much of what it loses in credibility through its anonymity, it recoups by (1) its detailed description of the subject, which corresponds in many ways with those given by more recent reports; (2) the creature's owner being named; and (3) the author being an eyewitness: "I have taken so much pains to exhibit nothing but Verity, that I laid by as suspicious the reports of my dearest Friends, till I had purchased my own satisfaction with going to be an eye witness." The General Report of a Monstrous Creature," reads the report "which was lately produced in St. Johns Street, beckoned my curiosity thither,

> where from one Mrs. Denman, at the White Hart, a Victualling House in that same street, I had the ensuing Relation. That about a week since (it being part of her lively-hood to keep Hoggs) it happened that she had a Sow brought forth seven live well-shap'd Piggs, which at that time she thought were all … but about half an hour after their production, to her great Amazement, she beheld her deliver'd (in a bag by itself) of a wonderful Monstrous Creature, which upon entreaty, she openly exposed to my view.

The description then given lists a variety of traits typically mentioned in other reports about hominosuids, to wit:

- The skin was "soft and smooth, and not so much as Cover'd with a hairy Down."
- "Its Body somewhat resembles a common swine but its hooves … turn up at the ends."
- "It would have puzzled the curious inspection of the most inquisitive Anatomist to decide its Sex."
- "The Ears … hung backwards and pointed towards the Nape."
- "The forehead being broad towards the Crown."
- "Instead of eyes you might have seen the plain blank of a whole Skin, which look'd as if Oblivious Nature had forgot to Chalk out the Eye-holes, which in lew of those glittering luminaries … were supply'd by a pair of sinewy Bunches stuft up with a grisly substance."
- The nose "seems rather to resemble the Trunk of an elephant" [and] "has but one Nostril."

Skokloster, Sweden. Swedish surgeon Olaf Acrel (1750, p. 17) stated that, "At Skokloster [Castle], some pigs were born seven or eight years ago, and three had almost human faces. Archiater [Nils] Rosén [von Rosenstein] exhibited one of these in the anatomy hall at Uppsala."

Naples, Italy. The *Bote von Tyrol* (Innsbruck, Sep. 16, 1819, p. 285; ||38pyjduf) reported a human-headed pig farrowed near Naples.

New Haven, Connecticut. In an account of the court trial of one George Spencer, held in New Haven, March 2, 1641 (Hoadly 1857, pp. 62-63), there is a description of a hominosuid, a cyclops.

> The 14th of February, 1641, John Wakeman a planter and member of this church acquainted the magistrates that a sow of his [which] he had lately bought of Hen[ry] Browning, then [with] pigge, had now brought among divers living and rightly shaped pigs, one p[ro]digious monster, [which] he then brought [with] him to be viewed and considered. The monster was come to the full growth as the other piggs for ought could be discerned, butt brought forth dead. It had no haire on the whole body, the skin was very tender, and of a reddish white colour like a childs; the head most stra[nge], it had butt one eye in the middle of the face, and that large and open, like some blemished eye of a man; over the eye, in the bottome of the foreheade [which] was like a childes, a thing of flesh grew forth and hung downe, it was hollow and like a mans instrum[ent] of generation [i.e., a frontal proboscis]. A nose, mouth and chinne deformed, but not much unlike a childs, the neck and ears had also such resemblance. This monster being after opened and compared [with] a pig of the same farrow, there was an apparent difference in all the inwards.

Spencer stood accused of having "beene actor in unnaturaeall and abominable filthyness with the sow," and witnesses repeatedly testified that he had admitted having done so (ibid., pp. 66-67). After the trial, having been found guilty, he was hanged, and the mother sow, too, was put to death (ibid., p. 72). It was nothing unusual for the animal "accomplice" to be executed in such cases. The motives behind this seemingly needless punishment of a brute animal were spelled out by the Jewish scholar Philo Judaeus (*Special Laws,* III, 8:46-50).

> And some persons prefer mules to every other kind of animal for the yoke, since their bodies are very compact, and are very strong and powerful; and accordingly, in the pastures and stalls where they keep their horses, they also keep asses of an extraordinary size, which they call celones, in order that they may breed with the mares; and then the mares produce a mixed animal, half horse and half ass, which, since Moses knew that its production was wholly contrary to nature, he forbade the existence of with all his might by a general injunction, that no union or combination between different kinds of animals should on any account be permitted. Therefore, he provided thus against those evils in a manner suited to and consistent with nature; and from a long distance off, as from a watchtower, he admonished men and kept them in the straight path, in order that both men and women, learning from these precepts of his, might abstain from unlawful connections. If, therefore, a

> man seek to indulge himself with a quadruped, or if a woman surrender herself to a quadruped, they shall all die, both the man or woman and the quadruped. The human beings, because they have gone beyond even the bounds of intemperance itself, becoming discoverers of unprecedented appetites, and because with their new inventions they have introduced most detestable pleasures, the very mention of which is infamous; and the beasts shall die, because they have been subservient to such iniquities, and also to prevent their bringing forth or begetting anything intolerable, as would naturally be the result of such pollutions. Moreover, those who have even a slight care for what is becoming would never use such animals as those for any purpose of life, but would reject and abominate them, loathing their very sight, and thinking that whatever they touched would at once become impure and polluted. And it is not well that those things which are of no use for life should live at all, since they are only a superfluous burden on the earth, as someone has called them.

Thus, behind the practice of executing animals in such cases, there was the belief that a hybrid might be produced, that the animal, though it was guiltless, might nonetheless bring forth something "intolerable."

Another New Haven case, a cyclops with asymmetric ears, is found in Dexter (1917, vol. 1, pp. 158-159), in a record for the general court session for January 15, 1652.

> The Governor acquainted the Towne that the cause of calling [them] together this day is aboute a monsterous pigg, [which] was brought forth by a sow of John Vincons: it was like a pigg in the body & leges, but [without] haire, the skin being white; the head something like a piggs head, but y[e] nether chapp [i.e., the lower jaw] something like the nether chapp of a man, one eare something like a piggs eare, the other like two little teates hanging downe, one great red eye in [the] face of it, and from the forehead a peece of skiney flesh hanging downe, hollow like [the] member of a man [i.e., a frontal proboscis].

Though nearly four centuries old, the two New Haven accounts just cited constitute robust evidence for actual existence of hominosuids because, given that these are court cases, it is hardly likely that the descriptions given of these creatures were false, particularly given the fact that it is clear from the account itself that they were seen by multiple witnesses. And yet, those descriptions correspond closely with those in numerous reports already cited.[12]

Havelberg, Saxony-Anhalt, Germany. Praetorius (1678b) recorded the farrowing of two hominosuids on August 10, 1663, at a place called Jederitz, in Havelberg.

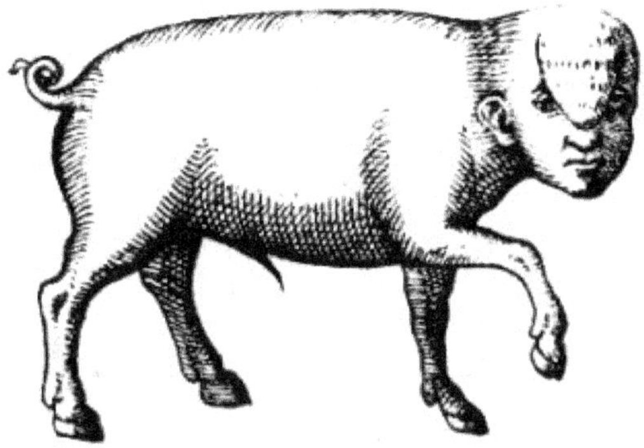

The Nicosia hominosuid (Schenck 1609, p. 113, fig. 85).

> *Who would undertake to resolve what species that monster was of, which is mentioned by Licetus (Bk. I, c. 3), with a man's head and hog's body? ... Had the upper part to the middle been of human shape, and all below swine, had it been murder to destroy it? Or must the bishop have been consulted, whether it were man enough to be admitted to the font or no? As I have been told it happened in France some years since, in somewhat a like case.*
>
> —JOHN LOCKE, *Essay Concerning Human Understanding*

Nicosia, Cyprus. Johann Georg Schenck (1609, p. 113) said a hominosuid with a human face was farrowed near Nicosia, Dec. 12, 1568, along with four normal piglets. From its forehead, he said, hung a growth like the comb of a chicken, which partly covered its eyes and nose (frontal proboscis?). He also said its ears were like a human's.[13]

Bornholm, Denmark. Thomas Bartholin (1657b, §80, p. 382) said, in translation, "In August 1648, on H. Knuts' farm on the isle of Bornholm, a sow farrowed, with three normal piglets, a misshapen one with a human head, a gaping mouth, closed eyes, and a wide forehead. ... The skin also was like a human's in lacking bristles. The hooves and tail were a pig's."

Hungary. Aldrovandi (1642, p. 430) said a woman in Hungary, in 1565, bore a "child with an elephant's head" (a likely hominosuid).

Hechthausen, Lower Saxony. Both Meyer (1679, p. 289) and Praetorius (1678a, 2.3) said that in 1678, at Klint (now in Hechthausen), a sow farrowed, along with five normal piglets, one with a human face, smooth skin over its entire body, and having, as well, two open and two "hidden" (perhaps skin-covered?) eyes.

The Brussels hominosuid (Paré 1641, p. 665)

Brussels. Paré (1641, p. 666) said that "In 1564 in Brussels at the home of Joest Dickpert in Warmoesbroeck Street, a sow bore a litter of six. The first was a prodigy with a man's face, arms and hands, and generally like a human from the shoulders up. But it had the two hind legs and rear parts of a swine, and the genitals of a sow. Nursing with the others, it lived two days, but was then killed, along with the sow, due to the horror people had of it. This prodigy is here shown most naturally, as it was in life"[14] The illustration in question appears above.[†]

Dillon, Slovakia. Praetorius (1677, p. 319) said that, in 1663, a sow at Dillon bore a creature with upturned hooves and a woman's foreparts.

Halle, Saxony. Dreyhaupt (1755, vol. I, p. 645) stated, in translation, that "On Easter, 1536, as Cardinal Albert was conducting high Mass at Moritzburg Castle,[‡] a remarkable monstrosity was cast down before him, hideous to behold. It stood on two legs, looked like an ape or a human, and its tongue hung down to its neck from a gaping mouth. Its eyes were covered with spotted blisters. The right ear was like a man's, the left like a pig's. Between the ears was a bald spot. The mouth and nose were apelike, the hind portion of its body, like a pig."

Xenogenesis. Holländer (1921, p. 237) reproduced a report about a woman giving birth to two piglets in Binszwangen, Germany, in 1574.

†. Pieter Bor (1621, *sub anno* 1564) gave largely the same account of this birth.
‡. Moritzburg Castle is in Halle. Dreyhaupt (ibid.) lists a piglet with a human head farrowed at Halle in 1523. But this date is likely based on a mistaken reading of Lycosthenes (1557, p. 528), who says a human-headed piglet was born Easter Day at Halle, 13 years *after* 1523 (i.e., in 1536). The Halle hominosuid was known as the Priesterschwein, or porcosacerdos, both of which mean "priest pig." Protestant Reformation propaganda said its head portended God's wrath because it was shorn like a priest's (cf. the Münchkalb, p. 63, and the Sea Monk, p. 943).

Three hominosuids that Friedrich Wilhelm Schmuck (1679) said were farrowed, June 20, 1627, outside a butcher's in Strasbourg. Schmuck lived in Strasbourg. Also see Anonymous (1627).

London. Sir John Hayward (1840, vol. 7, p. 107) writes that in the year 1562, "a sowe farrowed a pigge having hands and fingers like a man child" (also see Crawford 2005). An account of this creature published in a broadside that same year reads,

> This present yere of oure Lord God a thousande five hundred three score and two, one Marke Finkle a joiner dwelling beside Charing Crosse by Westminster, had a sow that brought forth one pigge onely, upon the seventh of Maye, beinge Ascention daye, the whiche pigge had … two fore feet, like unto handes, eche hande havinge thre[e] long fingers and a thumbe, bothe the thumbes growinge on the outsides of the handes.[15]

Lycosthenes (1557, p. 379)

No locus in quo. Lycosthenes (1557, p. 379) stated that a pig with a human head was born at the end of the reign of the Holy Roman Emperor Henry III. He gave a date of 1055 for this event but specified no *locus in quo*.

* Henry III, called the Black, or the Pious, reigned from ca. 1016 to October 5, 1056.

The Liège hominosuid (Liceti 1634, II, p. 183; ||q4w4yux).

Liège, Belgium. Various writers mention the case pictured above, Ambroise Paré (1641, p. 665) among them: "In the year 1110 a sow in the town of Liège farrowed a pig having the head and face of a man, and human hands and feet, but the rest was like a pig" (my translation).

* Sigebert of Gembloux (ca. 1030-1112) seems to have been the primary source for the Liège birth, which he says took place in the year after Philip I of France died, that is, 1109 (see Migne 1854, p. 430).[16] Sigebert can scarcely have been citing earlier authors, as he himself died just three years later. In 1109 he was living in Gembloux, about fifty miles from Liège. In the same passage, Sigebert mentions a quadruped chicken hatched that same year. That later writers refer to both this pig and chicken shows they are following Sigebert. The exact year of this hominosuid's birth became confused over time. Bartholomaeus Fisen (1642, p. 349) dates the Liège birth to 1118 A.D. In his *Anglica Historia* Polydore Vergil (11:6; ||uzv8wym) dates these two events (chicken and pig) to around 1110, whereas in *De Prodigiis* (1664, 3:96), he assigns the piglet to 1008 and the chicken to 1111. Sir John Hayward (1613, p. 303) says the birth occurred in the thirteenth year of the reign of Henry the First (i.e., 1112 A.D.). Lycosthenes (1557, p. 374) also lists this birth, along with the chicken quadruped, and dates them to 1018 A.D. Irenäus (1584) says 1109 A.D. and cites Sigebert. He also mentions the chicken and assigns it to 1111 A.D. The hominosuid Lycosthenes reports as having been farrowed in 1055 (see p. 196 of this book) may be based on mistaken reports about Sigebert's report.

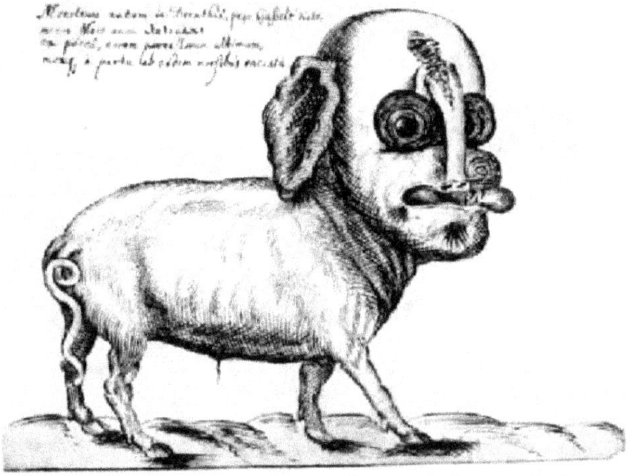

A hominosuid that, according to the caption, was farrowed along with nine normal piglets at Gasselte, Netherlands, in May 1661 (||s8whjmg).

Avella, Italy. Colonna (1592, p. 5). referred to an Italian hominosuid. In translation, he said, "To this may be added what would seem unbelievable had I heard it from anyone else. But because my dearest grandmother Caterina Pelegrina—who has such authority and learning and is so worthy of trust—told me this, I am persuaded it is certain. She told me that in 1528 when her father was alive and ruling over Avella, a monster farrowed by a sow was found there and brought to him. Its head was in all ways like a human's, but the rest was like the mother. The question arose whether some swineherd had impregnated her, so he ordered the matter to be turned over to a judge."

Mutation or Hybridization?

Genetically, some types of alterations are more easily produced than others. Mutations that result in omissions, such as limb deficiencies, merely require a silencing of the genes that govern a limb's development. When they are blocked, the limb fails to develop. In such cases, there is no need for new information. It's simply a matter of ignoring an existing set of instructions. But to alter pig genes directing head development so that the head instead develops to look like a human head, is not so simple. Where would the required information, the blueprint as it were, for a human head come from? Presumably, no such information is latent in the pig genome. How would a *random* mutation in a pig create, *de novo,* a plan for a human body part?

Left: The Caere animal. *Right:* The Tarquinii birth (Lycosthenes 1552).

A hominosuid with a pig body and human head and hand, as represented on the capital of a column in the Cathedral of St. Peter and Paul (Brandenburg an der Havel, Germany, ca. 1230 A.D.).

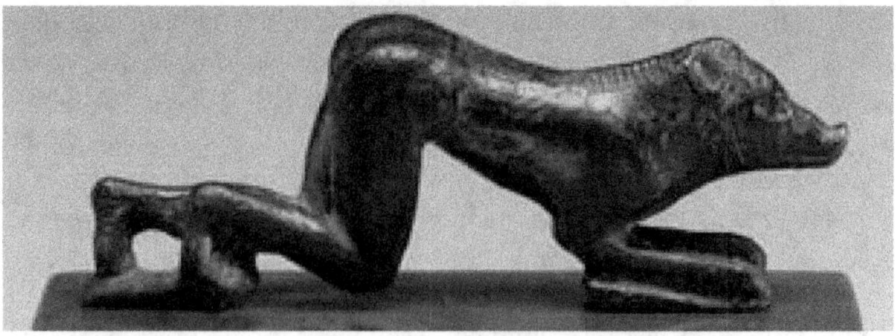

A hominosuid (Greece, 5th cen. B.C., Walters Art Museum; ||3y69wnau).

ANCIENT TIMES

Some of the earliest references to hominosuids occur within the context of the supernatural. Varaha, the third avatar of Vishnu, had a boar's head and man's body. Early images of this deity date to the 1st century A.D.

But there are also accounts that purport to be nonfictional. Livy (31.12) mentioned the birth, at Sinuessa, in Italy, in 198 B.C., of a pig with a human head. "According to reports," he said, "births of obscene offspring occurred in several places at that time. … At Frosinone a lamb was yeaned with the head of a pig, at Sinuessa, a pig with a human head, and in Lucania, a foal with five legs was born on public land."[17] Elsewhere (27.4), he referred to a previous birth of the same kind, a pig with a human face, at Tarquinii in 210 B.C. Julius Obsequens (§14) recorded a piglet with human hands and feet farrowed at Caere (now Cerveteri) in 163 B.C.

Two depictions of Varaha. *Above:* Bas-relief (Udayagiri, 4th century A.D.; ||y4sx4y5v). *Below:* Mughal painting (17th century; ||mprdwwxv).

Dio Cassius (43.2.1) stated that at Rome in 46 B.C., "a pig was born resembling an elephant save its feet." And Livy (27.12) mentions a child born with an elephant's head in 209 B.C. In this book, "piglets" and "children" reported as having elephant-like ears and trunks are classified as hominosuids. Mesopotamian omen lists specify the consequences of sexual interaction between humans and pigs (Freedman 2017, p. 72; Jastrow 1914, p. 52), and refer specifically to women giving birth to children with the heads of pigs, and sows birthing offspring with human hands and feet.

An illustration accompanying a manuscript report about a hominosuid farrowed in the Diocese of Speyer in 1576 (Wickiana Collection; ||bddhut3s).

HYPERTRICHOSIS

Plau, Mecklenburg, Germany. Johann Friedrich Wilhelm Nevermann (1839, p. 291), a physician practicing in Plau, described a case that he said came under his personal observation, that of an 11-year-old girl who bore on her sacrum what appeared to be a patch of black pig bristles, which prompted the classification of this case as a hominosuid.

INCERTAE SEDIS

Cairo, Illinois. The *Cairo Bulletin* (Mar. 12, 1869, p. 4c; ||yrmk47wm) reported a woman near Cairo giving birth to a nondescript with some porcine traits, but which did not fit the typical hominosuid profile.

> The entire surface of the body was red as crimson. The head, which, in general, was shaped like that of a human, presented no signs of hair, the extreme crown being as smooth as the face. The ears actually met behind the head, and the nose, very flat, covered fully one half the face. The arms were very slim and bony and were twenty-eight inches in length reaching eight inches beyond the "pedal extremities." The leg terminated not with a foot, but in something more in the nature of a hoof. The knots (for no word answers a better purpose) intended to supply the place of feet, were cloven like the hoof of a hog, and there was no visible sign of a knee or knee joint. There was continuous bone, like the femur, from the hip to the ankle joint. The fingers of the hand were in keeping with the length of the arm, and were entirely boneless, intertwining with each other like an equal number of strings or cords. The voice of the monster was sharp and shrill, more like the squeal of a pig, than any other sound to which it could be compared. Its cry was almost continuous from the moment of its birth until its death, on Tuesday.

NOTES

1. *Physica,* 2.3.3 (||y23f8hdn).
2. Indeed, some sources say it was born in Laos, not Myanmar.
3. Cyclopia and anophthalmia are common in telenothians.
4. In the original, the name of Wyatt A. Lunsford (1825-1887), a longtime resident of Crockett County, Tennessee (||yy95ppa6), was given as "Wyatt Lunceford."
5. Appendage deficiencies are frequent in telenothians.
6. Supernumerary body parts are frequent in telenothians.
7. Buffon and Daubenton (*Histoire Naturelle,* 1766, Tome 14, p. 392; ||y3ohy77m) gave a brief anatomical description of this creature.
8. Perhaps, Gustav Hartenstein (1808-1890), a professor at Leipzig University?
9. This is almost certainly Joseph Simon Fleischman (1881-1956), whose son, Joseph Fleischman, Jr., was born in Hawkinsville in 1910 (||y2advxn2).
10. Source of epigraph: ||yahw9dce.
11. The original French of the Bipemba report read, "Le jeudi 04 février dernier dans la commune de Bipemba, ville de Mbuji-Mayi, l'une des ville de la province du kasai oriental en R D Congo. Une jeune femme d'environs 22 ans s'est accouchée d'un enfant sans forme précise. Les témoins expliquent que l'enfant-monstre a la partie supérieure humaine et la partie inférieure animale (porc), deux courts bras terminés par deux petites mains à 8 doigts, deux courtes jambes terminées aussi par deux petits pieds à 4 orteils chacun. la cène [sic. probably "scène"] s'est passée à la maternité Meda au quartier Lubuebue, vers 12 heures locale. La cause de cette naissance tragique n'est pas encore connue et demeure sans explication."
12. The record of the courtroom examination of a suspect that took place February 1, 1646, in New Haven, Connecticut, mentions "a sow of Mrs. Lambertons pigging two monsters, one of them had a faire & white skinne & head [and] another [had] a head lik that of a childe" (Hoadly 1857, p. 295; ||yypgtdwt).
13. Another early case is listed in Chinese records of the Shanghai region, a pig born in the year 1600 with a human head and hands (Macgowan 1860, p.70; ||ldn5q97).
14. Translation mine. Comments by Helmont (1682, p. 286), who stated that the internal organs were similar to those of the mother, imply the animal was dissected.
15. Quoted in Lilly (1870, pp. 45-46).
16. Thus, Sigebert writes, "In parrochia Legiensi porca porcellum enixa est habentem faciem hominis" (Migne 1854, p. 235).
17. Translation mine. Livy mentions this same birth elsewhere (32.9).
18. *Colophon:* Paré (1641, p. 665).

Battle of the Centaurs (Artist: Arnold Bocklin, 1873).

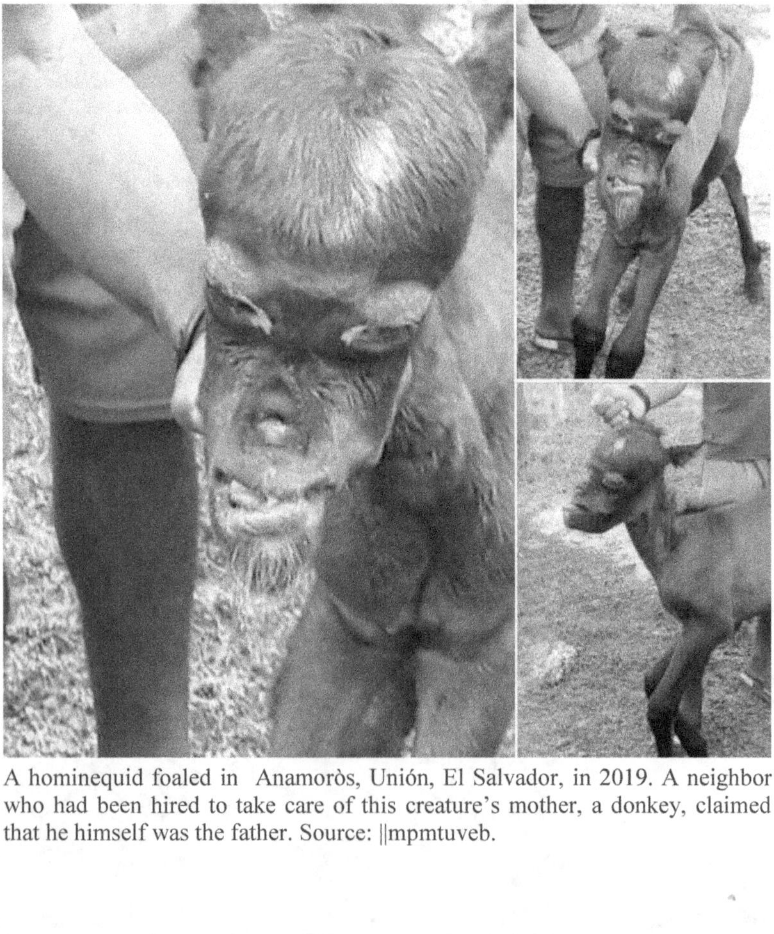

A hominequid foaled in Anamoròs, Unión, El Salvador, in 2019. A neighbor who had been hired to take care of this creature's mother, a donkey, claimed that he himself was the father. Source: ||mpmtuveb.

HOMINEQUIDS (ASS- AND HORSE-HUMANS)

Leviticus forbids women to copulate either with goats or with horses. It must be the case, then, that at one time such couplings were common.
—VOLTAIRE

Hominequids are among the few telenothians that have a common English name: *centaur.* Some of the oldest surviving depictions of centaurs appear on Mesopotamian boundary stones dating to about 1300 B.C. (Baur 1912, p. 2). What seems to be the earliest Greek portrayal yet found of such a creature, a terracotta statuette recovered at Lefkandi, dates to about 1000 B.C. Another Greek representation of a hominequid, in bronze, dates to the mid-8th century B.C. And Homer, writing in the same era, mentions centaurs.

Man and centaur, 8th cen. B.C.[1]

Among the Greeks and Romans, the notion of human-animal hybrids was widely accepted. In art, one of their gods, Pan, was regularly represented as human from the waist up, but as goat below. And he was considered a real being, so much so that the Roman scholar and writer Marcus Terentius Varro (116-27 B.C.) claimed that Pan had once been the governor of Spain.[2]

People accepted that Zeus, the lord of the gods, had seduced women in various animal forms, including an eagle, a swan, a ram, a snake and a dolphin. It was believed that, as a bull, he had impregnated Europa and carried her away to Crete, where she gave birth to the Minotaur, half man, half bull. In another version of the story, Pasiphae, the wife of King Minos of Crete, Europa's son by Zeus, lusted after a white bull sent by Poseidon. Again, the Minotaur was the result (see BOSHOMINIDS).

The Rape of Europa (Guido Reni).

Thomas Mann, in *The Magic Mountain,* went so far as to claim that "In the religions of antiquity, very often indeed, the sacred and the obscene were one." As Oliver Goldsmith put it, "fornication, incest, rape, and even bestiality, were sanctified by the amours of Jupiter, Pan, Mars, Venus and Apollo."[5] Apollodorus[3] said the first centaur, Chiron, was born when Kronos took a horse's form and made love to the nymph Philyra. Such behavior, seen in this sacred light, was permissible even to mortals.† Thus, Pliny stated without disapproval that Queen Semiramis "was so much in love with a horse as to have had sex with it."[4] To the Romans, says Salisbury (1994, p. 86),

> animals were not very different from people. They suffered the same emotions of love, anger, and jealousy; they had the same aesthetic appreciation for beauty and were able at times to set aside that aesthetic to appreciate someone from the highest motives of love. There was so little distinction between humans and animals that half-human/half-animal births were unremarkable.

According to historical accounts, Julius Caesar rode a horse with human feet. As Suetonius reported,

> He rode a remarkable horse, too, with feet that were almost human; for its hooves were cloven in such a way as to look like toes.‡ This horse was foaled on his own estate, and since the soothsayers had declared that it foretold the rule of the world for its master, he reared it with the greatest care, and was the first to mount it, for it would endure no other rider. Afterwards, too, he dedicated a statue of it before the temple of Venus Genetrix.§

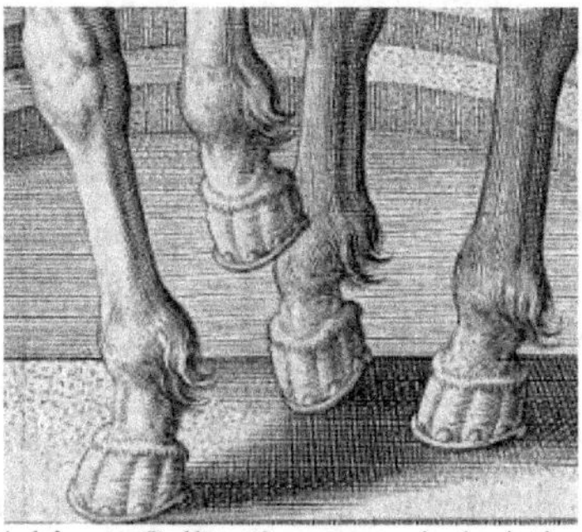

Adriaen Collaert's guessed depiction of the feet of Caesar's horse's (||nmyux4w).

†. Various authors document coition between human beings and equids, from early times to the present (e.g., Evans 1906; Devereux 1948; Krafft-Ebing 1906; Laws and O'Donohue 2008, p. 391; Miletski 2009; Valcuende del Río and Cáceres-Feria 2020). See, too, ||5n8w7nn5

‡. Rzaczynski (1721, p. 359) reported the reverse, a man with horse hooves, living in northern Poland, near Malbork, at "Insulae Mariaeburgenis." A video shows a teenager with hooflike feet living in Campos dos Goytacazes, Brazil (||4d4wcwxw).

§. Pliny (8:64) states that "It is said, also, that Caesar, the Dictator, had a horse, which would allow no one to mount but himself, and that its forefeet were like those of a man; indeed, it is thus represented in the statue before the temple of Venus Genetrix."

Julius Caesar's human-footed horse, as represented in Adlersflügel (1703, fig. 21). Compare to the description of the Bloomingdale hominequid.

Pliny himself, one of the most enlightened men of his era, was certainly a believer in centaurs. Elsewhere (ibid. 7:3), he tells of one sent to Claudius Caesar (Roman emperor, 41-54 A.D.) from Egypt preserved in honey and claims to have seen it himself.[6] In the same place, he mentions the birth of another such creature as if it were an historical event. It was born in Spain, he says, at Saguntum, in the year that city was destroyed by Hannibal (218 B.C.).

In his paradoxography, *On Marvels* (34-35), a later writer, Phlegon of Tralles, a literary freedman in the employ of Emperor Hadrian (reigned 117-138 A.D.), described an embalmed hominequid he found in the imperial collection. He says it was originally sent to Emperor Claudius, so whatever this specimen might have been, it seems it was the same one mentioned by Pliny. It was known as the Centaur of Saune, a place in Arabia where it was supposedly caught. Phlegon says it had a fiercer face than a human's, hairy arms and fingers, and a human torso that merged with a horse's body and limbs. He says other centaurs had been reported, but

> So far as concerns the one sent to Rome, anyone who is skeptical can examine it for himself, since, as I said above, it has been embalmed and is kept in the emperor's storehouse.[7]

St. Anthony meets a centaur (Master of the Osservanza, ca. 1430).

Somewhere between the days of Phlegon and modern times, belief in centaurs waned. But the pagans were certainly not the last to accept the idea of horse-human hybrids. In middle eastern art, the prophet Muhammed's steed, the Buraq, is usually portrayed as a centaur, often with wings. The early Christians, too, believed in centaurs. In his *Life of Saint Paul, the First Hermit*, St. Jerome tells how St. Anthony was setting out in search of Paul, when,

> All at once, he beholds a creature of mingled shape, half horse half man, called by the poets Hippocentaur. At the sight of this he arms himself by making on his forehead the sign of salvation, and then exclaims, *Holloa! Where in these parts is a servant of God living?* The monster after gnashing out some kind of outlandish utterance, in words broken rather than spoken through his bristling lips, at length finds a friendly mode of communication, and extending his right hand points out the way desired. Then with swift flight he crosses the spreading plain and vanishes from the sight of his wondering companion. But whether the devil took this shape to terrify him, or whether it be that the desert which is known to abound in monstrous animals engenders that kind of creature also, we cannot decide.[8]

Centaur battling wild beasts (mosaic, ca. 120 A.D., Hadrian's Villa, Tivoli).

Zirkle (1941: 487) notes that the existence of centaurs was accepted even in late medieval times. He mentions the Byzantine scholar Manuel Philes (ca. 1275-1345) as a believer and quotes Albertus Magnus (*De animalibus,* 22.2. 24): "The onocentaur,† as they say, is a mixed animal: for it has the head of an ass and the body of a man. Some are said to be found with a horse's body and with an upper human part."

Vincent of Beauvais

In the *Speculum Naturale,* a work written in the late Middle Ages, Vincent of Beauvais (1624, 19.97, 1435) stated that "The onocentaur is a monstrous animal of a biformed nature, that is, with the head of an ass and the body of a human being. His face bears horrible bristles and its hands are well-formed to carry out

†. The name *onocentaur* means "donkey-centaur." *Onos* is Greek for "donkey." The *Parallela Graeca et Romana* (Babbitt 1936) offers a germane anecdote: "Aristonymus of Ephesus, the son of Demostratus, hated women and used to consort with an ass; and in due time the ass gave birth to a very beautiful maiden, Onoscelis by name. So Aristocles in the second book of his *Strange Events.*" *Onoscelis* means "donkey legs."

> any task. He utters sounds, as if about to speak, but his untutored lips cannot form human speech. After those who pursue him, as the Experimentator says, he sends sticks or stones. The philosopher Adelinus says this monster was not created originally with the other animals but was sometime and somewhere produced accidentally through an adulterine mixed generation. But many contradict this opinion. [my translation]
>
> * The "Experimentator" was a writer now known only from fragments in the writings of later authors (||k6t696mw). Adelinus (ca. 639 A.D. - May 25, 709) was Abbot of Malmesbury Abbey and Bishop of Sherborne.

Such allegations referred not only to occurrences long ago and far away. Albertino Mussato, the Italian historian and statesman, in his account of the Holy Roman Emperor Henry VII, described the birth of a centaur as a real event. In translation, his account reads, "At this time [1254 A. D.], in the Veronese colony of Porcilio, a mare gave birth to a notable prodigy, namely a living quadruped with an

The Verona centaur (Schenk 1609)[9]

> equine body but the neck and head of a human being, which with its horrific gibbering so terrified the peasant at whose hovel it was born that, when he had seen it, he at once hacked it to death with a sickle. He then impaled its head on a sharp stake in front of his cottage where it was seen by passersby. Having been reported to the authorities, he was summoned and interrogated about the begetting of this creature, and about why he had killed it, with no proper authority, except in that he perhaps feared an attack. But in the end, he was absolved of any crime. And all others were cleared of suspicion given that the monster was born in the fields, which concealed the facts of the matter.[10]

Many later scholars cited Mussato. Thus, nearly four centuries later, in his encyclopedia of equine medicine, veterinarian Georg Simon Winter von Adlersflügel (1703, pp. 136-137) spoke of this prodigy as if it were something he believed. He even went to the trouble of providing a good quality illustration.

The Verona centaur (Adlersflügel 1703, fig. 22).

In his *History of Four-footed Beasts* (1608), Edward Topsell (ca. 1572-1625) discussed centaurs. He expressed the opinion that while it was possible for centaurs to be produced by "unclean" copulation, it was "unpossible that they should live long after birth."[11] And indeed, various other early writers reported centaur births as fact.

- The monk Ralph of Coggeshall said a creature found struck by lightning near Maidstone, Kent, June 26, 1205, had the head of an ass, a man's belly and "other portentous members far different from those of any single kind of animal" (Stevenson 1875, p. 155; Russell 1881, p. 363).
- Jobus Fincelius (1566b), a professor at the University of Jena, told how in 1554, before Easter, in the village of Recow near Cammin ("Rekow bey Camin") in Pomerania, a mare foaled a hominequid. Supposedly, its right forefoot was human, the other foreleg, a stump ending at the knee, and the rear feet bore horse hooves. Red lumps of flesh took the place of eyes, he said, and a big tuft of hair ("Dolle") replaced the tail. The mouth was like the bowls of two spoons pressed together. Unable to nurse, it died in three days and was buried on the edge of a field.

* Also see Lycosthenes (1557, p. 642) and Schenck (1609, p. 102). cf. the Vertaizon BOSHOMINID. Modern Rekowo and Kamień Pomorski, Poland. Recow and Cammin appear on old maps (e.g., ||4rfatmdr).

- Lycosthenes and Batman (1581) say a woman at Berwick-Upon-Tweed bore a two-headed hominequid with horse's ears on Jan. 5, 1579.
- The German parson Christian Lehmann (1699, p. 725) said a "boy" whose penis was like a horse's ("priapo equino") was born at a place on the Große Mittweida ("der Miepa"), a river in Saxony, in 1555.
- In 1661, Gaspar de los Reyes Franco,[12] a physician, said he had been told by Juan de Orduña, a Jesuit professor at the College of Carmona, that he himself had seen a creature with "the hands, feet and face of a human" that had been foaled by an ass. Orduña said a priest from the Guadajoz Valley, where the birth had occurred, came to ask whether it should be baptized, a question that sometimes arises with telenothians.
- Johann Christoph Bekmann, a professor at the University of Frankfurt, asserted that a foal with a human head had been born at Wittstock in Brandenburg in 1661.[13]
- Paullini (1686) communicated a manuscript compilation of curiosities collected by the monks Isibordus von Amelunxen and Alexander Insulanus at the Abbey of Corvey around 1200 A.D. Observation 46 of that compilation (ibid., p. 209) stated that "Herr Heinrich Himmel, a priest at Mihla [in Thuringia], told me in good faith that he knew a boy, a miller's son, who had been born with ears like a donkey's."[14]

A centaur (*Der Naturen Bloeme,* ca. 1350; ||yaxoqdoj).

Centaurs, as depicted in an early zoology text (Jonston 1650).

18TH-CENTURY REPORTS

And such reports persisted well beyond the seventeenth century. Douthwaite (1997) comments that "accounts of animal-human interbreeding abound in eighteenth-century natural histories and reveal a curiosity mixed with repulsion for hybrids of all kinds."

Jonas Lillequist (2006, pp. 159-162) documented a court case in which the accused claimed he encountered a human-horse hybrid: In 1706, a Colonel Anders Sparfeldt served as head of a Swedish military tribunal at Skara trying a soldier, Sven Jönsson, who had been away without leave. Jönsson maintained that during his absence he had had sex with a creature who was a woman in every aspect except that she had "shaggy legs and a mare's tail." By the end of the trial, Sparfeldt, who closely interrogated Jönsson, concluded this event had in fact occurred, and that the creature in question was the offspring of a man and a mare. Sparfeldt had a portrait drawn of this *tertium quid* after Jönsson's description (shown at right). Therefore, even in 1706, centaurs were still taken seriously enough to be accepted in court.

Jönsson's hominequid.

An erect centaur, much like Jönsson's (Kassite kudurru, 2nd millennium B.C., Jastrow 1915, pl. 53).

Indeed, animal-human hybrids were accepted by some of the greatest minds of the era. Voltaire commented (*Essay on the Customs and Spirit of Nations,* 1756) that,

> Nearly all of the ancient authors speak of satyrs. I do not consider their existence impossible. Leviticus forbids women to copulate either with goats or with horses. It must be the case then that at one time such couplings were common. And, so far as one can tell, it is to be presumed that monstrous species could have been born of these abominable amours.

Centaur relief on a Roman cup (Berthouville, 1st cen. A.D.; ||y5s48ha9).

Elsewhere he wrote,

> Was there indeed a race of satyrs, that is to say, were there girls who became pregnant by the same means that monkeys do, and who gave birth to hybrid offspring in the same way that mares give birth to mules and jumarts? All antiquity testifies to these singular facts. Many saints have seen satyrs—this is not an article of faith. So, the thing is quite possible, but it must have been rare.[15]

Voltaire
1694-1778

Gmelin
1709-1755

In an account of his travels in Siberia, the German naturalist Johann George Gmelin (1752, p. 460) said that in Yakutsk, in about 1740, locals brought him a fetus taken from a slaughtered mare. It had a human head, he said, but otherwise looked like a horse. The *Affiches des Trois-Évêchés* (Metz, Aug. 11, 1774, p. 2b; ||y668xmt7) said a colt foaled July 27, 1774, at Altroff (Moselle, France) had a human neck and head but soon died. And Desgranges (1798, p. 72, obs. 4) reported a woman at Vevey, Switzerland, giving birth to a nondescript with a horse hoof.

19TH-CENTURY REPORTS

Alexandria, Virginia. From the *Gettysburg Compiler* (Gettysburg, Va., Feb. 11, 1824, p. 2c; ||y2msxnpd):

> ALEXANDRIA, D.C. Jan. 30.—A monster is to be seen at the Alexandria museum. It was found yesterday in a public stable and is evidently what the advertisement asserts it to be, *half human and half horse ! !* It has the human head almost perfect, with only one exceedingly large eye[16] in the middle of the forehead, with two marks where eyes should properly be. Most of the joints appear like those of a human being.—There are many other indications of what we state, which we conceive unnecessary to particularize—nor does our time well admit of doing so.
>
> We had ocular demonstration last evening of what we have mentioned of this wonderful prodigy; and it will only be kept in its present natural state until to-morrow evening in order to preserve it for the museum, those of our citizens who are in the least doubtful of what we state can satisfy themselves.
>
> * Alexandria was part of the District of Columbia until 1846.

Anamosa, Iowa. From the *Des Moine Register* (May 2, 1887, p. 2a; ||8u9wuen4):

> ANAMOSA, May 3.—Special to the State Register.—Anamosa had the greatest wonder of the day in the shape of a colt that had but one eye, which was in the center of the forehead, about the size of a teacup. It had also nails like a human being. It died and will be sent to Chicago to be mounted. It is considered here the greatest wonder of the day.

Sharon Springs, New York. From the *New York Times* (Sep. 22, 1885, p. 5c; ||ycxcdd62):

> CANAJOHARIE, N.Y., Sept. 21.—Veterinary Surgeon Putman, of this village, was called to Sharon Springs [a village in Schoharie County, N.Y.] to attend a foaling case yesterday. The mare was a trotter of much prominence on Long Island, owned by Dr. Howard Green, of New York, and $2,000 had been refused for her. During the illness the horse and colt died. The upper part of colt's head from below the eyes resembled a human cranium, there being a place for the ears and eyes, and the soft spot on the top [i.e., the fontanelle], which is never found on the colt. The brain cavity was large and round, measuring 8 inches in diameter and 22 in circumference. The lower part of the head resembled the face of a pug dog, the mouth being dark colored. The length of the head was 13 inches. There were only three joints from the poll to the body, where there should have been 12, … The malformed head was taken to Utica to be preserved.

A painting, dated 1835, on display in the Musée Fragonard, an anatomical museum in Paris. The documentation accompanying this painting says the creature depicted was a hydrocephalic mare ("jument hydrocéphale") and that it was foaled January 21, 1821, at Huningue, Haut Rhin, France (||w9z3aa5s). The resemblance of its face to a human's is, however, unmistakable. The *Journal de Savoie* (Chambéry, May 11, 1821, p. 183; ||ycxyfvmy) reported this creature soon after its birth and, indeed, described it as human headed ("androcéphale"), not hydrocephalic. It would have been fourteen years old in 1835. The depiction is likely accurate, as the artist, Joseph Bordes (1773-1838), was a skilled portraitist (||2wzj8s3d).

New Britain, Pennsylvania. From the *Bradford Reporter* (Towanda, Pa., Jun. 30, 1859, p. 2; ||yx8rmtwr): "Samuel Brandt, of New Britain, Bucks County, has a mare that recently had a colt whose head resembled the human face, and has but one eye right in the center of the head.[16] It died shortly after its birth. The owner has preserved its hide."

* Jacob S. Brand (b. Middletown, Pa., 1858) was the son of Samuel Brandt (|y4xdjumh).

Nashville, Tennessee. From the *Nashville Daily Union* (Nov. 2, 1865, p. 3c; ||y3bcwfqe):

> We recently noted the discovery, at Nashville, Tenn., of a remarkable *Lusus naturae*. The monstrosity has a human head and features, while the body and legs are of the brute creation. It is called half horse and half man … To what class or order of conformation does this wonder belong? With our limited knowledge of animal monstrosities, we are totally unable to assign it a connective place with any known phenomena of nature. It belongs to none of the tribes or families enumerated and discussed by the historians of teratology but seems to be the actual representation of that species of deformity which had only a fabulous existence in and prior to the seventeenth century. In those days of superstition, the Minotaur, the centaurs, satyrs, dragons, tritons, sirens, and mermaids were regarded as creatures half human half animal, but this mythological belief is no longer shared by any but the hopelessly ignorant and credulous. The astounding fact is established beyond cavil that such a freak (the unity of man and beast) does exist in the form now on exhibition at Nashville. It surely is a subject for philosophical and medical study, and we hope some of the modern Hippocrateses, Aristotles, Plinys, Galens, or St. Hilaires will investigate it for the enlightenment of our own country and the world. This great phenomenal curiosity is in the possession of Dr. L. L. Coleman,[16] a prominent physician of Nashville. As a man of superior attainments in the sciences of anatomy and embryology, he treats it as the veritable wonder of the times. The attention of the medical world will be attracted to it and result in the most remarkable demonstration of fetal mystery.
>
> * In the Nashville report, note the manifest cognitive dissonance of the writer who both states that the existence this centaur "is established beyond cavil" and simultaneously states that "this mythological belief [in centaurs and other such composite creatures] is no longer shared by any but the hopelessly ignorant and credulous." Is this an example of potent taboo subverting an otherwise rational mind?

Portland, Oregon. From the *State Rights Democrat* (Albany, Ore., Jun. 20, 1890, p. 1c; ||mkpnwy4): "A FREAK.—There is now on exhibition in Portland a curious freak of nature, says an exchange. It is half man and half horse having limbs formed like a horse and a shaggy mane growing on the back of his neck. His head is like that of a colored man, with a very intelligent face. He can trot at a three-minute gait and converse freely with visitors."

Harrison, Kentucky. The *Bourbon News* (Paris, Ky., May 2, 1899, p. 5a; ||l4bbw7p) said, "On Frank Ammerman's place in Harrison there was foaled last week a colt which has eyes, eyebrows, forehead, nose and mouth like a human, the rest of the body being that of a perfectly formed colt."

Coventry, Vermont. Newport Express (quoted in the *Orleans Independent Standard,* Irasburg, Vt., May 17, 1867, p. 2b; ||y5wlcgh2): "The fabled centaur of old has actually been re-produced in our own time, in a head almost human, growing upon the body of a colt. Samuel Burbank of Coventry, last Thursday morning found a valuable brood mare of his dead. In seeking for the cause of death, a colt was taken from the body. The body of the colt was perfectly developed, but the head bore not the slightest resemblance to the equine species. The upper part of the head very closely resembles a well-formed and fully developed adult human head. The contour, the temples, the upper extremities of the cheeks, three crowns [?] upon the top of the head, the soft spot or opening in the skull as found in infant heads [i.e., the fontanelle], close, striking and unmistakable. There are two distinct eyes, much like the human eye, though larger, and these are separated from each other only by a membrane and placed in one large socket or cavity in the center of the head [i.e., synophthalmic cyclopia]. There is but little semblance of a nose, while the mouth somewhat resembles that of a colt, but is located more like a human mouth, and the protuberance below the mouth is as much like a chin as the under lip of a horse. The extremities of the ears are somewhat like that of the horse, and in fact is about the only resemblance to a horse. It is the greatest curiosity we have ever seen, and we doubt if a more singular and remarkable malformation has ever been produced. Dr. I. B. Kimball, of Coventry, now has the head in his possession, and is taking measures to preserve it. All wonder-loving people should see this unparalleled development of nature."

*A Samuel Burbank (1826-1905) lived in Coventry in 1867 (||y5655lfg, ||y4pssyc7, ||y53e7uht). He was a representative from Coventry to the state legislature (||y2d4s4f7). An 1870 census record for Coventry lists an Isaac Kimball, a physician, aged 44 (||y5yyjltt). Kimball lived his entire life in Orleans County (||y6py42jg).

Montgomery County, Missouri. The *Wetmore Spectator* (Wetmore, Ks., Oct. 20, 1883, p. 1g; ||2p85sv8c) said one of two "living curiosities" from Montgomery County that were exhibited at the Moniteau County Fair,

> was a colt with a human head, a most extraordinary freak of nature, the human cast seeming to run as far back as the fore-shoulder, where the lines of the colt set in. The location of the eyes and ears and the shape of the head were decidedly human in appearance.

> * The *Enterprise-Courier* (Charleston, Mo., Nov. 16, 1883, p. 3c; ||bdha9v3s) said the Montgomery animal was showing in Ste. Genevieve, Missouri, a month later.

La Grange, Missouri. The La Grange *National American* (quoted in the *Free Trader,* Ottawa, Ill., Jun. 6, 1868, p. 5b; ||yy76ej2x) said, "We learn that a large gray mare belonging to Robert Harrison, a farmer residing about five miles west of La Grange, gave birth to a colt, a few days since, having a face strangely resembling the human features divine. In every other part, except the head, it is not unlike any other young colt, and may be considered well-formed up to its ears, where the human-like face begins. The eyes are smaller than usual, and very expressive; the nose is flat, while the nostrils are thin, the nose having quite a ridge through its center. The mouth, lips, and chin are modeled after the human form and are quite perfect. The colt has no teeth, which is unusual, and cannot obtain its nourishment from its mother, but is fed milk with a spoon. Its breathing seems difficult and hard, and it is not very active, being loath to stand upon its feet. We have given this account from the description furnished us by a gentleman who saw the colt on Tuesday afternoon, and numbers of our citizens are going out to see it every day. It is thought that it will not live, but if it should, arrangements will be made to exhibit it over the country. P.S., since the above was in type, we learn the colt died on Tuesday night."

* A Robert H. Harrison (1828-1876) lived near La Grange in 1868 (||y5swtm9c).

Manila, Philippines. El Independiente (Las Vegas, N.M., Nov. 17, 1910, p. 3c; ||2x23vc8p) said, in translation, "On a farm near Manila a mare foaled a colt with the head of a monkey. As usually occurs with such freaks, the colt died in a few days, despite the efforts of the veterinarians. This freak will be exhibited in a museum in Manila." This monkey-headed animal has been classified as a hominequid, as human heads and monkey heads are, in comparison with horse heads, much alike.

Williamsport, Maryland. The *Williamsport Pilot* (quoted in the *Shepherdstown Register*, May 4, 1872, p. 1e; ||53dw6wuv) said, "One day this week on the farm of Mr. Graves, a few miles below town [i.e., Williamsport] in the direction of Two Locks, a mare gave birth to a foal which had a human head—at least in appearance. From a gentleman who saw this curious freak of nature we learn that the head of what should have been a colt bore a horrible likeness to the head of a human being. The ears small and setting upon the side, the top of the head, round and covered with a glossy suit of fine velvety hair, and the general outlines of the face bearing a strong likeness to humanity. The man-brute lived but a few hours, but the head has been preserved."

* A James T. Graves, born circa 1840 in Virginia, lived near Williamsport in 1870.

Topeka, Kansas. From the *Wichita City Eagle* (Wichita, Ks., Aug. 12, 1875, p. 2c; ||y6col5kp): "A Shawnee County mare gave birth to a well-developed colt with a human head. This is a BLADE story." The *Blade* was published in Topeka, the seat of Shawnee County.

Cabo Rojo, Puerto Rico. La Correspondencia de Puerto Rico (San Juan, P.R., Mar. 14, 1893, p. 3c; ||2hy4uub2) said, "A rare freak has appeared among the livestock of Mr. Juan Tornos at Cabo Rojo. A mare foaled a colt with round head without a muzzle, or really without jaws. The ears are round, the nose very pronounced, and the mouth, small. The fore feet have fingers instead of hoofs. And, finally, the freak's face much resembles that of a human being. In other respects it was normal for a horse. It lived only half a day."

Salem, Ohio. From the *Jeffersonian* (Findlay, O., Jun. 14, 1878, p. 3c; ||y2axoxol): "A curious freak of nature was on exhibition in Upper Sandusky [Ohio] last week. It was a colt with a head shaped like a human person's, the face and eyes resemble those of a man, except that the face was covered with hair, and the lower jaw projected like that of a natural colt. The forelegs were perfect to the feet which slightly resembled human feet. The monstrosity was foaled on the farm of A. B. Inman, near Whartonsburg. Owing to the imperfect development of the nasal organs it lived but a short time."

* An 1880 census record lists an A. B. Inman, a retired farmer, aged 78, as living in Salem Township, Wyandot County, Ohio (||y22h92va).

Macomb, Illinois. The *Wichita Daily Eagle* (Mar. 13, 1891, p. 2; ||jurb49j): "MACOMB, ILL., March 12.—A remarkable curiosity made its appearance in the world here yesterday morning. A mare belonging to Asher [Benjamin] Blount [1819-1899] gave birth to a colt about noon. The little animal was a perfectly formed horse, with the exception of its head, which was as near like that of a man as it could be without being human. The neck is rather long and slender; the cerebrum is round and about the size of a grown man; the ears are delicately formed, and in proportion to the size of the head; the mouth and nose, though in their proper place, are very much deformed, the two joining each other. The nose is like that of a horse, and the mouth is very wide. There are no eyes, and apparently no place for any.[17] The little animal lived but a few minutes. It will probably be preserved in alcohol and placed on exhibition."

Guthrie, Oklahoma Territory. From the *Wichita Daily Eagle* (Wichita, Kan., Jan. 29, 1892, p. 5b; ||y6qn92sj). "Yesterday morning when the Santa Fe train came from the south it brought one of the most remarkable monstrosities ever seen in this city. It was no more nor less than a colt with a human head. It was handled as if it was a human body, for the animal was dead, and taken to the undertaking establishment of Boaz & Gill when the artistic embalmer of that firm, M. H. Miller, proceeded at once to embalm it. Through the courtesy of that gentleman, an EAGLE man was given entrance to the room where the monstrosity was. It was laid out on a slab and was then undergoing the process of embalming. The body is of the shape of a colt, but the head is distinctively human. The body is devoid of hair and looks in almost every respect like the body of a Mexican dog.[18] It is of a brownish color, something like the color of sole leather. The head is very interesting. It is almost precisely the shape of the lower order of human beings and the eye has a look of intelligence about it that is really astonishing. The ears are not fully developed at the side of the head, but the outlines of the human ears are plainly visible. There are two small ears on the back of the head which bear a resemblance to horse's ears, only that they are very small. The top of the head is covered with black hair, which is almost as fine as silk, and that is the only suggestion of hair about the body. Logan County, O. T., has the honor of producing the monstrosity. It is now the property of Dr. Moore, who had a museum in this city recently, who bought it from its original owner, a farmer living near Guthrie. Mr. Miller will spend two days embalming it, and, when he has it completed, it will be sufficiently preserved to be kept on exhibition for a long time."

* A Wichita city directory for 1892 lists an undertaker named M. H. Miller (||yyshp9k5), likely Merritt H. Miller (1869-1894).

Ashland, Wisconsin. The *Asheville Citizen* (quoted in the *Daily Concord Standard*, Concord, N.C., Oct. 16, 1897, p. 4b; ||mdg36qp): "Dr. G. W. Barlow, a veterinary surgeon of Ashland, Wis., is here with what is termed "the modern centaur," which is to be kept on exhibition for a time. The freak stands about 18 inches high and has the body of a colt and the head of a human. The freak was born at Ashland, Wis., January 25, 1895, and lived two hours. It has been in Dr. Barlow's possession ever since its birth. He has a certificate signed by a number of business and professional men of Ashland testifying to its genuineness."

Fort Yuma, California. La Crónica (Los Angeles, May 29, 1878, p. 4b; ||4ze65yet) said, in translation, "Don Rafael Varela, of Fort Yuma, writes to say that one his mares recently foaled a freak. Its forelegs are like a human being's, with hoofs like nails. And given that the hind legs are longer than the forelegs, its muzzle is down by the ground when it walks."

Wellsburg, Missouri. From the *Warrensburg Standard* (Warrensburg, Mo., Jun. 1 1866, p. 1d; ||ytfkb4ab):

> WELLSBURG, ST. CHARLES COUNTY, MO. MR. EDITOR—I wish to inform you that on Tuesday, the 15th inst., my mare had a foal and it had but one eye,[17] and that was in the centre of the forehead; it was a large white eye, with a black heart in the centre; it had no nose, and the mouth and chin the same as a man, with a beard on its chin; the jaws worked exactly like a man's. Everyone round the country is coming to see it. It is the wonder of the age.
>
> Yours, respectfully,
>
> THOMAS LUCAS
>
> P.S.—I forgot to mention that the body, legs and tail have a perfect horse foal. It is three and a half feet [1.07 m] high.

* Wellsburg was the name of a post office in Cuivre Township (||3a389unt). Two men named Thomas Lucas, father and son, lived in Cuivre Township in 1860 (||ma33u43u).

Fort Wayne, Indiana. From the *Vinton Record* (McArthur, O., May 22, 1873, p. 1f; ||zwppue7): "A few days since, on a farm six miles from Ft. Wayne, Ind., on the Bluffton road, there was foaled a horse colt which was part human. It had the head, neck, breast and forelegs of a horse, and the rest of its body was human. It was foaled alive, and no doubt would have lived and been the greatest monstrosity ever created. The owner, however, killed the creature as soon as it was foaled."

The Madness of Titania (detail; artist: Paul Jean Gervais).

HOMINEQUIDS BORNE BY WOMEN

Doddington, Northumberland. In 1674, reports said a certain Jane Paterson bore at Doddington a nondescript with the head, mane and hooves of a horse. It was dubbed the Northumberland Monster (Morley 1892, p. 248, Vaughan 1674). Townspeople scalded it to death.

Henderson, Kentucky. In Furman (1880, p. 465), Dr. Thomas W. Taylor, of Henderson, reported that he witnessed a woman "delivered of a child stillborn having the head and complete ears of the jackass."

* Starling (1887, p. 787) said Thomas W. Taylor was "a highly educated physician."

Ciudad Mendoza, Veracruz, Mexico. El Correo (Mérida, quoted in *El Mensajero,* Phoenix, Az., May 27, 1939, pp. 1b & 3b; ||5xar2r6z) said, in translation, "A GIRL WITH THREE FACES AND THE EARS OF A BURRO.—The RESUMEN, the newspaper in Río Blanco, Veracruz, has reported that a girl with three horrible faces and the ears of a burro has been born at Ciudad Mendoza [6 km from Río Blanco]. The VANGARDIA, the newspaper in Orizaba, [which neighbors Río Blanco] said that the birth took place in the Colonia Obrera neighborhood of C. Mendoza; that the mother of the freak is Mrs. Paz Monrroy, that the girl had one face in front and, in back, two (one with the brow of a human, the other with, a burro's); that her body was like a girl's; that the back of the neck was fused with the shoulders; and that Mrs. Monrroy's condition was very delicate, so that she had to be taken to the hospital to recover. But the latest reports about this freak say that Mrs. Paz Monrroy, a manual laborer, was then in perfect health; that the girl died after three days, and that her cadaver, on exhibit at the hospital, was attracting multitudes.

Amsterdam. Bartholin (1654, II, p. 242) reported a woman giving birth to a daughter with the head of a horse in Amsterdam in 1637.

Nerchinsky District, Siberia. Der Humorist (Vienna, May 5, 1847, p. 428; ||y34xq3ol) said, in translation, "The journal of the Russian Ministry of the Interior, which in addition to births of monstrosities, reports other things such as violent deaths and various accidents, tells (in the Feb. 1847 issue) of a monstrosity 'that is so remarkable that one knows not what to think of it.' The wife of a Siberian farmer in the Nerchinsky District, Mariana Klioutchewa, 22 years of age, gave birth on the 23rd of October last year [1846] to a thing that bore absolutely no resemblance to a human being, rather it was much more like a horse, that is, it had a horse's head and hooves, a small tail, extraordinarily large eyes and 19 ribs on either side."[19] Horses have 19 ribs.

Richmond, Mississippi. The *Medical and Surgical Reporter* (1861, vol. 6, p. 318a; ||yxfjyspp) stated that, in the May 1861 issue of the *Nashville Journal of Medicine and Surgery* (an issue now seemingly unavailable), Dr. M. M. Davis, of Richmond, reported "a case known to him, in which a woman had been, during her pregnancy, frequently frightened by a horse. Labor came on in due time 'and the object of her labor was expelled lifeless.' Dr. Davis thus describes it:—'To the astonishment of the husband and all the attendants, it proved to be, instead of a child, something like the shape of a horse. Its head, ears, nose, neck, body, feet, and legs were all as much like a horse as if it had been sired and foaled by that species of animal.'"

* A. Dr. Moses Marion Davis lived near Richmond in 1860 (||5du28ya9).

The Nerchinsky and Richmond reports may be compared to three other, much older ones about women giving birth to equids:

- *Padua, Italy.* Gisbertus Voetius (1655, p. 924), rector of the University of Utrecht, said a woman bore a horse at Padua in 1625.
- *Messina, Italy*. In a letter to Paul Zacchias, the Italian physician Pietro Castelli stated that in 1634 a woman at Messina bore "a being that looked like a donkey" (quoted in de Ceglia 2014).
- *Glogow, Poland*. Friedrich Lucae (1689, p. 2237, misnumbered 2227) said a woman bore a donkey at Glogow, Jan. 7, 1672.

* The last several cases fall under the heading of *xenogenesis,* that is, a mother giving birth to offspring not of her own kind (see p. 1235). In addition, at Hildesheim, Germany, in 1580, a mare reportedly foaled two boys (Irenäus 1584; Franck et al. 1585, p. 1050).

Anderson, Indiana. The *Huntington Democrat* (Huntington, Ind., Dec. 20, 1888, p. 3; ||k2mxp5bh): "A few days ago a lady in Madison County gave birth to a child, the lower limbs of which were like the hind legs of a horse, having hoofs and hair. The freak was dead when born."

Wabash Township, Ohio. On August 12, 1891, according to reports, Emily Whited (1850-1935), wife of John Newton Whited (1844-1932), gave birth to an 8-pound, male hominequid at their home in Wabash Township. One source (|vx73m7d2) says they named their "son" Orlando and that he died eleven days later, on the 23rd. Among the many reports about this case, one of the most complete appeared in the *Buffalo Enquirer* (Buffalo, N.Y., Aug. 18, 1891, p. 2; ||5b843w87).

> GREENVILLE, OHIO, Aug. 18.—On Wednesday morning last Mrs. Whited, wife of J. N.[20] Whited, who resides on a farm near Yorkshire, this county, gave birth to a male child which is considered a monstrosity. It is part human and part animal, and it was born alive and is living and doing well.
>
> Your correspondent visited the residence of Mr. Whited, which is about 17 miles north-east from Greenville, and about four miles north-west from Yorkshire [in Wabash Township] … A call was also made upon Dr. N[oah]. D[aniel]. Berry [1854-1928] at Yorkshire, the attending physician at the time of the delivery of the child. He stated that the child is a freak of nature, and such a malformation as he has never heard of or seen before. Through the kindness of Dr. Berry THE ENQUIRER correspondent was permitted to see the infant and learn the facts concerning its birth … Mr. and Mrs. Whited are well-to-do and respected people in the community in which they live, Mr. Whited being a prosperous and successful farmer. They were married about 28 years ago, and Mrs. Whited's maiden name was Lyons. Mrs. Whited, prior to the birth of this monstrosity, gave birth to six healthy, perfect and well-developed children, all living, except one which was drowned.[21]
>
> This child is well developed and weighs about eight pounds. After birth the child seemed apparently dead, but by the careful management of the attending physician the child was soon breathing normally. On close examination it was found that the child's shoulders and shoulder-blades were slightly increased in size. In fact, the child was abnormal from head downward. The chest was somewhat rounded and on the barrel-shaped order. On the small of the back, or perhaps a little lower, there is a tumor about three and a half by four inches in diameter … claimed by medical men to be "spina bifida" … The great deformity in the child is that portion from the hips downward, which undoubtedly represents a colt or young horse. The size and shape of the pelvic bones or hips and legs of the child to a marked degree resemble the hind legs of a colt and this can be seen by the most casual observer. The legs of the child are set at a right angle with the body, and they have no backward movement, but they have some little

> forward motion. They cannot be straightened out and the child cannot walk upright, but would have to walk, if at all, upon all fours, just like a colt or horse.
>
> The thigh or leg of the child between the knee-joint and the hip is not round like a human limb, but somewhat flat and has the exact shape of the thigh or leg of a colt. At the knee-joint there is no cap, but there is a movement forward in the hip and knee, but very little backward. This joint and knee has the exact shape and resembles the knee and joint of the hind leg of a colt. The legs from the knee-joint downward cannot be bent backward on the thigh, which is also the case of a colt.
>
> There is also a deformity of the feet, which are twisted around. One is ball-shaped and somewhat resembles the foot of a colt, but in fact is what medical men call "club-footed." The other foot is not quite so badly deformed. ... The community has been considerably excited over the child, and hundreds have been to see it. [speculation omitted.]

Sidney, Ohio. The *Darke County Democratic Advocate* (Feb. 14, 1889, p. 1; ||3jd2rttv) said a child born at Sidney was "from its stomach up, a complete horse. Its arms stood out like the forelegs of a horse, and instead of hands were perfectly formed hoofs. The child lived two days."

Ghent, Belgium. Guislain (1842, obs. 28, p. 237), a doctor practicing in Ghent, described a "child" with a horse-hair birthmark.

Carthage, Indiana. Wilson Hobbs (1869, p. 709), a physician in Carthage, reported a child born with "a nevus as large as a man's hand ... covered with hair resembling that of a horse." A nevus is a birthmark.

* Both of the previous two cases (Ghent and Carthage) come under the heading of HYPERTRICHOSIS.

Cincinnati, Ohio. From the *Cincinnati Enquirer* (Nov. 10, 1880, p. 4c; ||yvnrawe8):

> At No. 785 Freeman Avenue, a couple of doors above Bank Street, there reside an aged and respectable German and his wife, who make a living by keeping a family grocery and provision store. The husband's name is Philip Volz. Mr. Volz is a large fleshy, good-natured, honest, temperate German, who is content to mind his own business and deal out to his customers all the commodities usually found in a grocery and meat store, save cigars, tobacco and liquor. The last three articles he does not keep, and he does not drink beer himself, because he says it does not agree with him.

His wife is a kind, pleasant old lady—probably sixty—and the people in the neighborhood call her "grandmother." She is very neat and clean in her appearance and is one of those sensible bodies that people like to talk to, either on business or merely for pleasure. With the exception of six years, when the couple and their family resided at Camp Washington, Mr. and Mrs. Volz have lived at Freeman and Bank for nearly thirty years. Formerly they occupied the northwest corner of the two streets and were in the saloon business. It is said that they own the property they now occupy as a dwelling and a store, a substantial three-story brick building.

This good couple are the parents of one of the most wonderful specimens of humanity that the world ever saw, and certainly the most wonderful that was ever known to exist in Cincinnati. It is a man having the head of a donkey on its shoulders. It most always happened that monstrosities born of either women or animals have died shortly after birth, if born alive at all, but this is an exception to the rule, for the subject of this article is alive now, and is thirty years old. An ENQUIRER reporter visited the grocery store in question last evening and had a lengthy interview with Mr. and Mrs. Volz. The reporter also saw the unfortunate being who was the cause of the visit. Before attempting a description of George Volz, for that is the name of the human mule, it is but fair that something should be said relative to its life and the manner in which it is taken care of.

Both Mr. and Mrs. Volz said to the reporter previous to admitting him to George's room that many unkind rumors had gained circulation relative to their treatment of their crippled son. They asserted that people reported that George was kept locked up and was never allowed to see or be seen by anyone except his keeper. This they emphatically denied. When George was young he was allowed to crawl out into the bar-room when he felt so disposed, and during the day he was permitted to remain all day in the dining room if he felt so inclined. There he could be seen by anybody and everybody that came to the house, which was a public one. His appearance was such, of course, as to attract attention and cause remarks to be made. The poor fellow, after he got older, was kept in a comfortable room in the rear of the place of business, especially since they embarked in the grocery business, for the reason that he was not a fit object for women to look upon.

The parents declared that they took the best possible care of their son and did not keep him locked up even when he was in his room. He is the only child they have at home now, the others being married or dead, and they provided a nurse for him who was in constant attendance. George has no desire to go out, and cares nothing about being left alone.

When the reporter was ushered into the presence of the man George, he could not repress a feeling of extreme horror at the sight. The scribe's flesh began to creep, and his hair almost lifted itself on end. Had he been alone his courage would not have allowed him to remain in the room a second. But being accompanied by Mr. and Mrs. Volz, the reporter used his every endeavor to overcome the feeling of indescribable terror that came over him upon beholding such a frightful spectacle. For a few seconds the reporter could not hear what was being said, or recognize what was being done, so terribly astounded was he, and he could do nothing but gaze with a wild fascination at the object before him. Gradually he mastered his gruesomeness and tried to talk, in the hope of shaking off the entirety of his fright.

George Volz sat in a large armchair near a bed under a mantel. By his side was a comfortable stove. The eyes of the cripple stared at the reporter in an unpleasant way, and George began to kick violently as though he were trying to get at the stranger, the while his hands, or what correspond to a human being's hands, pounded each other with frightful rapidity and apparent viciousness. The reporter finally became sufficiently at ease to make a close observation, and he saw that the face of the cripple was as near the shape of a donkey as could be, the only feature that at all approached the shape of the genus *Homo* being the nose. This was noticed to be more prominent than that of a genuine specimen of the donkey, but the outline was exactly the same. The mouth could not be told from a donkey's, and when it came to observing the upper part of the face and the head, the resemblance was exact. The head is very large and nearly square like the donkey's. It is covered with coarse black hair, as stiff and straight as ever seen on the animal referred to.

The eyes are the eyes of a donkey to perfection. But the most perfect resemblance to this animal are the ears. They stand straight up and run out to a sharp peak in the exact shape of the ears of the donkey. They are about three inches long but are not covered with hair. … The lower portion of the face is covered with a short growth of hair, which is sheared by the mother at regular intervals. The body of the cripple is also covered with a growth of long, stiff, coarse hair.

THE EXAMINATION

The reporter noticing the fact that the upper part of the head between the ears was much larger than the lower part, at first thought that this was due to the growth of hair. In order to satisfy himself on this point he advanced toward George to place his hands upon his head. As soon as he got within range, he received a couple of vicious kicks from one of the monster's feet. Mrs. Volz also advanced at the same time, and she also was kicked. The reporter, however, succeeded in getting his hands on George's head and found that the shape noticed was the shape of the skull and was not caused by the hair. All the hairs grow upward.

As the visitor stood looking at the object before him, it kept swinging its head from side to side, and the expression of the face was precisely that of the donkey. George cannot talk at all. He makes no sound that approaches to the human speech. George's mother informed the reporter that he had sufficient intelligence to comprehend enough to have his wants supplied, which were few. He expressed a wish for a drink or something to eat by signs. During the interview George several times became very much excited, and resumed his violent motions with his hands, and at the same time he contorted his features in a horrible way. Mrs. Volz offered him a glass of water, but he repulsed it, and it was not ascertained what he wanted. Anon he would become settled and recommence the swinging of his head. George's hands were mentioned above. There are really no hands, but the stumps of three or four fingers in the aggregate are attached to what would be hands if he were a perfectly formed human being. The shoulders are very narrow, and the spinal column is not sufficiently strong to enable George to stand up long at a time. He crawls when he wishes to move about. One of his legs is that of a human, but the other is shapeless. George is powerless to wag his ears. He eats the ordinary food that is served on the table. Mrs. Volz stated that his appetite was good usually, and that he was allotted a place at the family board. Sometimes he refused to eat this substantial food of an everyday meal, and he was then fed on cake or some other delicacy. The room in which he lived is on the ground floor, across the hall from the grocery. In it are George's bed and the bed occupied by his attendant.

HIS DRESS

George is kept attired in a sort of jacket so far as the upper portion of his body is concerned, and below that is worn a woolen frock or petticoat.

SHAKING HANDS

While in the room with George, the cripple's mother asked him to shake hands with the visitor. George held out his left arm, which was grasped by the reporter, but George made no effort to shake the reporter's hand, although he took advantage of the situation to launch a kick at him with his foot. When standing erect, George would measure about four feet eight or ten inches in height.

When questioned further relative to the intelligence of the cripple, Mrs. Volz said that he could distinguish between the picture of his dead sister, who died three years ago, and that of his sister who is alive. When asked where his dead sister is, he would point toward heaven.

Several physicians have spent a great deal of time in years past in endeavoring to improve the mental condition of the cripple, but in vain.

Mrs. Volz informed the reporter that during the terrible siege of cholera in 1849 here, she was encient with George. Her husband had the malady, one of her children died of it, and another had it twice. Everybody was deserting the city, and she became almost frantic at the terrors of her situation. They resided at Vine and Liberty streets then, and they owned a dairy. There was no one but her to attend to the dairy, and, being unable to do the work herself, she gave the dairy away to a man who agreed to take it and run it for nothing. It was during this frightful period, while she had to be among the animals belonging to the dairy, that her unborn son was marked [or she was impregnated by a donkey?]. She stated that the impression was out that she had been frightened at the sight of a mule running into a burning stable at the time their property burned, but this was erroneous, as George was born before the fire referred to occurred.

George Volz looks as though he might live for many a day, although he is not nearly so fleshy now as he was a few years ago. The term donkey is used above to assist in describing this cripple not in a facetious sense, but because he resembles that species of quadruped more closely than he does a mule or horse. If he could be procured as a subject of illustration in the Medical Colleges, he would doubtless be considered of great value by men of science. But his parents would never consent to anything of that kind, and application would be worse than useless, for it would only be annoying, if not cruelty, to the parents, who are naturally sensitive on the subject. It is fortunate that the poor cripple has such kind and careful parents to administer to his comfort, for that he is well provided for there cannot be a doubt.

* An 1880 Cincinnati census record lists a 27-year-old, disabled George Volz as the son of grocers Eveanna and Phillip Volz, aged 60 and 67, living at 785 Freeman St. (||4w5bfrny). The mother's maiden name was Eveanna Rosina Kern (b. May 7, 1820, Bergertshofen, Germany, d. Jan. 12, 1885, Cincinnati). The father's full name was John Phillip Volz (b. May 2, 1818, Achern, Ortenaukreis, Baden-Wuerttemberg, Germany, d. Mar. 27, 1891, Cincinnati). An 1860 Cincinnati census record (||bdddyhs5) lists several members of the Volz family, including a nine-year-old George, who would therefore have been born in the second half of 1850 or the first half of 1851, surely the former, if his mother was pregnant with him in 1849, as stated in the report.

Uruguay. Quoting an Uruguayan newspaper, the *Boletín Mercantil de Puerto Rico* (San Juan, P.R., Oct. 15, 1882, p. 3f; ||4xnen7be) said, in translation, "A lovely 16-year-old, married to a cowboy, has just given birth to a horrible freak. It's two feet long and about seven inches in circumference around its chest. The arms and legs are so thin that they look like pencils, in sharp contrast with what should have been its hands and feet, which have been converted into true horse hoofs by an intimate fusion of the digits. But the most striking thing about this freak is that, despite its extreme slenderness, it weighs 37.5 pounds, which the doctors say is due to the exaggerated development of its head, which has a lengthy upper jaw that makes it look like the head of a burro. The whole back of this freak is covered with hair, especially where the tail should have been. Otherwise, the child is just lovely!"

Williamsburg,[22] *New York.* The *Brooklyn Eagle* (quoted in the *Spirit of Democracy,* Woodsfield, O., Jun. 14, 1870, p. 2; ||4yvaujw6) said,

> The residents in the Eighteenth Ward have been recently much excited over the birth of a child so malformed that nothing like it was ever seen. On North William Street, between Meeker Avenue and Herbert Street, resides a German family by the name of Aufnacht. Nick and his wife Gertie live in a shanty on the rear of a lot surrounded by stagnant pools of water. These hardworking and worthy people were childless until the 3d ultimo, at which date their family was increased by the birth of a little girl.
>
> At that time the child was not so badly formed. Within the past ten days its head has been rapidly developing into resemblance of that of a horse. The forehead slopes at an acute angle, and the upper jaw overlaps the under one. The eyes are bright and hungry looking.—It has the nasal bone and cartilage extending down over the lower lip, and ending in an oval-shaped fleshy protuberance, which merely joins with the upper lip. The upper lip spoken of is double. The whole front part of the upper jaw is wanting, and looking into the mouth no palate roof is seen. The tongue is bifurcated, and gums of the bottom as well as the top double section in the mouth are flattened much more than in an ordinary child.
>
> It receives nourishment only by hand, being unable to nurse. It devours voraciously the food given to it and eats as much as could be stowed away by three children of its age. When born it was quite a fine child bodily, but within two weeks past the limbs have grown thin and the leg joints have swollen. The monster breaths stertorously, as an equine does. The sound made by it in respiring would deceive one into the belief that a horse was near.

Greenpoint, New York. Four years after the Williamsburg birth, the *New York Dispatch* (New York City, Aug 2, 1874, p. 5e; ||yc8h4k92) reported the birth of another horse-headed hominequid in the neighboring city of Greenpoint. The report reads: "In the wanderings of a DISPATCH reporter after items of news for the paper, he came across a phenomenon which shows that Nature can indulge in a freak occasionally. In Union Avenue, Greenpoint, a child was born on Friday of last week, presenting all the attributes common to all other children, with the exception of the head. The head is the perfect facsimile of a horse's head, though in diminutive form. This head has the nose, face, ears, and general contour of a horse, accompanied with a mane. The child was living twenty-four hours after birth and is very likely to continue to live."

20TH-CENTURY REPORTS

Lewiston, Idaho. The *Idaho County Free Press* (Grangeville, Id., Aug. 20, 1903, p. 1d; ||n2z83ho) said, "Half human and half horse, a being 35 years old [ergo, born ca. 1867] was discovered in the suburbs of Lewiston having the body of a man and the head of a horse. It stands about two feet six inches high, covered with hair over a foot long, without human intelligence or power of speech and eats grass like a horse. After much persuasion the parents, after being promised that the strange freak would be well cared for and its parentage not disclosed, have made arrangements with Lewiston parties to exhibit the freak. And in all probability, the people of Grangeville and vicinity will have an opportunity to see it at the coming street fair."

Rio Negro, Argentina. La Nación (Buenos Aires, quoted in *La Democracia,* Ponce, P.R., Aug. 31, 1901, p. 5d; ||mr2v5zdm) said the National Museum of Argentina, in Buenos Aires, had just received the head of a hominequid found in a village in Argentina's Rio Negro Province. Its discoverer was a Mr. Tomás Echegay. The body of the animal had not been preserved. The report said horsehair covered the entire surface of the head, but its facial features were just like a human being's in both their shape and their arrangement, with the exception of the lower jaw, which was very long, and the teeth, which were those of a horse, and nasal passages were absent. The overall size of the head was about that of a human being's.

Stetteldorf am Wagram, Austria. Der Tiroler (Bolzano, Aug. 24, 1905, p. 4b; ||cp344dsz) said, in translation, "A MONSTROSITY.—It is reported from Stetteldorf in Niederösterreich that on the 20th of this month a child was born with a completely deformed lower body. One of its feet is shaped like a horse's hoof and its joint bends towards the front."

Weston, Oregon. From the *Shaniko Star* (quoted in the *Daily Capital Journal,* Salem, Ore., Apr. 17, 1911, p. 2c; ||mr2rbb82):

> Over at Weston, Oregon, a freak colt on exhibition at Dr. Kennard's animal hospital is different from anything ever seen in that section. The mother of the unusual animal is owned by G. W. Staggs.
>
> The colt is the size of an average dog and has a well-formed body up to its neck. Here the resemblance to a young horse ceases, for the most part, as the neck, skull and nose are much more like those of a human being. The neck is round and slender while the skull is nearly round with but one eye, that being in the center of the right side of the head. The lower jaw resembles that of a horse, but the upper is more like that of a human. The only hair on the colt when born was on the upper lip.

Two centaur ads from the 1890s. *Left:* From the *Ottawa Journal* (Ontario, May 26, 1894, p. 8; ||y6rg7fq); *Right:* From the *Buffalo Times* (Buffalo, N.Y., Mar. 13, 1898, p. 10; ||y3c4mgoj).

Grand Forks, North Dakota. From the *Ward County Independent* (Minot, N.D., June 14, 1905, p. 9b; ||y6j6svg7): "Mr. Reddick, from Grand Forks, is at the Booth building in Minot this week with a most interesting exhibit of freaks. He has a hatted [*sic*] colt which weighed 202 pounds at birth. The colt seems to have a human head, covered with human hair. The skull appears to be like that of a man."

Redding, California. From the *Red Bluff Daily News* (Red Bluff, Ca., Apr. 2, 1915, p. 4d; ||2p982ukv):

> REDDING, April 1.—A colt with the features of a murderer stamped upon its deformed face was born early Wednesday morning at the ranch of John Lutz, the Stillwater stockman. The animal had but one eye[17] and that was prominently set in the middle of the forehead. There was no indication of nostrils, the colt taking its breath during its life of a few hours through a strangely formed mouth, the under jaw of which protruded nearly two inches.
>
> Persons at the Lutz farm say that the colt's revolting features bore an unmistakable resemblance to John M. Level, a cripple of unpleasing countenance, who slew Joe Pareenti on the Lutz farm last summer. Level is now in the penitentiary.
>
> The story is that Level was driving the deformed colt's mother when she was early in foal. Being a neurotic, he became wildly enraged at the mare, because she did not properly respond to his control of the reins, and, it is related, Level attacked the animal with his fists. He is then said to have choked her with his gorilla-like arms. It is evident that the features of the colt bear uncanny resemblance to the human, but distorted, malevolent and ghastly in their expression.

Thus, it is believed the colt was marked, receiving the frightened mother's impression of a brutal face, the remainder of the body being exceptionally well formed and developed.

In spite of the malformation, the animal lived until it was shot at 11 o'clock. Dr. G. C. Taylor, who was called from Redding, cut off the head and all afternoon the freak of nature was viewed by sightseers in this city who are interested in anything out of the ordinary, from a two-headed chicken to a six-legged calf.

Dr. Taylor says the freak is unheard of. He will send the head east for mounting.

Blooming Grove, Minnesota. From the *Herald-Advance* (Milbank, S.D., Jun. 14, 1895, p. 3d; ||y6fc9f6z): "WASECA, MINN., June 12.—Charles Fell[23] of Blooming Grove township is the owner of a two-year-old colt that is a monstrosity the equal of which is seldom seen outside of museums. The colt is well formed with the exception of its head, which very closely resembles that of a monkey. Mr. Fell said that when the colt was born, he did not think it would live long, but now it seems to be as lively as any other colt of its age, and nothing will apparently prevent its reaching a ripe old age. The colt is attracting numerous visitors to Mr. Fell's farm."

* Composites described as monkey-like would presumably also resemble a human.

HUMAN HANDS, HUMAN FEET

Camden, Virginia. From the *Times-Dispatch* (Richmond, Va., Aug. 26, 1909, p. 5g; ||y2tfqctw): "CAPE CHARLES, VA., August 25.—C. W. Lord, a farmer of Camden, has a few freaks of nature on his Jones Neck farm. A few days ago, a black mare gave birth to a mule colt that, instead of hoof feet, had four hands each with fingers. The colt was perfect in every other way and very healthy, but it could not stand up and died three days after birth."

Bloomingdale, Kentucky. The *Los Angeles Herald* (Jun. 7, 1897, p. 6d; ||hfxl2fh) said a "party near Bloomingdale reports to [t]his paper that a mare of Tom Adams has a colt whose body is half bay and half gray; that its feet look like human feet."

Spiš, Slovakia. Rzaczynski (1721, pp. 357-358) told how, in 1700 at Spiš, now in Slovakia, the prefect owned a "most noble and speedy horse," having on one leg "the sole, heel and toes of a human foot."[24]

SOME GENERALIZATIONS

Since nothing can exist that does not fulfill the conditions that render its existence possible, the different parts of each being must be coordinated in such a way as to render possible the existence of the being as a whole, not only in itself, but also in relation to its environment.

—GEORGES CUVIER, *Règne Animale* (1817, I, p. 6)

None of the reports in this chapter describe hominequids having a human body from the navel up joined to a horse body from the shoulders down, the standard form of centaurs in movie and myth, and such as shown at right. Instead, the case most commonly reported seems to be a human head and neck wedded to an equine body. But some reports describe an equine head and foreparts connected to human hind parts. Some report hands or human feet on a horse body, or hooves on a human one.

Centaur (*Heidelberger Schicksalsbuch*).

The incidence of cyclopean and anophthalmic births among hominequids is far higher than in non-telenothians, either horse or human. There were also several reports in which the offspring of a woman was so like an equid that it was simply referred to as such. Births of this kind, if they are not straightforward hoaxes, may perhaps be attributable to the creatures in question being hybrids that were so similar to their sires that no human features were apparent, or were so minor in comparison to the equine traits that the reporter failed to mention them (see XENOGENESIS, p. 1235).

But in such composites, it is not surprising that the various body parts, whether horse-like or human-like, tend to be arranged according to the standard quadruped pattern, especially in viable individuals. As Cuvier pointed out long ago, no organism can exist if its parts are not so coordinated as to permit that existence. Imagine, for example, the disruptive effects of concatenating two successive digestive systems, as the standard depiction of a centaur implies.

The Fall of Ixion by Hendrik Goltzius (1588).

MYTH

In one myth about the origin of the centaurs, Ixion was an evil king of the Lapiths who, when Zeus invited him to a banquet, lusted after Zeus' wife, Hera. Zeus soon became aware of Ixion's plans, made a cloud that looked like Hera, and tricked Ixion into making love to it. Having confirmed Ixion's guilt, Zeus bound him to a wheel of fire, which he then set spinning eternally through the heavens.

But the cloud nymph, Nephele, who Zeus had sent to deceive Ixion, did become pregnant. And so, Centaurus was born, a deformed child who, due to his disfigurement, could find no peace among humans and so spent his days among horses. There he mated with the mares and sired the race of centaurs, who lived in the mountains and ate raw flesh.

In later legend, at the wedding feast of Pirithous, the centaur Eurytion, tried to abduct and rape the bride. The centaurs were defeated in the ensuing battle and expelled from Thessaly. Thus, even in stories of their origin, centaurs were almost always associated with unbridled lust.

The "Monstrous Tartar." Many pamphlets and broadsides of the day depicted this hominequid with the neck and ears of a horse. He was supposedly an Ottoman archer captured by the Hungarian Count Nicholas von Serin at Bratislava (known in German as Pressburg) in February 1664. Royal Collection.

After his capture the "Monstrous Tartar" was supposedly held captive. In translation, the caption reads, "This is the tartar that Count Nicolas von Serin captured in 1664. Drawn directly from life by [the engraver] Copird in Pressburg." Pressburg was a former name of modern Bratislava, Slovakia.

ON THE VALUE OF NEWS REPORTS

The authority of those who teach is very often an impediment to those who desire to learn.

—CICERO, *De natura deorum,* 1.5.

Michel de Montaigne based his essay *Of Cannibals* in large part on the testimony of one of his servants who "lived ten or twelve years in the New World." He explained his reasons for trusting such information:

> This man that I had was a plain ignorant fellow, and therefore the more likely to tell the truth: for your better-bred sort of men are much more curious in their observation, 'tis true, and discover a great deal more, but then they gloss upon it, and to give the greater weight to what they deliver, and allure your belief, they cannot forebear a little to alter the story; they never represent things to you simply as they are, but rather as they appeared to them, or as they would have them appear to you, and to gain the reputation of men of judgment, and that better to induce your faith, are willing to help out the business with something more than is really true, of their own invention. Now in this case, we should either have a man of irreproachable veracity, or so simple that he has not wherewithal to contrive, and to give a colour of truth to false relations, and who can have no ends in forging an untruth. Such a one was mine; and besides, he has at divers times brought to me several seamen and merchants who at the same time went the same voyage. I shall therefore content myself with his information, without inquiring what the cosmographers say to the business.

Regarding the nature of hybridization, biologists stand in the same position of authority today as did Montaigne's cosmographers regarding cannibals in his day. Likewise, biologists have many preconceptions about hybridization and describe it accordingly. They discuss it at length, but "They never represent things to you simply as they are." On the other hand, from reporters, farmers, ranchers, obstetricians—people who are "plain and ignorant" when it comes to the phenomenon of hybridization—we get seemingly unvarnished descriptions of things they have seen. No doubt many biologists would propound on their opinions as to whether such things are possible. But in science, observation, not opinion, is sacred. Opinion, even expert opinion, is belief, not knowledge. Like Montaigne, "I would have everyone write what he knows, and as much as he knows, but no more."

NOTES

1. Public domain image (Metropolitan Museum; ||y5q5dsu6).
2. See Pliny, 3.1.8 (||yyp9un2n).
3. See: ||y6rvzlax.
4. *The Library,* 2.
5. Pliny, *Natural History,* 8.64. Translation mine. Original Latin: "Equum adamatum a Semiramide usque in coitum Iuba auctor est."
6. Compare the Egyptian satyr supposedly sent to Emperor Constantine (see p. 106).
7. Hansen (1996, p. 49).
8. §7; ||jjqbejf.
9. Image source: Schenck (1609, fig. 72).
10. Mussato (1536, liber 7, p. 52). My translation. Paré (1573, p. 482; 1641, p. 646), Liceti (1665, p. 183; citing Paré), Lycosthenes (1557, p. 438), Adlersflügel (1703, pp. 136-137) and Arcos (1786, p. 117) all dated this birth to 1254. Liceti (ibid., p. 184) cited Mussato for a second centaur at Verona, in 1315, but seemingly in error.
11. See: ||y6x589h4.
12. Reyes Franco (1661, 45.27, p. 325).
13. Bekmann and Bekmann (1751, p. 883).
14. Translation mine. Original Latin: "Bona fide mihi dixit M. Henr. Himmel, Ecclesiastes Mihlanus, se nosse puerum auribus asinis natum."
15. Voltaire (1768, pp. 120-121). Translation mine.
16. Leonidas L. Coleman (b. 1827 at Kinston, N.C., d. Jan. 1, Memphis, Tn.). See: ||y6th2jhf and the *Memphis Avalanche* (Jan. 2, 1876, p. 4c; ||4v4sf94).
17. Cyclopia and anophthalmia occur at elevated rates in telenothians (see p. 1211).
18. It seems that many human-animal telenothians are hairless, or nearly hairless.
19. *Der Humorist* (Vienna, May 5, 1847, p. 428a; ||y34xq3ol). Translation mine.
20. In the report, Whited's middle initial was given as M., but records (||vx73m7d2) show this was John Newton Whited (b. Jun. 16, 1844, Xenia, Greene Co., Ohio, d. Jun. 5. 1932, Piqua, Miami Co, Ohio). So this initial has been corrected here.
21. William Henry Whited (b. Nov. 2, 1873, d. Jul. 1, 1888 at North Star, O.).
22. See the *Daily Phoenix* (Columbia, S.C., May 25, 1870, p. 1a; ||yc6954e7).
23. A Charles Gottlieb Fell (b. Feb. 28, 1862, Rems-Murr-Kreis, Baden-Württemberg, d. May 30, 1948, Waseca, Waseca Co., Minnesota, USA) was living at Woodville Twp., Minnesota in 1900 (||yy9bemlr). Woodville is eight miles from Blooming Grove.
24. Translation mine. Original Latin: "cum planta hominis, calce ac digitis, in uno pede."
25. *Colophon:* Aldrovandi (1642; ||y3pvkvnc).

The suidovid Livy said was yeaned at Frosinone, Italy, in 198 B.C. (Lycosthenes 1557, p. 135).

SUIBOVIDS (PIG-CATTLE)

And so if, as you recently told me, an extraordinary lamb with a pig's head was born, then it is because a boar mated with a sheep.
—POLYDORE VERGIL, *Dialogues on Prodigies,* 3.26†

Call a mix of a pig and a bovid a suibovid. Bovids include cattle, sheep and goats. The mothers mentioned in such reports are usually bovids. Since Bovidae and Suidae are families of the same order, Artiodactyla, suibovids are not, strictly speaking, telenothians. However, as artiodactyl-human composites are so numerous, a chapter on suibovids is a suitable addendum to the section on homungulates.

SUIDOVIDS (PIG-SHEEP)

It's well documented that sheep and pigs sometimes will mate (videos: ||249styej). Indeed, even the ancient Akkadians knew that pigs and sheep sometimes do (Freedman 2017, p. 6). It's a common barnyard occurrence. Although many reports exist about such animals being farrowed by sows, a few say they can also be yeaned by ewes. The direct production of such a creature from a ewe mother seems to preclude the idea that animals like the one pictured above are simply a breed of woolly pig (the "Mangalitza") obtained without recourse to hybridization. So, a few reports about ewe-yeaned suidovids will be cited before discussion proceeds to animals produced with the cross in the more usual direction (sow × ram).

†. My translation. Original Latin: "Ac propterea nuper narrasti agnum suillo capite insignem natum, quod verres iniverit ovem."

Chambersburg, Pennsylvania. From the *Telegraph* (Harrisburg, Apr. 4, 1911, p. 2): "CHAMBERSBURG, PA., April 4.—J. Frank Huber, a farmer near town, has in his possession a freak which is attracting wide attention. A few days ago a sheep gave birth to an animal that is part hog and part lamb. The body is covered with wool with the exception of the legs, which are of the hog family. The head is part hog and part sheep. The animal has but one eye and one ear."

Monroe City, Missouri. The *Moberly Monitor* (quoted in the *Chariton Courier,* Apr. 3, 1908, p. 6b; ||kwmnat7) said, "Geo. A. Hampton, one of the large and successful farmers of east of Monroe City, has a ewe that has dropped a pig. The offspring had the face of a sheep, ears of a pig and was covered with wool back to the shoulders. From there to the tip of its pig tail, it was covered with bristles. The legs were those of a lamb, but the feet, those of a hog. In color it was red and black. The only record of a cross of that nature is the reverse. Several years ago the BREEDERS' GAZETTE reported a sow dropping a lamb."

Marshfield, Wisconsin. From the *Minneapolis Journal* (Sep. 24, 1902, p. 13; ||gtdvbw9):

> MARSHFIELD, WIS.—A farmer living six miles east of this place is the possessor of a freak of nature that can well be called half sheep and half pig.
>
> The combination is one year old and weighs 200 pounds. While it was born of sheep parentage and is covered with a thick growth of wool, its feet, legs and neck have every appearance of the swine family, and its grunt and squeal are proof that the animal is half hog. It is attracting much attention and will be brought to this city and put on exhibition during the street fair, Sept. 25, 26 and 27.

Julianges, Lozère, France. L'Union des Gauches (Mende, Jun. 30, 1929, p. 1f; ||492h7ukd) said, in translation, "JULIANGES.—A FREAK.—Pourcher Piere, of Mazet, found a young lamb in his sheepfold this morning that was certainly unique for its species. Its eyes and ears were under its head, and that head ended in a pig snout with large nostrils but lacked a mouth. Otherwise well formed, it soon died, as it was unable to eat."

Sachsenburg, Germany. Gottfried (1710, p. 1128) said a ewe yeaned a suidovid with pig foreparts and sheep hindparts at Sachsenburg in 1603.

Frosinone, Italy. Livy (31:6) recorded a ewe at Frusino (modern Frosinone) yeaning a lamb with a head like a pig in 198 B.C.

Foster Dwight Coburn
1846-1924

Mexico. In his book *Swine in America: A Textbook for the Breeder, Feeder & Student,* F. D. Coburn (1916, p. 63), a farmer and government official who served as secretary of the Kansas Department of Agriculture, said that a hybrid known as the cuino, a cross between sow and ram, was bred in Mexico. This assertion is corroborated by reports quoted below.

Huautla, Oaxaca, Mexico. In an agricultural journal, Roger de La Débutrie (1900, p. 513) reported on cuinos.

> In a recent letter … I mentioned the cuino, the product of a cross between a ram and a sow. Requests for particulars as to this animal have been received, and I give them with pleasure. The information, however, must be very incomplete, I know, and unsatisfactory on many points, but I beg to be excused on account of the great difficulty of gathering data in this country [Mexico], the inability of most people to give satisfactory explanations from want of knowledge or observation, and the lack of any attempt to experiment in any form.
>
> The cuino (pronounced queeno) is not considered a special race of hogs, but a hog of any race which has been improved, refined, by the introduction of sheep's blood. I do not know if the cuino is bred in all Mexico, but he is to be met with all over this State, in the mountain districts as in the extensively cultivated valley of Oaxaca.
>
> The ram used for crossing on sows belongs to the ordinary well-known, long-legged, light bodied Mexican sheep. He must be polled, and if not, he has to be dehorned for fear he should hurt the sows. No pains are spared in his training, which begins at an early age; in fact, as soon as the lamb is weaned. When able to eat, he is taken away from his mother and the flock, and from that time forth has to live entirely in the company of hogs, never being allowed to see any of his own kind. He is pastured and fed with the drove of pigs and sleeps with them at night. With such an education and training he cannot but believe himself a pig as well, and, when the time comes, has no objection to serving the sows. When the sow is in heat, he is shut up with her in a narrow pen and left as long as judged necessary. If he is slow in serving her, he is given an aphrodisiac composed of roast maize and salt, which always settles matters. This roast maize and salt is also used for this purpose in Mexico with other animals like the jack when slow in serving. The sow used to produce the cuino belongs to any race, but, as a rule, to the razor-back, which is the more numerous. There is never any difficulty of her accepting the ram when breeding time comes.

> The progeny is a pig—unmistakably a pig; he has the form and all the characteristics of the pig, but he is entirely different from his dam, if she is a razor-Back. The sow in that case is, as you know, long legged, flat-ribbed heavy boned, with an immensely long snout. She runs miles from the ranch, foraging in the bush, where she is too often the prey of the leopard or the puma. She can put away bushels of corn without any appreciable effect on her carcass; in fact, I need not describe her, she is well known in your Southern States. Her pig is entirely different; He is round-ribbed and blocky, his short legs cannot take him far from his sty, and his snout is too short to root. His head is not unlike that of the Berkshire. His body is covered with long thick curly hair, not soft enough to be called wool, but which, nevertheless, he takes from his sire. His color is black, white, black and white or brown and white. He is a good grazer and is mostly fed on grass, with one or two ears of corn a day, and on these he fattens quickly. I have seen the litter of a mule-footed sow bred to a ram; her pigs were either single hoofed or cloven footed, about half of each kind.
>
> The cuino reproduces itself and is often crossed a second and third time with a ram. One of my neighbors, who breeds cuinos on his ranch, told me that it requires two or three introductions of ovine blood to produce a highly improved race of cuinos. I asked him if by breeding the cuino a number of times over again to a ram the progeny would not have more of the form and characteristics of the ram. He answered that the progeny would always be hogs. Then I made the objection that, in such a case, he might, after a number of crosses, have a hog practically a pure blood sheep. To that he made no answer, only laughed.
>
> Be it what it may, the cuino is the most popular breed of hogs in this State of Oaxaca and became so on account of his propensity to fatten on little food. … I do not suppose any of the readers of *The Gazette* will ever want cuinos, having such excellent breeds of hogs, but they may like to know something about the ways of their Mexican neighbors and the extraordinary cross that produces the cuino.—Roger de la Débutrie

Zacatecas, Mexico. In a geography of Zacatecas published by the Mexican Ministry of Development ("Secretaría de Fomento"), Velasco (1896) lists among the principle domestic animals of the state the Cuino Hog ("Cerdo Cuino"), giving it the name *Sus hybridus*, and stating, too, that it is "derived from the union of a ram and a sow" ("Proviene de la unión del carnero con la puerca"). Zacatecas borders Oaxaca on the south.

Hidalgo, Mexico. Claudio Dunning (1906a, 1906b) a resident of Hidalgo, a state just north of Mexico City, wrote to both a livestock breeding periodical and to a veterinary journal to say that it was "common practice" for Hidalgans to cross sows with rams.

> Is it generally known that a ram can be crossed on a sow and not only reproduce pigs, but that the get will reproduce its kind? I have been handling stock all my life as a breeder but never heard of such a thing until I came to this country [i.e., Mexico], eleven years ago, and for a long time would not believe it and did not in fact until I personally saw it demonstrated.
>
> It is a common practice here among the poor people that want to improve their hogs. The cross is much larger than the common breed of hogs and large for any breed and they are slow to mature, not being grown until they are two and a half years old. They grow very tall, and their hair is very curly and fine but otherwise they are just like the hog. The ram used is taken away from the rest of the sheep when it is small and not allowed to be with them again but kept running with the hogs all the time.

Miahuatlán, Oaxaca. Otón Efferd (1903), a physician practicing in Miahuatlán, published an eyewitness report documenting this cross. In translation its title was, "Hybrid between pig and sheep, which is known as the cuino in the Central Plateau." In translation it read,

> When I arrived in the district of Pochutla, Oaxaca, a few years ago, the inhabitants informed me that certain pigs, which differed externally from ordinary pigs, were the offspring of a ram mated with a sow, and that they were called cuinos.
>
> At first, I laughed at the zoological impossibility of this claim, and by my incredulity lost some of my scientific reputation among the inhabitants of this region. But accumulating, little by little, information about the reality of these crosses, I gained the opportunity of verifying the facts of the matter with my own eyes.
>
> These crosses are only produced in regions with a moderate climate, since rams cannot stand climates that are too hot. It's extremely hot in Pochutla. So the closest place to Pochutla where they produce these hybrids is Miahuatlán, which is a three-day journey on horseback.
>
> In March of last year, I asked for leave and traveled to Miahuatlán. Very quickly, I convinced myself there that these crosses are almost certainly real.
>
> The thing that convinced me was the fact that there are many individuals who do business renting their rams to the owners of sows. The Indians are such sage observers of daily life, and so tight with their money, that it is difficult to conceive of one of them paying even a single penny to these ram owners for nothing.
>
> The economic motive for these crosses is that "cuinos" fatten more easily than do conventional pigs.

When I got back to Pochutla I communicated my observations to a dozen laboratories, in both the old and new worlds, inviting them to send a specialist to carry out experiments and research with regard to these revolutionary hybrids. But they all replied that this cross was an absolute zoological impossibility, that I was the dupe of Indian fables and that it wasn't worth the bother of sending a specialist to verify things that cannot exist.

Having lost all hope of convincing a zoologist, I decided to carry out studies of my own. I resigned my positions at Pochutla and Puerto Angel and moved to Miahuatlán forthwith.

Six weeks ago, my research into "cuinos" began. To conduct these experiments in such a manner as science truly requires would take six months. However, I am curious to know what the comments of my colleagues in the Central Bureau of Mexico might be with regard to these hybrids. People here say that these crosses are made in many parts of the Central Plateau.

If any of the readers of this journal would like to assist me in this research, which will, if the results are positive, certainly revolutionize the theory of evolution, it would be much appreciated. (*From the School of Medicine of Mexico*) [my translation]

Los Angeles. From the *Los Angeles Herald* (Oct. 3, 1908, p. 12; ||hp6fcx2):

Miss Lina is registered at Eastlake [Zoological] Park. Miss Lina is half sheep and half pig, and represents an industry peculiar in Mexico, from which country she hails. The man who is familiar with Sonora will easily place her grunts and bleats, but the tourist will find Miss Lina a most interesting study.

The cross between a ram and a sow is very common in some parts of Mexico, like the mulefooted or solid hoofed hog, but they puzzle the hog man of the corn belt, who is used to the Berkshire, Poland China or the red Duroc Jerseys. If [Zoo] Keeper Rice of the Eastlake Park zoo records the expressions of the eastern stock men who inspect Miss Lina, he will add considerably to his vocabulary.

At the Chicago stock yards this peculiar animal is known as a "cueno," [*sic*] but at Eastlake Park Miss Lina is registered as a jabolin.

"She looks more like a what-is-it," said Keeper Rice.

Miss Lina did not have a Pullman berth. She traveled in a biscuit box.

* Emmet Bedford Rice (1878-1939), an animal keeper at Eastlake Zoological Park, was a lifelong Los Angeles resident (||3cyj3k3e).

Nevada, Missouri. From the *Herald Democrat* (Leadville, Co., Nov. 30, 1921, p. 5b; ||ybfgqnny): "NEVADA. MO. Nov. 29.—Prof. S. A. Woltmor will be able to eat pork and mutton at one time when he kills an animal he now has in his fattening pen. It is a cross between a sheep and a hog, a present from

Colonel Goodnight of Goodnight, Texas, known as the Burbank of the animal kingdom. The animal is more "sheepy" than "hoggish" in appearance, but a close view reveals the ears and feet of the swine. The animal was produced so that it would have the fighting qualities of the hog and could protect itself from the attacks of the coyotes of the Texas prairies. It, however, is not vicious like the wild hog, but will defend itself with full strength when attacked. Its coat of wool is heavier and considerably coarser than that of the ordinary sheep. When fully grown the animal is expected to attain a weight of 200 pounds. The meat is said to possess a particularly pleasing flavor."

* Suidovids have dark red meat like a sheep's, not light-colored meat as in ordinary pigs.

Caruthersville, Missouri. The *Caruthersville Press* (quoted in the *Ripley County Democrat*, Doniphan, Mo., Aug. 7, 1908, p. 1f; ||kxukq78) said, "Tobe Baird, living north of this city, is a much-puzzled man. He has a fine brood sow that is the mother of eighteen of the strangest looking pigs that ever visited his barn lot. They are all coal-black and have long wool and short thin ears like sheep. Their noses are somewhat similar to other pigs, though he has never seen them root in the ground as other hogs do. He is inclined to think that a freak of nature has brought into existence a family of strange and mysterious beasts that have never lived upon the earth before."

Lincoln, Wisconsin. The *Marshfield Times* (quoted in the *Daily Tribune* Wisconsin Rapids, Oct. 1, 1902, p. 1) said, "One of the most remarkable natural curiosities ever discovered in this country was unearthed in the town of Lincoln this week and will be exhibited here during the carnival. The anomalous animal is a cross between a sheep and a hog, showing some of [the] characteristics of each, and has been named by the present owners a Schaf-Schwein. The curiosity was raised by John Anderson on his farm six miles west of the city and is a year old. It resembles a sheep more than a hog in build, but is much larger than the ordinary sheep, weighing about 300 pounds. The animal is healthy and lively, has the body of a sheep and the head and limbs of a hog and grunts like a hog, though covered with a heavy coat of wool. The Schaf-Schwein is now in the hands of a Marshfield syndicate and will be exhibited in a tent as one of the star attractions of the fair."

Middletown. New York. The *Binghamton Republican* (quoted in the Ogdensburg Journal, Ogdensburg, N.Y. Jan. 6, 1888, p. 4 ; ||keypr4k;) said, "A farmer near Middletown, has an animal said to be formed like a pig, but which has the horns and wool of a sheep."

Chenailler-Mascheix, Corrèze, France. Le Populaire du Centre (Limoges, Oct. 2, 1952, p. 2c; ||mr26ka6x) said, in translation, "CHENAILLER-MASCHEIX.—A sow belonging to Mr. Pouyade, at Nestève, has farrowed a truly curious piglet. The animal, which did not survive, had two mouths and the four legs of a sheep."

Bordeaux, France. Courrier de Tarn-et-Garonne (Montauban, Mar. 10, 1852, p. 2c; ||wc62u3hy) said, in translation, "A SHEEP-PIG.—A freak animal, which resembles at the same time both a pig and a ewe, the one by its snout and body structure, the other by its fleece, arrived day before yesterday at Bordeaux, where it was put on sale in the market on Rue Pont-Long. … This animal is completely covered in wool."

Ars, Creuse, France. L'Indépendant de Boussac (Mar. 17, 1892, p. 2c; ||bdj8ykvy) said, in translation, "ARS.—A FREAK.—A ewe belonging to Mr. Barre, one of Madame Dumiral's farmers, just gave birth to a remarkable freak with the body of a sheep and the anterior parts of a pig. The monster has one eye in the middle of the forehead."

SUIVACCIDS (PIG-COWS)

Rockford, Illinois. The *Evening Gazette* (Monmouth, Ill., Mar. 20, 1886, p. 2a; ||3d2xhama) said, "ROCKFORD. ILL., March 20.—A most extraordinary freak of nature is a species of a hybrid or cross between a hog and a calf, to which a cow belonging to S. Comings gave birth Thursday. In everything, but the head, which was that of a calf, it exactly resembled a hog. It was born dead, and the local scientists have taken charge of the carcass. It is thought that a hog which was confined in a pen adjoining the barnyard frightened the cow with this strange result."

* The *Chicago Tribune* (Mar. 19, 1886, p. 6; ||4upxukre) added, "It was born dead and weighed 40 pounds. Its hair was something between hog-bristles and cow's hair."

Clonkelly, Tipperary, Ireland. The *North Wales Chronicle* (Gwynedd, Wales, Jul. 12, 1856, p. 5b; ||4z3xeurn) said, "One of the most extraordinary instances of lusus naturae took place a few days ago on that portion of the celebrated moving bog of Dundrum, near Clonkelly. A fine cow belonging to a farmer in the neighbourhood gave birth to a calf, which had two mouths, one resembling the snout of a pig; each limb was double jointed, and what was still more remarkable, the animal had but one eye placed in the center of its forehead. The cow was ten days and twelve months in calf."

Decatur, Illinois. The *Decatur Herald* (Jul. 12, 1908, p. 1g; ||5n7kas37) said, "A freak, part pig and part calf in its appearance, was taken from a cow Saturday afternoon at the Danzeisen packing house south of the city. It weighed 128 pounds. There was also a perfectly formed calf fully developed, weighing 28 pounds. The freak was put on exhibition Saturday night at the Penny Arcade in the south side of Lincoln Square."

Janesville, Wisconsin. From the *Rhinelander Daily News* (Rhinelander, Wis., Dec. 2, 1932, p. 3b; ||2zt5twyx):

> JANESVILLE, Dec. 3 (AP)—A strange creature which died soon after birth has attracted many curious persons to the L.E. Tiffany farm near here.
>
> Tiffany describes it as a "palf", explaining that it has the hide of a calf and the meat of a pig.
>
> "Take a look at the mother," said Tiffany, pointing to a dejected cow.
>
> "Two years ago she had an ordinary calf. A year ago she bore twins. Now I get this critter, half calf, half pig."

Honey Grove, Texas. The *El Paso Daily Herald* (Feb. 5, 1901, p. 6a; ||ycfcuk9n) said, "A freak of nature was found by J. C. Yoakum, near Honey Grove, last week. It was half pig and half calf, the head and body being that of a calf, while the lower jaw, legs and feet were those of a hog. It was dead when found."

Lindley, New York. From the *Ithaca Journal* (Mar. 11, 1927, p. 6a): "CORNING, March 11.—What in years to come may be a side show attraction was born to a cow on the farm of William Welty, a well-known farmer in the town of Lindley. The form [*sic*] was brought to Corning to be stuffed, it having lived only about 30 minutes after birth. It was a hog's head, a calf's face, hog's legs, short tail and a calf's body, with very little hair."

Hautapu, New Zealand. The *Waikato Independent* (Jul. 21, 1927, p. 4; ||y3blnnyf) said, "There was born on Mr. Carwell-Cooke's farm, Hautapu, yesterday a monstrosity, in the shape of an animal that was apparently partly calf and partly pig. This freak, born dead, the progeny of a three-year-old heifer, presented a most remarkable appearance. The top part of the head was like that of a calf, but the bottom jaw, eyes and ears were those of a pig. There was no neck, the head being set direct on the body. The front legs had no shoulders, and the tail, like that of a pig, was attached to one of the hind legs. The strange animal possessed both male and female organs. A number of neighbors who saw the freak state that they had never before seen anything like it, and it was an object of interest to all who saw it."

St. Albans, Vermont. The *St. Albans Weekly Messenger* (Apr. 11, 1873, p. 7c; ||5n82z8t9) said, "A few days ago, a cow belonging "to Mr. M[erill]. L. Jeffords [1832-1899], surprised her owner, by making an addition to his stock which was a novelty in the animal kingdom. It was a combination of calf and pig, though it had neither hair nor bristles. The head, body and legs were those of a calf, but the ears, hindquarters and tail were those of a pig. This pig calf at its birth was alive and in good health. ... Mr. Jeffords thoughtlessly had the animal killed."

North Montpelier, Vermont. The *Argus and Patriot* (Montpelier, Vt., Feb. 27, 1878, p. 3b; ||26scpuyh) said, "A cow belonging to Samuel Edwards, of North Montpelier, recently gave birth to a monstrosity. The body and nose were those of a calf, and red [in] color. ... Both fore feet and one hind have three toes, and the other but two. The upper half of the tail is red, and like that of a calf, with the rest of it the color and size of that of a pig, and the hindquarters also resemble those of a pig. No indications of sex could be discovered."

Brand, Tirschenreuth, Bavaria. The *Limburger Koerier* (Heerlen, Netherland, Aug. 2, 1926, p. 5a; ||y7f7an6t) said, in translation, "A calf with the head of a pig has been born in Brand. It is still alive."

Gmina Trzebnica, Poland. The *Breslauische Sammlungen* (1722, p. 436; ||4t599a58) reported a cow calving a nondescript with a pig's snout in Massel (modern Gmina Trzebnica) near Breslau in April 1721.

Goshen, Indiana. From *Billboard* (Dec. 14, 1907, p. 35; ||zbk3dtt): "FOR SALE, CHEAP. The most perfect FREAK ever shown. Half Calf and Half Pig, and three distinct bodies on one pig head: four ears, twelve legs and three tails. O. F. WILLIAMSON, GOSHEN, IND."

Thiergeville, Normandy. *Le Petit Journal* (Paris, Mar. 27, 1873, p. 3d; ||4kwf96we) said a Thiergeville cow had borne a calf with two pig heads.

Draria, Algeria. *Le Radical Algérien* (Algiers, Mar. 11, 1892, p. 2c; ||3rj234v4) said a cow at Draria bore a calf with a pig head and five legs.

Lebanon, New Hampshire. The *Independent* (Elizabeth City, N.C., Oct. 28, 1932, p. 6; ||yyk6z3nq) said, "A freak animal, half calf and half pig, is attracting wide attention on the farm of Frank Sanville, at Lebanon, N. H. The mother is a 26-month-old heifer ... It has the legs of a pig, eight inches long, with the front feet pointing right, and the hind feet in the opposite direction. The neck and head resemble that of a pig, together with the tail, but the body is that of a calf."

Berkshire, Vermont. The *Green-Mountain Freeman* (Montpelier, Vt., Apr. 30, 1873, p. 2h; ||ypcsayb8) said, "A cow belonging to Mr. M. L. Jeffords, of Berkshire, has made an addition to his stock which is described as a combination of calf and pig, though it had neither hair nor bristles. The head, body and legs were those of a calf, but the ears, hindquarters and tail were those of a pig. This pig calf at its birth was alive and in good health."

Eastman, Georgia. The *Macon News* (Macon, Ga., Oct. 21, 1921, p. 1c; ||ymw3m3m6): "EASTMAN, GA.—W. J. Daniel of this city, has ... a calf born a few days ago with no tail at all and with one normal eye, the other eye being similar to the eye of a pig, it being very small and shaped like the eye of a pig."

Lectoure, Gers, France. In a section reporting events in Lectoure, *La Voix du Peuple* (Auch, France, Sep. 15, 1912, p. 4c; ||4pzc3k7t) said, in translation, "A FREAK CALF.—A cow belonging to Mr. Nouvion, has given birth to a calf that in no way resembles others of its own kind. Its mouth, due to a deformity of the lower jaw, gapes constantly and is open up to the ears. The front legs are reversed, and the animal looks like a boar. Though hand fed, it seems to be doing quite well."

SEXUAL IMPRINTING

Young pigs are sometimes nursed by cows. Thus, the *Daily Record* (Pioche, Nev., Apr. 30, 1876, p. 2b; ||bvhb7ha4) said, "At Eureka mill, Carson River, a pig has taken to sucking a cow. It is said the cow has developed a great affection for the pig and shows much uneasiness whenever it is out of her sight." Such behavior can lead to sexual imprinting, which, in turn, often results in hybridization.

WANTED, to Sell, Remarkable Freak of Nature, part Pig and part Goat. Alive. Aged Ten Months. Very strong and healthy. Price, £4. A sure fortune to small Showman. TOM NORMAN, 12, Suffolk-street, Birmingham.

Source: The Era (London, Jan. 28, 1888, p. 18).

SUICAPRIDS (PIG-GOATS)

Henderson County, North Carolina. The *Western Democrat* (Charlotte, N. C., Jun. 25, 1867, p. 3d; ||gqpd3ce) said, "A Mr. Farmer, of Henderson County, informs us that he had born on his place an animal that was evidently half pig and half goat. The head and shoulders were hog, but the other part of the body was goat."

Moulton, Alabama. From the *Birmingham News* (Birmingham, Ala., Aug. 1, 1898, p. 4f; ||52ehnhcw): "Farmer Joseph Plunkett, living [near] Moulton, is the owner of a most remarkable freak of nature, or, more properly speaking, of a number of freaks. The freaks are described by those who have seen them as five animals given birth by a sow. Two of the litter are half goat, half pig, with head and shoulders and miniature horns of a young kid. Two of them half pig, half goat, the head and shoulders of a pig, while from about the middle the hair changes to goat hair, and the rear limbs are plain goat, with short tail and cloven hoofs. One of the litter is a plain, simple pig."

Mount Olive, North Carolina. The *Goldsboro Messenger* (quoted in the *Eagle,* Fayetteville, N.C., Mar. 2 1871, p. 2g; ||46kz5ns6) said, "Mount Olive has a sensation in a novelty of a litter of pigs … said to be part pig and part goat. We have trustworthy information from several gentlemen (eyewitnesses) that a sow of Washington Wynn … near Mount Olive, recently had a litter of seven "quadrupeds," two of which are pigs and the others, goats—only their head resembling that of a pig and the bodies have all the appearance, shape, hair and tail of the goat. At last accounts, all of them were living and doing well. The owner, we learn, has already been offered one hundred dollars for the sow and 'brutes.'"

San Juan, Puerto Rico. La Correspondencia de Puerto Rico (San Juan, Oct. 4, 1894, p. 2c; ||593e6395) said, in translation, "A FREAK.—We have seen a composite animal, a mixture of goat and pig, which will soon be put on exhibit. The skin of the animal is a pig's … the forelegs are those of a goat, the hind ones, of a pig."

CERVIBOVIDS (DEER-COWS)

Bernhausen, Germany. Rommel (1689) reported seeing at Bernhausen a hermaphrodite cervibovid with deer hoofs. It was calved by a cow.

Paris. Nicolas de Blégny (1652-1722), physician to Louis XIV, described (in Planque 1762, p. 511) a deer-cow mix he saw in 1698 at the St. Germain Fair. Overall the animal, a male, looked like a bull with the muzzle and pelt of a deer. In addition, de Blégny said, in the middle of its back it had a well-formed deer leg, which was attached to a penis about a foot long, also like that of a deer, with two testicles. There was a cavity of considerable size in the top of his head, and the end of his tail was forked.

* Many other reports about cervibovids have been collected elsewhere: ||ycxbzwhf

A felhominid (Zakariya al-Qazwini ||ycks8c8v).

Carnhominids

Even among men, monsters are sometimes found that remind us partly of a human being and partly of an animal.

—Paracelsus
De animalibus natis ex sodomia[1]

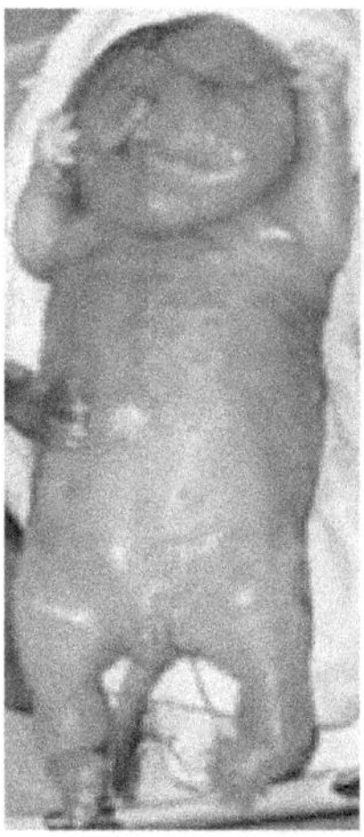

From a taxonomic standpoint, a carnivore is any mammal assigned to Order Carnivora. Examples are bears, dogs, cats, hyenas, weasels and mongooses. Seals, too, are usually assigned to Carnivora. Carnivora is, however, somewhat of a misnomer, given that not every kind of animal belonging to that order eats meat. Some, for instance, have a diet of insects, others, feed on fruits and other plant-derived foods. Giant pandas are members of Order Carnivora, but they eat virtually nothing but bamboo. At any rate, creatures mixing the traits of humans with those of carnivores will here be called *carnhominids*.

A felhominid reportedly born at Zürich (Wickiana collection MS F34, Bl 225r; ||4wrt8mn2). According to the report that accompanied this picture, it was borne by a woman at a hospital in Zürich, June 30, 1586, and had "a cat's head—the nose, eyes and mouth like a cat's—but the rest of the body, including the hands and feet, were like a child's" (my translation).

FELHOMINIDS (CAT-HUMANS)

> *The goddess Bast is usually represented in the form of a woman with the head of a cat.*
>
> —SIR WALLIS BUDGE
> *The Gods of the Egyptians*

Call any composite mixing the traits of humans and cats a *felhominid.* Nonfiction reports about such creatures are noticeably rarer than those of some of the more common kinds of HOMUNGULATES (e.g., BOSHOMINIDS or HOMINOSUIDS). They are sufficiently common, however, to establish that such births do occasionally occur.

Some of the earliest portrayals of felhominids represent Bast, a goddess worshipped in Egypt as early as the Second Dynasty, some 5,000 years ago. The form in which she was usually represented, a woman's body topped by the head of a lion or a cat, may have first been inspired by an aberrant cat-headed offspring borne by women. Quite a few records of women giving birth to such infants, dating to the modern era, exist. Thus, the following report describes a birth like the Zürich felhominid (depicted on the opposite page).

Bast (bronze, Egypt, Greco-Roman period, ca. 300 B.C.).

Lawrence, Massachusetts. From the *Lawrence American* (quoted in the *Orleans Independent,* Irasburgh, Vt., Jun. 10, 1859, p. 3; ||y4slcmxz):

> INFANT BORN WITH THE HEAD OF A CAT. One of the strangest freaks we have known recorded, occurred in this neighborhood last week, being the birth of a child with a well-developed body, perfect in its proportions and weighing ten pounds, but with a remarkable malformation, the head being formed in every particular precisely like the head of a cat. It lived but a moment after birth. The facts are vouched for by the attending physician, one of the highest standing in the city, and cannot be doubted.

Plau, Mecklenburg, Germany. Johann Friedrich Wilhelm Nevermann (1839, p. 291), a physician practicing in Plau, stated, in translation, that

"A woman in the first month of pregnancy went into the kitchen where a black cat ran between her legs and so frightened her that she fainted. I later delivered her of a child with a cat's head."

* Formerly, doctors widely believed that a mother's experiences could shape her baby in the womb (see MATERNAL IMPRESSIONS).

Owen, Indiana. From the *Indianapolis News* (Oct. 22, 1889, p. 1c; ||yapjde88):

> JEFFERSONVILLE, October 22.—Dr. G[eorge W.] Haymaker [1831-1901], of Charlestown, was called to Owen, Sunday, and assisted in the birth of a fearful monstrosity, a male child with a perfect cat's head. The mother was attacked some months ago by a vicious cat.[2]

London. British surgeon Joseph Hands (1879, p. 381) stated that "Mr. Catell—one of my pupils—saw December 3rd, 1853, the offspring of a woman, brought by herself—to fulfill the order of Mr. Furguson—to King's College. This child, in appearance, resembled a cat. The woman accounted for the malformation from the circumstance of her being frightened by this animal. Mr. F., whilst addressing his auditors, remarked that 'children were said to be angels, but this monster resembled a demon.'"

Budapest. Welt Blatt (Vienna, Aug. 26, 1875, p. 9b; ||yxwm4pge) said, in translation, "A woman living in Pest, the wife of a servant, gave birth to a boy whose head was exactly like that of a cat. Both mother and child are doing fine." ("Mutter und Kind befinden sich wohl.")

Milan. The *Wiener Zeitung* (Vienna, Apr. 25, 1772, p. 3b; ||k2c38es3), in a section covering current events in Milan, said, in translation, "Recently, a wicked woman living in St. Simon's Alley ["St. Simonsgasse"] has brought an extraordinary monstrosity into the world. Although, its entire body, indeed, had the usual structure of a human being, its head was like that of a cat. Fortunately, this horrible fruit of the womb soon expired."

Richmond, Virginia. In an account of the proceedings of the Richmond Academy of Medicine, Edwards (1876, p. 839) stated that Dr. George H. Bright, of that city, "delivered a woman of a dead child whose head in all respects looked more like that of a cat than of a human being."

Leoben, Austria. Deutsche Volkszeitung (Innsbruck, Oct. 28, 1926, p. 3a; ||yc75284v) reported that in a maternity ward in Leoben a woman had given birth to a stillborn nondescript with the head of a tiger in December 1925.

Antwerp. The physician Salomon Reisel[3] (1683, p. 274) describes a felhominid (see image, right) born Sept. 17, 1682, at Antwerp, "whose head and shoulders had skin, fur and whiskers like those of a cat, while the other parts of the body were like a human being."[4]

The Antwerp, felhominid (Schmuck 1683). Schmuck says this creature lived two hours after birth. For whatever reason, perhaps because they were trained to depict adults, sixteenth-century artists often depicted anomalous neonates as if they were older children, or even adults (as here).

Leiden. Thomas Bartholin (1616-1680), now primarily remembered for his discovery of the lymphatic system, recorded the birth of yet another felhominid. In a collection of unusual cases, Bartholin (1654, cen. II, p. 241) states that in the Netherlands, "At Leiden in 1638, near St. Peter's Church, a peasant gave birth to a child, which was in other respects normal, but with the head of a cat."[5]

Montpellier, France. Johann Helwig (1680, p. 12), who studied medicine at Montpellier,[6] said, in translation, "In 1633, a dead fetus was brought to Montpellier for investigation. It had been born in February of that year in a certain village in Provence. Of male sex, it had arms much longer than normal for its age and the genuine head of a cat, which in no way resembled that of a human being." He claimed to have seen it himself.

Pisarzowice, Silesia. Poland. Fincelius (1566b) claimed a woman at Schreibersdorf (now Pisarzowice) gave birth, on August 4, 1562, to a daughter with the face and claws of a cat. Irenäus (1584) also lists this birth.

HAS HISTORY BECOME MYTH?

All felhominid reports cited thus far describe beings that would have looked much like the cat-headed representations of Bast. One imagines that any composites of this sort, born at an early date in Egypt, would have been interpreted as avatars of Bast, just as births seemingly combining human and animal traits are seen as incarnations of deities in India today (see pp. 21-22). In *The Lord of the Rings,* Galadriel says, "And some things that should not have been forgotten were lost. History became legend. Legend became myth." Given the cases thus far cited, perhaps the Egyptians—with their stable civilization, which lasted more than three millennia, and their lengthy recorded history—remembered things that have now been lost, things that have now, as Tolkien put it, "passed out of all knowledge." Modern science, then, may be mistaking formerly remembered realities for myth.

Bast (bronze, Egypt, Greco-Roman period, ca. 300 B.C.).

In ancient times, Herodotus travelled over much of the known world, including Egypt, to question people about the past. He asserted that "Among the Egyptians themselves, those who live in the cultivated country are the most assiduous of all men at preserving the memory of the past, and none whom I have questioned are so skilled in history."[7]

But not all reports about felhominids describe creatures with cat heads and human bodies, as we shall see.

OTHER CONFIGURATIONS

Tirano, Italy. A manuscript report in the Wickiana Collection (||5dks2hm5) describes and pictures a felhominid borne by a woman at Tirano, June 6, 1568. It supposedly had a cat head, four human arms and a human thorax, four feline legs, and a feline tail and abdomen. The protruding tongue, it said, was as long as a finger, and the nose, human.

The Tirano felhominid (Wickiana Collection; ||5dks2hm5).

London. A pamphlet reports a cat giving birth to a conjoined felhominid that had in place of a head, a large cyclopean eye (Anonymous 1645, p. 5):

> In Lombard Street in London, in a Goldsmith's house, a Cat brought forth a monstrous Kitten, which was partly shapen like a humane creature, and the other part Monster like. The manner of the forme was thus: It had eight feet upon one body, and two tayls, his pawes were like a childes hands, and never a head, but one great eye placed in the body where the head should be.

Wetwang, Yorkshire, England. Chapman (1657) gave an account of "three strange and prodigious things like young cats, all speckled, which came from a woman dwelling at Wetwan [*sic*] in Yorkshire," but he provided no details, except to say that all three of these nondescripts died soon after birth.

XENOMELIA

Amsterdam. Hansen (1681, p. 51) described a specimen in an Amsterdam cabinet of curiosities. It was like a child, he said. but had atop its head the head of a cat. The owner supposedly saw it alive.

The Amolaradiema felhominid (Liceti 1634).

Amolaradiema, Italy. Fortunio Liceti (1577-1657) spent most of his long career at the University of Padua. A friend of Galileo, he authored a veritable flotilla of scholarly books, most within the fields of medicine and natural philosophy. His interests included the topics of generation, development, and teratology.

Perhaps Liceti's most famous production is his *De monstrorum caussis, natura et differentiis*, a remarkable tome that attempted to assemble all reports of abnormal human development. In it, Liceti (1634, pp. 193-194) reports and pictures a cat-human telenothian conjoined twin in which the human portion was parasitic (see COMPONENTS, p. 1197). This birth supposedly took place at Amolaradiema, near Padua, in 1629. Liceti was then living in Padua. In translation, his report reads,

> Recently (July 1629), in the district of Padua, near Monselice [about 17 mi (28 km) from Padua], in a village known as Amolaradiema, the wife of a cowherd of Lord Bernard Grifalconi (as reported in the writings of the most illustrious Lorenzo Pignoria, and via a verbal communication from Martino Sandelli,[8] famous for his learning and excellence) gave birth first to a daughter and, soon after, to a monster that was a complete cat, from the hind parts of which hung down two complete legs, as well as the partial hips, of a human being. This human appendage it dragged behind it. The monstrosity was brought to the city of Padua and here thrown into the Bacchiglione River, where it sank from sight.[9]

In the present context, it's interesting that Landauer (1961, p. 156) says Liceti "expressed the belief that the gestation period of women is sufficiently variable in duration to permit the occurrence of hybridization with animals. He saw in the fusion of fetuses of different genera, developing in the same womb, a source of monstrous development."

Ryazan Governorate, Russia. The *Leipziger Zeitung* (Leipzig, Germany, Jul. 29, 1826, p. 1956a; ||yhvvnnmr) said, in translation, "In Ryazan a farmer's wife lately bore a child with a cat's head and paws. Stunned, the unlucky mother killed it and buried it in the garden. But she was seen and reported to the police, who dug it up. The district physician put it in alcohol and sent it to the Tsar in Petersburg."

Moultrie, Georgia. The *Cincinnati Enquirer* (Aug. 30, 1891, p. 17; ||4xrur65w):

> The little town of Moultrie possesses a natural curiosity in the form of a woman who has the ears of a cat. She is the wife of a well-to-do merchant, and exceedingly sensitive on the subject of her peculiarity, which she very successfully conceals by an arrangement of her hair. Her ears are somewhat larger than an ordinary house cat's, but of the same shape, and are covered with a fine fuzzy hair, very much like an animal's fur. They lie close to her head, except when she is excited, when they protrude and work back and forth. She claims that her hearing is not keener for her possessing such a peculiarity, though her relatives and friends assert that this sense is abnormally developed in her, and that in disposition she is distinctly feline.
>
> Physicians who have examined her declare this deformity extends only to the outer shape of the ear and that the internal organ is the same as any other human being. With the exception of this strange feature the lady is very handsome and even intellectual, well educated and very refined in appearance. [speculation omitted]

Sodbury, Gloucestershire. The British surgeon Joseph Hands (1879, p. 381) stated that "I once knew a Mr. Wright, of Sodbury, who was known by the name of 'Cat-eyed Wright.' The pupils of the eyes of this gentleman were elongated like those appertaining to the domestic cat."

Naumburg, Saxony-Anhalt, Germany. Philipp Salmuth (1648, cent. II, p. 66, obs. 15), a physician in Saxony-Anhalt, wrote, in translation, "We see all too often how Nature can err in shaping fetuses. Indeed, I often think of Mittwoch, a boy in the Schulpforta [a school near Naumburg], who had a cat's paw for his left hand, a sad sight to see."

Rust, Michigan. The *Crawford Avalanche* (Grayling, Mich., Aug. 21, 1890, p. 3c; ||m59y6yj) stated that "Andrew Smith a farmer living in the above township [Rust had been mentioned] is the possessor of a cat that has human hands on her front legs, as perfect as can be, and while walking passes one over the other, which is undoubtedly a freak of nature."

* The *Helena Independent* (Helena, Mt., Jun. 24, 1891, p. 2; ||kvpv5gv) claimed "a cat with a human hand" appeared in a Boston cat show. See, too, the parallel cases describing a boy with a bear's paw in Jöhstadt, Saxony, a boy born with one dog's paw in Medina, Ohio, and the numerous JUMARTS with hoof replacements.

FELINE BODIES

Teresina, Piauí, Brasil. El Correo de Cantabria (Santander, Sep. 14, 1892, p. 3b; ||hfem4zjt) gives an account of a living felhominid on exhibit at a theater, probably in Teresina. According to the report this living female nondescript was "half woman, half lion" ("mitad mujer, mitad leon") and was caught in the wild. She was like a woman above the waist, per the report, but from the hips down, like a cat (presumably, a jaguar, given the locale) with a long tail. Judging from the description given, her general conformation would have been rather similar to the felhominid shown at right.

A felhominid
(Aldrovandi 1642, p. 25)

Guthrie County, Iowa. The *Carroll Sentinel* (Carroll, Ia., Mar. 18, 1891, p. 3c; ||y38b3o44) said, "A Guthrie County woman gave birth last week to a wonderful monstrosity having the body of a perfect feline and the head and neck of a human being. The body is covered with an abundant growth of hair the color of a Maltese cat and is fine and glossy as silk. This strange deformity is said to have inherited a ravenous appetite for rats and mice and will subsist on nothing else, and, when the appetite is not satisfied, will growl and spit like a cat when angry. The cause of the birth of this strange creature is said to be due to the fact that about ten weeks before the mother was confined, she was bitten on the arm by a vicious cat belonging to a neighbor, inflicting a wound that was not entirely healed at the time of birth."

* Carroll is the seat of Carroll Co., which adjoins Guthrie. So, the proximity of the *Sentinel* tends to substantiate this report, which names no witnesses.

HYPERTRICHOSIS

Paraguay. A felhominid known in nineteenth-century Spanish-language reports as "La Mujer Tigre" (tiger woman) allegedly had thick feline hair over her whole body. Born in Paraguay circa 1867, per reports, she died at the age of 24. (*Boletín Mercantil de Puerto Rico,* San Juan, Dec. 23, 1883, p. 3b; ||3abdatnp & Jan. 7, 1891, p. 3a; ||uhvr5k5m)

Minnesota (no locus in quo). The *Holton Signal* (Holton, Ks., Aug. 26, 1891, p. 6; ||y2b3c8cl) quoted a dime museum agent as follows. "In

Minnesota a short time ago we came across a little girl of ten years whose body was partially covered with soft, black fur of about the same length of the coat of the ordinary house cat. Her parents were very ordinary country people and could give no explanation regarding the singular occurrence beyond saying the fur was first noticed when the child was about six years of age, when it appeared as a sort of down, and since has grown very rapidly. With summer much of it falls out, as a cat's does in shedding and near the approach of winter it grows heavy again."

El Paso, Texas. The *Philadelphia Times* (May 30, 1890, p. 5c; ||264bjh2t) said "HALF CHILD, HALF PANTHER.—STRANGE FREAK OF NATURE THAT PUZZLES TEXAS SCIENTISTS.—SPECIAL TELEGRAM TO THE TIMES.—El Paso, May 29.—A singular freak of nature is attracting much attention in this neighborhood. It is the one-year-old child [ergo, born in 1889] of a Mexican woman living a short distance from town, which child seems to partake more of the nature of a wild beast than of a human. It has the curved claws of an animal of the feline race, sharp pointed teeth and short, coarse hair bristling all over its body, which is of a peculiar brown and tawny hue. When hungry or made angry the creature emits cries that are exactly like those of a panther and is fierce and unmanageable, scratching and clawing savagely If any attempt is made to meddle with it. It subsists almost entirely on raw meat, which it tears to pieces, growling and snarling like a wild animal at anyone approaching it while it is eating. During the day the creature seems to lie half asleep, coiled up in a bed of straw, but at night prowls about the house, and if allowed will make for the woods. It is possessed of incredible strength, leaping easily distances that a strong man would find impossible. The features of the strange being are good and its expression at times very intelligent and human and even melancholy, its eyes being large and speaking, but if noticed or approached at such moments will assume a mischievous, impish look and begin leaping and grunting. While perfectly able to walk erect it seldom does so, but prefers to run about on all fours, which it does with the greatest ease and activity. It has never spoken an articulate word nor attempted to, but its voice when it screams is extraordinarily powerful and ear-piercing. ... It is a female and is now about the size of a well-developed child of three years of age."

Dülmen, Germany. Franz Wesener (1818, pp. 31-33), Kreisphysikus at Dülmen, heard from midwives about a child that had been born with a cat's head. Investigating, he found a three-week-old girl with a patch of shaggy black hair on the right side of her face. The mother blamed this hairy nevus on a cat that frightened her while she was pregnant.

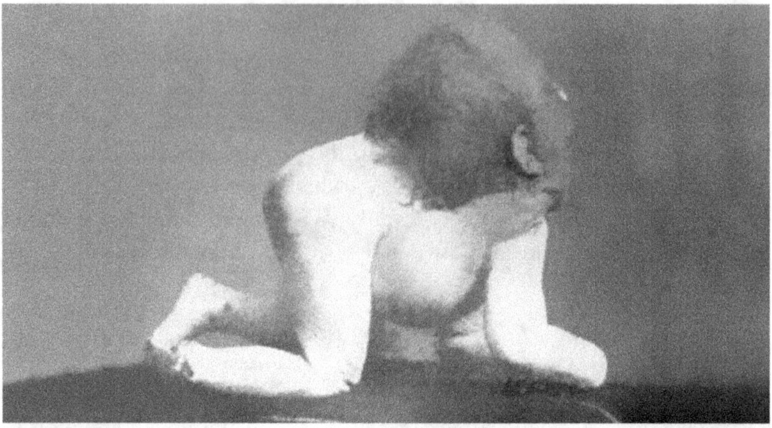

A nondescript borne by a woman in Pottsville, Pennsylvania, Aug. 20, 1889. Of it, Yocum (1891, pp. 155-156) said, "I will not describe the monstrosity very minutely, as the photograph … will explain it much better. The only deformity was the absence of the cervical vertebræ. The occipital bone resting on the lower portion of the dorsal vertebræ, being fixed at about an angle of 60 degrees. The other portions of the body is normal, with the exception of the face, which has a beastly, unnatural appearance. The weight is probably from seven to nine pounds. The specimen has been placed in the museum of the Hanemann Medical College. … Many persons who have examined the photograph say that the child resembles a lion."

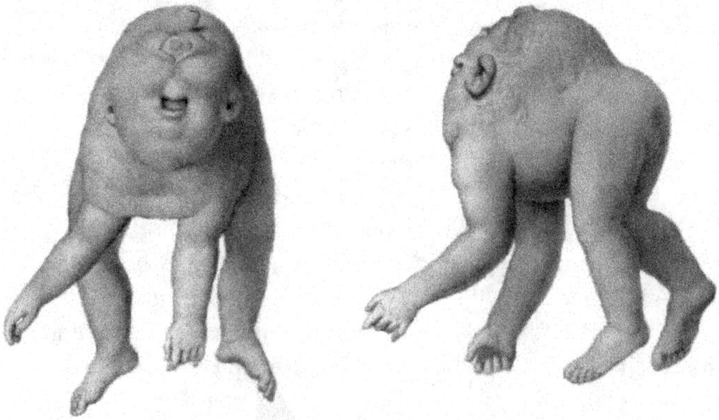

A cyclops with frontal proboscis born to a woman near Saint-Leonard-de-Noblat, near Limoges, France (Cruveilhier 1835-1842 pl. 6; ||fszj4zw8). Although the subject clearly resembles the Pottsville nondescript (pictured above), note that neither, despite Yocum's assertions, exhibits traits that are specifically feline. Both have therefore been classified as incertae sedis. An early pamphlet (Anonymous 1661) describes a birth at Ipswitch, England, on Oct. 24, 1661, which apparently resembled the births pictured here. Their structure links these specimens to the Ravenna carnhominid (see p. 274).

HOMOLEONIDS (LION-HUMANS)

Yet she, singing upon her road,
Half lion, half child, is at peace.
—WILLIAM BUTLER YEATS

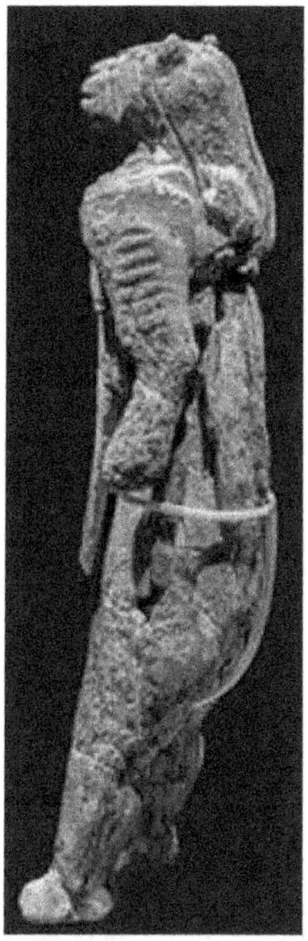

The Löwenmensch
(Thilo Parg; ||y63kms9p)

Call a lion-human composite a *homoleonid.* In 1939, at the Hohlenstein-Stadel cave in Germany, a figurine was recovered from sediments of Upper Paleolithic age. Carved from mammoth ivory, it depicted a human being with a lion's head. Carbon dating showed it was between 35,000 and 40,000 years old, making it the oldest known animal-shaped sculpture and one of the oldest examples, as well, of figurative art of any kind. Known as the Löwenmensch ("Lion-man"), this primordial statuette is now in the collection of the Ulm Museum in Germany.

But who would expect reports about a similar, viable being, living in the United States during the nineteenth century? And yet, on May 29, 1890, the *Greensboro North State* (Greensboro, N.C. p. 5; ||myww742) carried a notice about the death of Jay Hill, a 70-year-old "monstrosity, his head resembling a lion's more than that of a human being." And an expanded report appeared in the *North State* (quoted in the *Standard,* Concord, N.C. May 30, 1890, p. 1; ||mw2tsj7x):

> We have just received information that the life of one of the most remarkable men that ever lived in North Carolina terminated a few days ago at his home, about five miles north of Trinity College [near Greensboro].
>
> The name of this unfortunate being is Jay Hill, and at the time of his death he was nearly 70 years of age [therefore born circa 1820]. He has been an idiot from birth, but since infancy has possessed physical strength to a rare degree and was well formed in every particular with the exception of his head, which was so extremely ill-shaped as to give him more the appearance of a beast than a man.
>
> When but a child he was often so violent that it was necessary to chain him to the floor. Notwithstanding this unnatural existence he grew and strengthened day by day until his chain was not sufficient to hold him, and two or three times he broke loose and fled to the woods. During his brief liberty he was as violent as a tiger, and it was difficult to recapture and return him to his place of confinement. He was fastened with stronger

bonds, and with the exception of a few days he remained chained to the floor in the same room for about fifty years.

He had an insatiate desire for tearing to pieces anything that he could get, and at times it was impossible to keep him clothed. Flax was spun into coarse, strong threads and woven into cloth which was doubled and quilted and made into garments for him, but with his teeth and talon-like fingers he tore them into threads.

He has been known to tear to pieces a solid stick of hickory wood with nothing but his teeth and fingers. He ate as ravenously as a lion and could drink, without the slightest pain, boiling coffee.

On several occasions different members of the family narrowly escaped being killed by him. One evening a sister started to go out of doors. She passed too near him, and he sprang at her with great force, knocked her through the open door into the yard, so badly injuring her that it was not thought possible for her to live.

The writer has often at eventide heard his demoniac yells when more than a mile distant.

His father provided for him before his death, leaving property enough to maintain a wretched life through all these years. It was stipulated in the will that he should be kept on the old homestead.

Various early reports about homoleonids lack the Greensboro report's detail:

- According to Hannemann (1677) a woman at Buxtehude, Lower Saxony produced a stillborn fetus with the head of a lion.[11]
- Crouch (1682, p. 110) said a woman at Fenstanton, England, bore a creature that had "Mouth and Eyes like a Lion, which was both Male and Female" Sep. 23, 1580 (also see Anonymous 1580).
- Rzaczynski (1721, p. 350, citing Stryjkowski (1582, book 24) and Bielski (1597)) said a lion-headed babe was born at Brancovia, Poland, in 1527.
- Praetorius (1677, p. 302) said a child with an eye, paw and tail like a lion's was born at Salm an der Egra, Bohemia, in 1628.
- Lycosthenes (1557, p. 447) said a woman at Konstanz, Germany, bore a lion with a human head, in the year 1290.[12]

This photo shows that Lionella, a polydactyl stage artist, billed as a lion-woman and mentioned in previous editions of this book as a purported homoleonid, had no obvious leonine traits (Image: *Unu Revista Literara Lunara,* 1932, no. 46).

A homoleonid (Nineveh, ca. 640 B.C., Wikimedia, Carole Raddato ||bd6kv4ns).

The Guennol Lioness

Ancient civilizations frequently depicted homoleonids. The Guennol Lioness, a Mesopotamian limestone figurine with a lion's head and human body, is thought to have been created circa 3,000 B.C.—at about the same time as the invention of the wheel. The Sphinx at Giza is one of the oldest Egyptian monuments, dating to the middle of the third millennium B.C. It, too, may have been intended as a homoleonid, but in *Gods of the Egyptians*,[13] Budge says the form in which the goddess Bast was

> worshiped in the earliest times was that of a cat, and her identification with a lioness probably belongs to a comparatively late period.

Elsewhere, Budge (1902, p. 183) states that "The temple of Bast is at least as old as the time of Khufu, for this king's Horus name was found on a granite block which evidently formed part of the original building." Thus, felhominids are among the most ancient telenothians referenced in the annals of humankind. Nefertem was another homoleonid revered by the Egyptians.

* The twelfth-century scholar Gerald of Wales bemoaned the habits of a certain Parisienne, Johanna, who regularly made love to a lion (Dimock 1867, chapter 26).

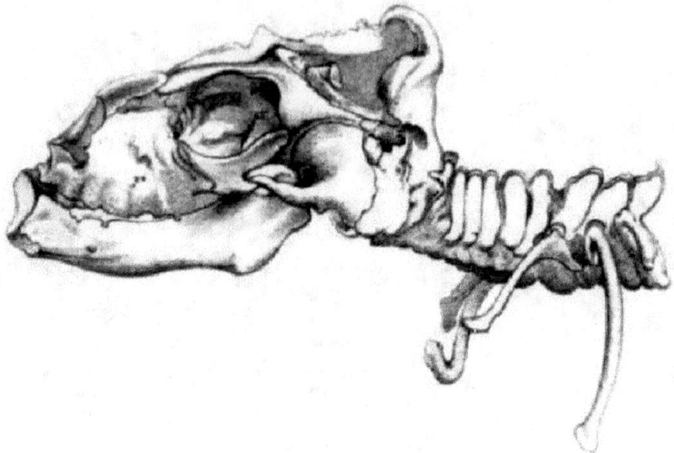

Skull of an infant born to a woman near Leipzig, Germany (Ahlfeld 1877, plate II, fig. 4). Ahlfeld (ibid., p. 160) said it was strikingly like that of "a domestic animal" ("fiel mir auf durch seine Aehnlichkeit mit dem Schädel unserer Haussäugethiere"). This case has been classified an incertae sedis carnhominid.

INCERTAE SEDIS

Basel, Switzerland. In translation, Lycosthenes (1557) said a creature with "a hairy head like a dog's, or indeed, more like a cat's or monkey's" was borne by a woman at Basel in 1556.

Maidstone, Kent, England. A broadside printed in London (Anonymous 1568) claimed that a certain Marget (sic) Mere gave birth, at Maidstone, to a nondescript with a mouth like a leopard's ("Libardes"). It lived 24 hours.

Stuttgart, Germany. Salomon Reisel[3] (1684, p. 272) reported a woman in Stuttgart birthing, May 22, 1682, a living, male offspring bearing a growth covered with fur like a gray cat's. This structure, he said, was like a scrotum attached to the back of the neck at its juncture with the head. It was larger than two walnuts placed side by side in their shells he said. Reisel suggested the cause might have been a neighbor's having placed a cat with fur of that color on the pregnant mother's neck (see MATERNAL IMPRESSIONS).

Brooklyn, New York. Samuel Hart (1860) reported five abnormal births, although he said the first four had feline heads, the actual descriptions he gave of these four neonates seem to fit better with the idea that they were HOMANURANS, as does the fact that a single physician delivered four such births in a single career, given that homanuran births are so common. Hart said these births occurred in 1837, 1843, 1845 and 1849.

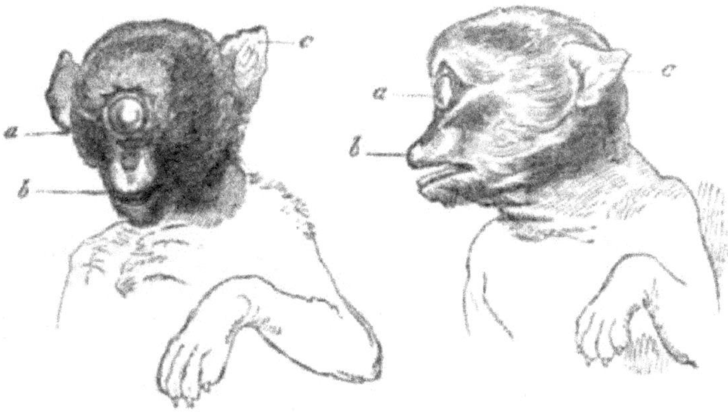

One of Westminster felpithecines (Atkinson 1833).

FELPITHECINES (MONKEY-CATS)

Westminster, England. In a letter to the *London Medical Gazette,* Westminster physician John Charles Atkinson[14] (1833) wrote,

> Sir, if you consider the subjoined extraordinary case of lusus naturae worth recording, you are at liberty to insert it in your journal; and should any skeptical person make application to your office, desiring to be further acquainted with the facts, you have simply to refer him to me, and he shall have ocular proof of the truth of my statement.
>
> On the 13th instant, a plumber's cat, in Horseferry Road, Westminster, at her *accouchement* brought forth three kittens, all exhibiting strong characteristics of the monkey tribe, as evinced by the general conformation of the head, the peculiarity of the hands or paws, the position and shape of the ears, the structure of the pelvis, and the facial angle. It would seem that either copulation had taken place between a monkey and cat [or that the deformities resulted from a MATERNAL IMPRESSION] From this digression I shall come more immediately to the notice of the accompanying sketches, which are given with great fidelity … Two of these kittens were furnished with no other eccentricities than what have been mentioned; but the third, a species of monoculus, forcibly brought to my recollection the mythological idea of the Cyclops: the single eye in this instance was placed in the middle of the forehead, as seen in fig. 1, a, with total obliteration of the nose.

Frankfort, Kentucky. The *Yeoman* (*Frankfort,* quoted in the *Louisville Daily Courier*, Mar. 22, 1850, p. 3b; ||2p8a5nva) said, "We saw the other day a singular freak of nature in the shape of a well grown kitten, which lived until it was three days old, having five legs, two well-formed tails, one eye and a head, lip and mouth precisely like the monkey."

WOMEN BIRTHING CATS

When a woman dreamed she had been delivered of a lion, this signified that the country in which she had the dream would be conquered by foreign nations.
—CICERO, *De divinatione,* 1.53.120

Old reports exist about women birthing cats, events that can perhaps be interpreted as births of catlike felhominids. The *Ipswich Journal* (Aug. 3, 1734, p. 2a; ||tdv6dcjr) reported the following.:

Andreas Elias Büchner†
1701-1769

> PARIS, AUG. 4, N. S. [the French date, per Gregorian calendar, which had not yet been adopted in England] There has been shewn here for some Days, at a Midwife's in Rue Aubribouche, a Monster which a Woman brought into the World two hours after she was delivered of a Daughter. That Monster is about seven Inches long, and is like a Lion, all his parts resembling his.

Büchner (1734, p. 576), referring to what seems to have been the same event, described this creature as a double-bodied, single-headed, conjoined-twin lion. He claimed it could be seen in that city, preserved in brandy, at the home of a midwife called Souches.

And Foresti (1535, p. 382) said a woman at Pavia bore a cat in 1471.‡ Stumpf (1548, pp. 240-241) said a woman at Sempach, Switzerland, gave birth to a lion in 1278.[10] A very early case appears in Paullini (1686), who published a manuscript collection of curiosities compiled by monks at the Abbey of Corvey. Observation III of that collection (ibid., p. 186) claims that in the year 874, a woman at Gadelheim, Hoxter, gave birth to a cat and, as a result, was burned at the stake.§

†. Büchner became president of the German Academy of Sciences (Leopoldina), in 1736.
‡. Both Franck (1536, p. 245b) and Liceti (1634, p. 216) mention the Pavia birth.
§. Also see Isensee (1843, p. 287).

NOTES

1. Translated in Waite (1894, p. 124). Original Latin (||sswpo3a): "Inter homines etiam nonnunquam monstra reperiuntur, quae ex dimidio hominem, ex dimidio brutem seu animal referunt."
2. For an obituary of Haymaker, see Graham (1901).
3. Salomon Reisel (1625-1701, in Stuttgart) worked as a city physician, court physician and personal physician in several locales (||m3x4bdxh).
4. Translation mine. Original Latin: "cujus caput & humeri quoad cutem, pilos & barbam felinis similes fuerint, reliquae partes humanae." The French physician François Planque (1761, p. 58), too, mentions the Antwerp felhominid. He cites the *Gazette de France,* without specifying a date
5. Translation mine. Original Latin: "*Lugduni Bat.* 1638. Femina vulgaris condicionis prope templum D. Petri enixa est foetum capite felino, caetera elegantem."
6. About Helvig's training, see: ||y4b2eld3.
7. *The History* (2.77.1; ||y4srm2ge).
8. Pignoria and Sandelli, two Paduan scholars, were friends of Liceti and Galileo. Both died of plague in 1631. Sandelli translated works of Galileo's into Latin (see Drake 1957, p. 84, and *Journal des Sçavans,* 1750, vol. 150; p. 136; ||opgsg7d).
9. Translation mine.
10. The *Nuremberg Chronicle* (ccxvii; ||ybyrek2k), first printed in 1493, says the same. This event was cited in later authors (e.g., Schenck 1596, p. 604; Delrio 1755, p. 148b).
11. Hannemann (in Bartholin 1677, vol. 4, pt. 1, observation 11, p. 37).
12. Sebastian Franck (1543, p. 228), too, mentions a homoleonid as being born at Konstanz, but dates it to the year 1274.
13. See vol. 1, p. 446; ||lrgcz2f.
14. Atkinson signed himself "J. C. Atkinson, 16, Romney Terrace, Westminster" (the address in an 1845 issue of *The London Medical Directory* for John Charles Atkinson, G.P. and Member of the Royal College of Surgeons of Edinburgh).
15. *Colophon:* Aldrovandi (1642, pp. 138-139; ||y2ghzh4q) said, in translation, that André "Thevet [1516-1590] claimed to have seen among the Indian merchants of the Red Sea an animal akin to a tiger called a Thanacht. In size, and in the conformation of its parts, it was not much different from a tiger, but the face was almost human, albeit the nose much flattened. It had forefeet, too, like human hands but the hind feet of a tiger. It had no tail and bore a dark coat of hair. And the head, ears, neck and mouth were human, and bore blackish, curly hair recalling an Ethiopian's. The rest was like a tiger." Thevet (1556, p. 52) said the merchants had brought it from Calicut (which lay in what is now India's state of Kerala). His depiction does not differ significantly from Aldrovandi's (shown below). Thevet traveled to the Sinai in 1552.

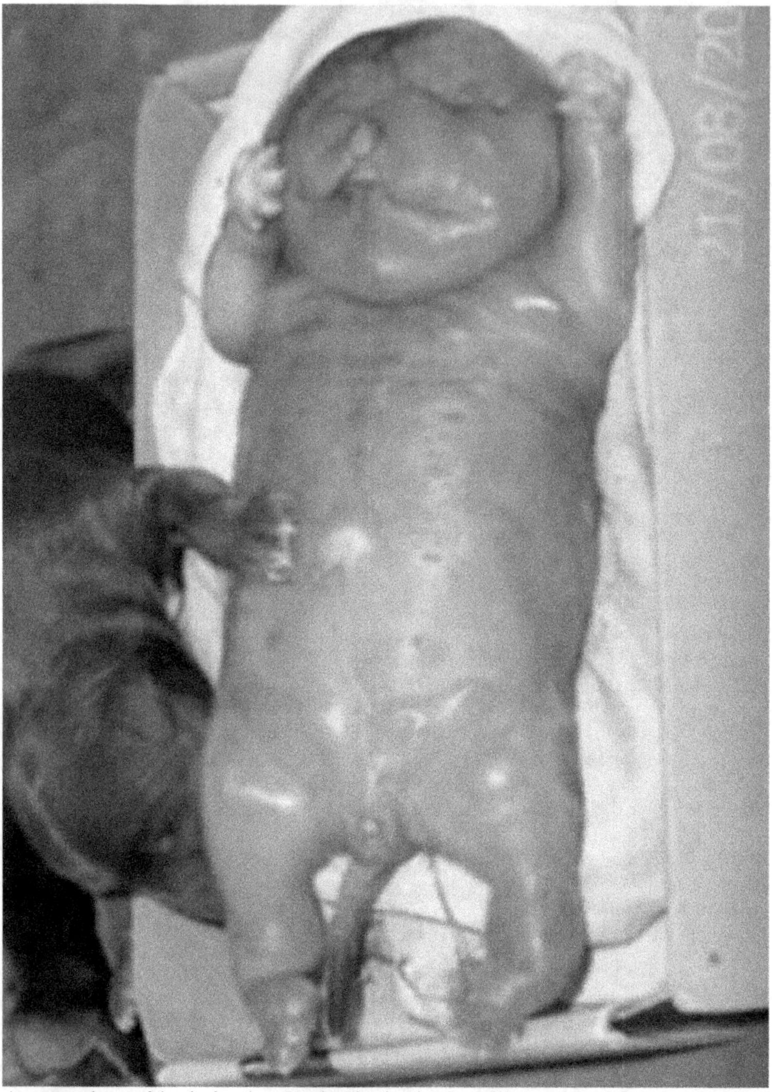

A cynanthropoid whelped at Umbaúba in the Brazilian state of Sergipe in 2007 (||yxzccq67). Note the tail and the dog-like paws, nose and mouth. A line of nipples is also visible. Human traits in evidence are the naked skin, general shape of the head and thorax, together with the way the appendages are attached. Nakedness is characteristic also of most BOSHOMINIDS. This individual appears to be anophthalmic, a condition commonly seen in telenothians.

Skull of a dog (left) compared to a 13-week fetus (right) miscarried by a woman (Haven 1897, p. 27). Dr. Joseph Haven (1897) described this fetus to the Chicago Pathological Society, calling it a "kynocephalous monster" (dog-headed monster). Haven (1886) said it was born in Chicago on Nov. 1, 1885.

CYNANTHROPOIDS (DOG-HUMANS)

If a woman gives birth to a child with a dog's head, the city will be in distress, and evil will be in the country.

—AN ASSYRIAN BIRTH OMEN[1]

This book attempts to include all serious reports about each different kind of telenothian, especially reports made by scholars. This policy was chosen in order to avoid at least one type of systematic reporting bias. The same is done here with reports about creatures combining the traits of dogs and human beings, that is, about *cynanthropoids*.

A cynanthropoid
(Schedel et al. 1493)

From ancient times to the present, creatures that mix the traits of dogs and humans have been repeatedly reported, often by sources that seem reliable.

Dr. Charles H. St. John (1843-1904) and his wife Eugenia were living in Kansas City in 1897 (||yyqteydg).

REPORTS

Kansas City, Kansas. Charles H. St. John, a Kansas City physician, reported the following case history (May 1897, 6(5), p. 148; ||ycfzuws4).

> Mrs. K------, a primipara, was delivered of a monstrosity in the month of March, 1897. It had the face of a pug dog without any development of the head and brain, with an imperfect right foot with five toes, being much the shape of a dog's foot. The period of labor was over twenty hours. There was an abnormal amount of fluid present. The sack had to be punctured twice before the liquor amnii escaped sufficiently to permit the birth of the child. Vertex presentation … sex female; perfectly developed, with the above exceptions; weight about six pounds; breathed but once after delivery as far as noted by the obstetrician. … The face of the child was a most perfect picture of the face of the pug dog that was and is now a pet of the household.

Goose Creek, North Carolina. The *Goldsboro Headlight* (Goldsboro, N.C., Aug. 3, 1893, p. 1g; ||2p8ap9fs) reported that a Mrs. Steve Watkins of Goose Creek, 25 years old, "gave birth to a child which died in a few hours. It had a dog's ears, its tongue lolled out like a dog's, it had hair on its neck and breast and its arms came out from its body where a dog's legs issue from the body."

London. In a list of cases seen in his practice, John S. Beale (1863), a physician practicing in London, included, "A child born with mouth and upper and lower extremities resembling those of a dog."

Warsaw. *Deutsche Volkszeitung* (Innsbruck, May. 27, 1927, p. 3c; ||bwvdz9rn) reported a woman at a Warsaw maternity hospital who bore a daughter with three noses and the face of bulldog but that was otherwise normal. The hospital's director had her head x-rayed and pronounced its structure canine. She was still alive after four days.

Franklin, Wisconsin. Dr. P. J. Hamilton (1849, p. 455), whose practice was in Franklin, said,

> I was called to see Mrs. R___, whom I found in labor with her fifth child. The pains were very strong when I arrived. The waters were discharged in a short time. I made an examination and was at a loss to determine the presentation. The pains being strong, I was not long kept in suspense. A child was born with a dog's head. In every other respect, it was perfect. The cerebellum and part of the cerebrum were covered with their proper membrane, the dura mater. A cartilaginous substance formed part of the os frontis. The ears were long, and lopped down, like a dog's. The eyes were very large, and in every respect resembled a dog's.

Separately, Hamilton (1848, p. 456) reported that a woman had given birth to a cynanthropoid in which "the superior maxilla was protruded far in advance of the other features of the face, and grinning like a dog. A friend of ours, at our elbow, who amputated the jaw, closed the hare lip and related to us the story, has the bone yet."

Cleveland, Ohio. The *Crestline Advocate* (Crestline, O., Dec. 10, 1896, p. 8c; ||bdfzkrxp) said, "A child was born recently in Cleveland and was perfectly formed from the neck down, but its head was shaped like that of a dog. Fortunately it did not live. [speculation omitted]"

Market Deeping, Lincolnshire. James Clapperton (1875), a physician practicing in Market Deeping, reported a child "born with a mark over the sacrum, which now resembles a dog." Whether the "mark" was a head, a tail, or some other canine part, he did not say.

Ravenna, Ohio. The *Cincinnati Enquirer* (1878, Nov. 18, p. 2; ||2p8hcnaf) published a lengthy story with the headline "THE DOG-BABY" from which the following excerpt is taken.

> In the jar on the table was the semblance of a human being, hanging suspended in alcohol, as is usual with such specimens. ... The entire length of the body is about thirteen inches. The superior and inferior extremities varied but little in length, a feature that would be extremely remarkable in a human being, but not at all among the four-footed animals. The face turned upward on an extremely short and barely discernible neck. The eyes looked downward over the cheeks. If placed in an erect position, the eyes would have turned heavenward; on all fours their direction would have been correct. The superior and inferior maxillaries were very prominent. The occipital bones were also very prominent, but the parietal bones appeared wanting, throwing the forehead back in close conjunction with the neck. The eyes are large, full, and of oblong shape. The nose is almost entirely missing, two small holes like a snout marking the place where it should be. The mouth is wide, and the jaws contain a full set of sharp teeth. ... The doctor says the child, or puppy, is at least at the seven-months stage. ... Investigation and inquiry at her house developed the fact that for nearly a year previous to her removal to Ravenna a large dog had been her constant companion, following at her heels and sleeping in her bedroom. The formations appearing like those of a canine on the monstrosity resemble this dog in appearance.
>
> * The doctors who conducted the delivery, Seth Lincoln Sloan (1858-1925) and Calvin L. Belden (1818-1894), both said they thought it was a dog-human hybrid.

Ottawa, Illinois. In a case report, Chester Perkins Hard (1883), an Ottawa physician (1829-1896), stated that

> On the night of September 8, 1883, I was called to the vicinity of Starved Rock, in this county, to attend Mrs. G., a German woman, in her second labor ... just at sunrise she was delivered of a very large female child, weighing 14½ lbs. The body was well formed and perfect, but the head was almost the exact counterpart of a bull pup and was the most hideous monstrosity I ever saw. The eyes were high

up on the forehead, large and round, and protuberant, and were two and a half inches apart, no eyebrows. The nose was long and flat, and continued to the mouth, with wide open nostrils; the distance from forehead to nose and mouth 3½ inches. The ears were small and long, standing out from the sides of the head one inch, and looked like the cropped ears of a stable dog; they were 3 1/2 inches apart. Back of the ears was a single tuft of coarse reddish hair, about one inch long. There was no back part to the head; or rather, no bones, but a simple sac containing a soft, semi-fluid mass. Taking the whole face together, the resemblance to a dog was most striking, and was at once remarked upon by all present, some of whom were anxious that I should dispatch it at once, but I allayed their fears by assuring them that the child would not live. It did live four hours, giving out faint moans from time to time. [speculation omitted]

* The classification of the Ottawa case is doubtful. From the description it might be better classified as a homanuran.

Mullinavat, Ireland. Dr. Norris-Cane (1872) described what may have been a cynanthropoid, stillborn at Mullinavat, August 9, 1872. "The occipital was a mere flat plate, of a triangular shape [as in dogs]," he said, "... and two canine teeth were found fully formed in the lower jaw. The facial angle closely resembled that of the canine species. The foetus was as large as a child four months old."

Brooklyn, New York. Samuel Hart (1860, p. 164, case 5), a physician practicing in Brooklyn, reported delivering a cynanthropoid.

The last case, gentlemen, which I shall present to you, and the last of the kind I have had the misfortune to meet with, occurred on the morning of February 1st, 1855. It was a fourth labor, and a full period. Uterine contractions were soon well established, and taking a seat by my patient, the membranes were found protruding and distending the vagina; they were at once ruptured, and a flood of waters followed as in the preceding cases. ... The presentation was puzzling and uncertain, exhibiting to the touch quite a different sensation from either of the other cases. From the extensive smooth, flat, rather than convex bony surface that could be reached, it was evident a head of some kind was presenting, and quite clear it was not that of my old acquaintance the cat. ... As the labor progressed, a new phenomenon appeared, which proved perplexing, and somewhat obscured the definite character of the presentation. A thin, moveable substance at the side of the part approaching, and about an inch in

> width, was discovered; this was soon found to be attached and pendant from it; but the child was soon born and dissipated all uncertainty. The head of this child closely resembled that of the dog, both in form and feature, and presented a most sad spectacle; in form, a dog's head upon a human neck and body; in every other respect perfect. The pendulous portion so troublesome previous to the delivery proved to be the ear, both of which, nearly two inches in length, together with the head and all the features, were strikingly canine in form. The child was quite small, and although it never respired, a slight movement or two of some of the limbs made it probable it could have been easily resuscitated, but this was not attempted.

* Samuel Hart (1796-1878) received baccalaureate and medical degrees from Harvard. He practiced in Brooklyn from 1828 until his death (see his obituary in the *Brooklyn Union,* Sep. 4, 1878, p. 2e; ||y3xlob8e). More about Hart: ||y2ndf5bv.

Elkin, North Carolina. From Elkin, the *News and Observer* (Raleigh, N.C., Feb. 14, 1897, p. 3c; ||2p954un8) reported that

> A woman of questionable character living in Yadkin County, seven miles from here, gave birth to twins one day last week. One was living and the other stillborn. The latter was in many respects like a dog. Its head, mouth and other parts had the distinct features of a dog and instead of arms there were dog's legs, with claws on the end. It is certainly a remarkable case, but it is said by people who saw it that it resembles a dog more than a human. The mother of this monstrosity is a loose character and has given birth to other illegitimate children. We learn that a physician has secured the dog child and has it preserved in alcohol. The other child is alive and doing well and has none of the features of its canine brother.

Danville, Kentucky. From the *Enterprise-Courier* (Charleston, Mo., Mar. 9, 1894, p. 1; ||2wcs82w5): "A white woman living twelve miles from Danville, Ky., has given birth to a monstrosity in the shape of a child with a well-formed head of a dog. It is not a mere resemblance, but an actual formation."

Człuchów, Poland. Adler (1855), the district physician (Kreisphysikus) in what is now Człuchów, reported delivering twins that survived 18 hours and "had oddly flat occiputs, twisted bones in arms and legs formed almost just as in dachshunds, and big, flat hands and feet, whose volar surfaces faced forward and lay in a plane with the abdomen when the children were laid on their backs" (my translation).

William Dickinson
1822-1894

Georgia. Physician William Dickinson (1884) said he saw "a publication in the papers in regard to a child in Georgia that had a dog's head. The name of the physician was given, and I wrote to him out of curiosity to know what he had to say about it. He replied as follows: "Dear Doctor—In regard to the account you speak of, I must say it is a great wonder indeed, but as to its being part dog and part man, I don't know what to say to you. I will just give you a short account of the case and let you judge. The child is from an unmarried woman,

> her first, a male child. It presented pedalic extremities first, and flexed upon themselves, and with a good effort I succeeded with great difficulty to deliver, and when I delivered it, I found it perfect male up to the head and face. It has no mouth, no chin, and just where the mouth and chin ought to be, there is a something extending out like a dog or fox, or something of that kind, and its ears are down, under its jaws, or where its jaws ought to be, on its neck. Its eyes are not square across as usual, but diagonal, ranging from outwards and upwards at an angle of about 45 degrees. The nose rested upon this, the human nose, and in the center of this strange nose or proboscis, is a little round opening about as large as a knitting needle. Now to elevate the face or head, and look it in the face, it does not look like a human. and has some appearance of an animal, favoring a fox about as much as the dog, but the features are not that of a perfect dog. I must say I never saw anything like unto it in all my days. It is one of the great wonders of the age. … I cannot learn who or what was its father, as she is not very communicative upon that subject.

Elbridge, New York. Elbridge physician Charles Ovid Baker (1879) reported delivering a boy with "a deformed hand, bearing some resemblance to a dog's foot. The fingers are shortened and joined together, and only a rudimentary thumb."

Edinburgh, Scotland. Dr. Andrew McLennan Thompson Rattray (1892), of Edinburgh, described his delivery of a cynocephalus: "The quantity of liquor amnii was very great. The face of the child exactly resembled that of a pug dog. Fortunately, it was stillborn."

Edenton, North Carolina. Edward Warren (1854, p. 348), a physician practicing in Edenton, wrote that "There is a negro boy in this town, aged about ten years, in good health, and quite well grown, whose countenance bears a remarkable resemblance to that of a fox. The likeness is so great that it strikes every observer at first sight and attracts the immediate attention of all who see him. But this

Edward Warren
1828-1893

> is not all: he walks and runs habitually on his hands and feet, like a quadruped and is more active than most boys of his age who use their limbs in the ordinary way. He is solitary in his habits, shy in his manners and of a cunning and roguish disposition. In almost every particular, some resemblance to a fox manifests itself; and when all the points of similitude are considered together, the likeness is most remarkable.
>
> As soon as my attention was directed to him, I became interested in the case, and instituted inquiries in regard to the experiences of the mother during her pregnancy.
>
> From her and from other reliable sources, I learned that when in that delicate condition, her master secured a living fox, which he chained in a situation where she was compelled to see it daily during the continuance of the pregnancy.
>
> The resemblance to the fox was plainly visible when the child was born and has continued to increase until the boy presents the appearance and peculiarities mentioned above.

* For more information about Warren see: ||y6ay53nk.

Alton, Michigan. In a case history, Dr. A. Ford (1888), a physician residing in the now-defunct village of Alton, stated that

> Some years ago, I was sent for to attend Mrs. I. The child was born before I arrived and only lived a few minutes. It was given to me, and I kept it for some time. Its nose, eyes and mouth were like those of a hound puppy. Its legs were bent forward like a dog's hind legs. Its arms were extended like a dog's forelegs. On the side of the hands were claws like a dog's even to the nail of the claw.

Augusta, Georgia. Milton Antony (1838, p. 77), a professor of obstetrics at the Medical College of Georgia in Augusta, reported that he himself "once delivered a finely developed child, in all parts except the head, which was, in shape, ears, eyes, colour of hair, &c. &c. the precise representation (according to the testimony of the mother of the patient) of a dog which had, to her very great alarm, in early pregnancy, seized the mother by her dress on her entering a yard in the country."

Milton Antony
1789-1839

Durham, North Carolina. The *Durham Sun* (quoted in the *Roanoke News,* Weldon, N.C., Mar. 17, 1892, p. 1c; ||y6q8c4m3) said,

> A child was born in Durham this morning the mother being a colored woman, and it is exciting the curiosity of some of our physicians. A SUN reporter saw it this morning and it beats anything in the way of a freak of nature he ever beheld. The head was a perfect formation of a human being, with the exception that it had rudimentary ears, no chin from the mouth and the jaws extending backwards as in the formation of a dog. It was pronounced to be eight months old [i.e., of eight months gestation] and was born dead. There was a thick head of long black hair extending down the neck to the body.
>
> Under the microscope the body showed that it was covered with a hairy fuzz and at the end [of the] spinal column there was a well-defined tail one-fourth of an inch long. The body was peculiarly shaped, not exactly resembling a human being. There were no arms, but something in the shape of hands or paws, growing from the skin, with no bones. The body, adjoining the legs, was very peculiarly shaped. There were no ankles. The feet projected from the knee joint.
>
> The formation and appearance of the body was a very peculiar one and the physicians who saw it expressed various opinions, yet all agreed they never saw anything like it before. The body was photographed and then placed in alcohol for preservation.

Watertown, South Dakota. From the *Mitchell Capital* (Mitchell, S.D., Jul. 30, 1897, p. 1c; ||y3pt2pch):

> WATERTOWN PUBLIC OPINION: PUBLIC OPINION secures this piece of information from sources that warrant the vouching for its accuracy, although names are suppressed for obvious reasons: There reside in the reservation country, northwest of Watertown, among others, a man and wife who command the respect of their neighbors. Recently the woman in question gave birth to a child whose face and general features were those of a dog. Its ears were likewise shaped like a dog's, while its hands were more like claws than like hands. Its crying was an imitation of the bark of a dog. Otherwise, the child was properly formed. It survived but a comparatively short time. Our informant says that the woman was frightened on account of a dog fight several months ago, and this incident is presumed to have wrought the deformity of her child. [see MATERNAL IMPRESSIONS]

Morgan, Vermont. The following two-sentence notice, which refers to an event that allegedly occurred in Morgan, is from the *Orleans Independent Standard* (Irasburgh, Vt., Jun. 29, 1869, p. 3b; ||hhsfbdw): "A lady in this town recently gave birth to a child with a dog's head. It was permitted to die."

Memphis, Tennessee. From the *Memphis Daily Commercial* (Jul. 27, 1890, p. 5f):

> A monstrosity with the face of a human and the body, legs and feet of a dog. Such was the startling discovery made by patrolmen [John T.] Thompson and [John] Doherty yesterday evening.
>
> In the lot surrounding the house on the northeast corner of Market and Second streets it was found. The house stands back in a neglected yard overgrown with long grass and weeds, and for many months the old rookery has been unoccupied.
>
> There in the tall weeds was a sack almost concealed from view. Upon opening the sack, they drew from it a bundle weighing about five pounds, neatly and carefully wrapped in a piece of linen.
>
> It would be difficult to decide whether human or dog, although the dog seems to be the predominating element of this physiological monster. A human, childish face, black silken hair, round head, and the body that of a dog, perfect.

* John Dougherty (d. April 6, 1918, Memphis, Tn.) was a patrolman in Memphis for many years (||yxk5hab5). A policeman named John T. Thomas appears in various Memphis directories during the 1880s and 1890s.

La Filiación (Goya, Los Caprichos, #57). The masked woman at right has much the same appearance as that described for Sarah Walls in the reports below.

Willet, New York. From *The Sun* (New York City, Sep. 23, 1888, p. 12d; ||y72y486j):

> In a lonely cabin on the banks of Otselic Creek [now the Otselic River], in the town of Willett, Cortland County, may be found a remarkable freak of nature, known throughout the country roundabout as the "dog-faced girl." Here, mother and daughter, named respectively Orilla and Sarah Walls, live alone and in poverty, gaining a precarious support by picking berries and gathering roots and herbs, and from gifts at the hands of the numerous visitors whom curiosity brings to their hut.
>
> The peculiarity of the dog-faced girl's features lies in the chin, mouth and nose, which protrude from the head and neck to such an excessive and unnatural degree, and in a shape strikingly like a dog's snout or muzzle. Otherwise, she is well formed and fully developed for a girl of fifteen years and is notably robust and active. But she is weak-minded, and the wonderful thing about her imbecility is its canine characteristic in voice and movement.
>
> Up to about ten years old the girl persisted in running about on all fours and in this posture she could outrun any child of her age. Her vocal utterances, also, were marvelously like the barking or howling of a dog. As she approached the period of womanhood, she learned to walk erect, and her voice lost something of its canine tone, but her features and utterances are still so unmistakably dog-like as to attract marked attention and wonder.

> She has a constant stream of visitors, including physicians and scientists as well as ordinary curiosity mongers, and these are told truthfully that her father who died shortly after her birth, was in no way malformed or peculiar. Her mother is good looking in feature and figure, and ordinarily intelligent. None of the dog-faced girl's visitors have as yet suggested any plausible theory of the freak that gave her to the world.

In another report about Walls, the *Binghamton Republican* (quoted in the *Chenango American,* Greene, N.Y., Sep. 27, 1888, p. 3c; ||17xmmof) said,

> One of the most peculiar freaks of nature that has been produced in any country lives with her mother in a small cabin near the banks of Bragg Pond [which was a sometime name of Echo Lake], about four miles from Willett, Cortland County. It is a girl about fifteen years of age named Sarah Walls, the daughter of a woman whom the denizens of the locality call "Rilla," but whose full name is said to be Orilla Walls.
>
> The girl excites much curiosity among strangers who happen to meet her in their fishing excursions, because of protruding lower features. Her mouth, nose and chin seem to be merged into the form of a dog's snout [see ORAL SUBSTITUTIONS], from which she has gained the sobriquet, "The dog-faced girl," and is rarely spoken of otherwise. Her laugh consists of a series of short yelps; her speech is almost unintelligible, while intellect is little above that of an idiot.
>
> This lusus naturae is said to have outgrown many of her canine instincts but sufficient yet remain to make her a veritable curiosity. When still a child, she often amused herself by barking and running on all fours through the woods. It was her delight to sniff at decayed stumps and with her hands tear away the soft wood in search of mice and chipmunks. Her speed through the brush is marvelous, requiring the best efforts of a man to keep her in sight.
>
> She and her mother eke out a scanty living by fishing and picking berries. The mother is now about fifty years old. Though very illiterate, she is social and good-natured, and she does not appear to think there is anything remarkable in her offspring. The girl is of medium size, well developed, and save for her dog-like face would be considered attractive.

* An admission form says Sarah Walls entered the Broome County Almshouse in Binghamton, New York, May 4, 1917 (||yym7pegs). Sarah would then have been 39, as a 1900 census record gave her birthdate as June 1877 (||y5kqdmah). The cause of her admission was likely the death of her mother, who last appeared in a 1915 New York census record stating that she was 77 and still living with Sarah (||y5hvfgbz).

A bull terrier.

Liège, Belgium. Dr. Paul Liebrecht (1876) described in detail a case like that of Sarah Walls. He reported that a certain individual, a 21-year-old bookbinder with the initials F. P., consulted him about a facial deformity. He looked, says Liebrecht, in all respects like an ordinary human except that his nose and upper lip "bore a striking resemblance to the muzzle of a bull terrier ("présentait une ressemblance frappante avec le museau d'un boule-terrier"). The rest of the face, he said, was normal including the hard and soft palate, as well as the shape of the alveolar arches. The case was so unusual, he said, that he showed "F. P." to other members of the medical faculty there at the University of Liège, including Auguste Swaen and Voltaire Masius, both of whom served, during their careers, as rectors of the university. Liebrecht successfully excised his patient's canine snout. In commenting on the case (p. 15), he said, in translation, "One must say, in particular, that this was a question of a deformity not only rare, but perhaps unique in its kind, not a simple arrest of development or a hypertrophy, but a malformation recalling, in a striking manner, a particular extraneous structure with a characteristic form."

San Diego. Cottew (1913), a San Diego, physician (b. 1869), said he delivered a woman "of a male child that had almost a perfect dog's head, the ears with the hair over them, and over the entire head were characteristic, and the expression of the mouth, eyes and nose were all those of a dog." It died soon after birth.

Troy, Ohio. From the *Miami Union* (Troy, O., Apr. 17, 1880; ||zaegzbu):

TROY, O., April 13.—A singular and somewhat horrible case of the birth of a monstrosity occurred at or near Nashville [a place], in [southwest] Miami County [Ohio], last Friday, April 9. Dr. John G. Senour, of Troy, was called to attend a lady in childbirth, and the result was a stillborn monster—a male child having a face resembling a bulldog, but otherwise fully developed. It seems that the mother at the period of two months gestation had received a fright caused by the shooting of a bulldog, and the child was delivered at seven months. The countenance is an almost exact imitation of a bulldog, with hare lip, flat nose, flattened eyes, large mouth, and the head set upon the shoulders with no development of neck. There was a peculiar hole in the back of the head, said to be at the same position where the bullet struck the dog, and on the back of the child is a mark like that left on the neck of the dog by the burning of the powder. Dr. Senour has the child preserved in alcohol, and it has been visited by many of our citizens, and is, altogether, a remarkable case of this kind.

John Gahagan Senour
1854-1904

Portland, Oregon. The *Lebanon Express* (Lebanon, Ore., Aug. 10, 1894, p. 3a; ||hmj6qut) said, "A child was recently born in Albina with the head of a dog." The city of Albina was consolidated into Portland in 1891.

Springfield, Ohio. From the *Cincinnati Enquirer* (Aug. 24, 1895, p. 9e; ||mpz6nzwr):

SPRINGFIELD, OHIO, August 23.—Considerable curiosity is being created over a midget girl baby of Mr. and Mrs. Homer McCoy, colored, which is now two months old, and weighs only 1¼ pounds. It has the peculiar features of part human and part animal. Its face resembles a dog and has that look characteristic of the canine race. The child is healthy, apparently, and physicians say that if nothing unlooked for happens it will live and grow to be a human monstrosity.

* An 1894 Springfield city directory lists a Homer McCoy living at 56 S. Yellow Spring St. (||2fujdkmw).

Marinette, Wisconsin. From the *Oshkosh Northwestern* (Oshkosh, Wis., Jan. 31, 1905, p. 1b; ||2p83fhr8):

> MILWAUKEE, WIS., JAN. 31.—A special to the [Milwaukee] JOURNAL from Marinette says: A human monstrosity half dog and half man, was born here. The child is living, and the mother seems as fond of it as though normal and insists on keeping it alive, regardless of medical advice that it be allowed to die. The body is covered with hair and has a long, projecting nose and low forehead. It is said that the mother was scared by a dog jumping on her about a month ago.

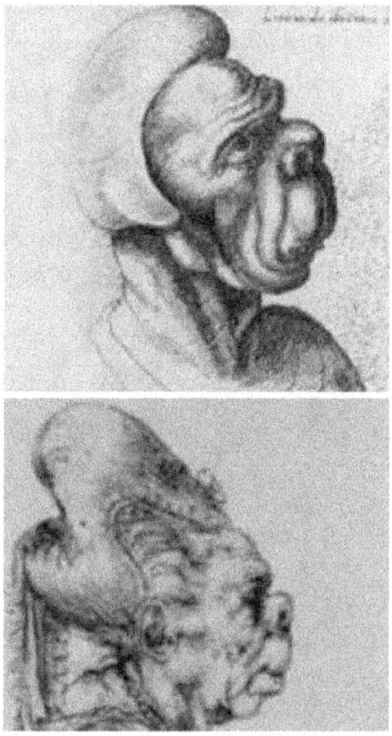

Two cynanthropoids (Hollar, ca. 1645, after da Vinci's *visi mostruosi*).

From the Chicago Tribune (Oct. 9, 1987): "Dear Ann Landers: My grandma was a midwife. She delivered a baby with the head of a Great Dane. It lived two hours and was buried secretly."

Cairo, Illinois. From the *Dodge City Times* (Dodge City, Ks., Aug. 8, 1890, p. 1a; ||heu8zrj):

> CAIRO, ILL., AUG. 4.—A lady named McLaughlin, residing on Twenty-first street, this city, recently gave birth to a child whose face and head was the image of a bulldog, the rest of the monstrosity retaining the normal condition and appearance of a healthy child. Sometime past the father of the child purchased a large bulldog, whose care he entrusted to the wife. About ten days ago the dog became vicious and frightened the woman so much so that she took to her bed. Yesterday the monstrosity was born. The child died a few hours after birth.

> * In 1900 a Margaret McLaughlin (1864-1910) and her husband Thomas Andrew Fuller (1855-1944) lived on 21st Street in Cairo (||yxt5rtb4, ||y6enhael).

Marietta, Georgia. The *Marietta Journal* (Sep. 6, 1883, p. 3d; ||3dr3xwtk) said, "A child was born in this county the other day having a head shaped like that of a dog, while its hands resembled dog's feet. It did not live" (Also see the *North Georgia Citizen,* Dalton, Ga., Sep. 13, 1883, p. 4a; ||4m7a6ktw).

East Poestenkill, New York. The *Troy Press* (quoted in the *Rutland Daily Globe*, Rutland, Vt., Jun. 12, 1873, p. 4d; ||y2o6hwo8):

> A CHILD WITH A DOG'S HEAD.—A very respectable lady residing at East Poestenkill, recently gave birth to a female child with a perfectly formed body and a dog's head. The child was born during a terrible thunderstorm. Some time ago its mother was badly frightened, while on the way to church, by two vicious dogs, one of which bit her. The child lived but fifteen hours.

Milwaukee, Wisconsin. The *Bossier Banner* (Bellevue, La., Jul. 27, 1867, p. 4) said, "A child was born in Milwaukee with a perfect dog's head."

Oshkosh, Wisconsin. From the *Appleton Post* (Appleton, Wis., Jul. 31, 1890, p. 7e; ||yvvpc6xa):

> A DIME MUSEUM FREAK
>
> OSHKOSH BABY WITH A DOG'S HEAD THAT SNARLS AND BARKS
>
> A dime museum freak of no mean dimensions has come to light in Oshkosh, says the Twin City NEWS. It is nothing less than a child with a dog's head or something that resembles the cranium of a canine more than it does that which usually goes with a human body. The child, or whatever it is, was born but a few weeks since and the parents are honest and respectable. They have endeavored to keep the infant from the gaze of the neighbors, but have been unsuccessful in the attempt, and now the strange appearance of the child is common talk. Certainly, the little one presents a strange and repulsive appearance and has undoubtedly received a birthmark from a dog, as the mother says that a month or so before the child was born, she was badly frightened by being attacked by a large Newfoundland dog. The mouth and nose of the child do not resemble those of a dog to any great extent, but they are very contrary to those of a human being. The ears of the child are very much like those of a bulldog, and are covered with a coating of soft, fine, reddish hair. The child does not cry like an ordinary infant but makes a whining noise like that of a dog anxious to escape from its kennel. The father and mother feel terribly mortified over the malformation of the child, which is their first, and they hope it will yet develop into something resembling a human being.
>
> Yesterday an agent for Kohl & Middleton's dime museum of Chicago was in Oshkosh and saw the child. He made the parents a liberal offer for the privilege of exhibiting it, but the offer was indignantly refused.

Bernard von Langenbeck
1810-1887

Kassel, Hesse. The *Berliner Gerichts-Zeitung* (Jul. 3, 1875, p. 3a; ||4wwxvuyd) said the well-known surgeon Bernard von Langenbeck was to operate on a ten-month-old infant born to a woman in Kassel. Reportedly, the right half of its face was that of a girl, the left, a dog's, and covered with long black hair, a description that suggests the patient was a gynandromorph (see p. 1231).

Fort Reno, Oklahoma. From the *Wichita Daily Journal* (Wichita, Ks., Jul. 11, 1890, p. 1a; ||y5lna2au):

> This morning a JOURNAL representative was the witness of a strange freak of nature at the union depot as the train from the south passed through for the east. The writer has heard of double headed cattle, six-legged horses, Siamese twins, children deformed in all sorts of manners, but he never before heard of or saw a baby with a face like a dog's and hands like paws and with every indication that the doglike characteristics extended even down under the clothing out of sight.
>
> A JOURNAL man was lounging about the depot as the train pulled in. A traveling man, an acquaintance, alighted and approaching the writer imparted the information that aboard the train was a curiosity such as would not be seen again in a lifetime. In the first car back of the smoker the reporter found a queer family. A man and a woman, evidently husband and wife, seated in seats turned facing each other and by the side of the father sat a very good-looking girl. On the seat next to the mother was an object which attracted the attention and excited the curiosity of everyone that saw it.
>
> This object, on whom the woman bestowed the most tender affection, was a child at least 2 years of age and a wonderful formation or rather a horrible malformation of nature it was. By permission the writer examined the child and found that its head, while human in formation, was adorned with ears like a dog's, the face was covered with a yellowish, taffy-colored hair, the lips curled back from the teeth like a bull dog's and the nose was flat

and flabby. As far as could be seen for the clothing, the body continued hairy and instead of prattling like babies usually do. This monstrosity whined like a little poodle and growled like an angry dog.

The reporter was horrified and interested and sought an interview with the freak's parents. They were rather loath to talk about the matter but the father at last said that his name was Leon Barst and his home on a ranch not far from Fort Reno [Oklahoma]. He had been employed there for several years, taking care of cattle, which were usually sold to the government for the Indians. [Here omitted is an account of an experience to which the parents attributed the birth of their doglike child, a fright the mother had had when she fell into the hands of Cheyennes and their dogs [see MATERNAL IMPRESSIONS].

They had tried to hide their secret and keep the strange looking baby concealed from view, but their secret had leaked out. An account of the affair had been sent to an eastern paper by a newspaper correspondent, and they were now on their way to New York, on invitation and at the expense of a museum manager, who wished to put the child on exhibition. They were poor people and the monied consideration was in their eyes very large, and therefore they accepted it. While stopping here at the depot full twenty or thirty people must have looked at the strange child. It was a wonderful sight.

Medina, Ohio. The *Medina Gazette* (quoted in the *Democratic Press* Ravenna, O., Nov. 25, 1869, p. 3d; ||ftrps9s7) stated that

> We are informed on reliable authority that on Friday of last week, in one of the villages of this county, a child was born with one of its hands shaped like a dog's paw. The arm from the elbow down is much smaller than the other, and the fingers are webbed about half the way up. The child holds the hand as a dog holds his paw when hurt.

XENOMELIA

Albany, Oregon. John Linsey Hill (1877), a physician practicing in Albany, reported that a woman with whom he was "well acquainted" had given birth to a child with "the exact counterpart of a dog's foot on its shoulder, and more remarkable still, this mark was thickly studded with hair resembling the hair to be found on the foot of a dog."

SEXUAL INTERACTION

Richard von Krafft-Ebing
1840-1902

Male dogs are promiscuous. They have been observed *in coitu* with a variety of animals, including cattle, sheep, pigs, cats, various primates, and even birds (e.g., with chickens and geese). Moreover, anyone who owns a male dog will be familiar with their "humping" behavior which amounts to an attempt to mount a human being. The German psychologist Richard von Krafft-Ebing was perhaps the first scientist to report on sexual relations between humans and dogs

Studies report the occurrence of such behavior, predominately involving women and male dogs, from early times right up to the present (Evans 1906; Krafft-Ebing 1906; Laws and O'Donohue 2008, p. 391; Miletski 2009, 2017). It has been documented in societies ranging from ancient Egypt and Mesopotamia, to India, Japan, Melanesia and various Native American cultures, to modern America and Europe (ibid.). It is due to such behavior that the Talmud forbids widows from keeping dogs (Abodah Zarah 22b; ||y5e2gsyv). Ancient Akkadian omen lists mention sexual relations between women and dogs. In a review of the subject of human-animal sexual interactions, Valcuende del Río and Cáceres-Feria (2020) state that

> Female bestiality is usually associated with animals linked to the domestic sphere, such as dogs [they cite Krafft-Ebing (1894, p. 562), Lucenay (1933) and Matté (2008)]; it is no coincidence, therefore, that in many cultures, a man engaging in sexual conduct with female dogs is considered unthinkable or grotesque [they cite Devereux (1948)]. In the north of Costa Rica, whereas sexuality with female pigs or donkeys is seen with a degree of normality, sexual relations between teenage boys and female dogs are rejected and ridiculed [Cáceres-Feria (2013)]. In Greece, dogs are prohibited for such practices, considered impure, and there is fear of the diseases that might be contracted [Anest (1994)]. Among the Inuit peoples, sexual relations with dogs were found among men and women alike. Male sexuality with female dogs was regulated: it had to be outdoors, never inside the home, and the animal had to be in heat [Laugrand and Oosten (2002)].

Romulus and Remus (Wenceslas Hollar, after Giulio Romano).

DOG-WHELPED CYNANTHROPOIDS

English, Indiana. From the *Cincinnati Enquirer* (Apr. 1, 1892, p. 5; ||3x6dahbh):

> ENGLISH, IND., March 31.—A pup with but two legs, the forelegs missing, one of those natural curiosities which occasionally shows itself, is in possession of Friedling Urich, foreman of the Bauer shingle mill at this place.
>
> Mr. Urich went about town this morning exhibiting the little creature and will doubtless raise it. Between the middle portion of the body and the head the monstrosity is shaped like a perfect man, and the head has some of the human shape. Science has laid down an excuse for human monstrosities and deformities, but how is this to be accounted for in the brute kingdom?

Torreón, Coahuila, Mexico. El Tucsonense (Tucson, Az., May 7, 1925, p. 1b; ||yyp5s2p7; also see: ||4ep7sch8) said, in translation,

TORREON, COAH., May 1.—A truly extraordinary event, which just occurred, has stirred all the residents of the barrio Torreón Viejo

It's nothing less than the birth of a dog with the face of a man, to which the similarity is amazing.

Osnabrück, Lower Saxony, Germany. Johann Friedrich Blumenbach (1813, p. pp. 4-5) said it had been recently reported that a birddog at Osnabrück had whelped a female nondescript with a human face.

Seneca County, New York. Byl (2002, p. 423) quoted an 1812 Seneca court record stating that a certain William Moulton

> should appear at the next court session and answer to the charge of "being guilty of the crime against nature commonly called sodomy in having communion with a beast commonly called a bitch." Moulton had been arrested on the said charge by a warrant issued by Justice of the Peace Osgood Church. ... Attached to this document is the evidence collected against Moulton, consisting of four depositions ... The first deposition is an eyewitness account given by a Levi Griswold who claimed to have stumbled upon him committing the said crime. ... Another deposition is from John Gandy, who testified that Griswold told him of Moulton's alleged crime shortly after the incident occurred. The last two depositions refer to a deformed litter of puppies that Moulton's dog reportedly mothered. Mattanach Wollet testified that Moulton's dog came onto his father's farm and gave birth to puppies the likes of which he had never seen. Allegedly these puppies "had large heads resembling a person's head, they had arms like a person's, and long straight hind legs, no hair on them nor tails, and on the side of their head they had small ears." This testimony, corroborated by his [i.e., Wollet's] father, was obviously an attempt to establish Moulton as the father of the puppies.

Cuba. From the *Charleston Courier* (Charleston, S.C., Jun. 21, 1855, p. 2):

> An extraordinary *lusus naturae* has been exhibited in Cuba. It consists of a pup of large size having its head, eyes, nose and mouth completely human. It had also an arm crossed over the breast, and the other suspended over the head. The hind legs, extremely diminutive and very neat, were placed perpendicularly.

Raleigh, North Carolina. A conjoined dog-human triplet was reportedly whelped by a dog in Raleigh. A report describing the Raleigh birth has been included in the section on CONJOINED DOG-HUMAN TWINS AND TRIPLETS.

LITTERS OF CYNANTHROPOIDS

Several reports allege human mothers bore multiple cynanthropoids in a single delivery, that is, to litters.

St. Louis. From the *St. Louis Post-Dispatch* (Mar. 25, 1881, p. 4; ||ycestyrc):

> The astounding rumor that a baby, half human and half dog, had been born in the neighborhood of Eighteenth and Morgan streets this morning was current upon 'Change, and was regarded as an absurd hoax. A reporter, however, went out to the place indicated, and after a great many inquiries was directed to an old two-story frame house, which had evidently seen better days, which is situated on Orange Alley, the thoroughfare running east and west between Franklin Avenue and Nineteenth Street. A trip was made up the shakiest flight of stairs which it has entered the human mind to conceive and, being careful of the weak places on a porch, the reporter at last secured admittance to a room which served as bedroom, parlor, kitchen and dining-room for a small, but interesting family … fellow-citizens. A large negro woman was lying on the bed.
>
> "I'm looking for Samantha Willis."
>
> "That's my name," said the lady in the bed turning away from the wall and grinning at the scribe.
>
> "I understand that you have had a child?"
>
> "Huh! Huh! I done hab free of um [I had three of them]."
>
> … Samantha looked to the reporter for more conversation. "Yes, I heard that," said he, "but they said that one of them had a queer head!"
>
> "Well, gimme a dime, boss, 'n I'll show you the chile; it is queer for sho'."
>
> The dime changed hands, the coverlet was thrown down, and a little … cherub about eight inches long was fished out of the intricacies of the bedding. He was clothed like Lady Godiva in the legend of Coventry, simply with purity, and at first glance there was little to distinguish him from the average … infant. A close inspection showed, however, that his head was exactly like that of a dog. It was covered with fur rather than wool, and the facial contour was just that of a young bull pup's. The nose and mouth projected almost an inch from the balance of the countenance, and the ears are long, pendulous and like a spaniel's. It has the power of moving its ears, which it does continuously, and is apparently a strong, healthy child, although Dr. Carstairs says it cannot live. The whole body is covered with very fine and short hair almost like the down on a peach, it is so soft and so almost imperceptible. The hands are all right, but the feet are malformed, there being a very perceptible dewclaw on the left ankle, and the whole foot is that of a dog's. The right foot is amorphous, simply

being a piece of flesh not differentiated into toes. [Limb deficiencies are commonly seen in telenothians] The other two children have none of the marks of the dog upon them except their ears, which are pendulous like their brother's, and their bodies are covered with the same downy fur. Mrs. Willis was attacked by a savage dog about three months ago and badly bitten in the side, and Dr. Carstairs traces the lusus naturae to that cause. [see MATERNAL IMPRESSIONS]

Ottumwa, Iowa. From the *Omaha Daily Bee* (Omaha, Neb., Apr. 18, 1882, p. 7e; ||h4gr7a6):

A woman at Ottumwa, last week, gave birth to five children, part human and part dog. The freak of nature is the subject of much wonderment and comment.

Arkadelphia, Arkansas. From the *Memphis Daily Appeal* (Memphis, Tn., Aug. 14, 1873, p. 1d; ||y4eo7aou):

ARKADELPHIA, ARK., August 13.—Mollie Holman, a negro woman living about one mile from here, whose husband died a year ago, gave birth on Monday night to five deformities, each weighing about three pounds. The face of these wonderful creatures most resemble a monkey, with the body and limbs of a dog, sparsely covered with hair. They were alive this morning and are fondly nursed by the awe-stricken mother. Our medical corps will visit the curious offspring this evening.

Sucre, Bolivia. In translation, the Spanish chronicler Pedro Cieza de León (1553, cxii) stated the following concerning events at a place ("Villa de Plata") near Sucre.

I have also heard for a fact that Francisco de Almendras, who lived near Sucre, caught an Indian woman and a dog committing a sin together. He had the woman burnt alive. And, besides this, I have heard from Lope de Mendieta, and from Juan Bortiz, and from others living near Sucre, that they heard from their Indians how an Indian woman in the province of Aulaga gave birth to three or four monsters sired by a dog, and that these survived several days.

Flora, Indiana. A report about twin cynanthropoids from the *Cincinnati Enquirer* (Aug. 21, 1894, p. 5e; ||4a8u4z6x):

LOGANSPORT, August 20.—A young woman of Flora has given birth to two monstrosities, which have created the greatest excitement in that place. The children, if such they may be called, have all the appearance and characteristics of the canine tribe, even to the voice and covering of hair. The mother is a very comely and fairly intelligent young woman.

* In addition to the cases above, the Litoměřice report (cited on the next page) also describes a litter of cynanthropoids.

A half human half dog (cigarette card, China, ca. 1910; ||mvz6stw2).

HALF HUMAN HALF DOG

Various women have allegedly given birth to cynanthropoids that were human from the navel up and dog from there down. The earliest such case that an extensive search has as yet identified dates to 940 A.D., a birth in Gascony, in southeastern France, which supposedly had "the form of a human being in its foreparts and that of a dog behind" (*Chronicon Tielense,* 1789, p. 62).

Rome. Ambroise Paré (1982, p. 67) writes that

> In the year 1493[6] a child [was born] having, from the navel up, upper parts similar in form and shape to the mother, and it was very complete, without Nature's having omitted anything; and, from the navel down, all its lower parts were similar in form and shape to the animal that was the father: which (just as Volaterranus writes) was sent to the pope who reigned at that time [i.e., Alexander VI].

The Roman cynanthropoid
(Liceti 1634)

The 1493 event received more attention than most such cases. Liceti (1634, p. 184) both described and pictured it (his illustration is shown above). In translation his account reads,

> At Rome, an unmarried girl bore a son who was half dog, human from the navel up, but from there down, like a dog. This unnatural event is recorded by Volaterranus, Cardano, Paré and others.

* The Roman birth was also mentioned by Bartholin (1657a), who stated that "At Rome under the pontificate of Alexander, a woman bore a half-dog."[7] Polydore Vergil (*De prodigiis,* 1664, 3:96) also refers to this birth.

Litoměřice, Czechia. Martin Schurig (1732, p. 61) states that in 1615 at Leitmeritz (now Litoměřice) in Bohemia,

> An event occurred in which a woman, who had had relations with an English hound, gave birth to five creatures that were human from head to navel, but from there down, like a dog, which shocked many who saw them. The mother was immediately imprisoned. [translation mine]

* The Belgian historian Aubertus Miraeus (1622, p. 224) gave an account of the Leitmeritz event that agrees with Schurig's. Schurig (1656-1733) was a German physician, remembered for his studies of anatomy of the sexual organs and of human reproduction.

Zeitern. Another report should perhaps be assigned to this section, given that it was allegedly dog below and primate above. Johann Georg Schenck (ca. 1560-1620), a German physician, reported and pictured an unusual birth, which, he said, had two heads, and arms without hands. Per his report (Schenck 1609, p. 20), this creature was borne, August 23, 1599, by a woman living in the village of Zeitern, which was at that time a mile from Brussels but has since been absorbed by the sprawling metropolis.

The Zeitern cynanthropoid
(Schenck 1609)

There is, however, some confusion regarding the claims of this report, as Schenck says the creature's heads and legs were like an ape's, and its body, like a human being's. And yet, his illustration shows legs like those of a dog, not an ape. Moreover, the faces depicted could just as well be human as simian. If anything, with their noses look more like human noses than like an ape's. It may be, then, that the general want of knowledge about apes at the time led Schenck to say the pictured legs looked like those of an ape rather than of a dog. There was a tail, he said, like that of a fox ("cauda vulpina"). Thus, his depiction can be interpreted, perhaps, as that of a creature that resembled a human from the waist up (albeit with deficient forearms and two heads), but that was like a dog from the waist down.

Dechard, Tennessee. Another report about a cynanthropoid being human above but canine below, appeared in the *Daily Leader* (Cleveland, O., Oct. 25, 1865, p. 2a; ||y4q96v8t).

REMARKABLE MONSTROSITY
A WONDERFUL NONDESCRIPT FOUND AT
DECHARD, TENNESSEE

On Tuesday morning, about seven o'clock, a wonderful nondescript was found lying on the top of a wood pile in the rear of the hotel at Dechard. The thing was in appearance, half human

> and half canine. The upper half of the phenomenon resembled the body and head of an infant a few days old. With the exception of the face and hands, this portion was covered with light curly hair, which, on top of the head, was fully two inches long. The hands and face were entirely exempt from hair, and were, in every way similar to an infant's. The lower half of this wonder corresponded exactly with the hind quarters of a dog, including a tail about twelve inches long, and covered with curly hair, similar to that on the head of the prodigy. It was alive when found and was regarded by the inhabitants with mingled wonder and horror. They were undecided whether it were not the holier work to burn the marvel in the flames. It was finally taken possession of by two army surgeons, who were stationed in the town.

Beaver Dam, Wisconsin. From the *Janesville Daily Gazette* (Janesville, Wis., Jul. 26, 1873, p. 1b; ||khzhsz9):

> The Beaver Dam ARGUS reports that on Thursday night a woman in that city was delivered of a monstrosity—half human and half canine, the head, shoulders, arms, and upper part of the body being a perfect formation of a child, while the lower part of the body was as perfectly in form of a dog.

Raton, New Mexico. From the *Fort Worth Gazette* (Feb. 4, 1892, p. 1d; ||yxogl73a):

> RATON, N. M., JAN. 27—The wife of a half-breed Indian living near Raton on the line of the Santa Fe railroad gave birth to a child a few days ago that is attracting much attention and exciting great curiosity. The legs of the child are exactly like the legs of a dog—in fact, the entire lower part of the body is the body of a dog. The flanks are joined together, and the legs and hips are covered with a thick yellow skin. The feet have five toes, but the nails are shaped like the nails of a dog, and under the foot there is a cushion such as a dog has. The child was ten days old before anyone knew anything about it, and then a man who happened to stop at the house discovered it. The father and mother are poor and illiterate and don't seem to care very much, in fact as the freak has been the means of bringing them quite a little money, they probably have very little regret. It makes a noise that sounds more like the whine of a pup than the cry of a baby. The head and arms were well formed and looked perfectly natural. The legs and

> feet were then shown. The legs were close to the body doubled up as a dog doubles up his legs when sitting on his haunches. The father says that six months before the birth of the child the mother had been attacked by a large shepherd dog and badly bitten. The fright threw her into convulsions and for three or four months afterwards she continued to have spells and she would fall down and, frothing at the mouth, bark like a dog and scratch at the floor with her hands and feet. For a few weeks before the birth of the child, however, she had appeared to be much better and had not suffered from an attack.

Wellston, Ohio. From the *Cincinnati Enquirer* (Jan. 27, 1893, p. 5; ||2whst6mf):

> WELLSTON, January 25.—Dr. W[illiam]. S. Hay [should be Hoy, not "Hay" (1856-Jun. 14, 1917)], President of the Southern Ohio Academy of Medicine, has in his possession a remarkable monstrosity in the form of a human dog. The creature was born on Monday, in company with six other pups, and lived but a few hours. The strange and unaccountable feature is the upper portion of its body, which is exactly that of a human being. It has a distinct human nose and upper and lower jaws, like those of a child. Instead of forelegs it has arms furnished with tiny fingers. The hind portion of the body is like a dog in every respect.

No locus in quo. An old book on midwifery (Anonymous 1830; p. 25) alleged the occurrence of such a birth and offered the picture at left, with the comment that "Some monsters are begot by a woman's unnatural lying with beasts; as in the year 1603, there was a monster begotten by a woman's generating with a dog; which monster, from the navel upwards, had the perfect resemblance of its mother; but from its navel downwards resembled a dog, as you may see by the accompanying figure."

In 1543, a woman and her dog were burnt at the stake in Avignon (see p. 301).

EARLY REPORTS

A monster in human form, with the face of a dog, has appeared in Lower Germany.
—MARTIN LUTHER[8]

Lyon. Dufieu (1763, p. 229) stated that "At Lyon, in 1757, I saw a monster with the head of a wolf, but with all the remaining members of the body like those of a human being" (translation mine).

* The Chevinay canisuid (p. 496) may have been this same animal, differently described.

Copenhagen. Thomas Bartholin (1673, pp. 56-57) reported that in Copenhagen, where he then lived, a child with a dog's head had been born, and that several such births had occurred there, though the parents, he said, fearing punishment, had tried to keep it secret.

Paris. Viardel (1674, pp. 241-246) gave an eyewitness account of how, in 1667, at Paris, in Saint-Germain-des-Prés, he delivered a child with a head like that of a fox. "I tried to encourage her," he said,

> telling her to bear up under the pain of expelling that horrible monster, which I drew forth by the end of its muzzle, aided by the lively efforts of the mother, who pushed it forth, only to have it die soon after. As in other births, the afterbirth followed. You will observe, however, that all other parts of his body were normally formed. It differed from other infants only with respect to the head, which was that of a beast. But such was enough to judge it a true monster. [translation mine]

Nierstein, Germany. The German polymath Eberhard Werner Happel (1690, anno 1681, p. 94) stated that on October 2, 1681, a woman at Nierstein gave birth to a cynanthropoid with a large dog's head.

Leiden, Netherlands. The Danish physician and anatomist Oliger Jacobaeus (1675, in Bartholin 1677, vol. 3, p. 97) reported a woman at Leiden birthing an intersex nondescript with doglike legs.

Bavaria. In translation, Christian Franz Paullini (1688, p. 49) said that in 1635 a Bavarian camp follower, after having had sexual connection with a mastiff ("molossus"), bore a stillborn cynanthropoid having "a head entirely like that of a dog, but hairless, with a lolling tongue, the legs of a dog in the place of arms, and a short tail; the remainder of the body being like that of a human."

Messina, Italy. Far to the south of Bavaria, a separate case was reported that same year. In an appendix of Paul Zacchias' *Quaestiones medico-legales,* the Italian physician Pietro Castelli (1574-1662) gave an account of a cynanthropoid born to a woman at Messina on December 26, 1635. Castelli (translated in de Ceglia 2014) thought it had been "deliberately crushed because it was a monster." Zacchias, who taught medical science and forensic medicine, had been consulted by judges in Messina regarding whether the birth could have been the result of hybridization between the woman and a dog. According to Castelli (translated in de Ceglia 2014), "It was shaped like a dog and its sacrum seemed very wide and full, but without a tail. The skin or, better, the cutis, was completely hairless, ruddy and highly tenuous, only in this was it similar to a human. Instead, its head, considering its shape and the position of the eyes, could be more easily compared to that of a bird than to that of a dog, although the ears were canine, the right one more oblong than the left one, sticking straight up. In place of the nose, it had a wide, pendulous membrane [frontal proboscis?] … It had a small, round mouth with the two lower incisors. Its front feet were reminiscent of those of a dog, but without nails; instead, the rear feet were truly monstrous: they were made of four oblong fistulas, of equal size and empty, but osseous, some of which had a sort of round plug at the end. Its abdomen was swollen and livid. I immediately eviscerated it, but, since the entrails were putrid and very fetid, I could not study them carefully. … Likewise, I was not able to recognize its sex, but it seemed to me to be a female. The length of the cadaver from its clavicle to the coccyx was a handbreadth, and from head to toe a handbreadth and a half."

According to de Ceglia (2014, p. 133), Zacchias, who at one time had some doubts about whether humans could hybridize with animals, became "more and more convinced and finally he would declare: 'I was previously of that opinion, and now, I persevere much more in the [opinion] that nothing can prevent it that from the mixture of human seed with beastly seed some generation can follow.'"[9]

Tonndorf, Thuringia, Germany. Fincelius (1566b) said a woman at Tonndorf gave birth on November 1, 1556, to a legless offspring of indeterminate sex with both the large ears and mouth of a dog.

* Binhard (1613, vol. 3, p. 147) also records the Tonndorf birth.

Avignon, France. After discussing the girl who at Rome in 1493 had engaged in coition with a dog and later bore a cynanthropoid (see p. 295). Liceti (1634, p. 186) referred to a like event at Avignon: In translation, he said, "In his *Miscellanies* [Hieronymus] Magius relates that at Avignon, in 1543, after a connection of like nature, a woman gave birth to a monster, having indeed the head of a human being, but the ears, neck, forelimbs, paws, sexual organs, and other parts of a dog. The woman actually admitted her pregnancy to have been initiated by the dog. And so, in expiation of her sins, by order of Francis, King of France [i.e., Francis I (reigned 1515-1547)], she was soon thereafter (on July 31) consigned to the flames, along with her dog lover." The account given by Magius (1564, p. 61) reads, in translation,

> In our memory, in 1543 at Avignon, from intercourse of a similar nature, a monster was produced with a human head, but with the ears, neck, forelegs and forepaws, genitalia and other parts of a dog. … Soon thereafter, then, on the 31st of July, by the decree of Francis, King of France, both the hound and the mother were consigned to the flames. Furthermore, many other, not very different, cases occur.

Benedetto Varchi (1834, p. 152), who also described the Avignon birth, stated that much of its body was "covered in long black hair like

> **None born out of lawful Wedlock can be an Heir. Nor a Monſter, *i. e.* one wanting Human Shape, *as having a Dog's Head*; but one having Fingers or Toes too many or too few, or Limbs diſtorted, is no Monſter.**

A passage in Coke's *Institutes* (7th ed., part 1, p. 11; ||hoe377d).

that of the dog that the mother had confessed to be the father" ("coperte di pelo lungo e nero come era il cane, col quale confessò poi essersi giacciuta quella tal donna che l'aveva partorito") and that "it lived long enough to be taken from Avignon to His Most Christian Majesty, King Francis, at Marseille" ("Visse tanto, che fu portato da Avignone a Marsiglia al cristianissimo Re Francesco"). Given travel times of the day and Marseille's being 100 kilometers from Avignon, Varchi is alleging the creature lived at least a few days. Referring to the Avignon birth, Riolan (1650, p. 834) said, "Indeed, not long ago we saw a woman punished in the same way for a similar crime at Paris" ("Simile scelus Lutetiæ non ita pridem vidimus simili supplicio castigatum mulierem").

Tuscany. Liceti (1634, p. 185) said, in translation, that "In the *Commentariorum urbanorum,* Volaterranus [1530, 32, p. 374] writes that in the reign of Pope Pius III[10] [Sept. 22 to Oct. 18, 1503], a girl in Tuscany, who had lain with a dog, bore a half-canine offspring, with four paws like a dog and a dog's ears, but otherwise like a human."

* Wikipedia describes the *Commentariorum urbanorum*, as "an encyclopedia of all subjects known at that time, prepared with great care." Volaterranus may have himself seen the product of the 1503 birth, since he was then living in Tuscany at Volterra, the town for which he is named.

Erfurt, Thuringia, Germany. Annals say that in 1521 a woman in the parish of St. George (S. Georgenpfarre), Erfurt, bore a nondescript with the head of a dog (Kruspe 1878, p. 38).

Rome. Zacchias (1688, p. 582a) said he had a diary that recorded a woman bearing a cynanthropoid in Rome's Rione Monti ("Regione Montium") neighborhood in the reign of Leo X (1513-1521).

Mennel's depiction of the Vienne cynanthropoid (Mennel 1503).

Vienne, Isère, France. Mennel (1503) stated, in translation, that "about the year 914, a monster, with a dog's head but other parts like a man's, was presented to Louis."

* Rolevinck (1518, lxviii) says this was the Louis who warred with Berengar I (i.e., Louis the Blind of Provence). Berengar blinded Louis in 905. So, this cynanthropoid would have been born before then and, likely, have been seen at Louis' capital, Vienne.

Paris. In the *Speculum Naturale*, written under Louis IX, Vincent of Beauvais (1624, 31.126.2392-2393) said, in translation, "Also, in our time, an animal was brought to King Louis, with a head mostly like a dog's, but in his other parts truly like a human. The legs were naked like a man's, as were his hands and arms. His neck, too, was naked and white, but his back, hairy. He both stood erect and sat like a man. He took roast meats and wine with gusto, which he put in his mouth correctly with his hands. For his size, his penis was large. And, he was eager to have sex with both women and girls. For he could tell men from women. When at ease, he was well-behaved. But truly, when angry, he became most discourteous and vented his rage upon men."

* Vincent's cynanthropoid dining with the nobility recalls Gerald of Wales' boshominid, who supposedly dined regularly with Norman nobles at Wicklow in Ireland (see p. 69). Vincent elsewhere (19.102) gives further details about this creature. As the *Speculum* was likely completed by 1250, the events described would have taken place before that date but after Louis' accession in 1226.

The Hague cynanthropoid.

WOMEN BIRTHING DOGS

> *It is not very long since I received a letter from a distinguished member of the profession asking me whether, in my opinion, I thought it possible for a woman to give birth to a dog.*
>
> —JOHN W. BALLANTYNE, M.D.[11]

Many reports about animals bearing progeny not of their own kind likely refer to hybrids, since hybrids sometimes show resemblance to only one of their parents. When that likeness is to the sire, it may well seem that the mother has borne the wrong sort of offspring, when in fact she has produced a hybrid (see XENOGENESIS, p. 1235). Such seems a likely explanation of the reports about women birthing dogs.

In a compilation of cases seen in his practice, physician Stalpart vander Wiel (1686, obs. 72) said that in 1677 at The Hague, a woman gave birth to a living puppy. He said (ibid., p. 315) that his brother, also a physician, had also seen a puppy borne by a woman. Wiel, like Castelli (see p. 300), said the skin was hairless ("pilis carentem"). The *Breslauische Sammlungen* (1720, p. 71; ||2p9f882b) mentions a Ukrainian woman birthing puppies in 1719. Heldvader (1625, p. 265) records another such case, in which a woman allegedly bore three. He states that in 1600, "A girl at Schwabstedt [in what is now Schleswig-Holstein] bore three puppies that died soon after birth. Supposedly, she had had relations with an English hound"[12] (my translation). Johann Christoph von Dreyhaupt (1755, vol. I, p. 646), in his account of his native Halle, says that at Moritzburg Castle in that city, in 1646, a woman was delivered of "what, to look upon it, seemed to be a dog" ("die wie ein Hund anzusehen gewesen.")[13] Franck (1536, p. 245b) likewise states that in 1471 a woman at Brescia, Lombardy, gave birth to a dog.[14] And a certain monk Ulrich claimed that in the year 1000 a shepherdess at Visbek in Germany had lain with a hound and later given birth to two puppies (Paullini 1686, p. 193). These reports exist. Some likely are, at least to some extent, true.

Friedrich Wilhelm Schmuck (1683, #2) published this picture of a two-headed, three-legged cynanthropoid with a wolf's tail, which he said died during birth, along with its human mother, in March 1682, at Turin.

CONJOINED DOG-HUMAN TWINS AND TRIPLETS

Zmigrod, Poland. The Danish anatomist Thomas Bartholin (1661, vol. 6, obs. 49, p. 278) said, in translation, "Reports from Zmigrod in Silesia [now in Poland] tell of the birth on April 22, 1652, of a horrid three-headed monster. Its father is a cobbler, Johann Stein, whose wife is named Catharina. The middle head was like a wolf's, faced upward, and had a lolling tongue. The right was blood-red and skinless, but human, though skeletal in form. The left face was clearly human, but a rounded, hard and fleshy mass was attached to the occipital region and to the other parts of the head. Lifted, it exposed on the forehead a small bone that stood up like a crown. The middle head was in the usual position, but the right and left were in fact placed at the armpits and turned to face each other. The hands, stripped of skin, were blood-red. Here and there, too, on the chest and thighs were dark red marks."

* The reported condition of the Zmigrod fetus suggests a decomposing stillbirth.

Raleigh, North Carolina. The *Maryland Gazette* (Annapolis, Aug. 7, 1800, p. 2; ||y2tac4f9) published an extract from a letter from Raleigh: "A most curious spectacle was exhibited in this city a few days ago. A female of the canine species was delivered of the most perfect *lusus naturae* that has ever been beheld in this part of the world, or I believe in any other. I will give you as accurate a description of it as I can. Its shape more resembles that of a child than anything else I can compare it to—indeed it appears to be a composition of the human and brute parts of the creation. It has three heads, viz. one on each shoulder, and another between them—the one in the middle is the exact representation of a human face, those on the shoulders no way differing from those of a dog. It has six legs, two of which stand upright on its back, and four tails. I forgot to mention that the middle head, instead of being covered with hair similar to that on the body, is furnished with black curly hair like that of a negro—and hands, instead of paws, are placed on the ends of those legs or arms (which ever they may be called) which stand upright on the back."

Leavenworth, Kansas. The *Leavenworth Daily Commercial* (Dec. 12, 1873, p. 4; ||y6zltuq3) said, "We have it from the best authority that a colored woman residing in that part of town known as Cincinnati, a day or two since gave birth to a monster. It was a child which, although otherwise perfect, had two heads like the heads of dogs. The strange and revolting monstrosity was, at last accounts, still alive."

Two pairs of conjoined dog-human twins, as depicted in Liceti (1665, p. 182). Those on the left represent the birth that supposedly took place in 1125 A.D., those on the right, the one in 854.[11]

Philadelphia. Randell C. Rosenberger (1896), a Philadelphia doctor, described a composite that recalls the Hesbaye dog-human. "A young lady," he said, "has a dog's body upon her back, the head between, and the paws resting upon, her shoulders."

Hesbaye, Belgium. Jesuit historian Bartholomaeus Fisen (1642, p. 357), in his eccleciastical history of Liège, *Historia ecclesiae leodiensis*, states that in the year 1125 a woman in the Hesbaye, a region of eastern Belgium, bore conjoined twins with two nearly complete bodies—one, a child's, the other, a dog's—joined at the rear so that they faced in opposite directions.

The Hesbaye twins (Schenck 1609, p. 74)

* Reports relating to the Hesbaye birth have perhaps been confounded with reports about similar, but distinct births. Authors differ about where and when the births in question occurred, and about whether there was one birth, two, or even three. Lycosthenes (1557, pp. 404-405) said one took place in Albania in 1126, perhaps mistaking *Haſbania* (Hesbaye) for *Halbania* (Albania). He also says such a birth occurred in the year 854 A.D. (ibid., p. 352). Thus, he writes, in translation, "At the end of Emperor Lothar's reign … a certain woman bore a

monster, that is, a kind of man and dog joined in a single body, joined at the back with the spines solidly merged." Two Holy Roman emperors were named Lothar, one in the ninth century, the other in the twelfth, which may have led different authors to assign different dates to these events. Alberic de Trois Fontaines (1698, p. 249) mentions two such births as having occurred in the year 1125, one in Hesbaye and one in Spain. The confusion here seems again to stem from the similarity of Latin names (*Hasbania* and *Hispania*). Gobelinus (1599, p. 223), too, records such a birth at about this time but gives no exact year or locale. Irenäus (1584) says, seemingly incorrectly, that Gobelinus mentioned two such births, and then refers to a third (citing Kaspar Goltwurm), with two heads—one human and facing forward, one canine, facing backward—as having been borne by a woman in 1175 A.D. (likely a mistake for 1125). Both Thomas Lirer (1751) and Johannes Wolf (1600, p. 321) mention such a birth "in the reign of Lothar." Liceti was sufficiently confident in the doubleness of these events to picture two pairs of conjoined twins in a single illustration.

A related canivulpid. Such births, in which two different kinds of organisms seem to be soldered together, are rare, but not unheard of. Thus, the Baron d'Arsac brought to Paris in 1873, from Poitou, a living conjoined twin composed of a complete female fox and a complete male bull terrier, joined along their sides (*L'Union Nationale,* Montpellier, Dec. 8, 1873, p. 3a; ||2s3evb5r; also see: ||9v8tsey3). The fox portion was supposedly identical to its mother, the terrier portion, to a bull terrier.

Lancaster, Pennsylvania. The *Lancaster Express* (quoted in the *Public Press,* Northumberland, Pa., Nov. 20, 1874, p. 3c; ||y4onhdqu) said,

> Yesterday there appeared at the office of District Attorney Rosenmiller a young woman who bore in her arms the most horrible monstrosity which we have ever seen.—She hails from Columbia [in Lancaster County] and visited the office for the purpose of prosecuting a recent lover for fornication and bastardy and for an attempted abortion. The object which the young mother bore in her arms was a child, apparently three or four months old, but without hands or feet. Its forehead is low, its head ill-shaped; but the most remarkable and horrible features of the monstrosity consist in this: Where the right hand should be there is a lump of flesh which closely resembles a dog's head, while the right ankle terminates in a projection that might easily be taken for a dog's leg!

* The Lancaster birth occurred on Nov. 13, 1874 (||y6xtxvu9). David Porter Rosenmiller Jr. (1841-1901) was mayor of Lancaster from 1884 to 1886 (||y3ugglzy).

FRONTAL PROBOSCIDES

The births in this section are tentatively classified as cynanthropoids

Andalusia. Aldrovandi (1642) pictured a creature, born in Andalusia in 1581, that from the neck down looked like a dog, though its digits were somewhat finger-like (see image at right). But the upper part of its head resembled a human being, especially the ear.

The Andalusian cynanthropoid. (Aldrovandi 1642, p. 378)

Traylee, Ireland. The *Freeman's Journal* (Dublin, Oct. 4, 1847, p. 1e; ||9vywx5th) said a Traylee newspaper had reported that "The pup of a spaniel in this town presents one of the most extraordinary freaks of nature we ever beheld. It is quite perfect, except the head. But here is indeed the wonder. From the top of the forehead projects

> what may be likened to an elephant's trunk, on the extremity of which are to be found the nostrils, just below that is to be found one eye, the only one the animal possesses, in the centre of the face, then the upper jaw is short and, though it projects a little, is furnished with no bone of any kind; The under jaw projects considerably, giving the profile something the appearance of that of an old person. This extraordinary animal was brought forth alive and ran until the superstitious fears of the owner and his neighbours induced him to drown it.

Bevergern, Nordrhein-Westfalen. Landois (1896c, p. 97) described a synophthalmic cyclops whelped at Bevergern on February 7, 1895. The presence of a frontal proboscis in this animal links it to the Andalusian and Traylee births.

Rangoon, Burma. A related report appeared in *Allen's Indian Mail* (Mar. 24, 1873, p. 270b; ||48xh6cr8), which said that "The RANGOON GAZETTE reports the birth of a curious monster. The body and legs are like those of a normal pup, but the head strongly resembles that of an elephant, being furnished with an unmistakable trunk."

Alice Elizabeth Doherty in infancy.

Petrus Gonsalvus and his wife Catherine (Artist: Joris Hoefnagel). It's said that this couple were the original inspiration for *Beauty and the Beast*.

HYPERTRICHOSIS

> *Tauron gives the name of Choromandae to a nation which dwell in the woods and have no proper voice. These people screech in a frightful manner; their bodies are covered with hair, their eyes are of a sea-green colour, and their teeth like those of a dog.*
>
> —PLINY, *Natural History,* 7.2

Just as there are many reports in which a doglike head replaces a human one, in many others dog skin and hair replace human skin and hair. Various individuals in whom the face was so affected have worked as sideshow performers. These qualify as CYNANTHROPOIDS.

Minneapolis, Minnesota. Alice Elizabeth Doherty, born in Minneapolis in 1887, had long facial hair like a sheepdog's. It may be that she had other canine traits, but none seem to have been recorded.

Tenerife, Canary Islands. Born ca. 1537 on Tenerife, Petrus Gonsalvus (ca. 1537-1618), known as the Canarian Werewolf, was a shaggy faced cynanthropoid of Guanche descent. The Guanche were an indigenous people of the Canary Islands.

Portrait of Barbara van Beck (b. ca. 1632, Augsburg, Germany, d. ca. 1668) by an unknown artist (||yyvajr8v). In his entry for September 15, 1657, diarist John Evelyn (1906, p. 123) wrote, "I saw the hairy woman, twenty years old, whom I had before seen as a child. She was born at Augsburg, in Germany. Her eyebrows were combed upwards, and all her forehead as thick and even as grows on any woman's head, neatly dressed; a very long lock of hair out of each ear; she had also a most prolix beard, and moustachios, with long locks growing on the middle of her nose, like an Iceland dog exactly, the colour of a bright brown, fine as well-dressed flax. She was now married and told me she had one child that was not hairy, nor were any of her parents, or relations. She was very well shaped and played well on the harpsichord." Evelyn's account is yet another piece of testimony indicating that cynanthropoids can be both intelligent and fertile.

Left: Stephan Bibrowski (1890-1932), known as "Lionel the Lion-faced Man."
Right: Fedor Jeftichew (1868-1904), known as "Jo-Jo the Dog-faced Man."

Wilczogora, Poland. Born in the village of Wilczogora in 1890, Stephan Bibrowski was about as hairy as was Alice Elizabeth Doherty. But unlike Doherty, he appears in photos to have had a doglike nose, so that his face closely resembled that of the Star War's character Chewbacca.

Irma Loustau, another individual affected by hypertrichosis. She was billed as the Panther Woman (Die Pantherdame).

St. Petersburg, Russia. Another sideshow performer with doglike facial hair, Fedor Jeftichew, was born in St. Petersburg in 1868. His father Adrian Jeftichew was similarly affected. Their case is interesting in that, if they were dog-human hybrids, it demonstrates that such composites are, at least in some cases, capable of producing offspring.

Ava, Burma. Another case which suggests that cynanthropoids can be fertile is that of Shwe-maong and his daughter Maphoon. In a record of the events occurring during an embassy from the Governor-General of India to the court of Ava, John Crawfurd (1829, pp. 184-187), a Fellow of the Royal Society, described a 30-year-old cynanthropoid named Shwe-maong on Nov. 4, 1827 (ergo, born ca. 1797). Crawfurd said he was hairier over his entire body than is a normal human being, but that his head and back bore thick hair like an animal Colonel Sir Henry Yule (1858) later met at Ava Shwe-maong's daughter. Though her face bore lengthy locks of dense hair, the rest of her body was mostly unaffected. "This was Maphoon," he wrote (p. 93), "the daughter of Shwe-maong, the 'Homo hirsutus' described and depicted in Crawfurd's narrative, where a portrait of her, as a young child, also appears. Not expecting such a visitor, one started and exclaimed involuntarily as there entered what at first sight seemed an absolute realization in the flesh of the dog-headed Anubis." Two of Maphoon's children were similarly affected. Crawfurd (1829) said she was two when he saw her in 1827 (ergo, born ca. 1825).

Shwe-maong (Crawfurd 1829)

Maphoon (Yule 1858).

Giessen, Hesse. Johann Salomo Ernst Schwabe (1787, p. 28) a professor of medicine at the University of Giessen, reported a woman bearing a fetus like a child from the neck down, but with a head like a dog's covered in a coat of doglike hair. The ears were not like a dog's, he said, but not like a human being's either. He also described its voice as being more like the whimper of a dog than a baby's cry.

Waco, Texas. From the *Dallas Daily Herald* (Dallas, Texas, Aug. 17, 1880, p. 1b; ||y4cxw8ru):

> WACO, Aug. 16.—Lottie Campbell, … twenty-one years old, yesterday gave birth to a monstrosity. It had the face of a child, but a nose that had no bone or bridge and only one nostril, a small round hole looking outward instead of downward, as in the natural organ. The hair began at the eyebrows and covered the entire head; was very long, straight and thick, resembling a Newfoundland pup. The jaws were also covered with long hair and flesh, and they hung down like a dog's upper lip. The spinal column was covered with very thick hair and was pasted like on a fat dog's back. The collar bone and chest were also covered with hair of astonishing growth for an infant.

* A Lotta Campbell, a servant, 18, lived in Waco in 1880 (||y3mn3w6g).

Most of the individuals thus far mentioned in this section had heads and bodies that were human in form but had heavy pelage covering their faces and heads. Doctors, however, have reported large nevi of doglike fur to occur on other parts of the body.

Grayville, Illinois. Dr. J. G. Stokes (1880), of Grayville, reported that

> On August 14, 1878, I attended a lady who gave birth to a very singular child. The child was a girl, large and natural, except the left arm and a small part of one ankle. The left arm, from near the shoulder-joint, down to the wrist, and almost encroaching upon the hand, presented all the appearances of a dog's leg. This anomalous arm at birth was covered with hair, one-half to three-quarters of an inch long, over the entire dorsal surface of the arm; but not so copious upon the inner surface; yet the hair covered the entire arm, and to all appearance was (and yet is) dog's hair. Its color is dark brown. The skin is a dark, bluish color, like unto a dog's and is furnished with bluish, nasty excrescences, scattered here and there over the inner surface of this arm. There was also an unnatural growth of the same nature upon one ankle, the size of a silver quarter.
>
> The whole unnatural condition of this arm appears to be an extra growth or covering, over a natural arm.

> The line of demarcation, or the beginning and ending of this strange covering is well defined. At the shoulder and wrist, it projects very perceptibly above the natural skin; and on first seeing it, one would suppose that if this extra covering could be removed (slipped off) as a glove, that a natural arm, skin and all, would remain. I measured both arms at birth and found the strange one to be three-quarters of an inch longer from the shoulder to the elbow than the natural arm; from the elbow to the wrist, they were near the same length. I again measured the child's arms when one year old and found the upper curve of this singular one one inch longer than the other, and the hair had grown to be fully one inch long. The hair looks like dog hair and is thickly set on the outer surface of the arm; so much so that the skin is completely hidden from view. As far as I can discover, the child is in every other respect normal.
>
> The child is large, handsome, and very intelligent; in the enjoyment of perfect health and has the perfect use of both arms. … I have been, and shall continue to watch the case closely, and am studying and investigating the cause of this (to all appearance) combination of human and dog.

Horley, Surrey, England. Charles F. Williamson (1890), a physician in Horley, said that on September 26, 1890, he delivered a "child" whose right thigh was entirely covered in doglike hair. "The colour was shiny black," he said, "and studded with long white hairs."

Stony Stratford, England. Dr. William Bull (1882), a surgeon at Stony Stratford, said a woman consulted him who was "reported to have always been passionately fond of dogs, and would nurse them for hours on her lap." In 1880, she had a daughter, who Bull described: "I found the greater part of the skin of the abdomen, right side, and the whole of the back, from the line drawn between the axillae above down to the anus below, and extending down the right thigh … hypertrophied, superpigmented, and covered with hair one inch in length." Pea-sized, or smaller, flecks of pigment were also present, but no aberrations of other kinds.

Upwell, Norfolk, England. W. J. Tubbs (1843), an Upwell physician, described a "girl" whose neck and torso were dotted with doglike hair.

Denning, Arkansas. Dr. J. C. B. Harrod (1904, p. 378), who practiced at Denning (||3ntj69xj), briefly described a "baby" with what looked like a four-inch-long patch of black dog hair on his belly.

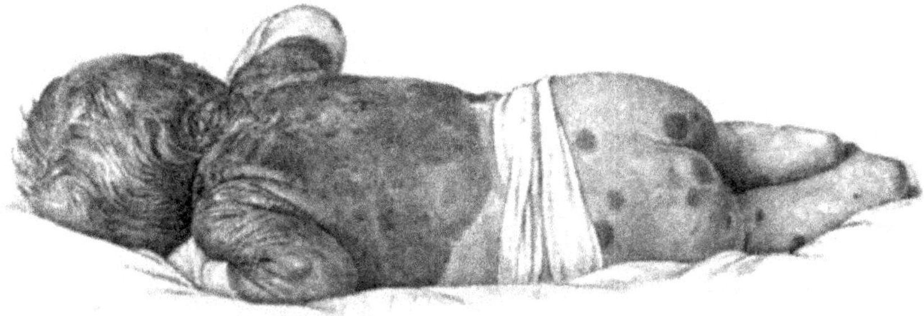

The Berlin homotherian (Joseph 1892).
Image: (Joseph in Mraček 1904, p. 545, fig. 9; ||9cmpdfr7).

HOMOTHERIANS (INCERTAE SEDIS HUMAN-ANIMAL COMPOSITES)

Some reports about hypertrichotic births do not mention dogs, but cynanthropoids are the commonest hairy homotherians borne by women.

Berlin. Joseph (1892) described an infant seen at his Berlin clinic. A deeply pigmented birthmark (nevus) covering most of the back and chest, as well as the upper arms. The nevus bore brown hair of varying length. It was sparse and short (less the 1 cm long) in some places but dense and long (4-5 cm) posteriorly. Joseph described it as similar to the pelt of a beast ("thierfellähnlicher"). So, this individual is perhaps best classified as an incertae sedis homotherian. Smaller nevi, some hairy, some not, dotted the body elsewhere. Within the nevi, the skin was chocolate brown, elsewhere white. No such markings were present in other family members. Although Joseph (1889) had previously described a similar case as having had bearlike fur, he made no such connection in the present case.

Sutton-in-Ashfield, Nottinghamshire, England. Waring-Curran (1869) described the right forearm of a "woman," a native of Sutton-in-Ashfield. The posterior part of the arm," he said,

> from the elbow to the wrist is covered with a sandy hair—the hair on the woman's head is quite black—fully an inch long, and very thickly set in a light brown hide or skin; the dorsal surface of the hand is also covered, except for the base of the thumb, with the same kind of integument and hair. The

anterior surface of the arm is made up of a soft fatty mass, from bend of elbow to wrist joint. It is impossible to find the flexor tendons through it; turning up the arm, this soft mass hangs pendant and wrinkled for three inches from the bones by its own weight, the fingers are small and crooked, and the arm is almost useless to the woman.

London. Sir William Lawrence (1830, p. 228) reported a case of hypertrichosis involving markings like those in the Berlin case but mentioned no resemblance to any particular animal. "I saw a curious looking child, a little time ago, a gentleman sent for me to see it as a curiosity. I think that a great part of the body was covered with hair. Over the lower extremities there were a great number of spots; one on the arm, and

Sir William Lawrence
1783-1867

others on the trunk of the body. The hair covering the child, in other respects, was of the natural texture, and indeed, these parts were not unhealthy.

*. Sir William Lawrence, 1st Baronet FRCS FRS (b. July 16, 1783, d. July 5, 1867), was an English surgeon who became President of the Royal College of Surgeons of London and Serjeant Surgeon to the Queen.

Trieste, Italy. La Verita (Waterbury, Conn., Feb. 5, 1927, p. 2a; ||4xtj5fwa) said in translation,

TRIESTE.—Forty-year-old Anna Lukas, wife of a postal service employee, gave birth to a perfectly healthy baby covered from the abdomen down to the legs in thick, long, fawn-colored hair, resembling fur. According to doctors, it is a very large pigmentary mole, which cannot be subjected to any dermatological treatment. The mother believes it was caused by her having wished for a lady's fur coat.

NORMAL VARIATION OR CANINE PARENTAGE?

There are, of course, many who deem human-animal hybrids impossible. These would claim that all the reports about cynanthropoids are (1) false or (2) refer to individuals that were in fact wholly human and were merely described as part dog because they had certain dog-like characteristics. In other words, they would assume that any supposedly canine traits reported in fact fall within the normal range of human variation and were mistakenly attributed to canine parentage. This sort of argument dates back to Aristotle, who wrote that in such cases,

Benedetto Varchi
1503-1565

> people say that the child has the head of a ram or a bull and so on with other animals, as that a calf has the head of a child or a sheep that of an ox … but they are none of the things they are said to be. There is only some similarity, such as may arise even where there is no defect of growth. [Barnes 1995, *Generation of Animals,* IV, iii, p. 1191]

But what would Aristotle have said about a birth such as the one Magius and Varchi say occurred in Avignon in 1543—a child with a human face, but otherwise mostly dog-like, though the offspring of a woman? And what would he say of Stalpart vander Wiel's hairless puppy, also supposedly born to a woman? Would he still have argued that such things fall within the normal range of human variation? Would not doing so alter the very meaning of the word *human?*

Saint with cynocephali (*Kievan Psalter,* 1397, artist unknown).

Cynocephali (*Livre des merveilles du monde,* 13th-century).

MYTH

Angamanain is a very large island. The people have no king and are idolaters, no better than wild beasts. And I assure you all the men of this island of Angamanain have heads like dogs, and teeth and eyes likewise; in fact, in the face they are all just like big mastiff dogs!

—MARCO POLO

Cynocephaly, the condition of having a human body and the head of a dog, is described in the lore of many different cultures. It was once widely believed that entire races of dog-headed men existed. Thus, Augustine (*City of God*, 16:8) expresses his puzzlement on the topic: "What shall I say of the Cynocephali, whose dog-like head and barking proclaim them beasts rather than men?" Pliny (*Natural History*, 7.2) said that in the mountains of India

> there is a tribe of men who have the heads of dogs and clothe themselves with the skins of wild beasts. Instead of speaking, they bark; and, furnished with claws, they live by hunting and catching birds.

St. Christopher

In Eastern Orthodox iconography, another Christian holy man, St. Christopher, was often represented as something less than wholly man, that is, as a cynocephalus.

Many medieval scholars seemed more concerned with how such creatures should be classified than with whether they existed. Thus, in a letter, the Benedictine monk Ratramnus of Corbie wrote, in translation,

> So, you would like to know what you should believe about the Cynocephali, whether you should deem them descendants of Adam, or count them as mere animals?[2]

Anubis (2nd century A.D.)

Indeed, the Romans believed that such births could be produced by human mothers. Thus, Phlegon of Tralles, the second-century author, says that in 65 A.D., "The wife of Cornelius Gallicanus gave birth near Rome to a child having the head of Anubis"[3] (that is, having the head of a dog).

And belief in the existence of dog-humans—and of wolf-humans—was rife not only in ancient and medieval Europe, but also in many other parts of the world. In *Myths of the Dog-man* (1991, p. 137), David Gordon White comments that

> Perhaps by virtue of a Mongol influence, the lupine ancestry myth is one that is widespread among the peoples of central Asia; but so too is the canine ancestry myth. Whether the latter is a variant of the former or an independent tradition that has become conflated with the former, as in the Mongol tradition, is an insoluble problem.

Cynocephalic gargoyle (Moulins Cathedral, France; ||ytc2rjxj).

Similarly, Nansen (1893, p. 272) noted that "myths of descent from dogs (or wolves, or bears) occur among many races, Aryan as well as Mongolian or American. They lie at the root of the mythology of many Indian tribes, who hold that the first woman took a dog to mate, and that they themselves are descended from this connection."[4]

In their book *The Mythology of Dogs* (1997), Hausman and Hausman make similar comments about the Ainu people of northern Japan: "The word 'Ainu' in Japanese means 'sons of dogs.' The respect accorded to these animals comes from a legend which states that Ainus are people descended from the union of a dog and a woman."[5]

The Egyptians worshiped various dog-headed gods, including the major deity Anubis. In Mesopotamia, the Sumerians revered the goddess Bau, who, under the name of Nininsina, was long represented with a dog's head. Even today, effigies of dog-headed men stand duty as temple demons in the wats of Thailand.

Anubis (*Book of the Dead,* ca. 1275 B.C.).

There are many traditions, too, that relate how women have been ravished by dogs. In the *Dhammapada* (verse 151), the Buddha utters a germane anecdote about Queen Mallika, wife of King Prasenajit (Bullough 1976):

> One day, Queen Mallika went in to bathe, and her pet dog came with her. And as she was bending to cleanse her feet, he mounted her, and she was not displeased. But the king observed this through the window. And when she came in, he cried, "O! Wicked woman! What did you there with that dog? Do not deny those things I saw with my own eyes!" The queen replied that she was only bathing and therefore had transgressed not. Then she said, "But that chamber is most odd. To those looking in from the window, anyone alone there seems to be two. If you believe me not, O King, I entreat you go into that chamber that I might look upon you through the window."
>
> So, the king did enter there. And when he came forth, Mallika asked why he had been transgressing there with a she-goat. He denied it, but she insisted she had seen him with her own eyes. The king was baffled, but gullible as he was, he agreed that that chamber was most peculiar indeed.

Such tales are legion. One Native American tale (from Boas 1917, p. 130) tells of a woman who produced a litter with a dog:

> THE DOG-CHILDREN
>
> Once there was a girl who lived near the mouth of the Fraser River. She refused all offers of marriage. At last, a strange young man visited her at night and lay with her. She wondered who he could be and made up her mind to mark him. She put red paint on the palms of her hands; and when he appeared the following night in the dark, she embraced him, leaving the imprints of her hands on his sides. She never went out of the house, but the next morning she went out to see if she could recognize the young man. The young men were playing, and called out, "Oh, see the girl! She has come to see us play." As she could see none of them with any marks, she went home. When near the house, she saw her father's large dog being fed by her mother, and on his sides were her hand-marks. Her mother said, "Who has been making a fool of the dog by painting his sides?" The girl was ashamed,

went in, and cried to herself. In due time she gave birth to eleven pups—five male and six female. One of the latter was half black and half white. The people were very angry. They beat the dog nearly to death and left the girl and her children to die. When they were gone, the dog became a man, and went into the woods, where he healed himself. The pups were hungry: therefore, their mother went at night with a torch to dig clams on the beach during ebbtide. On her return, when near the house, she heard the pups dancing, and singing, "She thinks we are dogs, but we are children." The black and white one was on watch and warned the others of their mother's approach; so, they all hastily donned their skins and kept quiet. She looked about and saw children's tracks where they had been dancing. She said, "It is strange that you are dogs, and still you give no warning, nor tell when strangers are around dancing." The following night, when she went out after clams, she put her robe on a stick, tied her torch to another one by its side, and hurried home. The pups thought she was still at the beach and kept on dancing and singing. She crept up stealthily, jumped over the one on watch, and seized the skins of the others before they could get them, and threw them into the fire. Thus, they remained children, while the black and white one remained a dog.

Now, their father returned in the form of a good-looking man and hunted for the family. He killed many goats and deer, and soon had great quantities of meat and fat. He put much fat into the caches of those people who had left some fish for the girl, and into the caches of those who had left nothing he put only bones. Now, the girl's grandmother pitied her and sent Crow with some fish for her. Dog-Man gave Crow fat to take back to the old woman, and by this it became known how well off the girl had become. Then the people all returned, and were fed by Dog-Man. The ten children of the girl grew up to be handsome people, and they married among themselves.

CYNOPITHECINES (DOG-APES)

> *However large the difference in their gestation periods, it is well known that a kind of dog can be easily bred from a dog and an ape.*
>
> GEROLAMO CARDANO, *De rerum varietate*[15]

Call animals mixing the traits of monkeys and dogs *cynopithecines*. Blumenbach (1776, p. 11) reported two cases of dogs whelping such animals. In the first instance, he failed to identify the primate sire, but his description suggests a Golden Lion Tamarin (*Leontopithecus rosalia*). In the second, he stated that the sire was a Diana monkey (*Cercopithecus diana*). In translation, he said, "As to whether monkeys and dogs can cross, I hesitate. They do seem very distinct. But I know of two dogs impregnated by monkeys, cases I deem a violation of the natural order. The first took place in Schwarzburg [Thuringia]. His Excellency [Christoph Gottlieb] Büttner kindly sent me a depiction of this hybrid. It was like a small domestic dog of a golden-yellow color, but in its eyes, ears, and shaggy mane, it differed from ordinary dogs and indeed in these respects resembled the father. The other, reported by an unimpeachable eyewitness, occurred some ten years ago in Frankfurt. A bitch, covered by a Diana monkey,[16] whelped a puppy that in its ferocity, disposition, humped back, and long tail, was exactly like its father.

J. F. Blumenbach
1752-1840

Christoph Gottlieb Büttner
1708-1776

Chester, England. In translation, Gerald of Wales (*Journey through Wales*, 2.11, ch. 11) related how in Chester,

> during our time [i.e., in the twelfth century A.D.], a dog was impregnated by a monkey and eventually produced a litter of whelps resembling a monkey in their foreparts, but a dog behind. At once, however, they were all killed by one of the peasant keepers of the inn where they were born, who was shocked at the strangeness of the prodigy and by their hideous double nature. He dispatched them with a staff he happened to have in his hand, but not without incurring the anger of his lord, who was indignant at this damage to his property.

Nairobi, Kenya. A video recorded in Nairobi in 2024 provides a clear view of an ostensible baboon-dog (||4cudu92z).

Johannesburg, South Africa. The *Sydney Morning Herald* (Apr. 13, 1901, p. 7g; ||y4tbfgx6) described an exhibited animal as

> a wonderful "freak" in the way of dogs. He seemed to me to be half dog and half baboon, he having the forequarters of the former and the hind of the latter. A black, white and tan, he had a good deal of the show head of a collie, and he is taller on his forelegs than his hind ones. He tapers off behind, and his short dock certainly gives him the appearance of the simian. M'Gregor [the ringmaster] told me he came from Johannesburg.

This description of the Johannesburg animal matches animals depicted in South African veterinary texts. For example, H. P. A. de Boom (Veterinary Laboratory, Onderstepoort) provided pictures of "baboon-dogs" (in Hutt 1964). Outside South Africa, this condition is known as "short-spine disorder" and is generally attributed to mutation, not hybridity (Hansen 1968). The viability of such animals is reduced, and males have not been reared, but females have reached maturity and produced offspring (Hutt 1964, p. 475).

* Hutt's comments about the viability and fertility of female baboon-dogs is consistent with Haldane's Rule, which states that "when one sex is absent, rare or sterile in the F_1 generation of an animal cross, it is always the heterozygous sex" (Haldane 1922). Thus, in a mammalian cross, if hybrids of one sex are more viable and/or fertile than the other, they will usually be female (among mammals, the heterozygous sex is usually the male).

A baboon-dog: A viable cynopithecine? Image: H. P. A. de Boom.

A pregnant baboon-dog (from Hutt 1964). This animal looks like an ordinary dog viewed in a funhouse mirror.

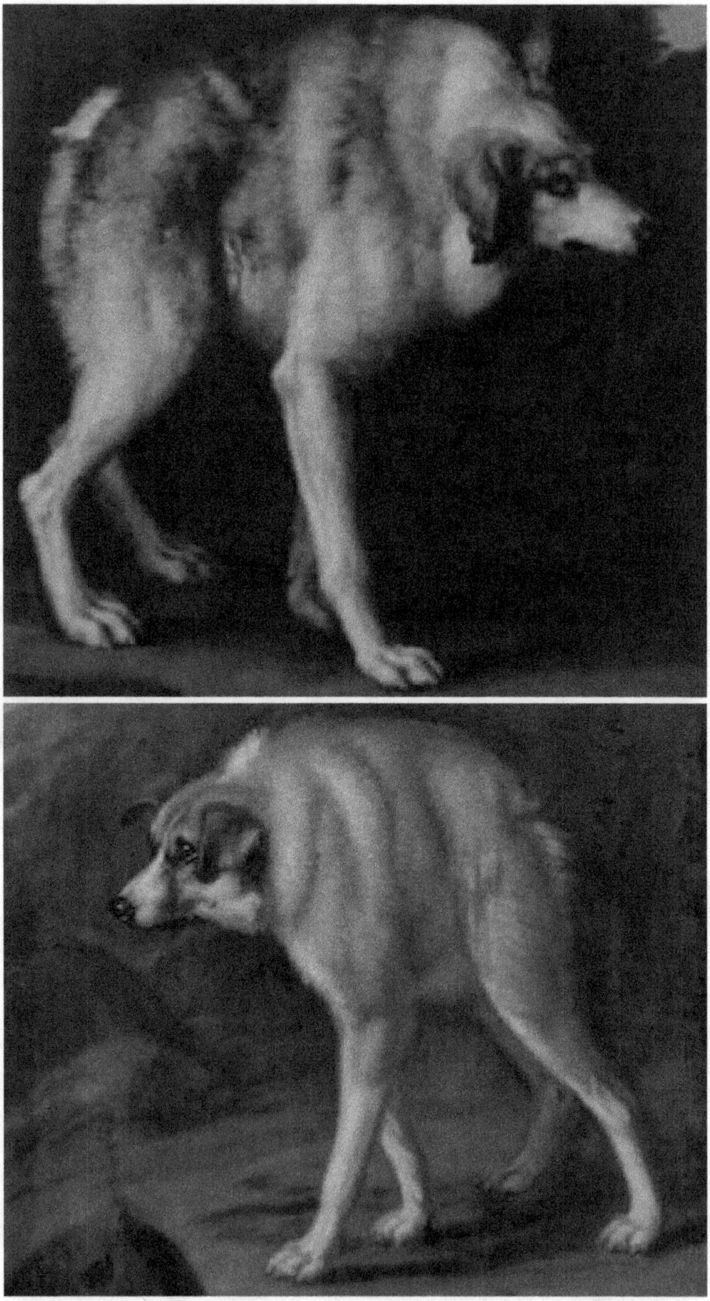

Baboon-dogs (D. K. Ehrenstrahl, ca. 1690; ||y6qvszme, ||y32gfnqs).

A comparison of a baboon-dog with a chacma baboon (*Papio ursinus*), also known as the Cape Baboon, native to South Africa. The Chacma Baboon occurs throughout Transvaal, where the pictures on page 262 were taken.

Indeed, one YouTube video shows wild yellow baboons (*Papio cynocephalus*) forming permanent, family associations with dogs (||loulxhf). *Papio* is New Latin for baboon, and *cynocephalus,* derived from Greek, means "dog-head" and refers to the fact that these baboons have doglike heads. *Papio cynocephalus* therefore means "dog-headed baboon."

Papio cynocephalus

Newcastle, South Africa. A notice in the *Baltimore Sun* (July 13, 1941, sec. 1, p. 5f; ||yxu6mul4) described a baboon-dog with hands. No ordinary dog has hands. But a cynopithecine might.

> ORIGINAL BABOON DOG DIES IN SOUTH AFRICA
>
> South Africa's original baboon dog is dead on the farm of Arthur Steel near Newcastle. Steel had three of these strange animals. The original was a freak in a litter of normal pups.
>
> Although a dog, it had many characteristics of the baboon. It could hold food in its hand, which resembled the hand of a baboon, and open a door with this hand, and sit and walk just like a baboon. Moreover, it could climb a four-foot wall with ease.

The kinkajou, an animal intermediate between spider monkey and olingo.

Kinkajous. Taxonomists once classified the Kinkajou (*Potos flavus*) as a primate, calling it *Lemur flavus* (Kays 2001, 2003; Kays and Gittleman (1995, p. 300).). But it is now assigned to Order Carnivora, though its diet consists mainly of fruit, particularly figs. Kays (1999) says its frugivorous habits are similar to those of spider monkeys. Its manipulatory abilities rival those of primates. Like spider monkeys, and many other monkeys sharing its Central and South American range, it has a fully prehensile tail but lacks an opposable thumb. The tails of other carnivores are not prehensile. And, like New World monkeys, and unlike other creatures categorized as carnivores, kinkajous are fully arboreal.

A spider monkey (*Ateles*)

Thus, while they do not exhibit any specifically doglike characteristics, they do combine traits of Order Carnivora, the order to which dogs belong, with those of Order Primates, and are therefore mentioned here in connection with cynopithecines.

* The nocturnally active Olingo (*Bassaricyon*) shares the habitat of *Potos* and is its closest ally among carnivores (Nowak 1999). It is therefore a likely candidate to be investigated in connection with the parentage of *P. flavus.* Unlike most primates, kinkajous are nocturnal.

NOTES

1. Hastings et al. (1912, p. 562b; ||o4ftz9u).
2. *Epistola de cynocephalis ad Rimbertum presbyterum scripta* (Ratramnus 1852, 1153). Translation mine.
3. *Book of Marvels,* Hansen translation (1996). Phegon says Aulus Licinius Nerva Silianus and Marcus Vestinus Atticus were consuls (65 A.D.).
4. An Inuit story tells of a woman who marries a dog and has hybrid offspring, called Adlets in the Inuit tongue (||y5cht9y8).
5. In certain cultures people still sometimes marry dogs. For instance, the parents of a Hindu girl recently arranged for her to marry a stray dog (||yxwebj8t).
6. Lynch (1662, p. 270) also mentions this birth. Note that the translator miscopies Lynch's "1493" as "1433" (ibid., p. 269).
7. Translated in Zirkle (1935, p. 56). Bartholin cited Johannes Lange (I. 2. *Epist. Med.* 9), who himself cites Volaterranus (Lange 1605, p. 377). Simeone Maiolo (*Dies caniculares,* vol. I, 42; ||n9djnbv) also mentions this birth.
8. Brunet (1844, p. 235). Translation mine. Also see Lauterbach (1872, p. 29, Feb. 13, 1538, *De spirituum apparitione).*
9. De Ceglia's translation of a passage in *Quaestiones* (Vol. III, Cons. XXII, p. 28).
10. The pontificate of Pius III followed immediately upon that of Alexander VI and lasted only 26 days (Sep. 22, 1503 - Oct. 18, 1503).
11. Source of epigraph: Ballantyne (1894, p. 140).
12. Nitzschka (1685, p. 465) made the same claim.
13. See too, Praetorius (1677, pp. 345-346).
14. Irenäus (1584) also mentions this birth.
15. Cardano (1565, p. 315). Original Latin: "Quamvis diuersitas temporis gestationis uteri accederet, constat facilius ex cane & simia canem posse procreari."
16. *Simia diana,* the name used by Blumenbach, is a synonym of *Cercopithecus diana.*
17. Colophon: A cynocephalus from an early edition of *The Voyage and Travels of Sir John Mandeville.*

An early depiction of an ape abducting a woman (engraving by John Chapman 1793, after J. B. Ihle, Wellcome Library no. 39699i). The caption reads, "The Orang-Outang carrying off a Negro Girl."

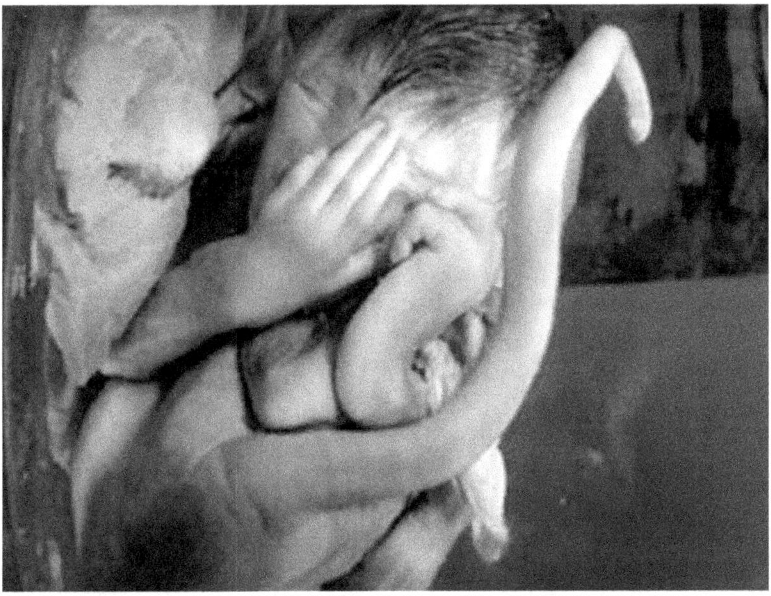

A homosimian with a long tail and handlike hindfoot (Museum Vrolik).

The only animal which possesses the least shadow of a possibility of ever being able to impregnate the female of our own species besides her own appropriate lord, would seem to be the ourang-outang of Borneo and the East.

—THOMAS W. BLATCHFORD

Observations on Equivocal Generation, 1844

HOMOSIMIANS (APE-HUMANS)

Call any creature mixing the traits of a human with those of a non-human primate a *homosimian.* Some of the reports about women bearing such offspring mention, or at least imply, that chimpanzees, bonobos or orangutans, were the sires. But others implicate monkeys, especially macaques and baboons.

Given that humans and all apes and monkeys are assigned to the same taxonomic order (Primates), homosimians are not telenothians. But many of the accounts about women being abducted and raped by apes are reminiscent of reports about women being abducted and raped by bears. HOMINURSIDS (bear-humans), covered in the following chapter, are in fact telenothians.

REPORTS

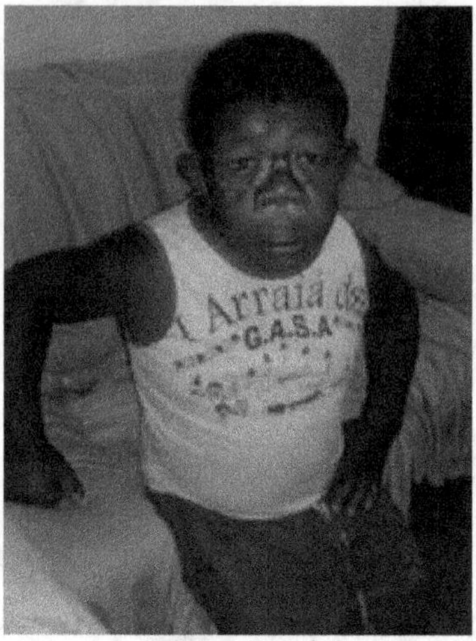

Augusto Dembo

Luanda, Angola. In January 2020, reports went viral about an Angolan woman claiming on her deathbed that she had had a long-term sexual relationship with a wild chimpanzee and that her 12-year-old son (ergo, born ca. 2008), Augusto Dembo, was a hybrid resulting from that relationship. But photographs cannot prove hybrid ancestry, and it seems Augusto has not been examined via genetic testing. Augusto was, according to reports, living with his uncle and aunt in Luanda in March 2021.

The Franciscan capuchin friar Antonio Zucchelli (1715, pp. 147-148) told, in translation, of Luandan women impregnated by apes.

> All the monkeys or apes found here differ from those in Brazil. Here they are of all sizes, small, medium and large. And because they are all extremely lustful, it indeed very often happens that the blacks have intercourse with these apes, They are impregnated by them and ultimately bring forth monsters. A few years ago, the same occurred here in the city of Luanda. Two black women impregnated in this way, when their time came, gave birth to monkeys. [The translation of this last sentence, unclear in the German edition, was based on the Italian original (Zucchelli 1712).] One of these fell to the lot of the Governor, which he plans to take back to Lisbon on his return to Portugal. The other was given to our brotherhood here in Luanda, which also lived for several years at the hospital before he died. Those who saw him tell me he truly had features more like those of a human being than a monkey.

* Zucchelli's "macachi grandi," which cannot here refer to a macaque (as macaques do not occur in sub-Saharan Africa), is here translated as "apes." The only large primates native to the vicinity of Luanda are yellow baboons (*Papio cynocephalus*), chimpanzees (*Pan troglodytes*) and gorillas (*Gorilla gorilla*).

* There are various other early reports from the Guinea region specifically referring to hybridization between humans and apes. Historically, the name *Guinea* referred, not to the modern country, but to the vast region adjacent to the Gulf of Guinea (all the countries between, and including, Ghana and Angola).

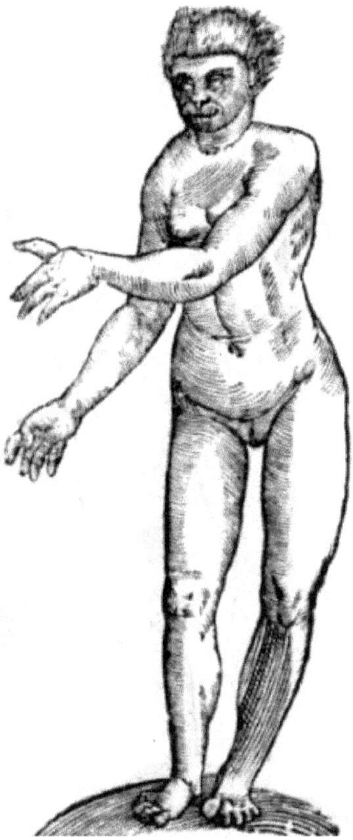

Guinea. Bogdan (1988, p. 25) said that in 1738 a colonial American newspaper carried an advertisement about a homosimian. The creature, it said, "was taken in a wood at Guinea; 'tis a female about four feet high in every part like a woman excepting her head which nearly resembles the ape" a description that matches a homosimian depicted by Aldrovandi (1642, p. 469; see image, right). The composite to which Bogdan referred was, apparently, the same reported in the *London Magazine* (Sep. 1738, p. 464): "A most surprizing Creature is brought over in the *Speaker,* just arrived from Carolina, that was taken in a Wood at Guinea: it is a Female about four Foot high, shaped in every Part like a Woman, excepting its Head, which nearly resembles the Ape: she walks upright naturally, sits down to her Food, which is chiefly Greens, and feeds herself with her Hands as a human Creature. She is very fond of a Boy on board,

> and is observ'd always sorrowful at his Absence: She is cloathed with a thin Silk Vestment and shews a great Discontent at the opening of her Gown to discover her Sex. She is the female of the Creature, which the *Angolans* call Chimpanze [*sic*], or the Mockman.

Bostota, Slovakia. The *Innsbrucker Nachrichten* (Innsbruck, Jun. 19, 1926, p. 8c; ||bmdfjjby) said, in translation, "HOUNDED BY NEIGHBORS TO HER DEATH.—BRATISLAVA, Jun. 18.—A farmer's wife in the Slovakian village of Bostota recently gave birth to a hairy apelike child, a frightful monstrosity with seven digits on each hand and foot. Because her neighbors accused her of sodomy she leapt with her child down a well. They both drowned."

Orange Park, Florida. The *Sun* (May 10, 2018; ||y858njn4) interviewed Gordon Gallup, a psychologist at the State University of New York at Albany, who said a viable human-chimpanzee hybrid had been produced during the 1930s at the Yerkes Laboratories of Primate Biology, then located in Orange Park. He claimed that

> They inseminated a female chimpanzee with human semen from an undisclosed donor and claimed not only that pregnancy occurred, but the pregnancy went full term and resulted in a live birth.
>
> But in a matter of days, or a few weeks, they began to consider the moral and ethical considerations, and the infant was euthanised.

Malaya. From the *Queensland Times* (Ipswich, Jan. 2, 1954, p. 2; ||lhu6l97; also see: ||mjbnznr & ||lxx8crb):

> HALF HUMAN, HALF APE ANIMALS SEEN
>
> SINGAPORE, Jan. 1.—Three hairy, fanged creatures, one female and two male, who appear to be half human and half ape, have emerged from the edges of a 150-mile tract of dense jungle in northern Malaya and scared the local inhabitants.
>
> A handful of people who saw the creatures described them as tall, powerfully built, and covered in hair, with unkempt hair from their heads hanging down to their waists.
>
> All had long dog-like fangs which could still be seen when they had their mouths closed.
>
> Both men and the woman had moustaches, and all were said to smell of animal.
>
> First to see the creatures was a Chinese girl rubber tapper who was surprised by them on the outskirts of a rubber plantation.
>
> She came running back to her village in hysterics and when the plantation manager, Mr. Sam Browne, finally got her story straightened out, he sent Home Guards men through the plantation on patrol.
>
> Corporal Wahib, who had previously shown his steadiness in killing terrorists in a jungle fight, found the three creatures on the banks of a river digging for tapioca with their bare bands.
>
> They caught sight of him, dived into the river, and hid under some overhanging scrub.
>
> The following day a Malay child playing by the river saw the creatures again.
>
> They were swimming.
>
> The child ran back to the village and sobbed out her story.

On December 27 a Tamil (South Indian) tapper on the plantation felt a pair of woman's arms steal around him while he was bending over the bole of a rubber tree.

He turned and saw the hideous fanged face of a female and her two companions and smelt their animal odour, so he burst clear.

His shirt ripped and he fell down the bank into some water.

The last he saw of the creatures, as he ran away, they were standing round and appeared to be laughing heartily at him.

Radio Malaya broadcast an interview with all witnesses in their own languages and all descriptions tallied.

The only point of difference was about the rudimentary clothes they were wearing.

Some said they wore bark loin cloths, others said skins.

All natives and European settlers in the regions are positive the creatures exist and say that, as they seem to be quite young, there are probably colonies of them back in the jungle which, in places, is impenetrable.

A subsequent story about the same event appeared in the *Barrier Miner* (Broken Hill, New South Wales, Jan. 5, 1954, p. 6d; ||53sx334k).

LONDON, MONDAY: British scientists reject reports that hairy half-human, half-ape creatures are roaming the Malayan jungle.

Malayan rubber tappers recently have claimed six times to have seen these creatures.

Malayan scientists hope to send an expedition to the Perak Jungle to look for the apemen.

In Britain Sir Richard Winstedt, who has studied Malaya's aborigines, said: "There's no chance of there being any unknown races left in the Malayan jungle."

Twenty-five years ago an Englishwoman reported seeing a bare-breasted ape-woman with hair down to her waist.

The Englishwoman's husband thought his wife was suffering from hallucinations. He sent her back to England.

But her description of the ape creature exactly fits the latest one circulated to Malayan police.

The police are now searching the fringe of the jungle for three ape-like creatures, two men and a woman.

Eyewitnesses said the men have darker skins than the woman.

They said none was wearing anything except a small piece of bark around the loins.

Memphis, Tennessee. Speaking of a case that came under his personal observation, Memphis physician James Madison Walton (1907, case 4) discussed the case of "Mrs. S., the mother of three well-formed and intelligent children. The fourth child, when born, resembled a monkey much more than it did a human being. Its body was covered with long black hair, especially on the back and arms. The face was shaped like that of a monkey, and covered with hair, the same as the hair on its head. The forehead was missing, and the mouth and nose were in the shape and location of those of a monkey. The body and limbs, and, in fact, all the features, resembling a monkey's more than it did a human being. The monstrosity lived only a few hours."

* A James Madison Walton, a physician, lived in Memphis in 1900 (||2jtu4d8n).

Pen Yan, New York. The *Buffalo Times* (Buffalo, N.Y., Apr. 20, 1903, p. 6; ||2cx8uj53) reported that a veiled nondescript had stopped off in Buffalo that morning at the railway station en route to Chicago. Among other things, the account said, "The freak came into the waiting room accompanied by a country woman. They had come from the neighborhood of Penn Yan, but just where, it was impossible to learn. The woman, when questioned, said she could say very little about the freak, as she was under a contract which prevented her doing it. She described the freak as about four feet high, with the body and legs of a human being. The head, neck and arms, however, are those of an ape. The woman said the head was that of a perfect ape and this was confirmed by the outlines of the head as seen under the veil. The head is covered with hair, and this hair runs down over the body. It is long, like that of an animal. The woman would say nothing about the birth of the freak; was silent on the matter of its parentage and also as to the place of its birth."

Wethersfield, New York. The *Star-Gazette* (Elmira, N.Y., May 26. 1894, p. 6d; ||yaxl4oh8) said that at Wethersfield, "a remarkably queer offspring has been born to Mr. and Mrs. Lewis Walker of that place. The body and face of the infant, it is said, is covered completely with a heavy growth of hair and in appearance it strongly resembles a monkey. The unfortunate freak of nature is attracting great interest, and many curious ones are anxious to get sight of it."

* Wyoming County resident Lewis Jacob Danforth Walker (1849-1913), and his wife, Susan (1860-1908), lived in Wethersfield in 1894. Records show they had a son, Boyd, born April 6, 1894, who was still alive in 1915, at which time he was working as a laborer on his older brother George's farm in Wethersfield. His World War II draft card shows Boyd was living at Arcade, in Wyoming Co., in 1942. The card bore his scrawled signature. Thus, whether he was an ape-human hybrid or not, Boyd Walker could write and reached at least the age of 48.

Budapest. The *Cincinnati Enquirer* (Oct. 4, 1891, p. 18; ||yckhbhby) said, "General O'Beirne has on his hands at the barge office a human monstrosity. Its name is Herman Ginsler, and it arrived with its mother and several brothers and sisters on the Kaiser Wilhelm yesterday. The father, John Ginsler, has been in this country two years. The family are Hungarians, and the unfortunate child was born near Buda Pesti, about ten years ago [ergo, born ca. 1881]. The child's face is ape-like in the extreme, and the head measures but fourteen inches in circumference. The hands and feet are abnormally large, and the natural position assumed is squatting upon the haunches. When it walks it hops along on hands and feet. General O'Beirne will demand a continuous bond of $5,000 before he permits the freak to land, as is required in the case of all imbeciles."

Parkersburg, West Virginia. From the *St. Louis Post-Dispatch* (Oct. 7, 1888, p. 6b): "PARKERSBURG, W. VA., October 6.—A woman of this city has just given birth to a hideous monstrosity. Its head, shoulders and part of its body exactly resembles an ape. It is said she attended a circus and became frightened at an old ape in the monkey cage."

Henderson County, Kentucky. The *State Sentinel* (Indianapolis, Ind., Oct. 31, 1883, p. 5a) said, "A negro woman in Henderson County, Ky., gave birth to a monstrosity one day last week, in the shape of a half human, half ape. From the waist up, it was a perfectly formed ape, while the lower extremities were human in every particular. It was born dead."

* Most descriptions of homosimians say they resembled apes in their upper parts and humans in their lower. Many hybrids do have a composite appearance in which the front part of the body is closely similar to one parent and the hind portion, similar to that of the other (as has been seen in many reports cited in this book). So it is in fact possible, at least in terms of the general rules of hybridization, that the creature described in the Henderson County report actually did have an upper half like an ape, but a lower one like a human being.

Honfleur, France. The *Rising Son* (Kansas City, Mo., Mar. 27, 1903, p. 2f; ||33m6yzym) said,

> A woman named Blondel, living at Honfleur, France, recently gave birth to a female child with extremities like those of a monkey. The hands and feet have only four fingers and toes respectively, but instead of nails they have long claws, while the skin hangs like a bag. It is not expected that the child will live.

* Some New World monkeys have claws (e.g., tamarins, marmosets, capuchins).

Wilmington, North Carolina. The *Wilmington Star* (quoted in the *Weekly Pioneer,* Asheville, N.C., Apr. 4, 1874, p. 1f; ||4ent7ft8) reported a case in which the sire was likely a capuchin, the typical organ grinder's monkey: "One of the most mysterious and unaccountable freaks of Nature that we have heard or read of in a long time has come to light in this community. Sometime during last week a child was born in that part of the city known as Brooklyn, the head of which in almost every particular resembled that of a monkey. The child was perfectly formed in every respect save the extreme length and slenderness of its arms, and formation of its head, neck and

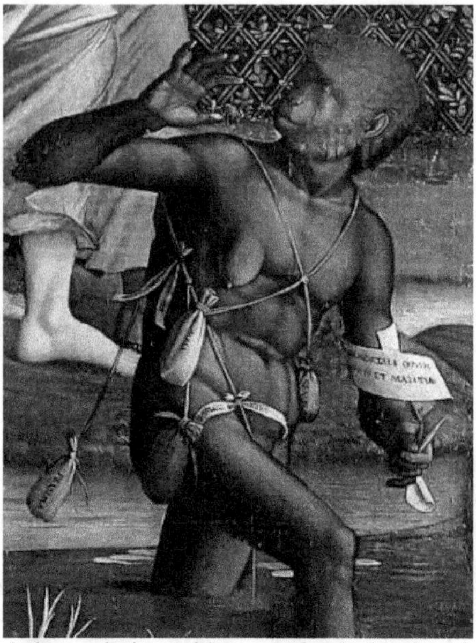

A homosimian with a monkey's head and human body. Detail from *Athena Driving the Vices from the Garden of Virtue* (Mantegna).

> face, and facial organs. The ears, eyes, nose, mouth, forehead and arms are indisputably those of a monkey in all the characteristics of form and feature. On the head is a formation of peculiar flaky skin, or scale (dark, light and brown colors), resembling the skull cap usually worn by an organ grinder's monkey, and it would seem was intended by nature to carry out this representation. The child was stillborn, and we learn will be preserved in alcohol for the benefit of science.

Vienna. The *Innsbrucker Nachrichten* (Nov. 10, 1863, p. 2493; ||3bub5d7h) said, in translation,

> An unusual monstrosity was born in Herr Professor Braun's clinic at the Vienna Maternity Hospital. A woman from Hietzing bore a daughter who had the front part of her head like an ape's. The chin, mouth and nose are not proportioned as in a human being. … The eyes and ears are normal.

Watertown, New York. Orson Squire Fowler (1844, p. 221) described a homosimian he saw in Watertown as "a child whose looks, actions, and shape of head, bore a close resemblance to those of the monkey."

Paris. The *Boletín Mercantil de Puerto Rico* (San Juan, P.R., Jul. 23, 1875, p. 3f; ||bddwbea4) said, in translation,

> MAN-MONKEY.—A singular creature has just been born in one of the poorest neighborhoods of Paris: a child-monkey, which seems to be perfectly healthy. This freak has been blessed with a long tail, and his body is covered with fine hair, red and silky. Like monkeys, this creature is four-handed ... He has four teeth already in his upper jaw and six in the lower, and this precious dentition prevents nursing, and he is being raised on a bottle. The father, a carpenter, was so shocked by the birth that he has left the house ... he cannot bring himself to accept the idea of being the father of a man-monkey, all the more so given that the doctors expect the creature to live. Still, he has no real reason to be upset. In this day, abundant in freaks and monsters, he will see that he has a fortune.

Churuguara, Falcón, Venezuela. El Constitucional (Caracas, quoted in *La Correspondencia de Puerto Rico,* San Juan, P.R., Sep. 2, 1908, p. 4c; ||mt7nctda) said a woman in Churuguara had given birth, on July 9, to a nondescript that looked like a monkey, with teeth, and hair 25 centimeters long covering its whole body.

Richmond, Mississippi. Dr. Moses Marion Davis (1861), of Richmond, Mississippi, gave the following account of the expelled uterine contents of a woman "who had taken a great fancy to a monkey, and miscarried." From its neck, he said,

> it had the appearance of a well-formed four-months male foetus, while its head, mouth, nose and ears resembled those of a monkey. Its left eye had no lids, but all the ball and membranes seemed to be as blue as indigo, all of which was covered by a thin transparent membrane. The right eye was not discernable.

* Dr. Moses Marion Davis (1836-1903) lived in Itawamba Co., Mississippi, in 1860 (||5du28ya9). Richmond in its early years was a part of Itawamba Co..

Lewis, Outer Hebrides, Scotland. In a report on lunacy in Scotland (Great Britain 1861, p. 249), Dr. Arthur Mitchell described a homosimian he encountered on the Isle of Lewis.

> I have never seen a better illustration of the ape-faced idiot than in this case. It is not, however, the face alone which is ape-like.

> He grins, chatters, and screams like a monkey, never attempting a sound, in any way, resembling a word. He puts himself into the most ape-like attitudes in his hunts for lice and often brings his mouth to help his hands. His arms are long. He has a constant tendency to drop on all fours. He grasps what he brings to his mouth, with an affenish [*sic*—from German *Affe?*] hold. His thumbs are but additional fingers. He tears his clothes with his teeth, and spits when angry. He has a leaping walk. He has heavy eyebrows, and short hair on his cheeks or face. His teeth are good, and his under jaw is large and round and greatly projecting. He is muscular, active, and not dwarfish. He sits on the floor in ape fashion, with his genitals always exposed. He has filthy habits of all kinds. He may be called an idiot of the lowest order; yet there is a mischievous brute-like intelligence in his eye. His head is not very small; its greatest circumference being 20.5 inches; but in its shape it strongly exhibits the ape form of abnormality.

Plancoët, Brittany. Maréchal (1757, pp. 231-232) said, in translation,

> Some years ago, a five-year-old girl [ergo, born ca. 1750] died here in Plancoët who looked and behaved exactly like an ape ["guenon" (French Wikipedia says *guenon* was once used to refer to a female ape, especially a chimpanzee)]. She never learned to speak, but did howl, just like a true ape. She was always scratching her thighs, would take what was given her to eat with both hands and lift it to her nose to sniff, and engage in all sorts of truly apelike behavior. She had difficulty holding herself erect and walked far more easily on all fours. Some gullible people claim that the girl was affected in this way only because her mother too attentively regarded a certain ape with whom she had been quite familiar.

* The Plancoët report mentions nothing about human traits. The only human connection, per the description, was that the individual in question was the offspring of a human mother. So there is nothing to indicate a composite character. However, it seems highly unlikely that any woman would choose to adopt an ordinary ape and pass it off as her child. Was this then simply a case of a hybrid closely resembling only one of its parents? Such hybrids, with a biased resemblance toward one parent, are known to occur in a wide variety of mammalian crosses (see XENOGENESIS).

Ulm, Germany. Ulm physician Eberhard Gockel (1700, p. 151) said, in translation, “A year or two ago [i.e., ca. 1698], I attended the wife of a military constable (“Steckenknechts”) who gave birth to an infant whose upper body was like that of an ape.”

Eberhard Gockel
1636-1703

Paris. Nicolas de Blégny (1686, pp. 96-97), personal physician to Louis XIV, said a discussion[1] was held in the Academy of Sciences in Paris about a woman who had borne a child that was hairless, but which was like an ape with respect to its face, arms, legs and trunk, with the exception of its hair. He said he had seen it himself.[2] Elsewhere, many reports in this book about likely human-animal hybrids mention the absence of a hair coat.

Nicolas de Blégny
1652-1722

Netherlands. The Dutch physician and anatomist Gerhard Blasius (1674, pp. 123-124) said, in translation,

> Aberrations worthy of consideration were present in a twelve-year-old boy [ergo, born ca. 1662], who was exhibited for profit by his parents. Scarcely two feet tall, with the face of an ape, he produced no articulate voice, only a harsh hissing and laughter. He was extremely hairy all the way down to his thighs. the hands were both so markedly swollen that they were lumpy and uneven. On the left side of the neck at the ear was a baglike structure where food collected before moving on to the stomach. The pregnant mother had been frightened by a baboon.

Gerhard Blasius
1627-1682

Wittenberg, Germany. The *Wittenbergsches Wochenblatt* (March 8, 1776, p. 76; ||y8j5wgvs) reported that on January 21, 1771, the wife of a day laborer bore a female nondescript with long fingers and toes, and the head and face of a monkey,[†] which died soon after birth.

Rochester, England. William Gregory (1741), a physician practicing in Rochester gave an account of a woman giving birth to a fetus "resembling a monkey," but gave no particulars, nor does he provide a picture. He said the birth occurred in September 1731.

Gdańsk, Poland. Rzaczynski (1721, p. 351), citing Lorenz Eichstadt, a physician practicing in Gdańsk, stated that on September 15, 1656, a woman there gave birth to a homosimian with an apelike nose, one eye and long arms that extended below the knees.

Trévoux, France. Le Brun (1714) describes a homosimian born October 18, 1713 at Trévoux, a suburb of Lyon. Briefly, he said it lived three days, had a flat nose, large canines, a severely cleft palate, long arms, legs and digits, and a hairy body. These features, together with the fact that the mother had the daily care of an ape ("singe"[‡]) belonging to a local resident, led Le Brun to the conclude that her offspring was in fact a hybrid.

Saxony. The German chronicler Christian Lehmann (1699, p. 726) recorded three early Saxon cases. He said the first, a boy like an ape with a monk's tonsure, was born in 1655, at Hermsdorf, the second, an ape-boy that died soon after its birth at Breitenbrunn, July 12, 1658, and a third, a nondescript with the head and a face of an ape, born at Elterlein, February 8, 1667. And in the *Breslauische Sammlungen*[§] (1720, p. 71; ||4y9ekxdf), a report from Dresden read, in translation, "In January 1719, here, in Scheffel Alley, a woman brought a stillborn monster into the world that was in all respects like an ape."

†. The term used in the German original was *Meerkatze,* which generally refers to any primate belonging to the family Cercopithecidae (i.e., Old World monkeys).
‡. French *singe,* like English *ape,* is often used loosely to refer not only to apes proper (i.e., gibbons, chimpanzees, bonobos, gorillas and orangutans), but to monkeys as well. So, if the Trévoux animal was in fact a hybrid between a human being and non-human primate, the type of primate is unknown.
§. *Breslauische Sammlungen* is a short title of what has been called the most important German-language scientific journal of the early eighteenth century (||ya4h3769).

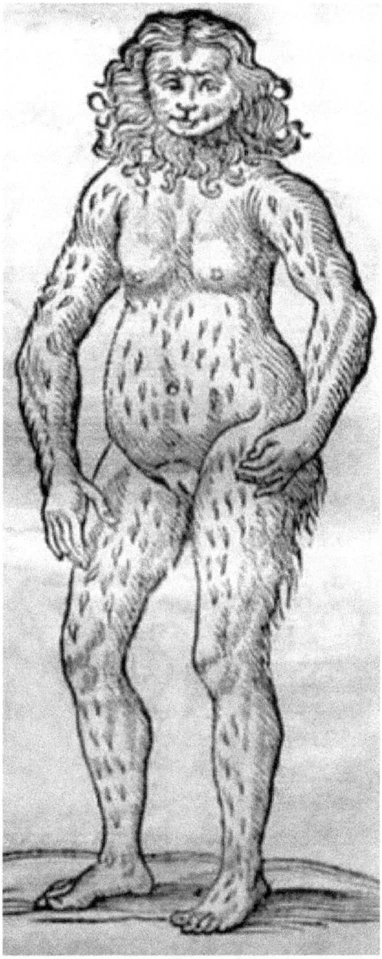

Of this picture, de Bondt (1658) said, in translation, "But see this wondrous monster with a human face, and who is also human in her manner, sighing and watering her cheeks with tears."

Java. Perhaps the first European to report an orangutan-human hybrid was the Dutch physician and naturalist Jakob de Bondt (1592-1631). He was also the first scientist to make a systematic study of the fauna and flora of Java.†

In 1627, de Bondt arrived in Batavia (now Jakarta), the capital of the Dutch East Indies, and for the rest of his life, "by the side of his official activities he worked indefatigably on Java's medical and natural curiosities" (Heniger 1986, p. 12).

Zoology texts often cite de Bondt as having given the first account of the orangutan. He does seem to be at least the first European to have used that word in print. But the individual he depicted (see his picture at right) cannot be mistaken for the animal bearing that name today. It was something different. The creature he pictures, though hairy, has a human face, and with the limbs, hands and feet were all in human proportion. He also said he witnessed one crying and shedding tears, things orangutans never do. So his "ourang outang" must have been some other creature.

Discussing the creature shown above, de Bondt wrote, in translation, "I saw some of these creatures myself, of both sexes, walking erect. In particular there was the female satyr shown here, who out of modesty concealed herself from strangers, hid her face in her hands (for it is accurate to use the word *hands*) and wept copiously, sighing all the while. She behaved also in other ways like a human being, so that you would say nothing human was lacking, except speech. Indeed, the Javans say they can speak but do not for fear they will be put to work. (What nonsense!) The name they give these creatures is *Ourang Outang*, which means “man of the woods.” And these, they insist, are born out of the wantonness of the native women who, with detestable lasciviousness, unite themselves with apes.”

William Leonard Hunt and side show star Krao Farini in 1883. Farini (ca. 1876 – April 16, 1926), known on stage as Krao, had thin hair over her whole body. Although she was often described as a missing link between humans and apes, the only other trait connecting her with non-human primates was the presence of cheek pouches. Neither humans nor apes have cheek pouches, but they are a characteristic feature of the Cercopithecinae, a subfamily of the Old-World monkeys, some of which are native to Laos. According to contemporary reports, Krao was found in Laos, living in the trees, along with other, similar beings to whom the use of fire was unknown.

An apelike homosimian supposedly borne by a woman at Stein, near Nuremberg, Feb. 20, 1739 (Holländer 1921, p. 273). Reports say this nondescript, a female, died two days after birth.

Saxony-Anhalt, Germany. In a book describing cases encountered during the course of his practice, Philipp Salmuth (1648, p. 66, obs. 15), a physician practicing in the cities of Dessau and Zerbst, Saxony-Anhalt, wrote,

> A certain nobleman, N. N., used to keep an ape, which would caper round his dinner table. This man's wife, while pregnant, often romped with this animal and later bore a child that was an ape down to the waist (this was the part of the ape she could see when he was prancing round the table) but from there down human, a sad sight. But it is sadder still that she is still nursing this monster. [translation mine]

Liguria. Sometime prior to his death in 1072, the eleventh-century cardinal Saint Peter Damian (1853, pp. 789-790) wrote about the birth of a viable homosimian. In the following he tells, in translation, of an Italian nobleman whose pet monkey became his wife's lover. "The following I heard from Lord Alexander,[†] the Pope, less than a month ago. He told me that Count William, who lives in Liguria, had a male monkey, commonly known as a maimo.[‡] His wife, an entirely shameless and wanton woman, used to play lewd games with

Peter Damian
ca. 1007-1072

this monkey, … embracing and fondling him. And he in turn showed signs of arousal. It was clear from its gestures that he was attracted to her naked body. So, her chambermaid said, "If it please you, ma'am, why not let him do as he likes? Then we can see what he really has in mind." What more can I say? She allowed it, and, shameful to say, united herself with this beast. This unheard-of crime became habitual.

One day while the count was in bed with his wife, this monkey, inflamed by jealousy, attacked him, savaging his rival with his claws and slashing him with his teeth. The count was mortally wounded. This innocent man, who had been faithful to his wife and given this animal his daily bread, had never expected either to harm him, as he had always been kind to both. But what a crime! The wife wantonly broke her wedding vows, and the beast rejoiced in his master's death.

And indeed, recently, a boy, almost full-grown, was brought to the pope, while I was in attendance. Twenty years old, he was unable to speak. And he looked so much like a monkey that he is known by that name [i.e., "maimo"]. So, the suspicion has arisen that a ruinous prodigy, I will not say a beast, has been reared in that house.

†. Alexander II (reigned 1061-1073), the only eleventh-century pope of that name.

‡. The animal in question is referred to as a "maimo," which may refer to a macaque, since the eighteenth-century naturalist Buffon refers to the Pig-tailed Macaque (*Macaca nemestrina*) as the "Maimon" (and since the French-language version of Wikipedia redirects the query maimon to the page for *M. nemestrina*). Given that most macaques are native to southeast Asia, the only macaque that would have been easily accessible to an eleventh-century Ligurian count would have been the Barbary Macaque, also known as the Barbary Ape (*Macaca sylvanus*), which is native to northwestern Africa.

The Messina homosimian (Bourriquant 1598).

Messina, Italy. Fleury Bourriquant [1600] told of a certain Hipolyta Biscontina, a 19-year-old girl giving birth to an ape human-hybrid at Messina. He said that under torture she confessed to having had repeated sex with a pet ape. He goes on to say that Biscontina was then publicly tormented with pincers in all the crossroads of the town before being burnt at the stake, along with the hybrid offspring and its simian alleged sire. In the illustration, note the tail, which would seem to preclude the idea of a chimpanzee-human hybrid. Bourriquant's description ("a large ape with a tail") suggests a baboon.[3]

A hamadryas baboon (*Papio hamadryas*).

Baboons. Reports about homosimians often refer to baboons. There are various kinds of baboons, but one, the Hamadryas baboon, has an especially close connection with humans. These animals were sacred in ancient Egypt (hence, their alternative name, Sacred Baboon), where they were popular as pets. Tomb paintings show them on leashes or playing with children. When temple baboons died, priests performed the Opening of the Mouth ceremony upon them, something usually reserved for humans. They were mummified and adorned with jewelry and buried in wooden or limestone coffins. In many cases they were worshiped after death as incarnations of the god Osiris. The mummified remains of a sacred baboon were found in the tomb of Tuthmosis II.

New Orleans, Louisiana. From the *New Orleans Crescent* (quoted in the *North State Whig*, Washington, N.C., Oct. 6, 1841):

> LUSUS NATURAE.—Yesterday morning we were invited by Dr. Rogers to see a child which had been born dead about three hours previous. It appeared to be well formed in all respects, except the head, which was that of a baboon.—There was no elevation above the eyes, and no frontal development—the top being perfectly flat.

> The mother is a slave belonging to Mr. Shall, of the City Hotel, who with a very commendable regard for science, has presented it to Dr. Mackie of the Circus Street Infirmary, but whether for dissection or preservation, we know not—probably the latter. There was a poile or hair upon the child such as covers a monkey; the eyes were large and seemingly blue; but the nose, mouth, and chin were baboon entirely.—During the period of gestation, a monkey which was kept in the yard, when angry, jumped upon the mother, and thence in all probability this singular freak of nature. The child was of the feminine gender, as large as newly born children generally.

Peru. In his book *Crónicas del Perú* (1553, cxi), the Spanish conquistador Pedro Cieza de León says that when he questioned the indigenes of Peru about the peoples inhabiting the interior, he was told, among other things, that they hybridized with some kind of large ape residing there. But he admits he is relying on hearsay. Moreover, there are no apes, *sensu stricto,* in South America. And the largest monkey would probably be one of the woolly spider monkeys (Brachyteles), which weighs only about 15 pounds at maturity. At any rate, the following is an English translation of the relevant passage.

> Deep in the mountains they say there are peoples so rustic that, having neither houses nor clothes, they wander about much like animals, killing for food with arrows whatever birds and beasts they can. These, they say, have neither leaders nor rulers and live in caves and hollow trees, some in one place, some in another. And so, they say, there are also some very large monkeys (though I have not seen them) that live in the trees there, whom, through the temptation of the Devil (who is always seeking how and where to make men commit the worst sins), they use like women. And they affirm that some of these give birth to monsters with heads and pudenda like those of men, but the hands and feet of a monkey. They say, too, that they have monstrous little bodies all covered with hair, and that in the end these grow up to look like the Devil their father, if it is true what they say. They also say that they cannot talk but instead have a bloodcurdling howl. I cannot confirm any of this, and only say what I have heard, but many from hearsay, claim that their women—and it gives me great pain to mention this—have given themselves to mules, dogs, stallions and other beasts. It may be that such is the case.

Antwerp. Thomas Fienus (1608, p. 119, §23), a professor of medicine at the University of Leuven, recorded another homosimian.

> There is at Antwerp a honest matron, who has the face and gestures of an ape. This is a consequence, as it is thought, of her mother having been very much delighted by an ape she kept at her house, and with which she had often played. One hears of many other examples, which I do not include here since they have not been confirmed by trustworthy authorities, and their proven truth has not been established. [translation mine]

A homosimian (Wenceslas Hollar, after one of da Vinci's *visi mostruosi*).

Azzo. A case publicized in the 1930s was that of "Azzo the Ape-man," an individual, found living in the Atlas Mountains, said to be intermediate between an ape and a human being. But from photos, the only trait connecting Azzo with apes was a small cranium. A small brain is, of course, a trait of apes. But it is also a trait of many other animals and does not in itself make Azzo a homosimian.

Oliver. Years ago, I was contacted, by the owners of a nondescript widely alleged to be a chimpanzee-human hybrid. I told them that in my opinion this individual, known as Oliver, had no obvious physical traits, not present in normal chimpanzees. linking him to humans. True, he spent more time walking upright than chimpanzees normally do. But chimpanzees can be trained to do many things that humans do.

Ivanov. During the 1920s Russian biologist Il'ya Ivanov attempted to impregnate chimpanzees with human semen but managed to carry out only three such inseminations (Rossianov 2002). So, his failure to produce any hybrids is not surprising. Hybrid crosses typically require more, in some cases many more, inseminations to produce a pregnancy than do matings between two animals of the same kind.

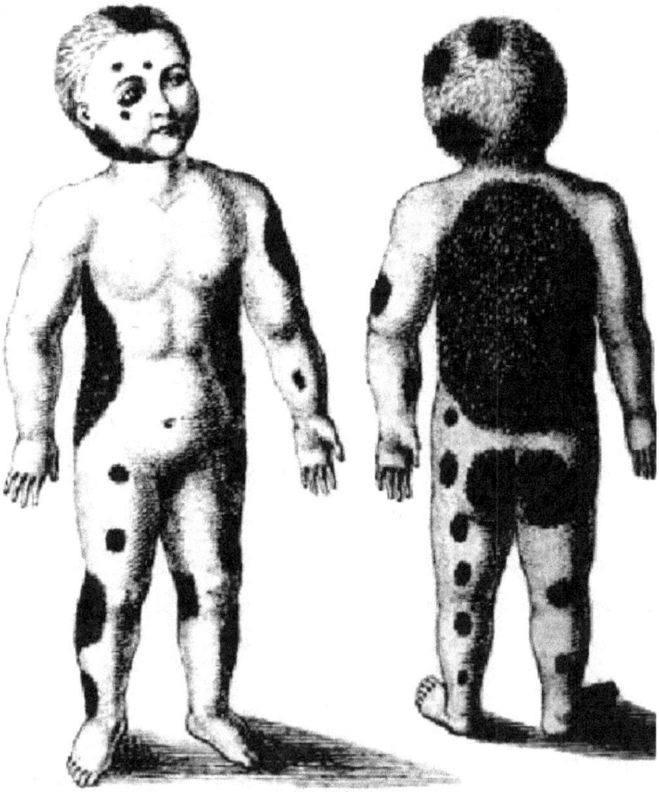

The Schönberg homosimian (Schönwald 1774).

HYPERTRICHOSIS

Over the years, various patients with large hair-moles, allegedly connecting them with apes, have been reported.

Schönberg, East Brandenburg. Schönwald (1774, pp. 565-566) said, in translation, "In January 1768, in the village of Schönberg near Soldin [now Myślibórz, Poland], the wife of Friedrich Neumann, a farmer, gave birth to a little daughter whose entire back from the neck down to the small was like that of an ape ["Affe"] and covered with the same kind of hair, which extended on both sides onto the flanks of the abdomen. In addition, there were various larger and smaller patches of the same nature, both on the child's body and on her face."

Milan. Drzewiecki (1891, p. 529b) referred to a subject in the Milan Anatomical Museum, like the Schönberg homosimian, supposedly the result of the mother's seeing an ape.

Buffalo, New York. In an account of an 1884 meeting of the Buffalo Medical and Surgical Association, Peterson (1885, p. 254) stated that Buffalo physician Samuel W. Wetmore "brought forward a little girl of Irish parentage, aged five years [ergo, born ca. 1879], who had a nævus covering most of her face from forehead to chin, and upon which there grew an abundance of hair. The nose was flattened, the eyebrows bushy and grown together, and, upon the whole, the child had rather a Simian appearance."

Bönen, Germany. Johann Hartmann Degneri (1742) described a girl born in 1732 at Bönen, who became extremely hairy, beginning at the age of three. He said the hair was like an ape's or some other hairy animal.

Nancy, Lorraine. Aimar Raoult (1902) reported a woman at Nancy who, supposedly as a result of contact with an ape, gave birth to a homosimian with a large nevus pilosus (the medical term for a hairy birthmark) and other smaller nevi. In translation, he said,

> In 1891, I myself saw, along with [Dr. Gaston Félix Joseph] Variot [1855-1930], a five-month-old girl whose father, mother and seven-year-old brother had no areas whatsoever of aberrant pigmentation. This infant has, on its right arm, a kind of black glove, covering the back of the hand and the forearm as far as the elbow. This region is covered with fine, silky, black hair and the skin itself is rough. The wrist on that side is smaller than on the other, and the bones, less developed. There are other black regions on the buttocks. The mother stated that she had, in the fourth or fifth month of her pregnancy, touched the hand of an ape. [see MATERNAL IMPRESSIONS]

Antwerp. Félix-Léonard van Camp (1847), a member of the Antwerp Society of Medicine, delivered a child, on December 5, 1846, that he said was marked on the lower abdomen with multiple large nevi resembling the skin of an ape.

* Félix-Léonard van Camp (b. Apr. 6, 1802, Sint-Gillis-Waas, East Flanders, Belgium, d. Mar. 1855, Antwerp?). Stevens (1855) gives an account of van Camp's life.

The London homosimian (Smith 1867).

London. Smith (1867) described a patient who "was admitted into St. Bartholomew's Hospital in 1865. She was at that time twelve years old (born ca. 1853). The left upper extremity and the greater part of the corresponding side of the trunk and neck were deeply stained with dark-brown pigment, from which grew an abundant crop of brown, harsh, lank hair, varying in length from one to two inches. The skin was rough and harsh; the arm was long, thin, and withered; the scapula was unnaturally prominent. In fact, the upper limb, shoulder and back bore a very strong resemblance to the corresponding part of a monkey." The Danish physician and naturalist Oliger Jacobaeus reported (in Bartholin 1677, vol. 5, obs. 103, p. 274) seeing, also in London, a girl whose body was nearly covered with hair and dark spots of various sizes, and whose mother was said to have begotten her with an ape ("ex congressu matris cum simia"). Jacobaeus was a professor of medicine at the University of Copenhagen.

Leiden, Netherlands. The Leiden-trained physician Gerard van Swieten (1754, p. 407) said he saw a patient who had long (1.5") shaggy dark hair covering her right hand and forearm. Her mother, he said, had been scared by an ape before giving her birth (see MATERNAL IMPRESSIONS).

SEXUAL INTERACTION

Samuel Purchas
ca. 1577-1626

Nearly all allegations of sexual activity involving humans and apes refer to women and apes, perhaps because a female ape is so much stronger than the average man. Some of the accounts refer to women voluntarily having sex with apes or being artificially inseminated. But there are also many stories about apes abducting women. Some of these are presented as fact, others, as fiction. In an example of the former, Maxse (1906) asserted that African pygmies appear

> as closely related to the chimpanzee monkey as it is possible for human beings to be, and the affinity is so far recognised by the chimpanzee of the present day that there are stories current of pygmy women being carried away by male monkeys and destroyed by their jealous wives.

And Proctor (1877, p. 421), in writing about chimpanzees, claimed "women in particular are often said to be carried away by these animals, and one negress is reported to have lived among them for the space of three years, during which time they treated her with uniform kindness, but always prevented any attempt on her part to escape."

As Proctor seems to imply, the victims in such claims of abduction are sometimes men. In fiction, Kala the she-ape rears the infant Tarzan. And there are many nonfictional reports about chimpanzees carrying infants off as prey.

Not all such accounts end in tragedy. Samuel Purchas published a collection of reports by travellers to foreign parts entitled *Purchas His Pilgrimage*. In one of them, Andrew Batell (1967) tells of his experiences in Angola. In a chapter on chimpanzees, which Batell called pongos, Purchas added a note (p. 55) which read,

> He [Batell] told me in a conference with him that one of these Pongos took a negro boy of his, which lived a month with

> them, for they hurt not those which they surprise at unawares, except they look on them, which he [the boy] avoided. [Staring at apes does provoke them (Goodall 1983).]

Dodds (2006) compiled various early accounts about marooned women impregnated by apes. Typically, in such stories the woman commits some transgression leading to her exile on an ape-infested island. There, she is whisked away to a cave and ravished by a brutal ape. She then gives birth to one or more homosimians. In the end, she is rescued by a passing ship, and her harrowing experience becomes known to the world.

Nicolas Gervaise (1700, p. 31) gave an over-the-top description of macaques assaulting women. In translation, he said,

> They [i.e., the macaques of Makassar] principally want women, and the first that catches sight of one cries out to his comrades. These surround her, throw her down, and subject her to a hundred outrages. They then strangle her and tear her to pieces.

In Africa, a widespread belief among indigenes that chimpanzees and gorillas were human has perhaps contributed to interbreeding of humans and apes. According to Savage and Wyman (1847, p. 426) all along the west coast of equatorial Africa, apes "are believed by the natives to be human beings, members of their own race degenerated. Some few, who put on a degree of civilization above the mass, will not acknowledge their belief in this affinity; such profess to view them as embodied spirits, the belief in the transmigration of souls being prevalent. They said that the *Enché-eko* or Chimpanzee has the spirit of a *Coastman*, being less fierce and more intelligent than the *Engé-ena* [or Gorilla] that of a *Bushman.* The majority, however, fully believed them to be men, and seem to be unaffected by our arguments in proof of the contrary. This is especially true of the tribes in the immediate vicinity of the locality. They believe them to be literally 'wild men of the woods.'"

Even today such interbreeding continues to be reported. According to one account (Ariyibi 2007), a wild chimpanzee was interrupted during an attempt to rape a female health worker in Ekiti State in southwestern Nigeria. Four previous rapes of local women by chimps were blamed on the same individual, who was hunted down and killed. And if one judges from reports, it seems clear that some women willingly engage in sex with chimpanzees. In various online accounts about the supposed chimpanzee-human hybrid Augusto Dembo, António Mendes, a doctor in the Angolan city of Luena, stated that "Sexual relationships between humans and chimpanzees are fairly common in the region." Indeed, reports about Angolan women hybridizing with chimpanzees have been appearing for more than three centuries, as was discussed above (see p. 334).

Orangutans, too, rape women. For example, in a review (||bddrzdxp) of Carole Jahme's *Beauty and the Beasts: Woman, Ape and Evolution,* Charles Taylor writes,

> Male orangutans procreate by rape and have been known to rape the women studying them ([orangutan researcher Biruth] Galdikas' cook was raped). This was almost the fate of Julia Roberts when she made a documentary at Camp Leakey in 1996. One male took a shine to her and grabbed her as she walked along a path. Luckily, a film crew was present, though it took five men to free her from the ape's grasp.

The rape just mentioned occurred when a wild-born male orangutan named Gundul, who frequently came to one of Galdikas's research camps, attacked a Dayak woman who worked there as a cook. Galdikas (1995, p. 294) said, "I had never seen Gundul threaten or assault a woman, although he often charged male assistants. The cook was screaming hysterically. I thought, 'He's trying to kill her.'" Galdikas told how she beat the orang and tried to pull him away, but then, she said,

> I began to realize that Gundul did not intend to harm the cook but had something else in mind. The cook stopped struggling. "It's all right," she murmured. She lay back in my arms, with Gundul on top of her. Gundul was very calm and deliberate. He raped the cook. As he moved rhythmically back and forth, his eyes rolled upward to the heavens. I was in shock. … Gundal let the cook go, stood up, and, soundlessly, moved off the feeding platform into the trees. It was over just like that.

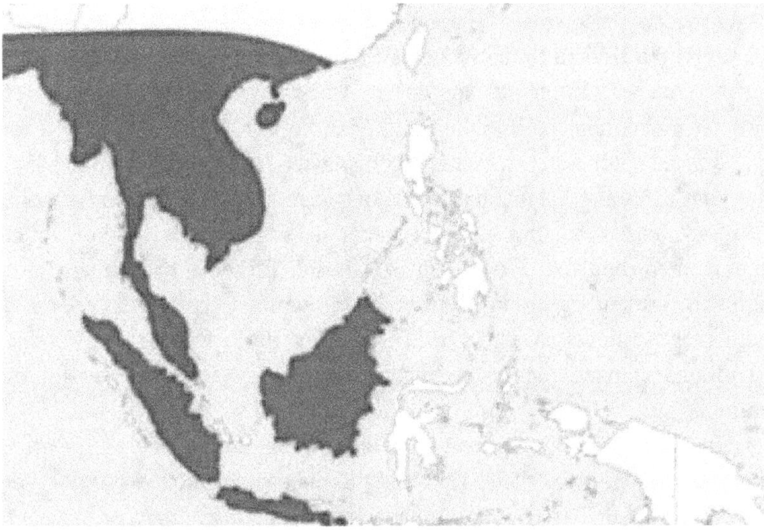

Former range of orangutans (red) in southeast Asia (based on Kahlke 1972).

The occurrence of such behavior in orangutans has been known for centuries. Honoré Gabriel Riqueti, comte de Mirabeau (1783, p. 150) refers to an orangutan at the Château de Chantilly that habitually violated women and had to be killed (although the word *orangutan* at that time might have referred to some other ape than *Pongo*). The second-century Roman author Aelian (*Characteristics of Animals*, 15.14), in describing the people of India, told how they brought various animals to their kings as gifts. But he said, orangutans†

> which are too fond of women, they do not introduce into their towns, but if they can contrive somehow to spring upon them, they put them to death, because they detest them as adulterers.

Orangutan populations declined by 97% in the twentieth century, But when these apes were more common, such encounters, too, must have been more frequent. Perhaps orangutan-human hybrids were, as well? But sexual encounters between humans and apes do not always involve women and male apes. Shaved female orangutans are exploited as prostitutes in Indonesia (Adams 2007).

†. Gossen (1935) identified Aelian's red ape as *Pongo*.

Indeed, orangutans were once far more widespread than today. Though now limited to the islands of Sumatra and Borneo, their former range covered much of southeast Asia and extended an as yet undetermined distance into the India (Kahlke 1972).

Various authors (Edwardes 1959; Masters 1962; Bagley 1968) say, too, that in Yemen, Hamadryas baboons, also known as sacred baboons (*Papio hamadryas*), have been popular as sex partners for both men and women, and that women in Sudan and Ethiopia bring them into harems. Ancient Egyptians, both men and women, had sexual relations with these animals (Masters 1962; Bagley 1968; Ramsis 1969). The Egyptians considered sacred baboons as among the most lustful of animals.

In Europe, baboons have long had a reputation for lusting after women. John Locke (1689) wrote, "if history lye not, women have conceived by drills" [i.e., have been impregnated by baboons]. Locke was likely referring to an event mentioned by Edward Coke (1648, p. 59) in the *Institutes of the Laws of England.*[†] He stated that the statute against buggery[‡] was extended to include women because "somewhat before the making of this Act, a great Lady had committed Buggery with a Baboon and conceived by it." The French artist and naturalist Jean-Baptiste Audebert (1799) said, in translation,

> Baboons are, in fact, the most ferocious of the monkey tribe. This is not because they are more brutish than other apes, but because they are more malicious. Those that one sees in the menageries let out horrible cries when any spectator dares to caress a woman in their presence. It would be dangerous, in fact, to let them have sticks or stones, with which these animals would not fail to wound those who excite their jealousy. The same cries are elicited if anyone touches their keeper. He can approach them without danger. The captive animals will permit him to touch them and even seem to encourage him to do so, by rubbing themselves against the bars, and by presenting him with their rear end. But if that same man approaches a woman, and touches her, the baboon won't stand for it. He leaps. He cries out. He jerks mightily at his chains and shows the same rage towards his master as towards a stranger.

†. The *Institutes* are recognized as a foundational document of the common law.
‡. Passed in the 25th year of the reign of HenryVIII (1533).

NOTES

1. The discussion Blégny mentions was recorded in Gallois (1674, vol. 1, pp. 143-192), which, however, seems to be more a theoretical discussion than a description of a specimen.
2. Blégny's homosimian is clearly the same as that described in the *Philosophical Transactions of the Royal Society* (1667, vol. 2, pp. 479-480). Blégny mentions that it also lacked thumbs. All primates have five digits on their thumbs, except for the spider monkeys of South America and the colobus monkeys of Africa, in both of which the thumb is small or absent. Woolly spider monkeys (genus Brachyteles), which are structurally intermediate (and thus likely derived from hybridization) between spider monkeys and woolly monkeys, also either have small thumbs or entirely lack them.
3. Bourriquant based his information on a Siennese original, now apparently lost.
4. *Colophon:* A homosimian lying on a tasseled cushion (Holländer 1921, p. 277). A woman reportedly bore it in the village of Mursellers, near Kimratshofen, Bavaria (Allgäu), Feb. 16, 1629. One account said it survived birth and was still alive.

An illustration by Friedrich Wilhelm Schmuck (1683, #5; ||z2y2sw9p) of a hominursid. In translation, his caption reads "On Captain Hochmuts' estate in Ingria [then a province of Sweden], June 22, 1682, a woman bore a malformed child with a head of hair looking like the style women are now wearing. It had many teeth, long, hanging ears, six or seven fingers on each hand, and legs like a bear, with which it scratched its ears and head as if it were a monkey." Supernumerary digits are often mentioned in reports about telenothians.

A woodcut from *Historia de gentibus septentrionalibus* (1555, p. 627), by Olaus Magnus, last archbishop of Sweden, in which he tells of a girl raped by a bear. According to the story, their son went on to found the Danish royal line.

Hominursids (Bear-Humans)

A bear lusted for a girl with wanton, erotic desire.
—Jean Tixier de Ravisi, *Officina*†

Many aboriginal cultures, both in northern Eurasia and North America, have thought bears can interbreed with human beings. From times primeval right up to the present, this belief has played a prominent role in shamanic ritual. The people of these northern cultures took bears as their totems and looked upon these dark and shaggy omnivores as divine spirits in corporeal form. Indeed, some cultures maintain such beliefs even today.

Haida bear totem.

The Haida people, native to southern Alaska and northern British Columbia, are an example of this tradition. Their choice of the bear as a

†. Tixier (1593, p. 78): "Amore lascivo et venereo puellam ursus adamavit."

totem refers to an old story about a Haida maiden who is so foolish as to laugh at the bears. Incensed at this outrage, a great pack of bears suddenly appears and kills all her companions. The chief bear then takes her as his mate. They have a daughter, half human and half bear. And later, when the daughter grows up, a hunting party from the tribe, thinking she is an ordinary bear, corners her in a tree. But, when they realize she is part human, they persuade her to come down and live among them. And now, according to Haida legend, this primal hominursid is the ancestor of all those entitled to wear the bear crest.

At least in some tribes, the bear was essentially a god. As Edward Tyrrell Leith (1893, p. 303) put it,

> The Bear and the Wolf are, according to the belief of the American Indians, children and consorts of the first woman, and in both forms the Great Spirit is conceived as embodied.

Such stories are widespread. In Eurasia, Russian fable tells of a woman who enters a bear's den, has sex with him, and later gives birth to a son. She names him Ivanko-Medviedko, meaning "Little John, the son of the bear." Human down to the waist, he has the lower extremities of a bear, much as satyrs are depicted as human above and goat below (see HOMINOCAPRINIDS).

In the Korean legend of Dan Gun, the progenitor of the Korean people, the god Hwan-in sends his son Hwan-ung to establish a new kingdom at Tae Baek Mountain in what is now North Korea. There he meets a tiger and a bear who want to become human. So, he gives them wormwood and garlic to eat and tells them to stay in a cave out of the sunlight for 100 days. The tiger could not remain there, but the bear did and became a woman. Hwan-ung then married her, and Dan Gun was their offspring.

The Greeks and Romans said an infant Atalanta was left exposed on Mount Parnassus but was taken and raised by a she-bear. And they had tales, too, about women mating with bears. Lemprière (1839) tells how Polyphonta, one of Diana's nymphs,

> devoted herself to celibacy and to a retired life, with such constancy, that Venus, irritated at the contempt she showed her divinity, inspired her with an unnatural passion for a bear, which the unfortunate nymph is said to have gratified. This was no sooner known than she was detested by Diana and pursued by all the wild beasts of the forest, but she escaped to her father's house, where she gave birth to twins, to whom she gave the names of Oreius and Agrius.

In early scientific literature, tales of bears abducting women continued to appear (Dodds 2006). Gesner (1551, p. 1086) said, in translation,

> A bear, who came down from the Alps of the Haute-Savoie, carried a girl off to his cavern lair and there raped her. Each day, he gathered fruits and herbs and brought them back for her to eat. And when he was away, he blocked the cave's mouth with a stone to prevent her from escaping. Eventually, however, when her parents after long search found his den, they were able to push the rock aside and rescue her.

Becher (1669, p. 242) said there were various stories about bears and monkeys coupling with women, and that from these unions proceeded offspring with a human figure, but shaggy and endowed with extreme strength. Gradually, these vague stories of early times were replaced with reports that named specific witnesses and locales.

REPORTS

Rome. Lycosthenes (1557) claimed that in the year 1282 a child had been born in Rome with a human head but the claws and shaggy coat of a bear. Aldrovandi later depicted this ursine composite (see figure, right). Fedeli (1654, p. 499), in his *De relationes medicorum,* likely was thinking of the same story when he said, in translation, "In Italy, of course, a woman, impregnated by a bear, bore a son who had a superior nature and founded a noble line."

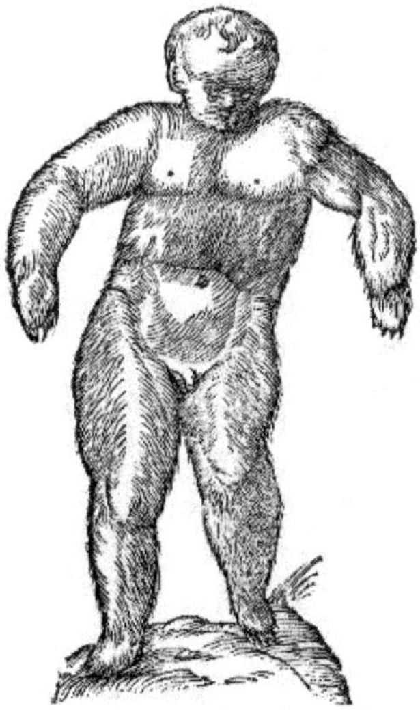

The Roman hominursid
Aldrovandi (1642)

Prague. Damascenus the Studite (see Boaistuau 1594, p. 35) told how at Prague a bear-skinned virgin was shown to Emperor Charles IV (reigned 1346-1378).

Poland. The Irish physician Bernard Connor (1699, pp. 133-134), who had traveled to Poland, told of a boy captured in the wild, whose skin was covered with hair, who could not speak and whose behavior was bear-like. Thus, he wrote, in translation, "Recently, in 1694, at the court of the now deceased King of Poland, John Sobieski,[†] there was a wild boy who was living with bears when he was captured by woodsmen somewhere in the neighboring confines of Lithuania and Russia. About ten years old, his skin covered with hair, he was frightful to see. He had neither reason nor speech, nor even

A bear nursing a child (Connor 1698)

a human voice. And he went about on all fours. Indeed, he had nothing in common with human beings except the mere form of his naked body.

Given that he did have at least the appearance of a human, he was baptized. But, separated now from his beast companions, he seemed unhappy. For he was always afraid and would try to run away, as if he were imprisoned. Eventually, by placing his hands against the wall, he was taught to stand, in the same way that infants are, and, as he gradually became accustomed to human food, he grew tamer. After a long time, he was able to speak a few words with a hoarse, inhuman voice. But even then, when asked about his former life in the forest, he could remember no more about it than we do of our days in the cradle.

The king himself, many senators and other trustworthy people assured me of these facts. And indeed, it is widely believed in Poland that children are sometimes reared by bears. For they say that if parents leave their babies close to the forest, or too near the fence, or carelessly out in a field, they may be devoured by a hungry bear, but that a lactating she-bear may carry the child away and raise it among bears. And then, years later, the child may return to human society when caught by hunters.

Rzaczynski (1721, pp. 354-355) gives a similar account of a boy captured among bears by Lithuanian hunters in 1657. Innumerable records attest to the odd adoptive choices mammals have made during lactation. Might a lactating bear choose to adopt a child?

†. Sobieski was the leader of the Christian coalition that decisively defeated the Ottomans at the Battle of Vienna, Sept. 12, 1683.

Connor's account recalls Tulp's Irish sheep-boy (see p. 112), also supposedly captured in the wild, and described as sheep-like both physically and in his behavior. Neither Tulp nor Connor speak of hybridization, but it is unclear how merely being fostered by an animal would cause a child to have a skull shaped unlike a human's, as in Tulp's case. And how would it make a ten-year-old boy have a "skin covered with hair"?

Dabbabi the man-bear (Zakariya al-Qazwini, Walters MS 659; copy of 13th-cen. original. ||44avb4fy).

The Magdeburg hominursid (Schelhammer 1691, fig. 34)

Magdeburg, Germany. The physician Günther Christoph Schelhammer (1691), who once served as Rector of the University of Jena, described a headless fetus borne by a pauper at Magdeburg on October 25, 1690. Its left arm was much larger than the right and ended, not in a hand, but in what appeared to be a bear paw with three claws. The right hand bore two fingers, and both feet, four toes.

Bad Langensalza, Thuringia. The Fall 1724 issue of the *Breslauische Sammlungen* (p. 358; ||yxhovpgj) says, in translation, that on November 12 of that year, a woman near Bad Langensalza, whose occupation was collecting firewood in a nearby forest, bore a child that "had ordinary bear's paws, and that from the rump to the neck was also formed like a bear, except that the skin was smooth and hairless. The face and the whole head, however, were formed as in a human being. ... Because it was born dead, it was wrapped for burial in the usual way and, since its face was human, no one regarded it as a monster, and it was interred in the customary manner."

Jöhstadt, Saxony. Christian Lehmann (1699, p. 725) said a boy at Jöhstadt was ordinary, except in having one hand like a bear's paw.

Strasbourg. Curtius (1762) said a woman near Strasbourg gave birth, in 1761, to a nondescript with a shapeless head, and feet with three digits.

* The Strasbourg case is tentatively placed here with hominursids, not because the fetus had reported ursine traits, but because Curtius said the mother had had, while pregnant, a disturbing experience with a bear.

Ghent, Belgium. Martou (1839) reported a certain Thérèse Peysan, 39 years old, birthing a bearlike offspring near Ghent in October 1839. It had been dead in her womb for several days.

Oschatz, Saxony, Germany. The *Dresdner Nachrichten* (Dresden, Apr. 20, 1863; p. 2a; ||2xpxap53) said, in translation, "We have been given, not only by word of mouth, but also many times in writing, information about the matter with which Oschatz and nearby places are now abuzz. They say a monstrous child has been born there, half human and half bear. For a thousand reasons, we have denied it all. But ever and again, they bring us this bear." And the editor went on to claim that all the reports were correct.

New Brunswick. From the *Democrat and Chronicle* (Rochester, N.Y., Apr. 16, 1885, p. 6; ||4jhrhu63): "Among the passengers of the day express yesterday, was a singular freak of nature. His name is David Kushlow, and he is a native of Canada. Instead of feet were claws like those of a bear, and each hand had five fingers and a thumb joined by a web similar to a bear's forepaws. The features of the face were very large and on the head was a thick growth of hair, similar to that on a bear. Kushlow speaks English and French and has been giving exhibitions at different places in the state." The *Western Sentinel* (Winston-Salem, N.C., Aug. 19, 1886, p. 5b; ||yxufckrh), too, reported a hominursid, likely Kushlow. "For two or three days past our citizens have thronged to see the 'man-bear,' which has been on exhibition in the Armory of the Forsyth Riflemen and has attracted no little attention. It is indeed a strange freak of nature and a curiosity such as has never before been seen in these parts. It has been proved by many leading physicians to be a genuine human being in the form of a bear. This strange being is a French Canadian and was born in the rural district of New Brunswick, is twenty-six years of age [ergo, born ca. 1860], is about four and one-half feet in height and walks on his all-four, as that of a bear, yet in other respects represents a human being. ... It has great muscular strength in both hands and legs." The *Hazleton Sentinel* (Hazleton, Pa., Feb. 8, 1886, p. 4; ||n54hyt2m), seemingly again describing Kushlow, said, "The eyes, nose and mouth are similar to those of a bear … His hair is long, forehead very low and he has very heavy eyebrows. It is impossible for him to stand erect, and he creeps around on all fours for all the world like a bear. The bones and joints of the body are more those of a bear than a human being. and he has five fingers and a thumb on each hand, webbed between each to the middle joint."

The illustration above accompanied the report below.

A report describing an individual of much the same appearance, presumably Kushlow, appeared five years later in the *Arkansas City Daily Traveler* (Apr. 16, 1891, p. 5; ||y2mzq4u3): "A rare opportunity will be afforded our citizens in general of seeing one of nature's greatest freaks—alive and living—that may not occur again in a lifetime. This strange freak of nature … is alive, white, weighs 153 pounds and is about twenty-seven years of age, and has flesh and power of speech, but is worthily termed a bear owing to the peculiar formation of his limbs—claws, six in number, on each of his fore limbs, webbed out beyond the second joints, while the nails split in the middle and hook at the ends. His head greatly resembles a bear, while a complete absence of patellas or knee-caps compels him to go upon all fours. The feet are also formed exactly as those of a bear, being entirely devoid of heels."

* The *Bennington Banner* (Bennington, Vt., Sep. 25, 1884, p. 3e; ||32yc5cxb) said "A hideous monstrosity, half man and half bear, arrived in town on Friday, and will be exhibited here during the fair." And the *Evening Telegraph* (Buffalo, N.Y., Apr. 24, 1885, p. 2b; ||yyt27mxk) reported, "GENESEO: A curious specimen of humanity is on exhibit at the Globe Hotel. In some respects, it resembles a man but in others, a bear, and seems to be half bear and half man." A hominursid, "Six-fingered Jack," depicted in the Chicago *Inter Ocean* (Mar. 9, 1890, p. 15; ||yywkyg74), may have been Kushlow.

Marble Falls, Texas. From the *Evening Report* (Lebanon, Pa., Jun. 13, 1901, p. 7; ||y3s5t4f5):

A young lady out in Marble Falls, Tex., was carried off by a "bear man" and returned home unhurt to tell the tale.

The Kickapoo Indians of this region have long believed in a great "bear man" who rules all the bears of the mountains. Miss Ramie Arland, the heroine of this thrilling adventure, believes her encounter was with the chief bear himself.

Ramie Arland is a pretty girl and the acknowledged belle of Marble Falls. She has always been a skeptic as to the existence of the "bear man." The remarkable story she tells has, however, gone far to convince the most skeptical.

The "bear man" in this case ventured almost to the back door of the Arlands, says the SAN FRANCISCO CHRONICLE. Ramie went out recently early one evening to gather her flock of sheep, which were grazing nearby. This was a common occurrence, and her absence was not noticed by the family until her mother heard her daughter scream wildly a short distance from the house. She rushed to the door. The screams were repeated but this time accompanied by the scream of a panther.

The mother seized a gun and rushed into the woods but could find no trace of her daughter. She returned to the house and, collecting a hunting party searched the woods all night. No trace of the missing girl could be found.

It was not until the next day that a hunter, wandering in the woods several miles from Marble Falls found Ramie Arland aimlessly walking about. He helped her home where she quickly recovered from her experience.

"I was walking along a narrow trail," she says, in telling her story, "when a large black bear suddenly appeared in front of me. He quickly turned away, when a curious looking animal, running on four feet sprang out of the chaparral into the trail. I saw at a glance that the monster in some way resembled a human being, and it flashed across my mind that I was confronted by the 'bear king' of the Kickapoos. It threw one of its long arms about my neck, glared into my eyes and uttered a horrible sound. I expected to be torn to fragments. The creature seized me and ran toward the mountains.

"It reached its cave at last and then left me lying on the ground. I tried to escape at once, but the creature struck me repeatedly on the head when I did so. I gave myself up for lost. Finally, however, he lay down to sleep. I waited fully an hour before attempting to steal away."

When the settlers and cowboys heard this strange story, they at once set out in the direction of the Moon Mountains for the purpose of destroying the monster.

It ground its teeth together, and while pounding its breast, it would roar and scream like a panther. It was now so apparent to the hunters that the thing was at least human in shape, that they hesitated to fire upon it. While they were deliberating it suddenly bounded with rage straight toward the astounded hunters. They were compelled to kill it in self-defense.

Camden, Tennessee. From he *Fort Worth Daily Gazette* (Aug. 3, 1889, p. 2e; ||y5mqcgww):

> CAMDEN,[15] TENN., July 30.—A young white woman near here recently gave birth to a half human and half bear, the resemblance to the latter predominating.
>
> The eyes are prominent and set far back in the crown of the head. A human nose in faint outline is seen in the center of the head. A prominent snout projects where the face should be, and from this a long tongue protrudes. The arms and legs are those of [a] human, but the feet and hands are those of an animal, except that the fingers and toes are perfectly those of a human. The creature was stillborn.

Montpelier, Vermont. The *Argus and Patriot* (Montpelier, Dec. 6, 1876, p. 4d; ||y2wd7jmz) said a local woman had borne a "child with a bear's head" that "died within a few hours of birth." No details were given.

Fredericton, New Brunswick. The *Montreal Gazette* (Jun. 16, 1879, p. 3a) said, "A child having a bear's head was born in Queen's County a short time ago. The mother was at Fredericton with her offspring during the visit of Barnum's exhibition, and the veteran showman, hearing of the curiosity, desired to purchase it, but nothing would induce her to part with it."

Milford, Michigan. The *Milford Times* (May 31, 1890, p. 1d; ||y7q8dsye) reported that "A freak of nature in the shape of a half man and half bear was on exhibition at the Central House last Friday."

Columbus, Georgia. The *Daily Times* (Columbus, Feb. 15, 1876, p. 4b; ||vkuzvbvf) said, "A negro woman living in the lower part of the city has given birth to a child, resembling a man and a bear. The likeness is said to be a blending one, while a portion of the child is the counterpart of the bear. [speculation omitted]"

Seattle. From the *Seattle Daily Post-Intelligencer* (Sep. 20, 1887, p. 3e; ||yaev5719): "A large party of Clayoquot Indians, from British Columbia, appeared in this city yesterday on their way to the hop fields.

Accompanying them is probably as curious a specimen of unfortunate humanity as ever was born. It is neither a man, nor a brute, but appears to be on the line dividing the one from the other. Considered as a human being, the being is a man; considered as a brute, it is a bear. It looks but little more like a man than a monkey does, except that the features are a little more distinct and there is not a coat of hair on his body. It cannot talk or walk upright. It crawls along on the hands and feet with the peculiar swinging motion of the bear. Its feet are at a very acute angle with the front of the leg and, when crawling, the hands move with an inward swing like the front feet of a bear. The expression is almost that of a bear. There is a peculiarly wild look about it, and the eyes are restless and sharp. Everybody instinctively calls it 'Indian Bear.' The Indians ... say it is 20 years old. But little attention appears to be paid to it by the Indians, and it wobbles around with about the aimlessness of an old dog, seeking a warm place, and eyeing suspiciously the approach of anyone not familiar with it. [an allusion to MATERNAL IMPRESSIONS here omitted]"

The *Wichita Daily Eagle* (Wichita, Ks., Mar. 13, 1890, p. 2e) referred to the same case: "One of the most remarkable freaks of nature ever known can be found near Vancouver, in an Indian settlement. The strange being is called Joe, and it is difficult to tell whether the animal is a bear or an Indian. The uncouth being is of prodigious size and strength, with limbs well formed, except that the ankles turn out, giving the feet the appearance of paws. The face and head are shaped very nearly like that of a grizzly bear. He never sits or stands like a man, but when at rest settles upon his haunches like a bear. When in motion he walks on all fours with the peculiar sidelong awkward motion of a bear. He cannot talk but gives vent to growls like a bear. When enraged he seizes the object of his wrath in a deathlike hug. Thus far he is to all intents and purposes a bear. But his parents are Indians, his associates are Indians, he loves red blankets and Indian trinkets, and his food is Indian food. Joe is well known throughout both British Columbia and Washington. Every year he goes with the other Indians to the Puana Valley, and the Snake and Columbia rivers in the hop picking season and proves himself to be the best hop picker of the tribe."

* A reference to, perhaps, the same hominursid from the *Dalles Daily Chronicle* (Dalles, Ore., Oct. 6, 1891, p. 2a; ||yeyuut3b): "A monstrosity in the form of a man born like a bear was on Tuesday's train from Roseburg [Or.] to Los Angeles. He is over 40 years old. His name is Phillips, and his parents, who are well off, are deeply attached to their unfortunate offspring. He is an idiot."

HYPERTRICHOSIS

Kętrzyn, Poland. Dr. Wilhelm Sommer (1885) described a bearlike hair-mole ("hairy nevus") on the arm of a man born in 1864 at Rostenburg (today, Kętrzyn). It was 30 centimeters (12 inches) long and stretched from the right shoulder to the elbow, nearly encircling the upper arm. The mole's deeply pigmented skin was covered by six-centimeter-long, black, pelt-like hair. Elsewhere, the hair was brown, the skin, white. Sommer (ibid, p. 408) said the skin of the mole was the color of brown, roasted coffee beans ("der Farbe braun gerösteter Kaffeebohnen"). Appealing to the theory of MATERNAL IMPRESSIONS, Sommer said the mother had been frightened by dancing bears during the first month of her pregnancy.

Canal Fulton, Ohio. Lewis F. Slusser (1852, p. 346), a Canal Fulton physician, reported delivering on February 28, 1852, a woman of a clawed female infant who bore abundant hair on her arms, legs and back. He attributed this nondescript's appearance to the mother's having been repeatedly disturbed by a bear belonging to a neighbor.

Green Bay, Wisconsin. Physician W. N. Brennan (1868) reported that "Natalie Logan, a half-breed Indian woman, a native of Green Bay, Wisc., aged 23 years, a widow, was admitted in the Sister's Hospital [St. Louis], Dec. 27, 1867 [ergo, born ca. 1844]. She suffered from [tuberculosis], of which she died Feb. 14th, 1868. She was also afflicted with epilepsy. The complexion of her face, the greater part of her body and limbs, was that of a half-breed Indian. She had a nævus maternus, or mother's mark, covering the whole trunk posteriorly, extending from the occipital region to the inferior nates, and from side to side. It covered the right mamma and reached midway on the right thigh and to the upper third of the left. It likewise reached midway on the arms. The parts so marked were uniformly black; not so black as a negro's skin but resembling more the rough skin of a Mexican dog [the skin of a black bear, *Ursus americanus,* is gray; ||2p8h95ef]. There was an unusual though sparse capillary growth an inch to two inches long on the entire discolored part; on the back below the waist, particularly, there was a thick growth of hair quite bear-like, two inches long. She was intelligent, and very adverse to exhibiting the "bear-mark," as she called it, ... She stated that the milk from the discolored mamma always sickened her child." [speculation omitted]

Housgield, Norway. Danish anatomist Thomas Bartholin (1661, p. 41) reported a boy, born in the village of Housgield, "whose hair precisely recalled a bear's both in color and form" ("cuius pili prorsus ursinos pilos refererunt & colore & figuræ similitudine"). Housgield no longer exists.

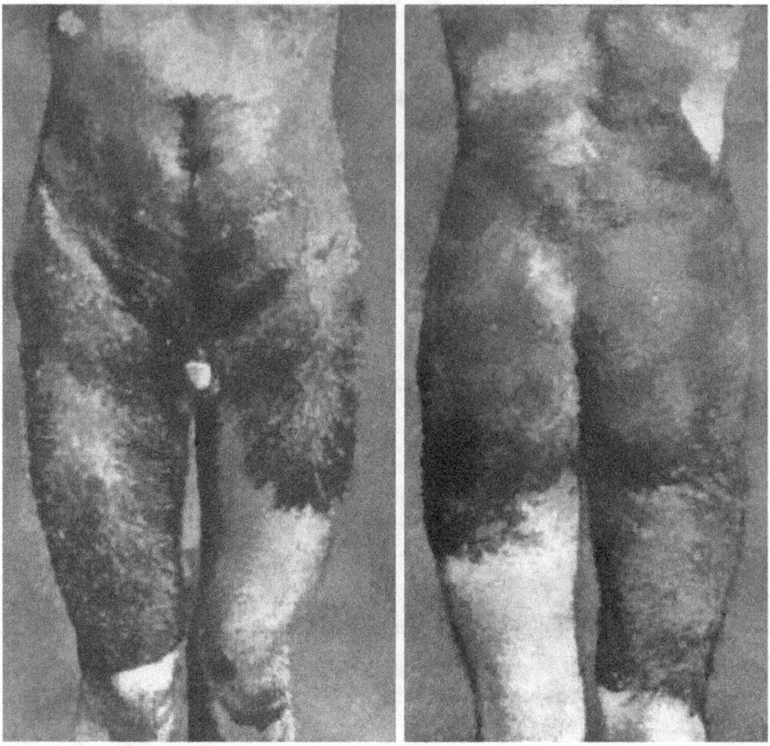

A patient with bear-like hair over much of his body (Joseph 1889, figs. 1 & 2).

Berlin. Joseph (1889) described a 22-year-old patient (pictured above) in Berlin with a skin condition recalling the Green Bay hominursid (p. 373). His scalp hair was blond, and his eyes, blue. But a bear-like pelt covered his lower back and belly, pubic region, buttocks and thighs. In these areas, the skin itself was pigmented, either dark brown or black. Forty coin-sized spots of a like nature dotted his body.

A Conjoined Bear-human Twin

London. A handbill dating to circa 1700 described a male "Child born with a Bear growing on its Back alive" being shown at London's May Fair (Ashton 1882, p. 263).

* Morley (1859, p. 323) cited this case and said his source was a book of handbills "once owned by Sir Hans Sloane, into which, I think, it was Sir Hans himself who pasted Handbills about some of the natural Prodigies which interested London from the days of Charles the Second to those of Queen Anne" (ibid., p. 315).

An early American hominursid, "Arctopithecus." Edward Topsell (1658, p. 16) said that its head was "like unto a child's" and that "one of them was given me by a Frenchman, which I kept alive for six and twenty dayes."

BIGFOOT ORIGINS: BEAR-HUMAN HYBRIDS?

From forests of the Northern Hemisphere, innumerable reports tell of mysterious hairy humanoids nicknamed *Bigfoot*. So widespread are the reports that different places dub them differently: *Sasquatch* in the Northwest, *Skunk Ape* in the Southeast, *Alma* in central Asia, *Yeti* in the Himalayas, *Mande Barung* in India, *Ban-manush* in Bangladesh and *Yeren* in China. So, what are these creatures? Do they even exist? The reports in this chapter suggest one possibility is that bigfoots might be bear-human hybrids. This theory would explain why reports of such creatures are largely limited to regions where bears occur. It also accounts for why different reports describe them differently—hybrids are notoriously variable and matings between different bears and different humans might be expected to produce different results. This notion addresses, too, an objection biologists raise against claims of the existence of megafauna cryptids of all kinds, that is, the implausibly large numbers required to maintain a breeding population. If bears and human hybridized sporadically and locally, the observed sparse distribution of sightings would be explained. No breeding population would be required. Under such an assumption, then, all the local variants of such creatures—Sasquatch, Skunk Ape, Alma, Yeti, Yeren, etc.—would simply be hybrids produced by various kinds of humans mating with whatever kind of bear happened to be locally present.

Killing of the Amberg hominursid (detail from a crude picture accompanying a report in the Wickiana Collection, Ms F 35, Bl 179v; ||2wtdkbks)

XENOGENESIS

A manuscript report in the Wickiana Collection (Ms F 35, Bl 179v; ||2wtdkbks) records a woman near Amberg, Bavaria, giving birth to a bear, or to an offspring much like a bear, in 1587. Supposedly, it was killed soon after birth, which the picture depicts.

Homoprocyonids (Raccoon-Humans)

Leave nothing but the nothings that belong
To this bare soul, let all men judge that can
Whether it be an animal or a man.
—William Butler Yeats
Parnell's Funeral

Both a bear-human and a raccoon-human fall under the heading of carnhominids. But the evidence for the existence of mixes of the latter type is more tenuous than for the former. In this book we will term such creatures, combining the traits of a raccoon with those of a human, *homoprocyonids*.

Henry Knox Stratford
1821-1915

Mendon, Massachusetts. Chicago physician Henry Knox Stratford (1881, p. 149) wrote that "Another most extraordinary case came under my observation many years ago. I saw a boy in the town of Mendon,

> Massachusetts, that had a face which was the exact counterpart of that of a raccoon. [Speculation on the discredited theory of MATERNAL IMPRESSIONS has been here omitted.] He would roll a small stick or plate around the palm of his hands for hours. His parents endeavored to keep him out of everybody's sight at all times; but he often got away and ran up a tree, which he would climb with extraordinary rapidity.

* Henry Knox Stratford was born in Mendon in 1821 and died in 1915 in Chicago.

Taylor, Indiana. From the *Memphis Daily Appeal* (Memphis, Tn., Jan. 24, 1880, p. 2c; ||h4rc8or):

> A MONSTROSITY
>
> SICKENING DETAILS OF THE BIRTH OF CHILD THAT IS HALF HUMAN AND HALF COON
>
> The Kokomo (Ind.) DISPATCH, of the twenty-second, publishes a highly sensational article on the recent birth in that county of a well-developed, sharply defined monstrosity, being half human and half

coon. The truth of the statement is vouched for by several of the best farmers of this county, who saw the *lusus naturae*, and whose testimony cannot be impeached. At the very urgent request of the parents of the monstrosity, their names will, for the present, at least, be withheld from publication.

About two weeks ago a woman living in Taylor Township, this county, about five miles from Kokomo, gave birth to twins, one of which was a well-developed boy and the other a hideous monster, half human and half coon. The regularly formed child is now living, but its twin monstrosity lived only a few hours after birth. Everyone who saw it at once remarked the striking resemblance to a coon. Its face was pointed and looked like a coon's face. It had four feet, resembling claws, on which were great sharp nails. It had a well-defined tail four inches long. It had no eyes, and its arms and limbs looked like the limbs of the animal it so strikingly resembled. Its body, or trunk, alone bore marks of human nature.

Our medical men are greatly excited over the matter, and great public interest is manifested in this rare and wonderful freak of nature. [speculation on the theory of MATERNAL IMPRESSIONS is here omitted]

Louisville, Kentucky. Thomas Brady Greenley (1887, p. 258a) reported attending, during the course of his Louisville practice, an infant who had "a deformed hand resembling the foot of the coon." "The baby," he said, "is called Cooney."

Thomas Brady Greenley
1818-1909

Allen, Indiana. From the *Chicago Tribune* (Dec. 16, 1871, p. 6c; ||y5xm9o9c):

> The Fort Wayne GAZETTE of the 11th says: "A few days ago a child was born in Allen township, in this [i.e., Noble] county, whose form displayed one of the most remarkable freaks of nature of which we ever heard. The child combined the appearance of a human being and a coon, the entire upper part of the body being an exact counterpart of the coon. The child was still-born, but the mother is convalescing."

HOMOPHOCIDS (SEAL-HUMANS)

Eudemus asserts that a seal fell in love with a man whose habit was to dive for sponges, and that it would emerge from the sea and consort with him where there was a rocky cavern.

—CLAUDIUS AELIANUS
On the Nature of Animals, IV.56

Rysum, Emden, Netherlands. Matthias Tiling (1670, p. 90), a professor of medicine, reported that in 1666 he had attended a woman in the seaside village of Rysum, who often went whaling to Greenland, and who experienced extreme pain and swelling during pregnancy. She eventually died, and autopsy revealed that she carried in her belly a large, hairy fetus clearly similar to a seal ("fœtus monstrosus figura externa cani marino plane similis ac multis pilis refertus in ventriculo ejus"). He said he saw the fetus himself.

Mattias Tiling
1643-1708

* In past centuries women, dressed as men, often served on Dutch ships (||mrxzacap).

Boufarik, Algeria. Le Petit Colon Algérien (Algiers, Aug. 7, 1884, p. 2d; ||yc4javbz) said, in translation, "Thursday, August 28, Mrs. D__., the wife of a carpenter in Boufarik, gave birth in that town to a monstrosity having all the appearance of a seal. The monster was stillborn. The delivery was performed by Mrs. Costa, a midwife, assisted by Dr. Séguy. The little monster was interred at once."

* Boufarik is about five miles from the Mediterranean. The only phocid occurring along the coast of Algeria is the Mediterranean Monk Seal (*Monachus monachus*).

Lewiston, Maine. From the *Sun-Journal* (Lewiston, Dec. 19, 1882, p. 3; ||ys8mczws): "A woman on Lincoln Street gave birth to a monstrosity a few days ago. The lower portions of the body were shaped like those of a well-formed child, while the head and upper parts resembled the young of a sea-lion. The creature had a hairy head, neck and muzzle. After it was a day or two old, it barked and spat. It was a hideous object. The woman nursed it for a while. A great many people saw it and shuddered at the repulsive spectacle. … the mother stopped nursing the monstrosity and it died."

Jersey City, New Jersey. From the *Camden Daily Courier* (Apr. 21. 1885, p. 1c; ||y6455h5f): "JERSEY CITY, April 21.—Mrs. Linder, the wife of William Linder, of Jackson Lane, Greenville [a part of Jersey City], gave birth last week to a female child which has a head like that of a sea lion.

The head is small and round and covered with soft, silky red hair. The child has no eyes, but there are cavities in the head. Its under lip is that of a seal. It will not take nourishment at the mother's breast but will drink milk and always seems hungry. Another peculiarity is that it does not cry like a human child, it cries resembling the faint bellow of a sea lion. Mrs. Linder spent last summer in Staten Island and made frequent visits to an aquarium in which were several sea lions … The child died yesterday."

* John W. Linder and his wife Rose lived in Jersey City in 1880 (||y3u49ged). Staten Island's commonest seal is *Phoca vitulina.*

Greenville, California. The *Chico Record* (Chico, Ca., Jan. 15, 1913, p. 1; ||5n7ypa7c) said a Greenville woman bore an 18-pound cyclops with three seal flippers and a human foot. It died just after birth.

Tranøy, Norway. Simon R. Kilbal (1839), provost of the Norwegian Church for the island of Senja, described a homophocid who reached maturity. Named Brede Sivertsen Normann, he was born April 7, 1805, at Tranøy, a fishing village in what is now the municipality of Hamarøy in Nordland. In English translation, Kilbal's report, which is itself a German-language translation of a Norwegian original (Berg, *Eyr: Et Medicinsk Tidsskrift,* 1835, vol. 10, pp. 293-299), reads

> The freak Brede Sivertsen Normann was born April 7, 1805, to parents Sivert Normann [ca. 1754-1824] and Johanna Margaretha Brede [ca. 1779-1837; see ||ypvt5zkj], on the Tanden farmstead in the parish of Hamarøy, Salten, Nordland, and baptized in the church of Hamarøy that same year. He was the only one of four siblings to be born a freak. The midwife was so frightened by the child's unusual shape that she lost all composure to lend her support to the mother giving birth …
>
> His gruesome form horrified all who saw him. The head, which was huge and bent to one side, ran to a point, ending in a fleshy knob about an inch wide. This lump was moist and hairless, though profuse hair—stiffer than is usual—covered the rest of the head. The forehead, which was quite flat, was divided in two by an opening about an inch wide from the nose upwards. This opening made it particularly noticeable that the skin dented inward towards the brain when the slightest pressure was applied, and that the skin on the forehead bulged when the lump on the crown of the head was pressed. Only as the boy grew did a substance form in the opening, which made it less yielding to the pressure. But this cartilaginous covering was still flexible when the finger pressure was applied. The eyes were exceedingly large and extended in an oblique direction up toward the temples. They were brown, and their gaze sharp and fiery. The nose was thick and blunt, a fleshy or gristly prominence, somewhat flattened. The nostrils had grown together almost completely, so that he could not breath

through his nose, which kept him from nursing. The mouth was prominent and unusually broad, the lips slightly overhanging, the teeth prominent and broad. The chin sat close to the mouth. The neck and arms were of ordinary human form, but the fingers were united by skin that was at first delicate, but that contracted, as the boy grew, and bound them together so tightly that the four fingers formed a point, so that the nails grew together, one into the other, and came to look like a single immensely large nail. From the roots of these nails a moisture constantly flowed, which stank horribly. The thumbs, which seemed only half-grown, were nail-less and merged with the palm of the hand, so that the fusion of all the fingers gave it the shape of a flipper. His chest and abdomen were normal and fairly well proportioned, though there was noticeable arching of the back and hips. The color was natural. His feet lacked toes, but five evenly spaced, little nails were present. The soles of his feet curved inward, which prevented him from using normal footwear. So he usually wore soft skin boots ("komager"). When he was washed, his body gave off a profuse perspiration that foamed the water so much that when his head or clothes were washed, the water foamed as if there were soap. His voice was hollow and harsh, piercing marrow and bone, It came usually as a howl or inarticulate shriek.

As to diet, he ate ordinary human food; but though equipped with strong teeth, he had a hard time chewing his food, especially hard things like rusks, bread crusts, and the like. Such foods therefore had to be dissolved in water or milk. The thought was that this was due to the opening in his forehead, which seemed to offer too little resistance to the pressure of the jawbone due to its mobility, so that it gave way when chewing hard food for lack of a point of support. …

He was much attached to his mother and expressed great anxiety in her absence., And if some time passed without her return, he would look anxiously in the direction from which he expected her to come, and his uneasiness lasted until her return, at which he would then show great joy. He was very afraid of fire and water, but he seemed to be more comfortable with the latter than with the former. For when the fire burned fiercely on the hearth, and no one was present, he always quickly put it out, and when children—from whom he generally seemed to withdraw—came near the fire, he drew them away from it at once. With a peculiar kind of dismay he therefore looked at the fiercely blazing heap of fuel on the hearth. He could hardly do any work but sometimes brought wood or water. For he was much hindered by the structure of his hands. If, however, he noticed that any work needed to be done quickly, he would scoot about anxiously and encourage exertion with his horrible, piercing cries. As to his mental and linguistic faculties, he possessed little of the latter and was very poorly endowed with the former. His mother is said to have noticed that he could now and then pronounce individual words, e.g., "Mother, food, drink, yes," he is said to have sometimes (once) said his thoughts and opinions even. But his brothers and sisters, who interacted with him every day, said that he could not speak, but made only inarticulate noises to indicate his hunger, thirst, and the like, to those around him. Generally, he expressed discontent and fear by howling or shrieking. … He seemed to have an idea of right and wrong, and what his mother had prescribed for him was sacred law to him. He was happy when you gave him a friendly look; but if he noticed anyone seemed afraid of him, he left at once. He was very brave. But when someone played a joke on him, he became cross, and the unfortunate creature would then sullenly take refuge in his mother, on whose sympathy and protection he seemed always to rely. Virtually helpless, even when he was older, he had to be dressed and undressed, to be provided with food and drink, and to be taken care of in all other ways, almost like a tender child. Yet he seemed, with his good-natured compliance, to encourage the patience of those around him, and to lighten their toil.

In his nineteenth year, when his father died on April 14, 1824, things became even more difficult. As the corpse was being carried out of the mourning house, and a psalm was sung, the lad suddenly fainted; and thereafter suffered a stroke that left him paralyzed. He had another three months later, and repeatedly thereafter, until he at last he died May 31, 1830, at the age of 25 years.

The mother of this unfortunate creature could offer no reason for his birth, other than perhaps the fact that once, while she was pregnant, she saw a dead seal that her husband had suddenly thrown down next to her. But she could not say what impression she had of the event. [see MATERNAL IMPRESSIONS]

Witnesses who saw the boy as a child, or when he was an adult, think he most resembled a seal with respect to the skin between his fingers and in the structure of his head, with its sharp and protuberant muzzle, and the shape of his forehead and eyes. It was also said that the mother herself, immediately after the birth of the child, had declared that nothing else could be to blame than the sudden appearance of the seal, which her husband had laid dead at her feet and which had been terrible to see, in that its head and torso had been badly injured by several blows.

The popular belief, that such a birth must always be regarded as God's punishment of some sin, was expressed, too, in this case, although the parents towered over the common folk in education and carefully hid the child during its first year from the public eye. Only later was this monstrosity seen by those who happened to visit the place. It is impossible to say just what progressive changes took place in this monster with the increasing age—which would no doubt have shed light on a fascinating question: To what extent could one call this creature *homo-phoca* [man-seal]? It remains for me to say that this report is based on the testimony of the mother, siblings and other credible witnesses and is extracted from the [Norwegian-language] report[4] of Pastor Berg of Hamarøy. Finally, I must also add that only the introduction of this report, and certain additions from other credible witnesses, are from my pen.

A harbor seal (*Phoca vitulina*).

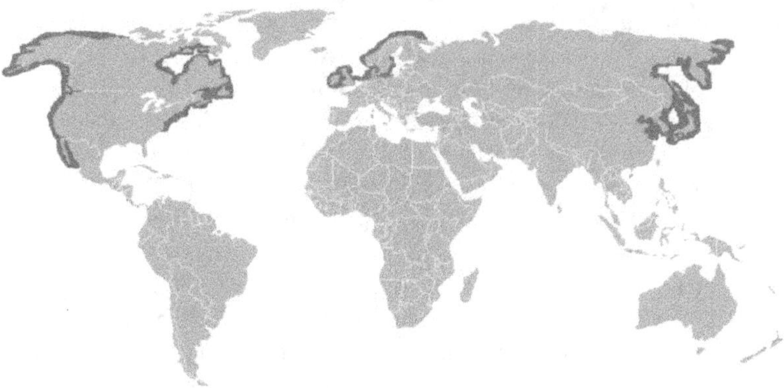

World distribution of the Harbor Seal (*Phoca vitulina*), the most common seal both in Norway and on the East and West coasts of the United States.

Venice, Italy. In 1741, a woman in Venice bore Marco Cazotte, who generally is usually described as being affected by phocomelia, a word that means "seal appendages,' which prompted classified as a homophocid. However, his appearance recalls that of the Buena Vista HOMINANSERIDS, four births that occurred in a locale where ducks are present, but no seals. He was just 80 centimeters (~31 inches) tall. Cazotte is pictured and described by both Vrolik (1849, pl. 77) and Verneau (1877). He lived to be 62.

* Cazotte's skeleton is in the holdings of Pierre and Marie Curie University.

Liverpool, England. The enigmatic creature described in this report, though billed as the "Talking Fish," seems best classified as a homophocid. The physical description given in the report, however, leaves much to be desired. From the *Liverpool Herald* (quoted in the *Berkshire Chronicle,* Reading, England, Jan. 22, 1859, p. 6d).

> In our last week's issue we called the attention of the public to an exhibition, then, and now, going on in Elliott Street of a marine wonder—namely, a talking fish. Although the animal had been exhibited more than a month, people were naturally incredulous as to the genuineness of the announcement on the bills; in fact the whole thing was looked upon as a hoax of so contemptible a character that a sneer or laugh were the only replies to the question "have you seen the talking fish?" For the purpose of exposing what we considered to be a wretched imposition, we entered the exhibition room, and the result was, we beheld the most extraordinary amphibious animal ever produced, and, in justice to its owner, made known our impressions. The consequence is that the creature is now the talk of Liverpool, for thousands have acted upon the axiom that "seeing is believing." At the same time we meet with many well-informed persons who ridicule our statements, because they have not seen. We are not at all surprised at this, for nothing but ocular proof can satisfy the mind where such extraordinary powers are stated to exist in one of the fish species. The intelligence of the creature is far higher than anything yet known in the animal world. It appears perfectly to comprehend the language addressed to it, and its fine expressive eye watches with lively interest every movement of its exhibitor. Its length is apparently about 22 feet in the confined space where it is kept, but it has the power of protruding the neck

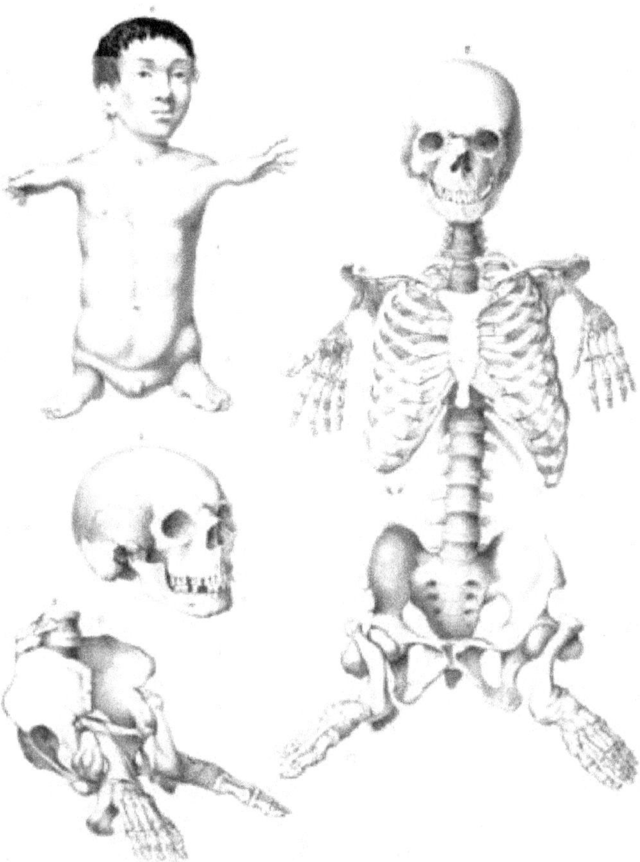

Above: Marco Cazotte, external appearance and skeleton (Vrolik 1849, p. 77). *Below:* Skeleton of a Mediterranean monk seal (*Monachus monachus*). Image: Wikimedia, Catlemur (||mr3dep6t).

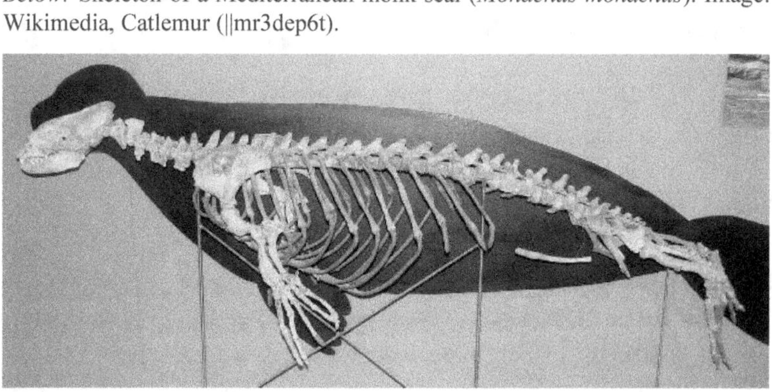

and shoulders considerably, which is seen when it erects itself, at the bidding of its keeper. Besides distinctly pronouncing the words "mama" and "papa," it can now call for "John," the man who attends upon it. In fact, we verily believe, from its tractability and high instinct, that it will yet be taught to speak several words and perform many antics that would now be deemed impossible. It is as docile as a tame dove and yet possesses all the power of a ferocious animal. We omitted, in our former notice, to observe upon the singular construction of the fins, or hands, on both sides. These will bend, and develop a hand, with joints like the human hand, wrists, and elbows. At command it presents either the right or the left to its keeper. At night it reposes on damp boards and will exist for days out of the water. On its passage to Liverpool it was without food of any kind, for it consumes nothing but fish, and yet sustained no injury. Its breath is, singular to state, perfectly sweet, and its docility such that ladies and children may approach it without the slightest trepidation. A very short acquaintance makes it familiar with a visitor, and it evidently appreciates kind words. All this we state from personal experience of the habits of the animal. The day previous to our first notice of this extraordinary creature Signor Cavana disposed of it to a gentleman of this town, who gave a considerable sum for the wonder. As a curiosity it excels all living things that have yet been submitted to the public, and we have no doubt, when it is exhibited in the metropolis, thousands upon thousands will be eager to see it. As a pecuniary speculation, it must prove a California to its present owner: as a wonder of nature, it stands alone, unapproachable in every respect. We have not overstated any particular we have detailed, and we feel confident all those who view the animal will give assent to the truthfulness of our remarks.

Bisceglie, Italy. According to Dutch physician Stalpart van der Wiel (1687b, obs. 35), Elizabetha Nastasia Consiglio, a resident of Bisceglie, a town on Italy's Adriatic coast, went down to the shore to wash clothes, and was there so impressed by the creatures she saw that she later gave birth to an infant whose "whole body, except for the head alone, was wrapped in what seemed to be the skin of a seal, which felt hard and rough to the touch" "totum corpus, excepto solo capite, veluti phocae pelle obducta, contactuque dura erant, & scabra."

Bernardo Consiglio
(Van der Wiel 1687b)

Van der Wiel's depiction of this "boy," who was christened Bernardo, is reproduced here. Other details seem to connect Bernardo with seals. Van der Wiel, in translation, said Consiglio, with his unusual skin, was little affected by cold even when naked, and that he "preferred to such a remarkable degree to dine on fish that even after eating them repeatedly, he still desired no meat. Indeed, I and other doctors who examined him noticed he exuded the smell of fish."

Although Consiglio has been tentatively classified as a homophocid, depictions show that his appearance was very similar to that of Edward Lambert and that of Lambert's various male descendants, all of whom have been classified as HOMERINACEIDS (see p. 412). Perhaps a clearer picture with regard to these cases will eventually emerge.

* When exhibited at The Hague in 1683, Bernardo Consiglio was about ten (van der Weil 1687b, obs. 35). So his birth year would have been circa 1673. Bekmann and Bekmann (1751, p. 876) say he was 14 in 1689. Morley (1859, p. 324) says he was also exhibited in England in the 1680s.

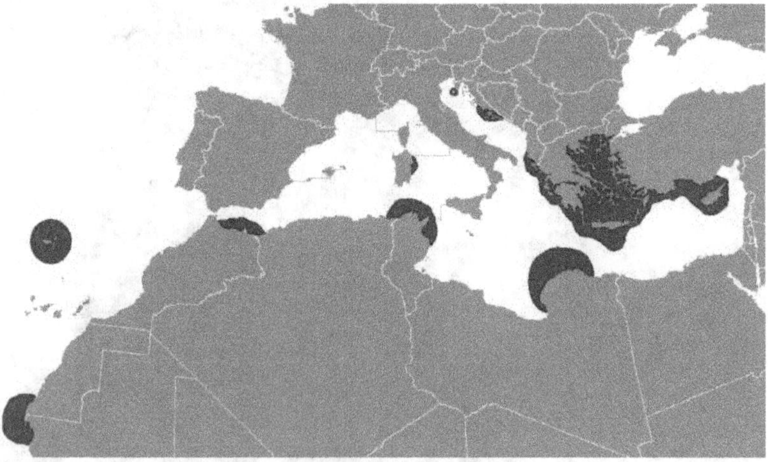

Modern distribution of Mediterranean monk seals (Wikimedia, Esculapio; ||3wakcxhd). They were once present throughout the Mediterranean. Considerations of distribution and abundance suggest harbor seals likely sired most homophocids, BOVOPHOCIDS, CYNOPHOCIDS, CHELONIPHOCIDS and PHOCOPISCIANS, as well as those TRIPLIANS with phocid traits. But the Italian homophocids (Venice and Bisceglie) and the El Kala TRIPLIAN, given their *loci in quo,* would involve the Mediterranean monk seal (*Monachus monachus*).

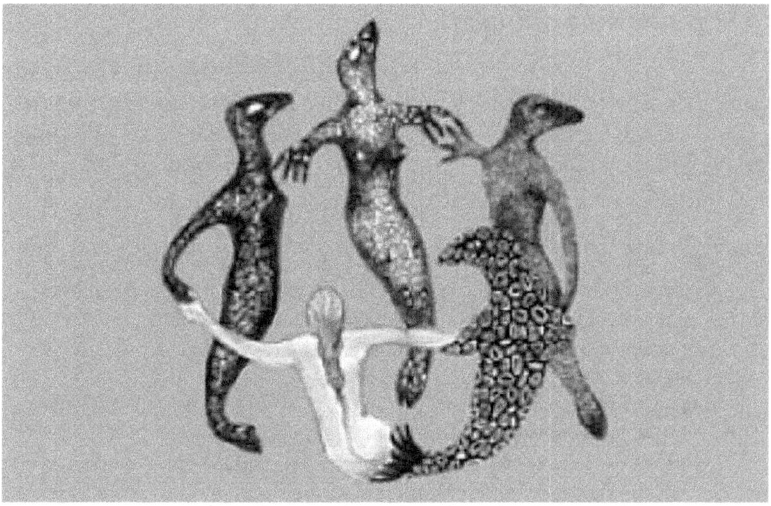

Dancing selkies (Artist: Edward Fuglø; ||yck42mdx).

SELKIES

I am a man upo' da land.
I am a selkie i' da sea.
An' whin I'm far fa every strand,
My dwelling is in Shöol Skerry.
—THE GREY SELKIE OF SULE SKERRY

In Norse and, especially, Celtic, mythology, selkies are beings that look like seals but that can shed their skins and take on human form. Monaghan (2004) states that "Some scholars remark upon the claim of seal ancestry by certain coastal families—in Ireland, the Coneelys, Flahertys, MacNamaras, Sullivans, and many families on Achill Island; on the Hebrides, the MacPhees; on the Scottish mainland, the MacCondrums; on the Isle of Skye, all fair-haired people." She says, too, "Marriage between water fairies and humans was apparently very common, if folklore that claims seal ancestry for many coastal families is to be believed" (ibid.). Mackinlay (1893, p. 5) stated that in the Shetland Islands, "A family living within recent times was believed to be descended from a human father and a mermaid mother, the man having captured his bride by stealing her seal's skin. After some years spent on land this sea lady recovered her skin, and at once returned to her native element. The members of the family were said to have hands bearing some resemblance to the forefeet of a seal."

CYNOPHOCIDS (SEAL-DOGS)

Calgary, Alberta. The *Calgary Herald* (Jun. 30, 1941, p. 15; ||subm2af) said, "A cocker spaniel body and a face like a seal's is the make up of a puppy owned by Mr. and Mrs. Percy Tydeman, 416A Fourth Street East." The account said it was the only unusual member of the litter and could not nurse but lapped milk with its "unusually large tongue."

Stockton, California. The *Fresno Bee* (May 6, 1936, p. 3Bc; ||t48gfb3) said,

> SACRAMENTO, May 6. … John F. Goldner, former newspaper man, to-day exhibited a freak which may cause considerable research.
>
> Queenie, a pedigreed police dog owned by Ramon Auntinteco, Stockton barber, gave birth to a litter of puppies.
>
> One of the offspring was born with a perfectly formed head and neck of a seal, with the remainder of the body that of a dog.
>
> Until a few weeks before the birth of the oddity, Queenie was a guard for seals on the beach at San Francisco, Golner said.
>
> The part-seal-part-dog lived a week and died due to the lack of vital organs in the throat.

Keyport, New Jersey. The *Home News* (New Brunswick, N.J., quoted in the *Chicago Eagle,* Oct. 3, 1896, p. 11a; ||mm2sz6j2) described an animal caught off Keyport in Raritan Bay: "The animal is 6 feet long and 3 feet in diameter at the center, tapering to a tail, on the end of which is a bunch of heavy fins. The head is large and resembles that of a huge pug dog. The animal has a coat of fine silky hair and growls like a dog."

* Seal-bovid mixes have also been reported (see PHOCOBOVIDS).

Le Thillot, Vosges, France. L'Express de l'Est et des Vosges (Épinal, Jul. 28, 1929; p. 3g; ||bddmef52) said, in translation, "LE THILLOT.—A FREAK.—M. Peureux Charles, Rue des Ballons, says his dog whelped a litter last Friday and killed one of her offspring that resembled a seal in every respect." This report amounts to an allegation of XENOGENESIS.

A LUTRAHOMINID (OTTER-HUMAN)

Szczecin, Poland. Fincelius (1566b) said a woman at Stetin (now Szczecin), on Epiphany (Jan. 6), 1554, bore an intersex nondescript with paws and claws like an otter's. According to the description given, the anatomy of this composite was jumbled, with a mass attached to the head resembling misplaced internal organs and an arm in place of one of the ears. The arms and legs lacked elbows and knees.

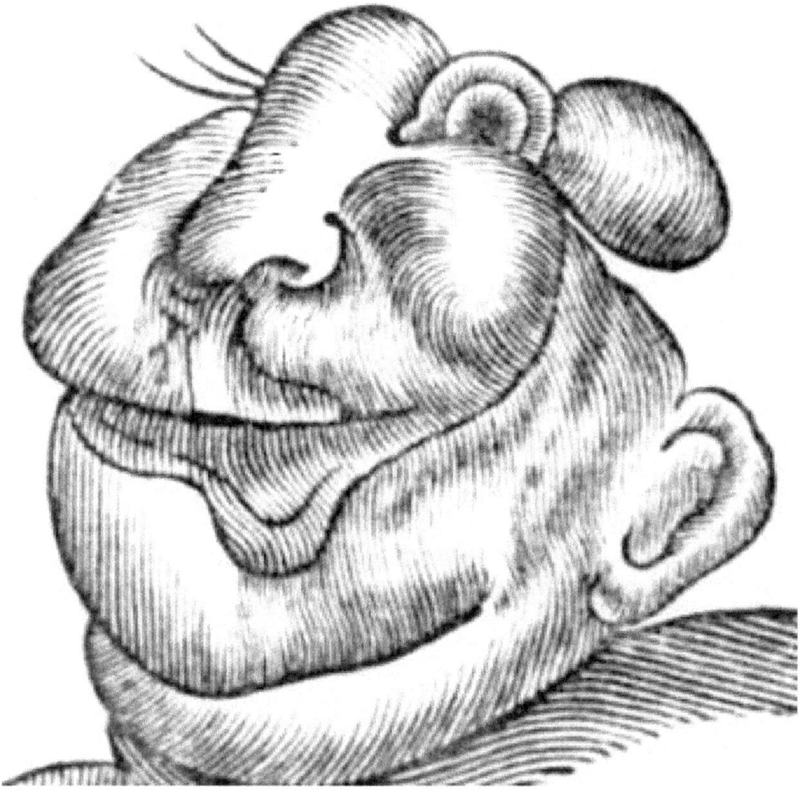

The Zerbst laghominid (Kühne 1728, p. 614).

Glirhominids

In some people we observe the upper lip in a manner slit or divided so as to resemble the upper lip of a hare.

—LORENZ HEISTER
A General System of Surgery

Any creature combining the traits of a human being with those of a glirid (rodents or rabbit-like animals) is a *glirhominid* (/glĭr-HAHM-ĭn-ĭd/).

The Vaduz laghominid, as depicted in a manuscript in the Wickiana Collection (||yfab8nch). According to the text accompanying this picture, the wife of a certain Christian Lampert gave birth, at Vaduz, to a creature with the upper body of a hare, but with human lower parts, on November 11, 1577 at Vaduz, which is now the capital of Liechtenstein. This description matches that of the Livorno laghominid. All reported laghominids were the offspring of women. Judging from reports, there is a strong tendency in these composites for leporid traits to be more expressed toward the cranial pole of development and human traits, toward the caudal.

Laghominids (Rabbit- and Hare-Humans)

> *These and other such things, the offspring of human beings, have been created by the ingenuity of Nature as toys for herself and marvels for us.*
> —Pliny, *Natural History,* 7.2

Cracow, Poland. Miechowita (1521, p. 358) stated, in translation, that on Oct. 22, 1494, "near Cracow, an unfortunate woman bore a monster with the ears and neck of a hare, and which had only an intestine in place of a stomach. It breathed through a gaping mouth."

* In 1494, Miechowita was living in Cracow.

The Cracow laghominid (Schenck 1609)

Vienna. The *Pustertaler Bote* (Sep. 30, 1853, p. 2a; ||45b5tse2) said a woman in Vienna had borne a stillborn daughter with a hare's head.

Zerbst, Germany. Kühne (1728) said a Zerbst woman, November 8, 1722, bore a brainless female with a whiskered hare-like face. Its fur and single eye were leporine. It lived 34 hours and was baptized.

Cœuvres, Aisne, France. L'Observateur des Pyrénées (Pau, May 23, 1847, p. 4b-c; ||4p465nzr) said a woman at Cœuvres bore a viable male fetus with hair, eyes, upper lip, forehead as in a rabbit. Two ears looked human, a third, like a rabbit's. The eyes lacked lenses.

Vaduz, Liechtenstein. A manuscript, in the Wickiana Collection (||yfab8nch), dating to 1578, states that in 1577, a woman at Vaduz bore a laghominid with the upper parts of a hare.

Livorno, Italy. The Lancet (Apr. 17, 1897, p. 1128; ||425p9vbt) said, "From Leghorn comes the announcement on the 8th inst. of the birth of a monster, the mother being a poor water-carrier of the city, and her offspring resembling from the head to the umbilicus a hare, and from the umbilicus to the lower extremities a normally developed infant. Dr. Botlari, who assisted at the birth, reports that the monster is alive; in place of eyes it has two cavities, its mouth and nose are similar to those of a hare."

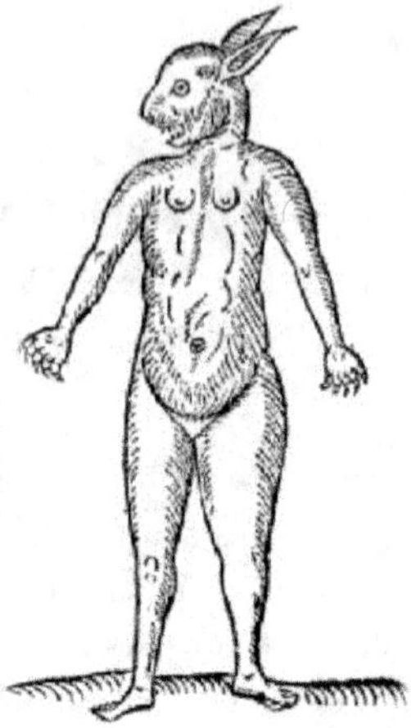

The Ottringham laghominid

Ottringham, Yorkshire. A pamphlet (Anonymous 1595) reports that on April 9, 1595, Elizabeth Rowling, a woman at Ottringham bore a nondescript with the head of a rabbit and hands like a mole's. According to the report, it lived and was otherwise like a normal female child.

Nuremberg, Germany. Johann Helwig (1680, obs. 5, p. 15), a Nuremberg physician, reported, in translation, that

> In 1641, Kunigunda, the wife of A. K., bore a monster with fore parts, including the head, like a flayed and torn hare. The skull was small and rounded. The mutilated ears, which replaced the usual small ones, were as long as a hare's and stood above the forehead. There was no tongue. The arms and legs were short stumps, there being only a single deformed foot. It was a male.

Tavistock, Devon. From *Berrow's Worcester Journal* (Worcestershire, Mar. 27, 1828, p. 4; ||mruvcn32): "The wife of a husbandman residing at Tavistock, named Downing, was on Monday last delivered of a male child, having six fingers on each hand, and six toes on each foot. The surplus toes and fingers are connected with the small fingers and toes. The child has a monstrous hare lip. The forehead has a striking resemblance to the upper part of the head of a hare."

Paris. From the *Athens Post* (Athens, Tn., Feb. 7, 1851, p. 3b; ||7ukse95w): "FREAK OF NATURE.—A lady of distinguished rank in Paris, lately gave birth to a child, which, though otherwise well formed, was totally without a brain [i.e., anencephalic]. The young creature, like the sensitive plant, experienced the liveliest emotion upon being touched. The top of the head was completely flat, which, joined to a long and narrow jaw, gave to the face the resemblance of the head to a rabbit. It lived twenty-two hours, to the great astonishment of the doctors."

London. In *The Lancet,* Edwin Child (1868), a London doctor, reported the birth of a laghominid. The relevant passage reads, "the child … was dead, and from the peeling of the skin I conclude it had been so for some time. … It was a fine, full-grown male child, formed naturally as to its body, with the exception of the nails on the thumbs, which were like those of a rabbit. On examining the head I was surprised to find the parietal, frontal, and part of the occipital bones wanting; at the space corresponding to, but larger than, the anterior fontanelle, was the brain, entirely denuded of skin or membrane not even being covered with arachnoid. There was a little hair over the eyes, not elsewhere. The eyes, palate, and tongue were similar to those of a rabbit."

Lausanne, Switzerland. The Swiss physician Jean-Baptiste Desgranges (1798, p. 70, obs. 2) reported, in translation, that a woman in Lausanne "gave birth at term to a girl half of whose face was covered with what appeared to be the skin of a rabbit, the upper lip was cleft, and there protruded from the angle of her jaw what appeared to be a rabbit's ear."

* Jean-Baptiste Desgranges (1751-1831) was a member of the Accademia delle Scienze di Torino from 1792 (||mwm84pv5).

Königsberg. The chronicler Andreas Angelus (1598, p. 13) said that in 1593, on Michaelmas, a woman at Königsberg (now Kaliningrad) bore a laghominid with an ear like a hare's. And in the *Historia Prussiæ,* Joannes Leo (1726, p. 435) said another, on January 21, 1566, bore one that had not only hare ears, but also four harelike paws.

Port Appin, Argyll, Scotland. Telling of "a young girl with an oblong patch of soft dark hair on the left cheek" that had come under his observation, physician Archibald Mackay (1891), of Port Appin, stated that "The colour and character of the hair on the patch closely resembled the fur of a very young hare."

Eisfeld, Germany. Balthasar Chilian (1683) said, in translation, "On May 4, 1680, between 11 and midnight, after a difficult labor, the wife of Eisfeld butcher Georg Schwencken, an upright woman named Elizabetha, who in six years of marriage had borne three healthy children ... was delivered of a 30-week fetus with a monstrous double face. It was a female. The distended head, neckless on the shoulders, was covered behind by a fleshy appendage, mangled, pendulous and blood-filled. The distended cranium, shaped like a hare's, contained a small brain. ... The forehead was narrow, almost nonexistent, its upper border, edged with fine black hair. There were three hare-like eyes—all open, one with two pupils in the middle of the forehead, and two others, one on each temple—and two narrow noses. The pair of hare-like mouths had a human tongue and cleft palate. Both ears were wide and floppy. Hair covered nearly the entire body. Between the cheeks was a navel-like opening. Below the diaphragm, an enlarged liver hid the small intestine. The other internal organs were normal."

Paterson, New Jersey. The *Globe-Democrat* (St. Louis, Mo., Mar. 28, 1877, p. 2; ||u9h37rw4) reported twins born in Paterson with hare-like noses, lips and chins.

York, Pennsylvania. From the *Centre Democrat* (Bellefonte, Pa., Nov. 29, 1860, p. 2c; ||yzfpvjja): "One of the most singular freaks of nature we have read of for a long time, occurred in the neighboring town of York lately, the particulars of which are given in one of the papers of that place, as follows: "A married lady gave birth to a child, possessing, generally, the forms and whims of a Rabbit. The features bore a close resemblance thereunto; and at the hind part of the head there was a short stumpy tail, precisely in imitation of the small pet animal. The child squeaked with the notes of a Rabbit. It was so formed that it could partake of no nourishment; and during the five weeks of its career on earth, it was necessary to pour, gently, a little milk down its throat. It has but little flesh, and that blue, resembling that of the innocent quadruped to which species it might have been classified to belong, instead of the human family."

La Tour-du-Pin, Isère, France. L'Express du Midi (Toulouse, Nov. 27, 1891, p. 2b; ||2m7trn3c) said a woman in a village near La Tour-du-Pin had borne a nondescript with the head of a hare, and that the head was covered with rough hair ("poils rugueux"). It was unable to eat.

Hamm, Germany. Der Wanderer (Vienna, Oct. 5, 1823, p. 3b; ||3exkjcmt; ||bddm9vhf) said a woman at Hamm bore a rabbit-faced anencephalic.

Edenton, North Carolina. Edward Warren (1853, p. 348), an Edenton physician, reported a woman had recently come under his observation who gave birth to a child that "bore so striking a resemblance to a rabbit, that the most casual observer could not fail to discover it at a glance" (and that a woman of her acquaintance afterward did the same).

Ghent, Belgium. Martou (1839) reported a woman in Drongen, a suburb of Ghent, birthing a girl with a harelike lip and two large incisor teeth.

Mattersdorf, Austria. The *Innzeitung* (Innsbruck, May 7, 1866, p. 3b; ||4k2w2wnh) said, in translation, "In Mattersdorf, near Vienna, a Jewish merchant's wife recently bore a child whose upper body had the complete appearance of a hare, while the lower was normal."

Weiltingen, Germany. Gockel (1700, p. 155) said a Weiltingen woman bore on St. Martin's Day (Nov. 10), 1685, a six-fingered son with harelike mouth, lips and voice. Baptized, he died soon after birth.

HOMORODENTIDS (RODENT-HUMANS)

They looked at one another, amazed. This thing they had never really believed in was coming true.
—JOHN STEINBECK, *Of Mice and Men*

Call any creature mixing the traits of both humans and rodents a *homorodentid*.

Albany, Oregon. In a paper read before the Linn County Medical Society, Albany physician John Linsey Hill (1877) claimed to have delivered a woman of a homorodentid with the foreparts of a squirrel. His account reads, "Another lady, whose mind, during pregnancy, was more than usually exercised about a pet squirrel, gave birth to a child with a squirrel head, squirrel hands, and the forearms, in shape and general appearance, were identical to the forelegs of a squirrel. It soon expired."

John Linsey Hill
1845-1919

Ten Broeck,[3] *Alabama.* From the *Daily Appeal* (Memphis, Tn., Dec. 6, 1884, p. 2d; ||hlorn43):

HALF HOG, HALF HUMAN
REMARKABLE LUSUS NATURAE DISCOVERED AT CHATTANOOGA, TENN.
A CHILD WITH MANY OF THE PECULIARITIES OF THE GROUND-HOG
[Special to the APPEAL.]

CHATTANOOGA, December 5.—"It is half ground-hog and half human." This remarkable utterance reached the ears of the APPEAL correspondent to-day as he entered an upstairs room of the Francisco House [a Chattanooga hotel]. In the room were gathered a mild-mannered, thoughtful-looking lady, and near her stood two gentlemen, who were introduced as Judge Brandon of the Probate Court of DeKalb county, Ala., and Capt. Frazier, of the same county, and a few feet from her was the wonderful monstrosity to which allusion was made. The child is in all respects the most remarkable creation the reporter ever witnessed, a most marvelous lusus naturae, a personification of the close affinity that can exist between brute and human.

The story of its birth is simple, and the cause of the freak is clear. Nine years ago, when enciente, Mrs. [Sallie] Lacey was shocked and terrified by seeing a ground-hog killed in her presence and witnessed its death struggles. The scene deeply affected her [see MATERNAL EFFECTS], and when her child was born it possessed the chief characteristics of the animal she saw killed. It will be ten years old on the 22nd of February. Its height is about three and one-half feet; its weight, about sixty pounds. It is hearty, vigorous and sound in all respects. It is a male.

A description of the remarkable creation is almost impossible, the features of beast and human being so closely allied. The observer is first struck by the appearance of the head. It is elongated, terminating about the forehead in a point shaped much like a horn. The back of the head is perfectly straight, from the neck curving gradually to the point. The forehead is concave and is fully six inches in height. The head is of extraordinary size, measuring fully twenty inches from ear to ear. The eyes are very large, the sockets are set very deep in the skull, but the eyes are full and protrude an unusual distance. The eyebrows slant and the sockets are triangular in shape. The mouth is also oddly shaped, being small and circular. The teeth are clustered and irregularly set, forming two or three rows. The roof of the mouth is concave. The body is perfectly white and has normal shape but is without shoulder blades. There are no joints in the hands or feet, nor any fingers or toes, but nails grow where the fingers would protrude.

The boy is very active and playful. While the reporter was present it was continually running about the room, throwing a small stick into the air. Its tongue is almost constantly out, and it utters a hiss and guttural noise peculiar to the ground-hog. The boy cannot talk and has no perceptive faculties. It never suckled, but was nourished by milk, administered in cups. Up to the age of three years it would walk on all fours like a hog but now stands upright. It makes no effort to speak and seems totally devoid of any human faculties. When it is frightened, it does not cry out but hides and crouches in any dark place.

The boy will eat anything given him but does not chew his food; his mother said it would choke to death at every meal did she not soften its food and give it in limited quantities; it usually takes milk. He sleeps naturally but will never sleep during the day. He is very docile and never becomes at all violent or vicious. He is obedient, like any animal that has been trained, and seems to have a deep affection for his mother.

The mother is about fifty years of age. She is a very quiet, matronly lady. None of her other children are deformed. She lives on Sand Mountain, near Ten Brook [today's Ten Broeck], Ala. The child will be placed on exhibition in this city as soon as arrangements are perfected.

The following report about the same individual later appeared in the *Spectator* (Staunton, Va., Mar. 4, 1891, p. 2e; ||5utnwv9j).

A GROUND HOG CHILD

ONE OF NATURE'S MOST SINGULAR FREAKS EXISTING IN TENNESSEE

A Chattanooga letter to the PHILADELPHIA TIMES says:

One of the queerest and most wonderful freaks in the world was seen in Chattanooga yesterday by the TIMES correspondent.

Only five like it have ever been born according to the best medical authorities and this is the only one that has ever lived. The freak is known as Jack Lacey, "the ground hog child." At first sight of the half human being a person is struck with the peculiar animal motion which the child keeps up as it stands on its feet; also with the two bow shaped hickory twigs about four inches long, which it plays with and balances on its peculiar stub-like hands. The twigs have been worn slick by its constant handling, and he is never without them when awake.

The wonderful monstrosity was born in the wilds of Sand Mountain, DeKalb County, Alabama, on Washington's birthday in the year 1875. Consequently on the 22nd of this month [i.e., February, the month in which the letter was written] the freak will be sixteen years old. The boy stands three feet seven inches and exactly twelve inches of his height are taken up by his abnormally shaped head. He has no cheek bones, but he possesses one more bone in every limb than a human being. The pupils of his eyes are twice the size of those of ordinary persons and a prominent Chattanooga physician who examined him says it receives five times as much light in his eyes as a human being. At night it can see just as well as any wild animal. His feet and hands are solid bone with no joints [stump appendages are commonly seen in telenothians]. The fingers and toes are knitted together and run out to points and are covered with one broad nail.

The birth of the freak is accounted for as follows: Its mother, seven months before its birth was soundly sleeping one night when some mischievous and thoughtless boys threw into the bed a ground hog they had captured. It bit the woman and frightened her terribly and left its impression upon her unborn babe. [See MATERNAL EFFECTS]

Mrs. Cynthia Lacy, the mother of the freak, says that the monstrosity was her eleventh child and that she was forty-two years old at its birth. Her other ten children were well-formed and grew up strong and healthy. She is a native of Jackson county, Ala.

One of the greatest peculiarities of the freak is its method of feeding. His mother, whom he never allows to get farther than two feet from him, chews up all his food and puts it in his mouth. Jack removes this food with his right stub hand and plasters it on his left hand. He then uses his hands in again putting the food in his mouth, as a plasterer uses his trowel and board in doing a delicate piece of work. The curious animal being will not wear shoes and has never chewed or spoken a word in his life. The only sounds it ever emits are those of a ground hog.

Lebanon, Warren County, Ohio. Dr. Edward Bruce Stevens (1881), a resident of Lebanon (||8u7na5zf), made the following report.

> I was called January 19th, to attend a young unmarried woman in premature confinement. The patient supposed herself about five months advanced, which I presume was not far from correct. The labor was tedious and proved to be a breech presentation. After the birth of the body I had more than usual difficulty in the extraction of the head, especially in view of the age of the foetus. After the birth was completed, however, upon examination, I found the hair on the head very abundant and that it extended down evenly and uniformly over the whole forehead, a compact matted mass continuous with the eyebrows. This peculiar development of hair gave the head a very singular expression—very closely resembling a squirrel. She made a very satisfactory explanation by claiming that when she was only a few weeks gone in pregnancy, she had been greatly startled by a pet squirrel that had attempted to bite her! Upon further examination I found two teeth in the lower jaw, the two lower incisors, quite well developed, and showing outside of the gum.

Vincennes, Indiana. Another report about a squirrel-human appeared in the *Indianapolis News* (Mar. 25, 1881, p. 2e; ||yxww2r5h).

> A child has been born and is now living at Vincennes, which is described as follows: The general appearance of the child is striking. Its shape is that of the ordinary human, excepting that the head, hands and feet are those of a flying squirrel. The fingers are webbed, and from under the arm along the body to the leg is a flabby piece of skin which gives to the animal the peculiar name by which it is known. The head is shaped somewhat like that of a squirrel but is devoid of any unusual capillary adornment.

The following, eyewitness report about the Vincennes birth, in the *Cincinnati Enquirer* (Mar. 31, 1881, p. 5; ||3w53dvtu), provides additional information and identifies the parents.

> VINCENNES, IND., March 30.—The account of the flying-squirrel monstrosity which was published in the ENQUIRER of last Friday did not appear in any other metropolitan sheet, and not in the local press till the following day; consequently, the publication in the ENQUIRER was hailed with a shout of incredulity, although it was eagerly perused and freely commented upon. The vagueness of the description and the fact that I "heard" the story from a third party may have induced your readers to refuse to place any confidence in the account. Since then I have interviewed the father of the child and am enabled to give an accurate description of the lusus naturae.
>
> The parents of the child are Mr. and Mrs. George Kleinklaus,[1] who live in the southern part of the city. The father is employed at a foundry in

this city and is as jolly a fellow as one would care to meet. They have two children, a boy and a girl, as fat as ever two youngsters could get. On the 24th of February last Mrs. K. gave birth to the object which forms the basis of this article. It was almost lifeless, and the midwife had her hands full in the effort to resuscitate the kid. It rallied, however, and was treated as other babies usually are. The matter of its deformity was kept silent, but leaked out, as such things always will. In my interview with the father he told me all that the public cares to know about the wonder. The child is now over a month old. While it is not very healthy, it is not in an alarming condition. It weighs twelve pounds, and has not increased one iota, either in weight or stature, since it was born. The forehead is shaped like a squirrel, with a human nose, but a decidedly inhuman upper lip. I asked the father if it had teeth. With a shake of the head he replied: "If you'd put your finger in his mouth once, you'd find out." Conclusion drawn from this: It has teeth. A growth of hair one inch long covers its head and meanders down onto the body, over which it extends—a colored fuzz. In front of its ears is also a fuzz, which reminds one of a young American's incipient beard. The boy's complexion is a light manilla. The second and third fingers on each hand and the relative toes on each foot are webbed to the tips. The others are webbed only midway. On the fingers there is a noticeable absence of the second joint, and the tip joint is crooked like a claw. The great toes on each foot are double.[2]

If the description stopped here, the child would be enough of a curiosity, but regard for candor compels a full description of this strange child. In my former dispatch I mentioned the wings of the child. An incredulous public heaped invectives upon the head of the poor, luckless wight of a newspaper-fiend for such an apparent breach of the truth. The father confirms the statement. From the elbow underneath each arm to a line circumferential with the navel, on each side of the body, extends the transparent piece of skin, exactly like that of a flying-squirrel, and which confirms the suspicion that the child is a rough counterfeit of that animal. This completes the description of the infant. The babe is not a repulsive-looking object by any means, and its mother is just as fond of it as if it had taken the blue ribbon at a baby show. One other peculiarity is the voice, which is unlike that of the conventional baby. It resembles the peculiar sound made by a squirrel when it is seized hold of by a strange hand.

London. F. H. Daly (1869), a London physician practicing in Dalston, reported to *The Lancet* on his delivery of a homorodentid in the early hours of June 4, 1868. "The child was dead," he said, "and evidently

had been so for some days. … When the mother was made comfortable, I took the child into another room and examined it. It was a most horrible monster. The body and limbs were natural; but there was no neck, the

> head being placed immediately between the shoulders, with the face upwards. The resemblance to a rat was most striking. The maxillae were prolonged, and terminated in an exact snout, the nose, mouth, and tongue being precisely like those organs in the rat. There was no rotundity of the cheeks, of prominence of the forehead; but a prolonged cone-shaped snout, looking upwards, attached directly above the sternum. ... she informed me that at the time of the conception, and for about three months afterwards, she lived in a house infested with rats. To use her own language, "Wherever she turned she saw a rat."

Sandusky, Ohio. In an excerpt from the *Coshocton Democrat* (Coshocton, O., Apr. 3, 1888, p. 2h; ||5fr5ycdj; datelined Sandusky, Mar. 26), a reporter said he went to a doctor's office to view an unusual specimen.

> ... upon arrival there the doctor took from a box, which was filled with jars containing "specimens" preserved in alcohol, a jar carefully wrapped in a thick cloth, which was removed by the physician and the jar held up to the reporter's view.
>
> "Great heavens! What's that!" exclaimed the reporter.
>
> "That," said the physician, "is a specimen for which I would not take $1,000. It is a fully developed male child with a perfect rat's head and tail. I was at this birth myself, and after much persuasion and a pledge of secrecy as to the name of the parents I was promised the monstrosity in the event of its death. It lived two weeks, and I then obtained possession of it, and have had it ever since. [speculation here omitted] When alive it squeaked like a rat, its black bead-like eyes snapped, and its tail moved restlessly to and fro. It had two half developed tushes protruding from the lower jaw and overlapping two tiny teeth in the upper jaw and ate solid food with avidity.

Arkansas. The *Holton Signal* (Holton, Ks, Aug. 26, 1891, p. 6; ||y2b3c8cl) quoted a dime museum agent as follows.

> In Arkansas, I found a mother with a baby that bore upon its little shoulders the furry head of a mouse. The teeth were long sharp and pointed, the ears were tiny and the sense of hearing very acute. The fur was perfectly white and very short, although heavy enough to conceal the skin. The joining of the neck to the body was really wonderful. From the delicate neck there was a graceful and gradual enlargement until it extended to the beginning of the shoulders. The body was perfectly formed and was without a blemish or contortion. The mother was a young country girl of some little learning and would not consent to any proposition I made to allow me to exhibit the then four-months-old mouse-babe.

Helmstedt, Lower Saxony. Moebius (1730) reported various observations made by his professor, surgeon Lorenz Heister. One (obs. 20, p. 38) involved an infant brought to Heister at the University of Helmstedt. In translation, the report reads, "In April, a three-month-old infant was brought to our professor, which had been born with certain excrescences, one next to each ear, in front of the tragus.

Lorenz Heister
1683-1758

> They were shaped just like the ears of a mouse and covered in short hair but were much larger and thicker than a mouse's ears, in length and width, about the size of the last joint of an adult's little finger and having a cartilaginous supporting structure. Therefore, since the parent's wanted this prominent deformity removed, our master snipped them off near the skin. The wounds then quickly healed after an application of wound ointment and plasters.

> * Cats with similarly placed, extra, rat- or mouse-like ears are pictured on the Internet (||2xreethp, ||xbxua8jx).

Lebanon, Pennsylvania. From the *Daily News* (Lebanon, Pa., Feb. 10, 1892, p. 1f; ||4ck7c97w):

> Mr. John Haulman [1864-1925], an employee of J. C. Hauer Sons, who resides at 211 South Tenth Street, on Sunday shot a rat at his home, which has created intense excitement in that neighborhood and his home was visited by a large number of people last evening to get a glimpse of it.
>
> The rat is black in color, the body is natural, but the face presents somewhat a human expression, the hind legs are formed like the legs of a human being, while the front legs are shaped like the hands of a human being. He has placed the rat in a bottle of alcohol, and it is quite a curiosity.

Houston, Texas. From the *Philadelphia Times* (quoted in the *Press and Banner,* Abbeville, S.C., Jul. 17, 1889, p. 6g; ||kegbb25):

> A CHILD-FACED RAT
>
> A singular freak of nature is on exhibition at the rooms of the Natural History Club, in Houston, Texas. It was brought in yesterday by a German family named Schweinfeldt, living in the suburbs of this

city, who tell the following remarkable story in connection with the strange creature:

A few months ago they were aroused one night by a shrill scream of pain from their year-old baby. Rushing to the cradle nothing was seen or heard, but the next morning, while bathing the child, the mother observed two red spots on the arm near the brachial artery, looking as if they had been punctured by a needle. The arm swelled a good deal and was still very sore. In about a week the baby was found dead in its cradle bathed in blood. The jugular vein had been bitten through.

The physician who was called in, on seeing the small but fatal wound, which consisted of a hole the size of a darning needle might have made, and hearing the history of the swelled arm, immediately said that both bites had been inflicted by a rat.

After the baby's burial the Schweinfeldts naturally determined to rid their house of the dangerous rodents and consequently traps of every fashion were placed about. Many were caught and drowned. One night, several months after the death of the baby, a rat was heard running about its narrow prison, and simultaneously the crying of a child was heard near by. The head of the family procuring a light, rushed to the place whence the cries seemed to come. To his astonishment it proceeded from the rat-trap, in which could be seen one of those animals.

Taking up the trap he examined the rat closely and was further amazed to find that the creature's face strongly resembled that of a human being, while yet it retained the characteristics of a rat. It cried piteously and so much like a hurt child as to be easily mistaken for one when out of sight.

It is this rat which is now on exhibition at the Natural History Club's rooms. Its eyes are somewhat larger and more human-looking and have more distinctive lids than are usual. The nose, however, is the most remarkable feature, being decidedly marked and prominent, with swelling nostrils. The mouth is small and has unmistakable lips, but the teeth are long, keen and rat-like. The feet show a slight resemblance to the human hand, although the nails are curved like claws.

Dr. Pinning, President of the Natural History Club, and a noted naturalist, agrees with the Schweinfeldts in thinking this must be the offspring of the rat that killed the baby, and the phenomenon is due to her milk being formed from the child's blood which she sucked.

CONJOINED RODENT-HUMAN TWINS

Glogow, Poland. German physician Matthäus Gottfried Purmann (1684, p. 169) reported a conjoined mouse-human twin he said he saw himself in 1667 at Glogow (now in Poland). The passage in question reads, in translation, "She gave birth to an otherwise normal girl with a thing resembling a mouse on the right side of her back, and it was so like a mouse that nothing, not even the hair, had been forgotten."

M. G. Purmann
1648-1711

London. Nearly, two centuries later, in a lecture on moles, Sir William Lawrence (1830, p. 228) reported a remarkably similar case. "Sometimes a part of the skin is brownish, or reddish, and thickly covered with hairs, like the coat of an animal. A little time ago, I remember there was a boy in the foul wards. I had him stripped to examine

Sir William Lawrence
1783-1867

> some syphilitic eruption. There was a part of the chest, about two inches one way and one inch the other, of light brown hair, like the coat of an animal. I looked at it, and said, 'What is that on the chest?'—'Oh, sir,' said he, 'it is a mouse.' I said, 'it is not a mouse—it has no tail.'—'Ah,' said he again, 'it had a tail, but somebody cut it off.'"

Buffalo, New York. Herman E. Hayd (1858-1944), a Buffalo physician, reported a third such case (of a mouse on the back of a child) as having occurred in his practice (see Peterson 1885, p. 255).

Market Deeping, Lincolnshire. James Clapperton (1875), a Market Deeping doctor, reported two cases, one, a child with a growth on "the throat, which now resembles a rat in shape and colour," and another with a mass on "its arm, which now distinctly resembles a mouse."

Madison, Georgia. The *Madisonian*[3] (quoted in the *Wrightsville Recorder*, Wrightsville, Ga., Dec. 15, 1883, p. 2; ||2haudkat) said,

> The strangest malformation that the writer ever saw was a double-headed child, almost twenty-four hours old. It was exhibited by Mr. T. W. Jones, of this county, who stated to us, that a mulatto woman living on his place gave birth to the monstrosity on last Wednesday at 12 o'clock m. He was passing the cabin about the time the birth was given, and his attention was attracted by the unusual excitement among the inmates of the cabin, who were intensely alarmed over the occurrence. Going in to ascertain the cause of the excitement, he was shown the above monstrosity by the sister of the woman who had given the unnatural birth and told by her that it breathed several times after birth. It was placed in a small box and interred, but Mr. Jones, thinking that this would be a valuable contribution to medical science, persuaded the mother to let him exhume it and preserve it in alcohol. After gaining her consent he immediately disinterred the double-headed child, put it in alcohol, and the following day brought it to Madison.
>
> Upon examining it, we observed a well-formed body, with two distinct and well-developed heads, each head having all the natural features. The ears are elongated more than human ears should be though medical men say that natural proportions are an uncommon thing in such cases. Each mouth had two well-developed front teeth. The eyes in each head were well-formed, and the heads were covered with a black, silky growth beginning at the top of the eyebrows and continuing down the shoulders, sides, arms and legs. But the strangest feature about the formation—it had two spinal columns, when a front view showed only one body—the placenta was between the shoulders, and the umbilical cord was attached to this instead of the navel; it was of double gender, with the male predominant, and about 12 inches in length. It is a puzzle to medical men, as it possesses both human and animal features. Twins were evidently intended by nature, but by some unaccountable freak, they were joined before birth, producing the malformation above.
>
> The woman that gave birth to this monstrosity is a mulatto of usual intelligence, married and of good general health. This is certainly the strangest formation we ever saw and will doubtless create interest in some museum of anatomy.

* Whether the birth described in the Wrightsville report should be classified as glirhominid is uncertain, but the presence of two large teeth in each mouth is consistent with such a grouping, as is the extensive hair coat on the body and limbs.

London. London surgeon Francis Mason (1877, p. 354a) described a case in which an individual with the appearance of a human being had a nevus (birthmark) like the hairy skin of a rat. He said this homorodentid, which he pictured in a woodcut (his figure 25, shown here), "came under my notice about three years ago. The mother attributed the deformity to a fright occasioned by seeing a rat [see MATERNAL IMPRESSIONS].

Francis Mason
1837-1886

Toruń, Poland. Seger (1672) said a man in Toruń (now in Poland), had a nevus like the skin of a mouse, with fur, on his right temple.

Boy with a rat hair nevus
(Mason 1877, fig. 25)

Sevenoaks, Kent, England. Thompson (1874), reporting a hairy nevus on the leg of one of his patients, commented that "certainly, if the skin of a dormouse had been fixed upon it, it could not have been more perfect, both in regard to texture and to colour."

New York City. Dr. Gurdon Buck (1876, p. 165), of New York's St. Luke's Hospital, author of the first American textbook on plastic surgery, stated that

> F. J., aged thirteen, a native and resident of the city, with fair complexion, light hair, and of a good constitution, has from birth had an abnormal growth of hair upon the left half of the forehead, covering the entire surface between the left eyebrow and the hair of the scalp, above as well as toward the left

Gurdon Buck
1807-1877

The New York City homorodentid (Buck 1876, fig. 68).

temple. The hair of the left eyebrow is much coarser than that of the right, while the abnormal growth of hair is short, soft, and fine, like that of a mouse-skin; both are darker colored than the hair of the scalp. The surface of the skin covered by this growth of hair, after being shaved clean, was found to be of a tawny hue, like that of the scrotum, and slightly elevated. (*See* Fig. 68)

NOTES

1. At the time in question, a George Kleinklaus, a foundry worker, aged 32, was living with his wife Elizabeth, 26, at 363 Bayou St., Vincennes (||9um3xuua). The mother (Elizabeth Osho, ca. 1854-1910) had borne two other children (||4e3399x4). The *Enquirer* (Mar. 25. 1881, p. 2; ||5hv7rv7p) had reported Mr. Kleinklaus bringing home a flying squirrel.
2. Supernumerary digits are commonly seen in telenothians.
3. The *Atlanta Constitution* (Nov. 27, 1883, p. 2; ||mshr5f3x) said that on Nov. 21, a woman near Madison bore a two-headed creature with black, silky hair "down the shoulders, sides, arms and legs" and with unnamed "animal features." The *Virginian* (Norfolk, Nov. 29, 1883, p. 1; ||3vc6xn7a) said an M. A. Peteet had the subject preserved in alcohol at his Madison drugstore. The *Sentinel* (Hazleton, Penn., Nov. 28, 1883, p. 1a; ||4h6ha7bh) said the birth occurred in eastern Morgan County on the Oconee River near Park's Mill. The *Atlanta Journal* (Nov. 26, 1883, p. 3) and the *Weekly Echo* (Savannah, Ga., Dec. 2, 1883, p. 2; ||mwmjycs5) also ran reports about the Madison birth. A Marshall Andrews Poteet (b. Jan. 22, 1843, d. Feb. 16, 1893), a druggist, lived in Madison in 1880 (||2p8vkum2).

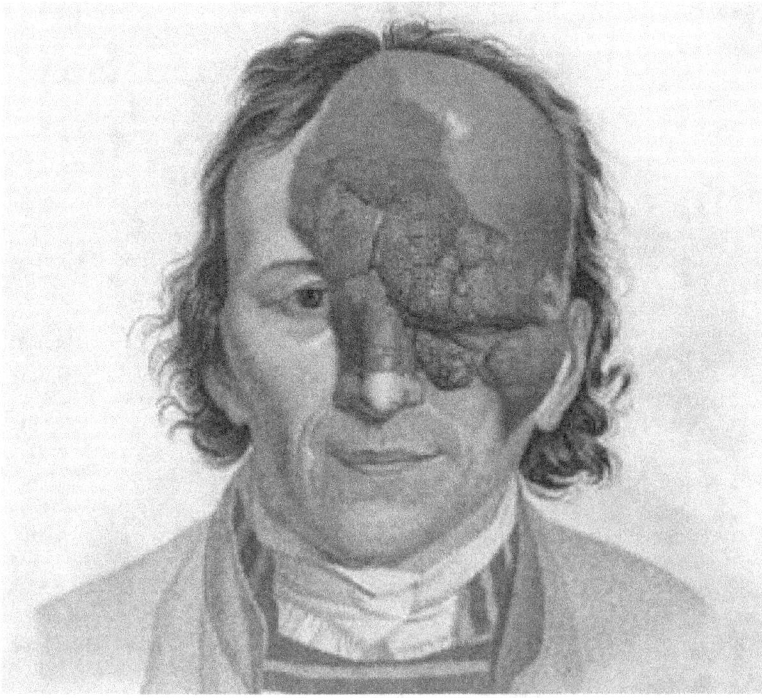

The Meaux homotalpid (Alibert 1832).

Telinsectivores

Call any animal combining the traits of an insectivore with those of some other kind of animal a *telinsectivore*. All reported human-insectivore composites seem to be the offspring of human mothers. The two insectivores mentioned in such reports are hedgehogs and moles.

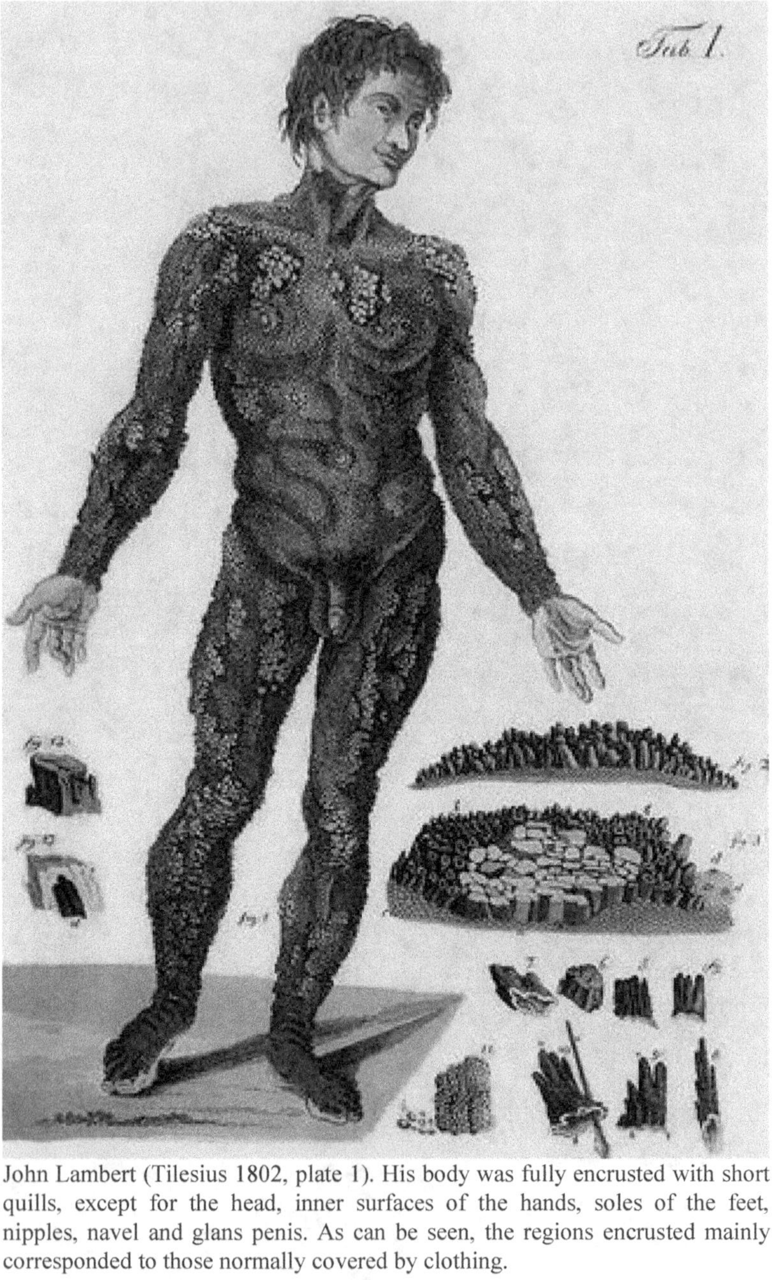

John Lambert (Tilesius 1802, plate 1). His body was fully encrusted with short quills, except for the head, inner surfaces of the hands, soles of the feet, nipples, navel and glans penis. As can be seen, the regions encrusted mainly corresponded to those normally covered by clothing.

HOMERINACEIDS (HEDGEHOG-HUMANS)

Animals of new kinds can arise when different species mate.
—THOMAS AQUINAS, 3.1

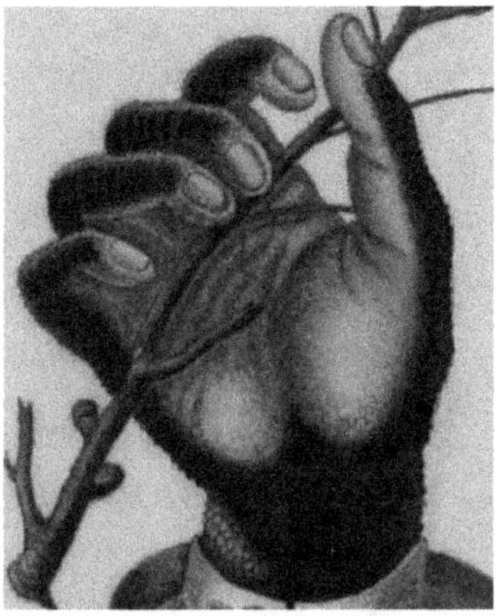

Edward Lambert's hand
(Edwards 1760a, plate 212)

Call a composite mixing the traits of both a human and a hedgehog a *homerinaceid.* Reports about such mixes seem all to relate to either Edward Lambert, born at Brandon, Suffolk, in about 1716, or to his similarly affected male descendants. Morley (1859, see pp. 315 and 324) says Lambert was shown at London's St. Bartholomew's Hospital sometime before 1727. And Machin (1731) reported that he was exhibited to the Royal Society:

> A Country Labourer, living not far from Euston Hall [which is about six miles from Brandon] in Suffolk, shewed a Boy (his Son) about Fourteen Years of Age, having a cuticular Distemper, of a different Kind from any hitherto mentioned in the Histories of Diseases.
>
> His Skin (if it might be so called) seemed rather like a dusky coloured, thick Case, exactly fitting every Part of his Body, made of a rugged Bark, or Hide, with Bristles in some Places, which Case covering the whole excepting the Face, the Palms of the Hands, and the Soles of the Feet, caused an Appearance as if those Parts alone were naked, and the rest cloathed. It did not bleed when cut or scarified, being callous and insensible. It was said he sheds it once every Year, about Autumn, at which Time it usually grows to the Thickness of three Quarters of an Inch and then is thrust off by the new skin which is coming up underneath.
>
> It was not easy to think of any Sort of Skin, or natural Integument, that exactly resembled it. Some compared it to the Bark of a Tree; others thought it looked like Seal-Skin; others like the Hide of the Elephant, or the skin about the legs of the Rhinoceros; and some took it to be like a great Wart, or Number of Warts uniting and overspreading the whole Body. The bristly Parts, which were chiefly about the belly and Flanks, looked and rustled like the Bristles, or Quills, of an Hedge-Hog, shorn off within an Inch of the Skin.

> His face was well featured, and of a good Complexion, if not rather too ruddy; and the Palms of his Hands were not harder, or in worse Condition than is usual for Workmen or Labourers. His Size was proper for his Age; his body and limbs straight, and, excepting as to this Deformity, well shapen.
>
> This rugged Covering gave him no Pain or Uneasiness, unless that sometimes after hard Work, it was apt to start and cleave and cause a bleeding. And notwithstanding the unusual Disposition of his Humours to form so strange an Integument, his natural Excretions were said to be in the ordinary Course and Manner, without any thing remarkable attending them.
>
> The Father knew of no Accident to account for this distempered Habit. But said that his skin was clear at his Birth as in other Children, and so continued for about seven or eight Weeks, after which without his being sick, it began to turn yellow, as if he had had the Jaundice; from which by degrees it changed black, and in a little time afterwards thickened, and grew into that State it appeared in at present. That he has been in Health from his Birth, and hath no Sickness at the Season when he sheds it. He further said that his Mother had received no Fright, to his Knowledge, whilst she was with Child; and hath born him many other Children, none of which have ever had this, or any other unusual Distemper or Deformity.

In the year 1755 Lambert, by then billed as the Porcupine Man, was shown in London, with one of his sons, Edward Jr., in the same condition. The English naturalist George Edwards (1760, p. 3) said that

> I then saw them both and examined them very carefully. The father, whose name is Edward Lambert, was born at Brandon in Suffolk and is now about forty years of age, a good looking well shaped man of a florid complexion and when his body and hands were hid seemed nothing different from other people, but he was covered with slender protuberances of a wart-like appearance, some flattish at top, others concave, some pointed like cones, others irregular, standing pretty close together like bristles in a brush … grown to the thickness of common-twine-packthread, which formed a rugged covering all over him, except his face and head, the palms of his hands, the insides and tips of his fingers, and the soles of his feet, which alone were free. Their color was dark brown, or rusty black; and they were so stiff and elastic, that a hand drawn over them produced a rustling noise. In some places they were more than half an inch long, and the general surface appeared pretty equal, though they gradually diminished as they approached the ends of the fingers and toes, where they disappeared.

Remarkably, Edwards says Edward Sr. had six sons, all of whom had the same hedgehog-like coat. But Edward Jr. was the only such offspring still living in 1755. This surviving encrusted scion was reportedly born in 1746, also at Brandon. He, too, was exhibited to the Royal Society, in 1754, and later worked as a gamekeeper for Lord Huntingfield at Heveningham Hall in Heveningham, Suffolk. Edward Jr. was the father of seven daughters, none of whom were encrusted, as well as two sons, John and Richard, both of whom were (Telesius 1802). This third generation of homerinaceids was born at Leiston, Suffolk. Telesius seems to have been the last investigator to publish a report about the Lamberts, other than Steinhausen (1828, ch. 1) who provided no new information.

In 1801 John and Richard traveled to Germany and exhibited themselves there in various cities. It was then that Telesius examined them and prepared detailed illustrations of John. Richard was 13 (therefore born in about 1788) and John was 22 (ergo, born circa 1779). Telesius also told how John had assumed his father's job as gamekeeper and said John's wife was pregnant.

Classification. The Lamberts were often described as porcupine men. But no porcupine is native to Britain, nor would any woman handle one. Hedgehogs, however, are common there—and often handled.

Fertility. If the Lamberts were, in fact, hybrids, it did not deter them from producing progeny. Edward Lambert Sr. and his wife produced six offspring. Moreover, according to Telesius (1802, p. 5), one of those offspring, Edward Jr., went on to produce nine offspring. And one of those nine, John, reportedly succeeded in impregnating his wife. Thus, the Lambert case is one of many suggesting that hybridization can lead to genetic exchanges between distantly related organisms.

Y linkage. As of 1802, all male descendants of Edward Lambert Sr. had become encrusted by the age of three months, but none of the female descendants ever did (Tilesius 1802, p. 5). This fact suggests encrustation was a dominant sex-linked trait among the Lamberts. So, did they perhaps carry a Y chromosome of hedgehog origin?

* A late mention of what was likely one of the Lamberts appeared in the *Leeds Intelligencer* (Leeds, England, Apr. 19, 1834, p. 4e). Campardon (1877, p. 400) describes what was likely one of the Lamberts, exhibited in Paris in 1770.

A hedgehog (*Erinaceus europaeus*)

A CANERINACEID (HEDGEHOG-DOG)

Saint-Quentin, Aisne, France. The *Messager de Gand* (Ghent, Feb. 12, 1837, p. 3c; ||r3redk5p) reported the whelping at Saint-Quentin of an animal with (in translation). "all the traits of a hedgehog. The head, the muzzle and the feet show no trace of the canine race."†

OVERINACEIDS (HEDGEHOG-SHEEP)

Meaux, France. Le Roy (1879, p. 63), librarian of the Société d'Agriculture, Sciences et Arts de Meaux, said one of the preserved specimens in the society's collection was a lamb rolled in a ball like a hedgehog ("un agneau ramassé en boule comme un hérisson").

XENOGENESIS

Meyrueis, Lozère, France. Le Courrier de la Lozère (Mende, Mar. 13, 1910, p. 3d; ||am33uhh7) reported that a ewe at Meyrueis yeaned together a hedgehog and lamb, and that both survived.

†. Also see Guislain (1842, p. 257, obs. 53).

The shovel-like hand of a mole (*Talpa europaea*).

HOMOTALPIDS (MOLE-HUMANS)

How many things are judged impossible before they actually occur?
—PLINY, *Natural History*, 7.1.7

Winnsboro, Texas. M. B. Pollard (1912, p. 1137b), a physician practicing in Winnsboro, said had personally observed an homotalpid (i.e., a "child born with hands and feet like those of a mole").

Belsericourt, France. Thibault (1761) said, in translation, "On March 25, 1758, at Belsericourt, near Noyon the wife of Antoine Fortin, a winemaker, being in labor, sent for my assistance. … she gave birth to a child, living, but extremely deformed, in that it had a long head consisting only of a tuft of hair, a brain and some supporting bones (maxilla, sphenoid and ethmoid). The arms were no more than three inches long, with hands of the same size and shape as a mole's. The legs were just four inches long, and the feet, like the hind paws of a mole. The legs, instead of extending normally and articulating with the ischium, extended directly out to the sides making a right angle with the body."

Meaux, France. The French dermatologist Jean-Louis Alibert (1832, pp. 803-805) described a gardener named Delaitre, and nicknamed "La Talpe" ("The Mole"), who was born at Meaux in 1756. A large growth covered much of Delaitre's face and Alibert (ibid., p. 804) said it was covered by "shiny, black skin that perfectly imitated the skin of a mole" (translation mine). Alibert (ibid., p. 803) noted that the growth emitted a disagreeable odor. Moles, too, have a bad smell that makes them distasteful to most predators. He fathered 13 children. Delaitre is pictured on p. 409 of this book.

* Alibert (1832, p. 805) said the growth felt like turkey skin. Delaitre's appearance does recall that of certain HOMOGALLOPAVIDS.

Wangenheim, Germany. Johann Oelmejer (in Horst 1660, p. 543a), a pastor in Wangenheim, reported a woman bearing, on August 7, 1624, a precocial, mole-like offspring with four feet and black hair.

Kirkintilloch, Scotland: William Whitelaw (1864, p. 168), a physician practicing in Kirkintilloch, reported a four-year-old girl with the "resemblance of a mole" (i.e., *Talpa europaea*) on her right calf. His description is, however, brief and nebulous.

Paris. Étienne Geoffroy Saint-Hilaire (1824, p. 453) detailed the mole-like traits of a nondescript that was the offspring of a woman.

A TALPISUID (MOLE-PIG)

Adams Township, Monroe, Ohio. The *Spirit of Democracy* (Woodsfield, Ohio, Apr. 24, 1877, p. 3b; ||4yceerc3) said, "Mr. Peter King [1859-1880], of Adams Township, informed us on Monday, the 16th inst., that a sow of his had given birth to a pig that had perfect hind legs while its fore legs were those of a ground mole from the knee joint to its feet, each of which had five distinct toes with a nail or claw on each."

A TALPICANID (MOLE-DOG)

Quimper, Brittany. Molinet (1684) reported receiving from Quimper a "dog" with the feet of a mole. The size of a weasel (*Mustela nivalis*), it had a small trunk (frontal proboscis?), he said, but no mouth or eyes. Supposedly, it lived three days, nursing through its trunk.

TELOCHIROPTERIDS (BAT COMPOSITES)

Lyle's Flying Fox, *Pteropus lylei* (||5cyyhuz6)

CYNOCHIROPTERIDS (DOG-BATS)

Animals that plainly combine traits of canids and bats include the various fruit bats of the genus *Pteropus*. These animals are called flying foxes because of the clear similarity their heads and necks bear to those of foxes. The range of *Pteropus* is approximately parapatric with both those of vulpid foxes (*Vulpes*) and of bat-eared foxes (*Otocyon*). These contacts occur in eastern Africa, southern Asia and Australia. Approaching the size of real foxes, flying foxes are far larger than ordinary bats.

* Parapatric contact between two different forms of life is often a sign of ongoing hybridization in those geographic regions where they interface.

Grey-headed Flying Fox
Pteropus poliocephalus

Sinaloan Mastiff Bat (*Molossus sinaloae*)
(Juan Cruzado Cortés Wikimedia ||53c9cvvf)

Other bats seem to qualify as cynochiropterids. In giving the first description of the genus *Molossus*, Étienne Geoffroy Saint-Hilaire (1805, p. 153) wrote, in translation, "*Molossus* bats are easily recognized by their fierce look and by the whole aspect of the face: their big head and very large muzzle have caused them to be compared with dogs and to be designated by the name *molossus* [Latin for an ancient mastiff-like breed of dog]."

CAMELOCHIROPTERIDS (BAT-CAMELS)

Bats of the genus *Hypsignathus* have camel-like heads. Moreover, they are in parapatric contact with camels in central and western Africa.

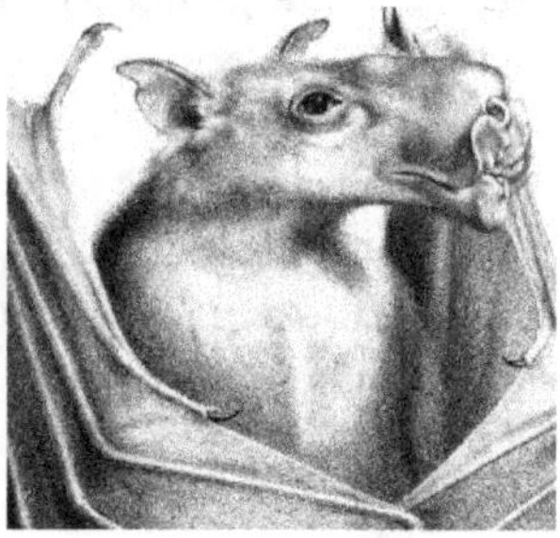

Hypsignathus

A HOMOCHIROPTERID (BAT-HUMAN)

Paroo River, NSW, Australia. The *Daylesford Mercury* (quoted in *The Leader,* Melbourne, Vic., Jun. 4, 1864, p. 4d; ||5cjt3xtx) described an aboriginal "woman" living on the Paroo River whose "arms, from the wrists, were connected to her body, just below the ribs, by a kind of web or skin, just the shape of the wing of the bat." Sugar Gliders (*Petaurus*) may also be considered in connection with this report.

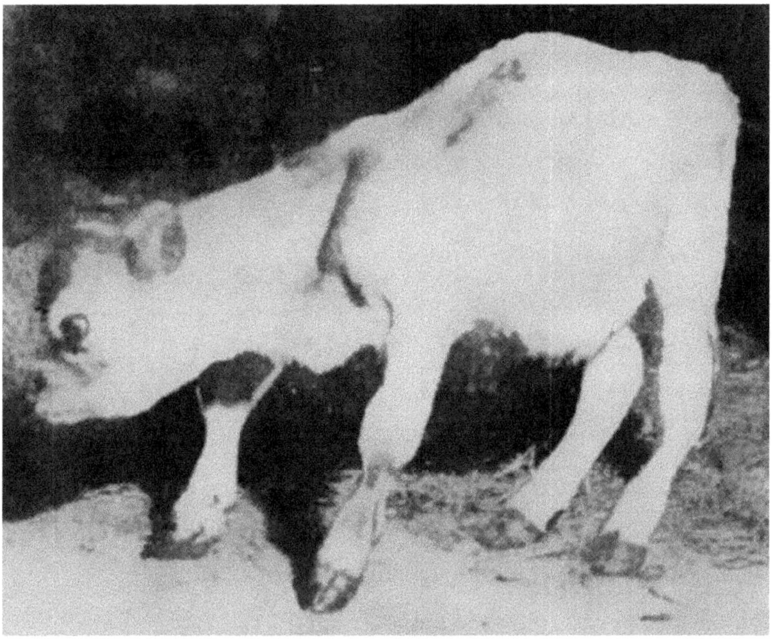

The Guysville canibovid (see p. 433).

Carnungulates

He that knows where to direct his view, sees much in a little time.
—SAMUEL JOHNSON

Any animal combining the traits of a carnivore with those of an ungulate will here be termed a *carnungulate*. Among the many reports about carnungulates, nearly all refer to ungulate mothers. If these creatures represent hybrids, in nearly all cases the sires participating in the cross would have been either dogs or bears. But a few cases implicate feline sires. Carnungulates with canine traits are especially common. Ungulates mentioned in reports about carnungulates include goats, sheep, horses, pigs and, most frequently, cattle.

There is a tall long sided dame
(But wond'rous light,) ycleped Fame
That, like a thin camelion, boards
Herself on air, and eats her words;
Upon her shoulders wings she wears
Like hanging-sleeves, lin'd through with ears,
And eyes, and tongues, as poets list,
Made good by deep mythologist,
With these she through the welkin flies,
And sometimes carries truth, oft lies,
With letters hung like eastern pigeons,
And Mercuries of furthest regions;
Diurnals writ for regulation
Of lying, to inform the nation;
And by their public use to bring down
The rate of whetstones in the kingdom.
About her neck a pacquet-male,
Fraught with advice, some fresh, some stale,
Of men that walk'd when they were dead,
And cows of monsters brought to bed.

—SAMUEL BUTLER, *Hudibras*

The Katuje canibovid.

CANIBOVIDS (DOG-COWS)

A dog atop a cow
A cow agog
Cow-dog
—CHRIS MILLAR

Call any animal combining the traits of a dog with those of a cow a *canibovid.* The mothers of canibovids always seem to be cows, often, heifers. Though most canibovids would presumably be F_1 hybrids, they are highly variable. Some, such as in the Katuje animal pictured above, have long, wolf-like snouts. More have flat faces like a bulldog's or pug's (see image at right). Many have dog paws and carnivorous teeth. Many, too, are viable, that is, they long survive birth.

A viable canibovid born at Sémeries-Zorées (*Le Grand Écho du Nord de la France*, Apr. 4, 1937, p. 7; ||mr3x9pnk).

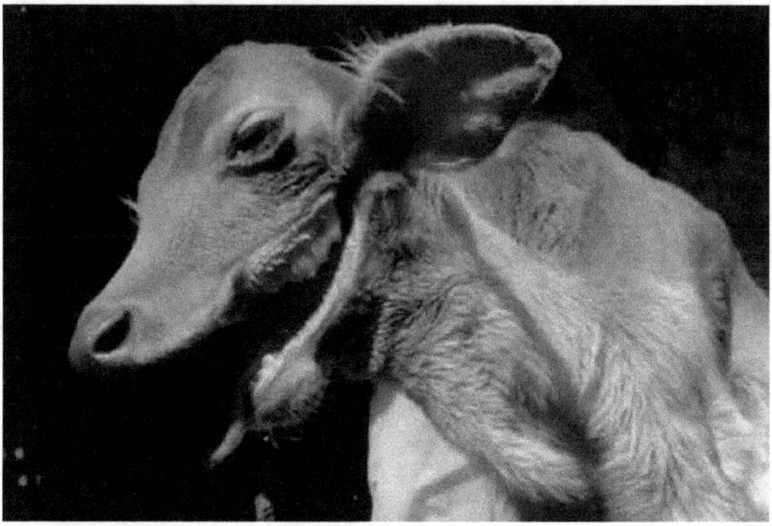

The Katuje canibovid.

REPORTS

> *It being produced by Linaria seems no less monstrous than a calf with a wolf's head being borne by a cow.*
>
> —CARL LINNAEUS[1]

There are many reports, stretching far back in time, about animals combining canine and bovine traits. But we can begin with a fairly recent one documented by clear images.

Katuje, Madura, Indonesia. According to Internet reports, a cow in the hamlet of Katuje on the island of Madura off the northeastern coast of Java, calved an animal with canine traits on February 2, 2011. The reports described it as having a wolf's head. But as no wolves are present on the island of Madura, and as photos show it looked quite doglike, it can be assumed that any canine parent involved in its production would have been an ordinary dog.

The owner's testimony suggests that the Katuje canibovid was also a dog-like in being an omnivore: "What is even more strange," he said, "is that this calf will not eat grass. It only drinks milk and eats snack foods like those eaten by humans." It seems to have been entirely viable, except in that it was, due to its aberrant mouth structure, unable

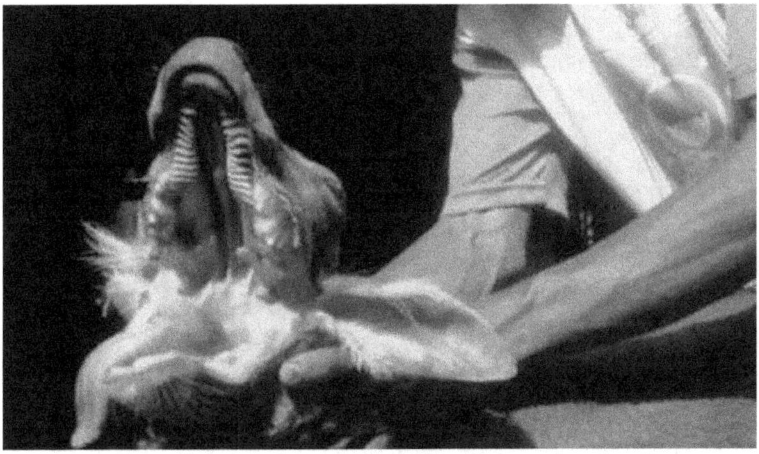

The oral cavity of the Katuje canibovid (note dental pad).

to nurse on its own. The owner had to nurse it by hand. As photos show, it had a huge gape, as do many carnivores. Most herbivores have small mouth openings. But in this animal, the mouth reached the ears. it also had a lolling tongue like a dog's. The muzzle, too, was very much like a dog's, and not at all like a cow's. The loose fur on the breast was also dog-like. Moreover, in the photos the erupting molars resemble the pointed carnassials of a canine, not the flat grinders of a bovine.

The dental pad of a cow.

It had a dental pad, which is not surprising in the offspring of a cow, since cattle and other ruminants all have this feature. The lower incisors were cow-like. The lower jaw was much shorter than the upper, and the mouth, agape. Mismatched jaws are common in canibovids (and certain other telenothians, e.g., JUMARTS, GALANATIDS and PISCAVIANS).

* The reports identify the mother of the Katuje animal only as a "cow." But Indonesian farmers generally do not keep European domestic cattle (*Bos taurus*). So, she would likely have been either a banteng (*B. javanicus*) or of the Madura breed (*B. javanicus* × *B. indicus*). The mother had given birth to three normal calves in previous years.

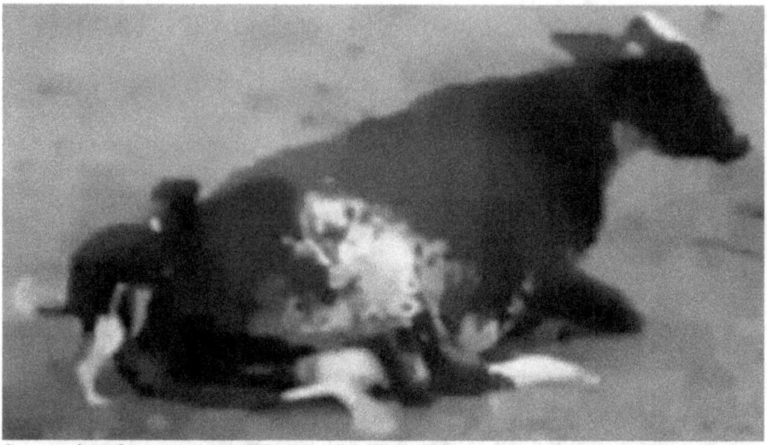

Screenshot from a video of a small dog (left) mounting a cow (||y6817goj). It is well established that male dogs will sometimes mount recumbent cattle, which is clearly relevant to the question of whether a bovine-canine cross might be possible. In general, mixed mating between small males and large females requires that the female be of a passive temperament. Here, cows fit the bill.

Perham, Minnesota. The *Donaldsonville Chief* (La., Dec. 18, 1912, p. 12b; ||gwmmy5j) said, "A dog-faced calf was born on the farm of Henry Krekelberg, near Perham, Minn., recently. The animal bore a startling resemblance to a dog, but unfortunately the carcass was not preserved. It had the long, narrow muzzle of a dog, droopy ears like a water spaniel, long lips and the teeth on the lower jaw were of the canine variety."

* Henry John Krekelberg (1877-1936), a farmer, lived near Perham in 1912 (||y4clfkef).

Montolivet, Seine-et-Marne, France. L'Avenir de Seine-et-Marne (Jun. 18, 1890, p. 3c; ||y5pe8v7r) said a cow, belonging to a Mr. Lelongt-Frutel at Montolivet, calved a tailless, viable animal with short ears. Neither male nor female, it had a calf's body but a dog's head and legs.

Étreux, Aisne, France. From *L'Aisne Nouvelle* (Saint-Quentin, May 6, 1952, p. 5c; ||2p98f4nd) said, in translation, "A FREAK CALF.—Mr. Dubois Jean, a cattle dealer, has an extraordinary calf, which is a few days old. It's head is like a dog's. The muzzle is flat; the ears are small and pointed. A tail is lacking. The extremely short legs end in feet shaped like dog paws. It never moos, but it does emit plaintive cries."

Marigné-près-Daon, Maine-et-Loire, France. Le Tourangeau (Tours, Sep. 22, 1897, p. 3c; ||4taryzc2) said a cow at Marigné-près-Daon calved a viable animal with the head and chest of a dog but otherwise like a calf. According to the report, it panted continuously like a dog.

Dormans, Marne, France. L'Écho du Nord (Lille, Dec. 6, 1901, p. 1b; ||4fzaby7x) reported a cow calving a tailless, viable animal at Dormans that was in all ways like a dog, except for its upper jaw, feet and hair.

Gersthofen, Bavaria. The *Tiroler Land-Zeitung* (Jun. 25, 1910, p. 8a; ||2p82eyta) said, in translation, "A MONSTROSITY, Jun. 23.—On a farm in Gersthofen a cow had to be slaughtered and when she was opened, a calf was found that had a complete dog's head ["vollständigen Hundeskopf"], double ears, no eyes and a dog's tail. The fetus was dead."

Lévis, Quebec. From the *Saturday Budget* (Quebec City, Jun. 4, 1892, p. 3; ||yyxlwud4): "There is at present on Mr. Joseph Samson's farm, Levis, a calf recently born, which by a freak of nature bears an immense head, the shape of a large dog's. The animal is viewed by large crowds every day. It has two rows of immense teeth."

* Records (||y67ho672) list brothers Joseph Honore Samson (b. Mar. 10, 1870, at Lauzon, Lévis) and Joseph Ulric Samson (b. Feb. 23, 1866, also at Lauzon).

Maroa, Illinois. The *Decatur Daily Review* (Decatur, Ill., May 14, 1935, p. 18; ||y5d8cx3s) said, "A dog-faced calf is reported to have been born on the farm of W. R. Clifton near Maroa. According to Mr. Clifton, the calf's hound-like ears droop down over its sharp dog-shaped head instead of standing erect. The calf seems to be otherwise normal. It is a shorthorn Jersey heifer and is more than a week old."

* William Redic Clifton (1878-1959), a farmer, lived near Maroa in 1940 (||y37kmhs4).

Lymington, Hampshire. Lymington butchers took from a cow a fetus with feet like dog paws (Thomas 1666a, 1666b).

Germany (*no locus in quo*). Citing Martin Friedrich Seidel (see note, p. 84), Bekmann and Bekmann (1751, p. 883) recorded "a calf with a dog's snout and no lower jaw," born Dec. 8, 1667.

* Haller (1774, vol. 1, p. 562), too, mentioned a calf with the head of a dog.

Advance, Fulton County, Ohio. The *Fulton County Tribune* (Wauseon, O., Dec. 14, 1917, p. 5c; ||byxhzc7h) said, "Harvey Reynolds brought in a freak calf's head this week which he obtained from the farm of Mr. Onweller, north of Advance. The head was shaped like that of a hound dog, ears set low and the under jaw entirely missing."

* A Frank E. Onweller (1863-1918), a farmer, lived north of Advance (||yu9kj9pb).

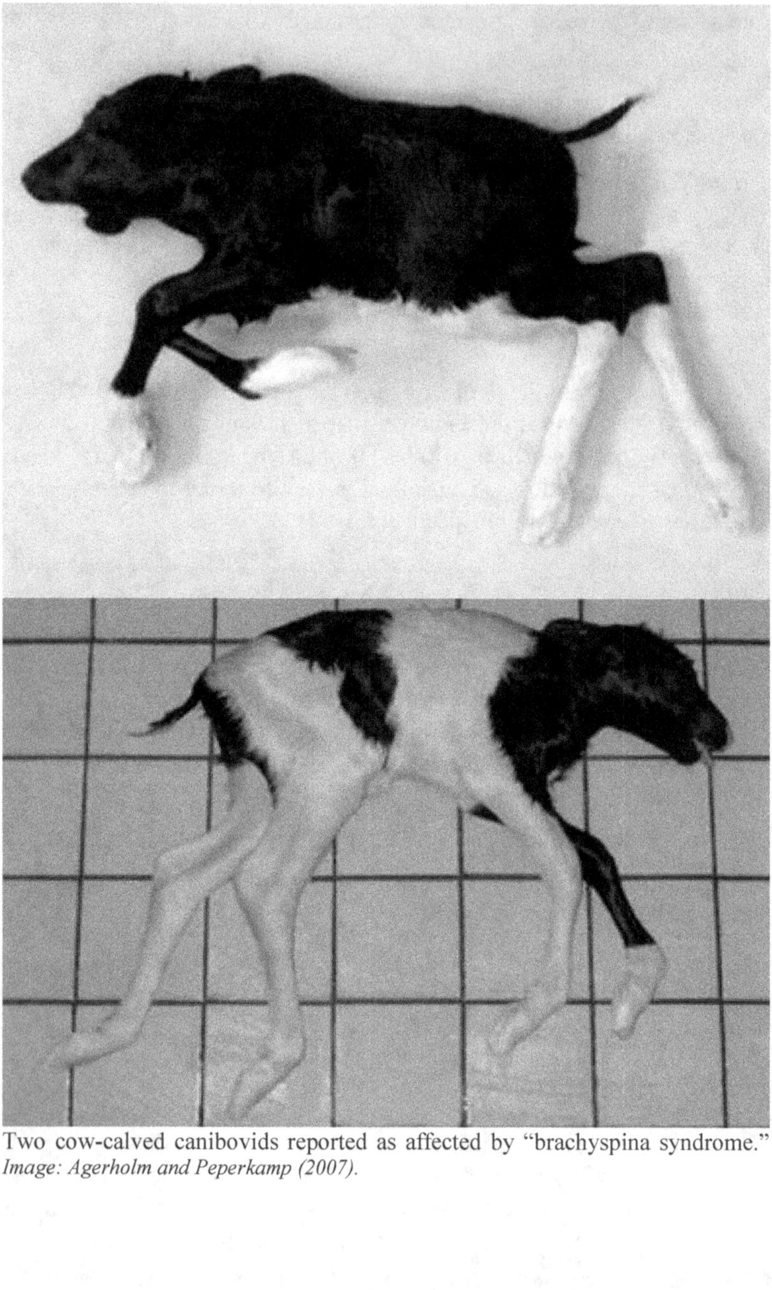

Two cow-calved canibovids reported as affected by "brachyspina syndrome." *Image: Agerholm and Peperkamp (2007).*

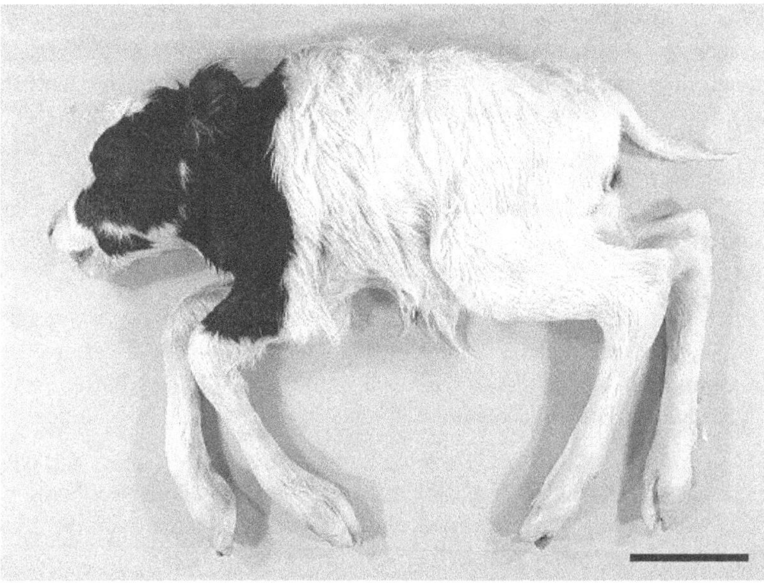

A cow-calved canibovid, described as a calf with "brachyspina syndrome." Bar is 10 cm, which indicates the overall length (nose to tail tip) was about 60 cm (23.6 in). *Image: Agerholm et. al. (2007).*

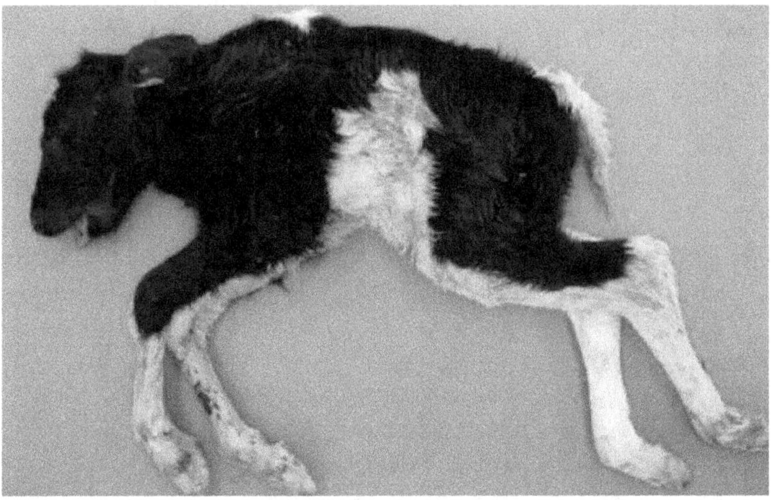

A cow-calved canibovid, reported as a calf affected by "brachyspina syndrome." *Image: Agerholm and Peperkamp 2007.*

Bréhain, Moselle, France. Le Messager des Charentes (Cognac, Feb. 9, 1894, p. 2a; ||bdfxu75e) announced the calving at Bréhain of an animal mostly like a calf, but with dog paws, and that further had the manner of a dog when eating. It was much smaller than an ordinary calf.

Durmenach, Haut-Rhin, France. L'Est Républicain (Nancy, Sep. 18, 1903, p. 3; ||bdewb28t) said, in translation, "A freak of teratology was just born in the stable of the White Horse Inn at Durmenach. It's a calf with the face of a dog, a single eye and just one nostril. It has no tail."

Vervins, Aisne, France. Le Réveil de l'Aisne (Chauny, Jan. 14, 1903, p. 3d; ||5xh6bp5f) said a tailless animal, calved on the farm of a certain M. Lombart near Vervins, was alive and dog-like in all ways but its calf-like hide. The head, face, teeth, body, legs and paws were all canine.

Marcenat, Cantal, France. Le Réveil de Mauriac (Mauriac, Feb. 4, 1899, p. 3b; ||yc866khd) said a cow had calved an animal with a perfect hunting dog's head ("un veau avec une tête parfaite de chien de chasse") at Marcenat the previous year.

Parishville, New York. The *Ogdensburg Journal* (Feb. 19, 1914, p. 6a; ||ydy3myep) said, "A remarkable calf was born recently on the farm of W. B. Christy about six miles from Potsdam. The calf had the head of a dog with the hide and hoofs of the ordinary calf, although its hair was a trifle long. It lived but a short time. The dam of the freak animal was an ordinary dairy cow in an apparently normal condition."

* In 1910 William B. Christy (b. 1875) lived in Parishville by Potsdam (||y3ru5zwz).

Terville, Moselle, France. Le Lorrain (Metz, May 17, 1905, p. 2c; ||2kfasbzh) said that on May 13 a cow at Terville had calved an animal that much resembled a dog and died soon after birth.

Saint-Thurial, Brittany. L'Écho de l'Orne (Mortagne, Jun. 26, 1926, p. 1d; ||yenkxkaf) said a tailless animal calved at Saint-Thurial had the head of a dog.

Barre, Vermont. The *Vermont Union* (Lyndon, Vt., Dec. 28, 1900, p. 4; ||y2bl976d) said, "The dog-faced calf belonging to H. A. Rugg of Barre is a thing of the past. It died Monday night and Fred Harriman of that city has taken it to his farm for burial."

* A 1910 census record lists a Herbert A. Rugg at Barre (||y6aeeffr).

Woodford, Vermont. The *Manchester Journal* (Manchester, Vt., Jun. 22, 1882, p. 3c; ||5cbcfvub) said a cow at Woodford had calved an animal "one-half ox and one-half canine."

Flagg Spring, Kentucky. The *Cincinnati Enquirer* (Jul. 14, 1896, p. 12b; ||yyhjk3zn) said, "G[ustave]. A[ugustus]. Gaiser [1861-1945], a farmer living near Flagg Spring, Ky. … is the possessor of a wonderful calf, which is a veritable freak of nature. The calf was born about 11 days ago and is a combination of the bovine and canine species. The head is perfectly white, and in shape and form is the counterpart of a shepherd dog. The face is also canine—particularly the nose and tongue. The hind parts and tail are also those of a dog. The legs and stomach are those of a calf. … Mr. Gaiser owns a shepherd dog.

The Flagg Spring canibovid (*St. Louis Post-Dispatch,* Jul. 26, 1896, p. 29; ||y5a5dwgw).

The Heart, Victoria. The *Argus* (Delegate, NSW, Jul. 23, 1914, p. 1a; ||yxpm9uuh) said, "At the Heart, near Sale, a half-bred Jersey cow gave birth to a lusus naturae, the head and neck of which resembled a calf, while the body and legs were those of a collie dog."

Villiers-Louis, Yonne, France. Le Petit Troyen (Troyes, Jul. 19, 1884, p. 3b; ||yc3vnatw) said a cow at Villiers-Louis calved a healthy short-tailed animal of indefinite sex with the head of a dog.

Cuincy, Nord, France. La Lanterne (Paris, Apr. 28, 1903, p. 1d-e; ||mpsnbayx) said a cow on François Godin's farm at Cuincy calved an animal with the head, paws and tail of a dog. The teats were in two rows, as on a dog. The Museé de Douai acquired the specimen.

Northwest Fork, Delaware. In a report datelined "GEORGETOWN, DEL, May 5," the *Washington Times* (Washington, D.C., May 5, 1922, p. 23f; ||ymtk2zhw) said, "Charles [Lamon] Terry [1863-1939], of Northwest Fork, reports the birth of a calf with a dog's head."

Shimerville, Pennsylvania. The *Morning Call* (Allentown, Oct. 25, 1915, p. 2; ||2s3scmsw) said, "J. Rinker, of near Shimerville, is exhibiting at the Broad Street stables a calf that has a head shaped like that of a dog."

Villeneuve-d'Ascq, Nord, France. L'Indépendant (Cambrai, Jul. 8, 1906, p. 3a; ||3fahz42w) reported that a cow in Flers-Breucq, a neighborhood of Villeneuve-d'Ascq, calved a short-legged, tailless, dog-headed animal with eyes the size of a five-franc piece (i.e., 37 mm. in diameter).

Rue, Somme, France. L'Intransigeant (Paris, Oct. 29, 1935, p. 9g; ||6jw2vmb2) said the Hospice de Rue bought a cow that dropped an animal with a dog's muzzle and ears, short legs, and a coat like a cocker spaniel's.

Massiac, Cantal, France. Le Facteur (Ussel, Apr. 5, 1913, p. 2b; ||ykhy2a8r) said a cow bought in Massiac bore a calf with a hunting dog's head, long drooping ears and dog eyes, but a calf's muzzle and body.

Phoenix, Arizona. From the *Arizona Republic* (Phoenix, Mar. 26. 1899, p. 4d; ||y46hphxg, ||ny3nfpwc): "Mr. O. H. Christy has lately deliberately destroyed what some men would have looked upon as a fortune. A few days ago, one of his cows dropped a calf which was a horrible monstrosity. It was perfectly formed as to every part except the head, which was that of a wolf. The ears were long and sharp and the lower one much the longer. The lower lip hung down about three inches and the jaws were filled with small, serrated teeth [which are characteristic of carnivores, not herbivores]. On account of the conformation of the mouth this strange beast was unable to suck. It had to be thrown down and milk was poured down its throat from a bottle. This kindly attention was resisted. The canine resemblance was carried beyond the shape of the head. The noise made by the calf was at times precisely like the whimper of a pup, and at others like the whining of one. The sight of it was revolting at times, and Mr. Christy reviewed it each succeeding time with increased disgust. Then he began regarding it with a sort of horror, which grew more pronounced as the animal seemed to thrive and give promise of living. His aversion to it thrived in equal ratio and at length he killed it."

* In 1899, an Orlo Huston Christy (1848-1913), a livestock dealer and farmer, had a ranch three miles west of Phoenix (||y3jplnoz, ||yygccvja).

Abbott, Mississippi. From the *News* (Tuscaloosa, Alabama, Aug. 30, 1929, p. 6; ||y4z9peo7): "SCOOBA, MISS., Aug. 29—Born this week on the farm of Hendon Smith, near here, the calf of a fine Holstein cow has a body and the upper part of its head shaped like a calf, but its mouth and the lower part of its head is shaped like a dog, and it has no tail at all."

* A Monroe Hendon Smith (1901-1974), a farmer, lived near Abbott, an unincorporated place in Clay County, north of Scooba in 1929 (||y2o3x3y4).

Hermival-les-Vaux, Normandy, France. La Petite Presse (Paris, Jul. 28, 1887, p. 1c; ||38y6ehen) said a dog-headed calf, with basset hound legs and a short tail like a terrier's, was on exhibit at Hermival-les-Vaux.

CALF GROWLS AND EATS MEAT.

Strange Freak Named Gyp, With Face Like a Dog, Pet on North Reading Farm.

DOG FACED CALF, NAMED GYP, ON NORTH READING FARM.

NORTH READING, May 4—Living on a farm here and enjoying the best of health is a freak calf. It has a perfect dog-face and its owner, Nick Preble, has given the little cerature the name "Gyp." The animal has already begun to attract much attention, and large numbers of people have visited the farm for the purpose of looking at him. The calf is a very intelligent animal, and although but a few months old Mr Preble says he can do most everything but talk. Its teeth are undershot, like those of a Boston terrier, and its horns are just commencing to sprout.

It has course hair and no tail, its ears resemble those of a dog and he sits up on his haunches like a canine. The front legs are short while the hind ones are long, resembling in some degree a kangaroo.

Gyp also has a habit of growling like a dog, and he takes great delight in eating meat and milk.

A report about a meat-eating canibovid living on a farm in North Reading, Mass. Calved by a cow, it enjoyed "the best of health," growled and had not only sprouting horns but also doglike ears and teeth. (*Boston Globe,* May 5, 1908, p. 3, ||yyyxn5cr)

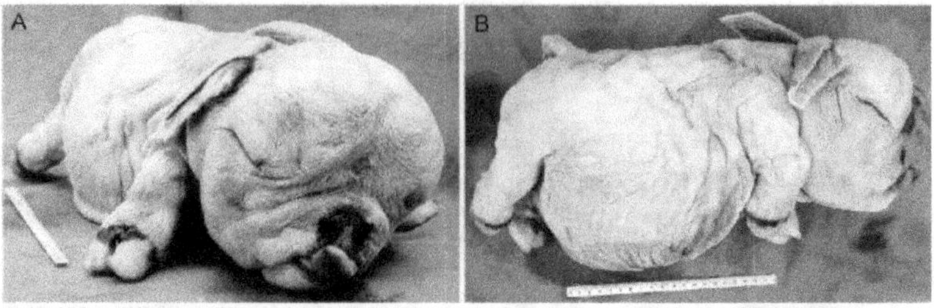

A bulldog-like canibovid calved by a cow. Note the presence of hooves.

BULLDOG CALVES

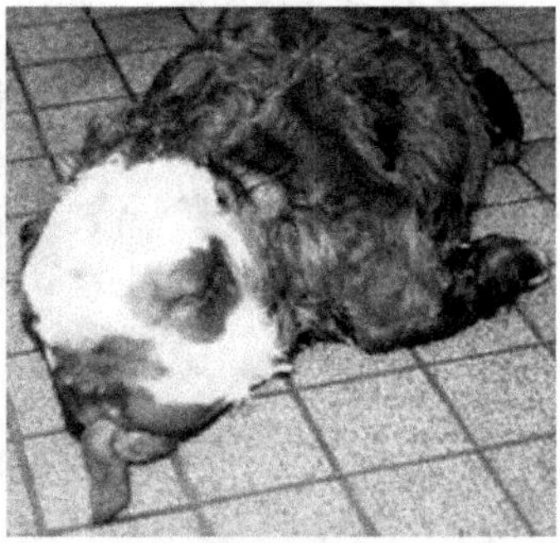

A bulldog calf
Armed Forces Institute of Pathology

The name "bulldog calf syndrome," is used to refer to subjects like those shown above and at right. Bradley (2018, p. 122) says "Bulldog calves usually are aborted before the seventh month. The animals are grotesquely malformed with protruding tongue; thick, rotated and abducted limbs; absent hard palate; and umbilical hernia and protruding viscera. The trait appears to be incompletely dominant."

But no mutation producing this phenotype has been characterized. In such a commercially important animal, why is this the case? Reports about canibovids often mention doglike faces and teeth, dog legs and paws. They speak of whining, tail wagging, snuffling and eating meat. Such statements speak against the idea of a resemblance produced by a mutation. Why would a single mutation produce, in a calf, an entire suite of canine traits? On the other hand, a cow-dog hybrid would be expected to have a wide variety of characteristics linking it with dogs.

Crescent City, Florida. The *Crescent City News* (quoted in the *Columbia Breeze,* Columbia, Ala., Aug. 22, 1901, p. 10b; ||3h9u5k5h) said, "H. F. Johnson has a freak calf which, though dead, has been preserved perfectly by taxidermy, the work of Mrs. J. P. Cowburn. The animal's head resembles that of a bulging bulldog, and its body is thick and long, mostly stomach. Its tail seems to grow out of its back and its legs hang from the sides of its body. … The calf was born in Grovesdale [now Huntington, near Crescent City] and lived but a short time."

Guysville, Ohio. The *News-Journal* (Wilmington, O., Jul. 16, 1935, p. 1; ||yc5zjtzj) pictured a canibovid calved at Guysville (their image is reproduced on p. 419 of this book). The caption read, "Business has picked up briskly at the hot-dog stand in Guysville, O., where crowds have come to see a half-calf, half-dog born to a two-year-old Jersey cow on the dairy farm of R. E. Calvert [b. 1877]. The animal has the face and forelegs of a bulldog and the ribs and hindquarters of a calf and whines and wags its tail in the fashion of a dog." Another report, in the *Coshocton Tribune* (Jul. 10, 1935, p. 1f; ||2p8uw4td) read, "A freak calf with the bowed forelegs and undershot jaw of an English bulldog was born on the farm of Roea E. Calvert near here. The body is that of a normal calf, but the tail wags."

Round Hill, Virginia. The *Baltimore Gazette* (in the *Staunton Spectator,* Jul. 6, 1880, p. 3; ||22nuzadt) published a letter from Round Hill describing a canibovid of the bulldog type. "On the 2nd instant an animal was brought to light in this community in which the scientific world, will doubtless take considerable interest. It is no less a lusus naturæ than the offspring of a cow which has only the feet and body of a calf, the balance of its members being those of a dog. Its head, mouth, nose and ears are precisely like those of a well-bred bull-dog, as is also its hair, which has the peculiar smell of a puppy, and its tail is curved and held up like a dog's. Its legs have the shape and peculiar "set" of a bull-dog's and the eyes are popped out to a degree that would fill the heart of a connoisseur with admiration. It eats fat meat and other food that a dog would relish, but drinks milk also. Its mother is a cow owned by Mr. Lunceford, which had given birth to well-developed calves previously. It was taken from the cow as soon as discovered, and its owner, in disgust, gave it to a young man in this place called [Joseph Lee] Lodge [1863-1953]. The animal is creating a sensation in this vicinity and was visited on Sunday last by people from all over the county. Its present owner has been offered $1,000 for it if it lives to be six months old. It is now about the usual size of a calf of its age."

Saint-Denis-en-Val, Loiret, France. La Presse (Paris, Nov. 10, 1893, p. 3c; ||yc42bbej) said a cow belonging to a farmer in Saint-Denis-en-Val had calved an animal the size of a piglet, with legs so short that one could easily take them for those of a basset hound. The rump was that of a dog with a cropped tail, the hide that of a calf. Per the report, the head from the front was just like a bulldog's, including the teeth. It was nursing and full of life.

Nouy, Indre-et-Loire, France. L'Avenir (Tours, Jan. 1, 1937, p. 3g; ||yfp8rnjx) said a cow, in the stable of Mrs. Fesland Lussault at Nouy, had had in late December a calf with the true head of a bulldog, long fangs, short ears, a bulldog's tail and short, twisted legs.

Sherman, New York. The *Westfield Republican* (Westfield, N.Y., Feb. 24, 1915, p. 2c; ||y8kbjh5v) said, "Charles Hein, who lives near the Sherman town line, had born on February 16, 1915, a black heifer calf that has a bulldog face. Every feature is exactly like that of a bulldog, the teeth and jaw being formed like a dog, and it puts its tongue out like a dog. It also makes a noise like a dog instead of a calf."

* A Charles Jacob Hein (1854-1938), a farmer, lived near Sherman (||mr2fn6cp).

Massena, New York. The *Commercial Advertiser* (Potsdam Junction, N.Y., Apr. 3, 1917, p. 4e; ||y9yrgsyt) said, "A cow belonging to G. A. Smith of Massena, gave birth to a freak calf recently. The body was nearly normal, but the legs were less than a foot long and the head closely resembling that of a bulldog. The calf lived only a few moments after being born."

* A Gustavus A. Smith, a dairy farmer, 62, lived in Massena in 1910 (||yy5oaps8).

Red Rock, New York. The *Eagle-News* (Poughkeepsie, N.Y., May 22, 1917, p. 4; ||2s3dtbst) reported a two-year-old Holstein heifer calving a bulldog-like canibovid in the hamlet of Red Rock in Columbia County. It had a protruding lower jaw, and the report said it "is about bulldog size and its tail is on its back, about six inches from the spot where nature intended tails should be located. The ears are those of a calf, the legs those of a dog. In color it is brown and white, the coat resembling that of a calf."

Château-Thierry, Aisne, France. Le Réveil de l'Aisne (Feb. 19, 1905, p. 2d; ||yks5fkv9) said, in translation, "M. Lecointre-Tranquille, a farmer in the village of Saint Martin [now a Château-Thierry neighborhood], has had since Monday a strange freak. One of his cows bore a calf which had, in fact, a bulldog's head. Its ears look cropped. It has no tail. It sits on its hind legs, which are only 25 centimeters [10 in] long and end in cloven hooves."

Artannes-sur-Indre, Indre-et-Loire. Le Tourangeau (Tours, Dec. 2, 1906, p. 3b; ||muuyenme) said, in translation, "ARTANNES.—A freak calf can be seen on the farm of M. Gallais at Aubardière. Now 15 days old, it has the head, the eyes, the ears, the tail and the paws of a bulldog."

Bellport, New York. From the *South Side Signal* (Babylon, N.Y., Nov. 25, 1899, p. 2; ||lluq97t): "A monstrosity in the shape of a calf with the head of a bulldog was recently born to a cow on the farm of William Mackintosh, at Bellport. Dr. W. B. Kelly, a Patchogue veterinary surgeon, who had charge of the case, says the calf was perfectly formed except as to its head and legs. The head was shaped like that of a bulldog, with the round face and the eyes of a calf. The legs were short and stumpy, thick set and shaped like a dog's leg. The monstrosity died soon after it was born."

* Bellport and Patchogue are now part of Brookhaven. A William Macintosh, a farmer (b. 1833), lived in Brookhaven in 1900 (||y452b38k), as did a William B. Kelly (b. 1868), a veterinary surgeon (||y3tftm4f).

A canibovid reported by Louis Griveaux (1904), veterinarian at Chalon-sur-Saône, Saône-et-Loire, France. Hydrocephalic and stillborn, it had short (17-20 cm) cartilaginous legs ending in hooves. The ears, too, were short.†

Margaretville, New York. The *Catskill Mountain News* (Margaretville, Dec. 6, 1912, p. 1; ||meqbqvd) said, "Supt. Chas. Knapp of Furlough Lodge [the summer home of financier George Jay Gould (1864-1923), a few miles south of Margaretville] brought a calf's head to this office Wednesday that resembled in nearly every particular the head of a bulldog. The lower jaw was undershot and carried protruding teeth just like a bulldog. The nostrils were down on either side of the jaw. It was necessary to take a second look to be sure that it was not the head of a bulldog that he had."

* Charles Gorse Knapp (1845-1910) was a Margaretville native (||m24x59vs).

Avrilly, Normandy. Journal d'Évreux et du Département de l'Eure (Évreux, Jan. 18, 1913, p. 3b; ||mr3k4t27) said a viable calf with a bulldog's head, jaws and forepaws had been calved two weeks earlier at Avrilly.

Dekalb Junction, New York. The *Farmer* (Malone, N.Y., Apr. 15, 1903, p. 1; ||kdjtmua) said, "Butcher Sawyer, of Dekalb Junction has a freak calf. The head resembles a pug dog [which resembles a bulldog's]. The legs have hoofs but resemble a dog's in other respects and are about as long as a pug dog's. The ears and tail look as though they had been trimmed, while the eyes are as large as those of a cow."

* An Ira W. Sayer, a butcher, lived in Dekalb Junction in 1910 (||y4geplhk).

†. In its general structure, the Chalon animal resembled the Cortland canibovid (pictured on p. 1196). Two other, viable, short-legged canibovids with bulldog heads were reportedly calved in France, one at La Celle-Saint-Cyr, Yonne (*L'Informateur*, Montereau, Feb. 13, 1909, p. 2a; ||kv3avf53), the other at La Capelle, Aisne (*Le Grand Écho du Nord de la France*, Lille, Feb. 23, 1912, p. 3f; ||5d2zh77p).

Ogdensburg, New York. The *Ogdensburg Journal* (Mar. 19, 1895, p. 4e; ||y9wqzwqw) said, "A wonderful freak of nature appeared on the farm of Seymour Stoddard of Water Street. It was a calf which has a pug dog's head and nose, with a long, lower jaw and a short pug neck and a dog's tail and legs, and measures fourteen inches from his tail to his head and is as thick set as any pug dog, but his feet have a divided hoof."

* Records show Seymour Lyman Clark Stoddard (1843-1914), a farmer, lived his entire life in St. Lawrence Co. (||y39xs3dd). Ogdensburg is in St. Lawrence Co.

* Many reports describe canibovids with legs that would be short for a calf. At a meeting in Paris of the Royal Academy of Sciences, Buffon (1744, p. 15) exhibited a specimen with no legs and a head like a pug's that made up most of its body.

Gilboa, New York. The *Amsterdam Recorder* (quoted in the *Daily Leader,* Gloversville, N.Y., Jan. 26, 1889, p. 3; ||n2apogb) reported a canibovid that was viable and almost just like a dog but for being hairless and hoofed:

> Gilboa, a place not many miles south of Amsterdam, has had several mild sensations during the past six months. The latest novelty that has set tongues wagging is the birth of what a youngster there has termed as "The bald-headed puppy calf on the Stryker farm." The monstrosity is certainly enough to wonder and think and talk about ... It was a gentle meek-looking cow, belonging to Andrew Baldwin, that gave birth to the queer looking object that is a bulldog in form, with the exception of its feet, which are formed like those of a tiny calf. Its upper jaw contains no teeth [cattle lack upper incisors] and there is not a particle of hair on any part of its body. It has a genuine dog's tail, which it wags when pleased. Its whine is peculiarly like that of a cur.
>
> * An Andrew S. G. Baldwin (1827-1913), a farmer, lived in Gilboa (||y6br3o5p).

Saint Albans, Vermont. From the *St. Albans Daily Messenger* (Aug. 26, 1893, p. 4c; ||y2r4nydf): "Joseph Goron, the taxidermist, has just set up a monstrosity in the shape of a cross between an English bulldog and a Jersey calf. It was dropped by a full-blooded Jersey heifer and lived a few hours. Its head is an exact reproduction of that of a bulldog with cropped ears, the short square nose, and protruding, massive underjaw. It is covered with hair like a calf and of the same color as the usual Jersey. Its tail is merely a tuft of hair something like that of a rabbit, while its feet are cloven hoofs something smaller than those of a calf. The eyes and teeth are those of a calf, but it is in other respects a bulldog. The animal measures two feet nine inches from tip of nose to tail and stands 18 inches high. This was the second calf-dog born in the same neighborhood in Franklin [County, Vt.]."

* An 1895 Saint Albans city directory lists a Joseph N. Goron (||y2uhz2mq).

Woodstock, Vermont. The *Herald and News* (West Randolph, Vt., Jun. 27, 1907, p. 3; ||ybqq63w3) said, "H. L. Jones of Woodstock is the possessor of a curious freak of nature. The monstrosity was born a calf but is apparently half bull terrier dog and of mixed sex. The animal has the coat, color and hoofs of a calf. Its ears are cropped like those of a sporting dog. It has a "screw" tail, a smooth tongue, its limbs are short and muscled, and shaped much like those of a dog; but the startling feature is the short, wrinkled nose, the undershot jaw and full lower teeth, the habit of dog-like sniffling for scent, and the cropped ears which give it a very uncanny aspect. The mother is a two-year-old heifer ... The calf seems healthy and likely to grow up. Mr. Jones has already been offered a large sum for the freak, provided it lives and thrives for a few weeks, enough to give promise of maturity. On account of its doglike jaws and teeth, the calf is fed from a bottle, to save the mother injury."

* Harry Leonard Jones (1865-1909), a liveryman, and Henry Herbert Daniels (1838-1913) both died at Woodstock (||y69we36f, ||y5oahyzq).

Guiscard, Aisne, France. La Gazette du Centre (Limoges, Feb. 21, 1883, p. 2e; ||2mf5jjxs) said a cow of a M. Dubois at Guiscard bore an animal with a calf's body and pelt but a bulldog's head and teeth. The legs, it said, were a calf's but ended in dog paws, the tail, too, that of a bulldog. It was initially healthy but, lacking an anus, died after three days.

Mutation or a canine sire? The *Journal d'Évreux et du Département de l'Eure* (Évreux, Jun. 1, 1895, p. 3b-c; ||y46axduy) said two cows in the same herd at Landepéreuse, Eure, had both recently calved offspring with bulldog heads. And *Le Tourangeau* (Tours, May 25, 1902, p. 3b; ||3key6bz2) said an exactly similar case occurred at Bléré, Indre-et-Loire, in 1902. Two cows in each herd being impregnated by a canine sire reasonably accounts for these two double events. Random mutation does not.

Boulange, Moselle, France. The *Gazette de Lorraine* (Metz, Apr. 28, 1886, p. 2d; ||9bmf2zv6) said, in translation, "Correspondence from Boulange, dated April 26, reports, "Today, at 1 p.m., the people of Boulange were able to view an extraordinary freak. A cow belonging to M. Jean-Pierre Penry, a landowner in Boulange, gave birth to a bulldog. Except for the four feet, which are like those of a calf, the body (particularly the head and tail) has the same form as a dog. It weighs 35 kilos [~77 lbs]. [It] was preterm and stillborn when taken from the cow."

Saint-Vaast-Dieppedalle, France. The *Journal d'Évreux et du Département de l'Eure* (Évreux, Mar. 16, 1898, p. 3d; ||mwy3nr2u) said a cow of Mrs. Grancher in Saint-Vaast-Dieppedalle bore an animal with "the legs of a basset hound and a bulldog's head" (translation mine).

Richford, Vermont. From the *Middlebury Register* (Jan. 14, 1916, p. 3c; ||zqbhp2z): "A freak calf was born recently on the W. B. Locklin farm in Richford. The calf had a bulldog's head, ears and tail. The nose and mouth were like those of a small pug dog, the tongue was about one-half inch wide or half round and protruded from its mouth about three inches and turned up to the forehead. The body was large for a young calf; the legs were about 10 inches long."

* A Wallace Byron Locklin (1855-1936) lived at Richford in 1916 (||yyem7eml).

Middlebury, Vermont. The *Middlebury Register* (quoted in the *Ticonderoga Sentinel,* Apr. 10, 1902, p. 8a; ||kyyjaq2): "C. Mathews has a freak in the shape of an animal with the body of a calf, the short head, short ears and protruding under jaw of a bulldog, carnivorous teeth and a curly tail. The monstrosity is weak in its forelegs, and as the cattle are afraid of it, the owner feeds it as a 'bottle baby.' A large number of curious people have been to the Mathews farm to look at it."

* A Charles James Mathews was born at Middlebury in 1855 (||y2oqvxdl).

New Haven, Vermont. The *Banner and Reformer* (Bennington, Vt., May 28, 1903, p. 3; ||z3t8t6o) said, "A two-year-old heifer owned by E. G. and A. W. Norton of Vergennes recently gave birth on their farm in New Haven to a monstrosity. It is about as large as a year-old bulldog and the head, tail and body closely resemble those of that animal. The legs are short like those of a dog, but instead of claws this freak of nature is provided with hoofs. The body is perfectly formed and the skin, which is covered with hair of a dark brown color, is like that of a calf. It was born dead, but measures have been taken to preserve it, and the Messrs. Norton are thinking of having it mounted."

* Brothers Eugene Galen Norton (b. May 1, 1859, d. Sep. 10, 1913) and Arthur William Norton (b. May 8, 1867, d. Mar. 11, 1939) lived in Vergennes (||y4zxmpt3; ||yyb9mnzh).

Brinon-sur-Sauldre, Cher, France. Le Bonhomme Limousin (Limoges, Jun. 9, 1907, p. 2b; ||56kn3ku9) said a cow belonging to M. Charles Michoux at Brinon-sur-Sauldre calved a composite with the body of a calf but having the head, ears, tail and paws of a bulldog. Drinking six or seven liters of milk a day, it was in excellent health.

Waterbury, Connecticut. The *Waterbury Evening Democrat* (Mar. 22, 1907, p. 4d; ||yen4zxuu) described a canibovid calved by a cow owned by local farmer Hans [Seth] Rasmussen (b. 1858). "The calf, which was from a Holstein, had only one nostril and had a face and head exactly like a bulldog." It died two days after birth.

South Reading, Vermont. A report in the *Spirit of the Age* (Woodstock, Vt., Apr. 10, 1878, p. 3; ||y2d3bqnn) about a stillborn canibovid: "SOUTH

READING.—Henry Austin had a cow which the other day dropped a singular monstrosity for a calf. It looked more like a bulldog than a calf. Its head with blunt nose, mouth with tushes [i.e., fangs] and ears were like those of a dog. From the forehead to the nose was 1½ inches. The legs were also like a dog but had hoofs. In fact, about the only appearance of a calf about it was the feet and hair. It was dead when found."

* A Henry S. Austin (b. 1814), a farmer, lived in Reading in 1880 (||yxr8fyg2).

Mineral Point, Pennsylvania. From the *Altoona Times* (Altoona, Pa., Apr. 5, 1906, p. 3; ||y32omx9p):

> A calf with a dog's head is the queer freak of nature which for the past ten days has been causing wonderment among the farming population of Mineral Point. The strange object is the offspring of a cow belonging to Charles Page … and it has been visited by a large number of people.
>
> The abnormal creature has the body of a calf except that its front legs are of such a character it can use them scarcely any at all, and it lies on a pile of straw practically helpless, except that it is able to shift itself about a little. The strange part, however, is the head, which is exactly like that of a pug dog, even to the turned-up nose, and it has a dog's ears and other features. Moreover, instead of drinking as a calf usually does, it laps up the milk like a puppy. It is fed several times a day from a saucer.
>
> * A Charles E. Page (1876-1916) lived at Mineral Point in 1900 (||yxw3q6v3).

Fourges, Normandy. The *Journal d'Évreux et du Département de l'Eure* (Évreux, Dec. 3, 1885, p. 2c-d; ||yc2cfpvc) said a cow at Fourge, belonging to a M. Pierre Désiré-Géroult had calved a tailless animal with "with the head, eyes, and fangs of a bulldog" (translation mine). It was otherwise like a calf, said the report, and still alive.

Fumeçon, Normandy. The *Journal d'Évreux et du Département de l'Eure* (Évreux, Mar. 16, 1898, p. 3d; ||mwy3nr2u) said a cow at Fumeçon calved "a sort of bulldog, resembling a calf in nothing except its hair" ("une sorte de boule-dogue n'ayant du veau que le pelage").

Beaucoudray, Normandy. The *Journal de la Manche et de la Basse-Normandie* (Saint-Lô, Mar. 7, 1908, p. 4d; ||5296r83s) said, in translation, that a farmer by the name of Verdière at Beaucoudray had a cow that had borne three weeks earlier "a calf with the head, paws and tail of a bulldog. The rest is like other calves. The animal appears viable."

Chassigny, Haute-Marne, France. *L'Écho de l'Yonne* (Tonnerre, Aug. 6, 1905, p. 3a; ||5bwxwsw2) said a cow belonging to Emile Bonin, a farmer at Chassigny, calved an animal with the head of a bulldog, sharp teeth, short ears, paws with claws and a tail less than 10 cm long.

Montreuil-en-Caux, Normandy, France. Le Progrès Libéral (Toulouse, Jul. 21, 1883, p. 1; ||43p9mc59) said, in translation, "A freak calf was born July 14 in the stables of M. Edouard Nouvel, a farmer at Montreuil-en-Caux. This calf had the face, the head and the ears of a mastiff ["chien-dogue"]. Its eyes are frightful, the size of hen eggs, and bulge from their sockets. The legs are short, from the foot to the knee rounded like those of a calf, but broad and flat from there up to the stomach. ... The animal is male and has a white head. The rest is a pale red. The teeth are rather sharp. This calf, when standing is 47 cm (~18 in) tall. Despite his abnormal form, he's viable and nurses excellently well."

* In 1900 Harris C. Boyes, 34, owned a harness repair shop in Wray (||yy87gsjm).

Wray, Colorado. The *Wray Rattler* (Apr. 20, 1901, p. 3d-e; ||y8rvamvw) said, "A freak calf was exhibited at the Boyes harness emporium this week. The head and legs are as nearly the shape of those of a bulldog as one could imagine. The nether jaw protrudes, while the nose is very much dished. ... It only lived a few hours."

* In 1900 Harris C. Boyes, aged 34, owned harness repair shop Wray (||yy87gsjm).

Moranburg, Kentucky. The *Evening Bulletin* (Maysville, Ky., May 8, 1901, p. 2c; ||yc5y7xc6) said, "Mr. William Marsh, of Moranburg,[2] reports the birth of a monstrosity on his farm a few days ago in the shape of a calf. It was perfectly developed in every respect, except that its head resembled that of a dog more than the head of a calf. The ears were very small, and back of them was a development or formation of the skin that could be rolled over the head somewhat like a hood."

* Farmer William A. Marsh (1847-1919) lived in Moranburg in 1901 (||yx8vrbpv).

Fort Worth, Texas. The *Fort Worth Daily Gazette* (Jul. 8, 1887, p. 4e; ||y585vmxu) said, "A GAZETTE man was shown ... the premature offspring of a cow on the place of Mr. Joe Born, living near town. ... its head was a perfect facsimile of a bull dog's, even to the ears and grinning teeth. The rest of the body was formed exactly like that of an ordinary calf, but the contrast between the face and the body was almost startling. The freak was dead when found ... It was about two feet in length, of a pretty fawn color."

Hamburg, Iowa. The *Fireside Guard* (Centralia, Mo., Feb. 21, 1874, p. 2; ||y5aaw5jk): "Hamburg, Iowa, has a monstrosity ... half calf and half dog."

Tosny, Eure, France. The *Journal d'Évreux et du Département de l'Eure* (Évreux, Jun. 2, 1897, p. 3c; ||449rdsxu) said a cow at Tosny, belonging to a M. Picard, gave birth to a calf-dog ("veau-chien") with the head of pug and the paws of a basset hound.

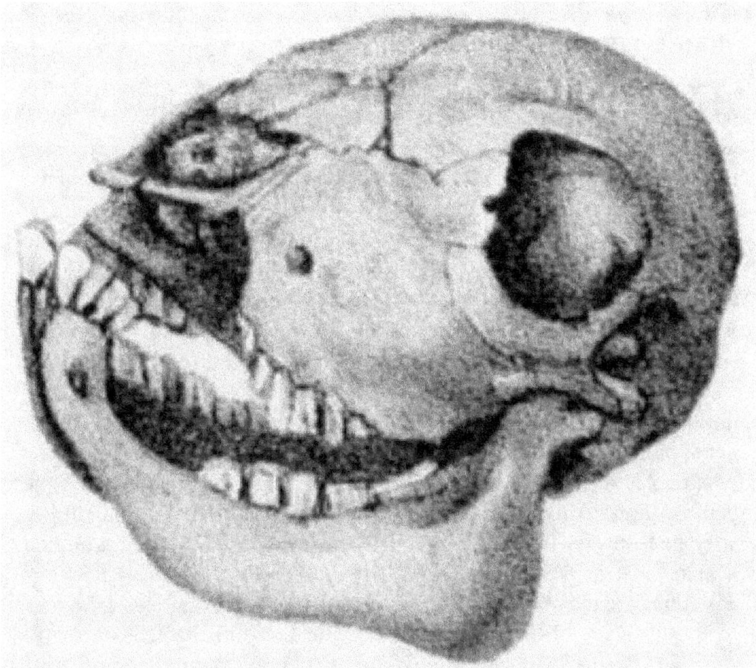

Skull of the Wrentham canibovid (Jackson 1847 pl. 8, fig. 20).

Wrentham, Massachusetts. Jackson (1847, p. 52) said an animal calved in Wrentham, Mar. 15, 1839, "was regarded in the neighborhood as a 'calf changed into a bulldog.'" He described it as follows, "The head was full and rounded, the line of the face was abrupt, the muzzle was short and square, and the under lip was somewhat projecting. The extremities were very short, but of full size … the anterior being turned directly backwards, the posterior backwards and inwards. The measurements were, from the muzzle to the root of the tail twenty-seven inches, and to the top of the head seven inches and three-fourths; circumference of body twenty-six inches; extremities six inches in length, the anterior being six inches and three-quarters, and the posterior seven inches in circumference. … The feet, … with the carpus and tarsus, are well formed [i.e., hoofed and normally shaped for a calf], but … the scapulae, humerus, radius and ulna, femur, tibia and fibula [consist] of very short, thick, and irregular bones, measuring from one inch and a fourth to two inches in length."

Oxford, Maine. From the *Lewiston Evening Journal* (Lewiston, Me, May 12, 1915, p. 12; ||y9uwua9j):

> Over at E. J. Boucher's taxidermy rooms in Auburn [Maine] is … a specimen … brought to Mr. Boucher to be mounted by Dr. J. H. Glover, veterinary surgeon in Oxford. The animal is in many ways like an ordinary calf. It is long-haired, black and white, with the white star on its forehead. The funny little tail is black and white also. The head is abnormally large and most pronouncedly like a bulldog's. The jaw is heavy and projecting, the eyes are small and bulging, not meek and mild like the ordinary gentle bossy. The ears in size and shape are like a calf's, but are set far back on the head, near the base of the skull, instead of high on the forehead. There are no traces of horns. The teeth are those of the cow species. The neck is extremely short, unlike the long, graceful one found on most farm animals and the head, bulldog-like, rests close to the shoulders. The hips are also those of a dog and the hind legs are stretched straight out behind with the gamble [app. a misspelling of *gambrel*] joint reversed. The animal lies stretched out, its forepaws extended and its hind legs straight behind it, like any dog might lie when tired. About as queer looking as anything, are the split hoofs at the base of the short dog legs. The legs to the ordinary observer are those of the dog but the feet present a startling contrast. The calf-dog would stand low from the ground. This specimen was born alive and lived for some three hours.
>
> * An Edmond J. Boucher, a taxidermist, lived in Auburn in 1910 (||yxmnxz7v). A James H. Glover, a veterinary surgeon, lived in Oxford in 1915 (||y65mgggk).

Laredo, Texas. The *Laredo Times* (quoted in the *Victoria Advocate,* Victoria, Tx., Feb. 13, 1925, p. 2; ||qyjqa4o) said,

> Yesterday afternoon John P. Dabbs telephoned to the "freak editor" of the TIMES to come posthaste to his dairy near the Holding Institute at Fort McIntosh and he would behold a freakiest freak that would furnish the gorm for a good story. And soon we were hastening to the Dabbs Dairy, aboard a Detroit Grief [apparently an old slang term for an automobile]. Reaching there, John Dabbs pointed to something prostrate on a gunny sack and … There lay an animal cold in the grip of death that, had it lived, would have had a hard time to find association with any other animal. The portion of the animal including its forelegs, and shoulders, head, nose, mouth, eyes, ears and neck were those of a British bulldog, while from the shoulders southward it was in the form of a heifer calf. We looked again and again. It was no illusion.
>
> With the consent of Mr. Dabbs, the freak animal was kidnapped and brought to Serrano's Studio and there photographed and then

> taken to Muter and Ince and turned over to Taxidermist Muter to be mounted. Talking about seeing pink elephants, legless camels, beakless pelicans and silk alligators during an attack of the d.t., you should, in your truly sober senses, see this half-and-half bulldog calf. … And the best of the thing is the camera backs us up for we have the photographs of this freak for a magazine story.

> * A John P. Dabbs (1887-1958) was living in Laredo in 1920 (||y4towvsh). A William R. Muter, a 51-year-old taxidermist, lived in Laredo in 1920 (||y2nh3vqv).

Hartford, Connecticut. From the *Hartford Courant* (Apr. 2, 1900, p. 8):

> A cow owned by Oliver Antrim of upper Windsor Avenue, Saturday night gave birth to a calf that in all characteristics resembled the English bulldog, having the large square head of that species of dog, "overshot" under lip, short stout legs, its tail also resembled the dog's. The monstrosity was as large as the average calf, and was fat and sleek, but had apparently died from lack of nourishment after it was born on account of its short legs.

> * In 1900, an Oliver Smith Antrim (1824-1906) was living at 1016 Windsor Ave. in Hartford (||y99dbv92). Note that this case *sounds* like XENOGENESIS (see p. 1235). More likely it represents a canibovid at the canine end of the spectrum. *L'Express du Nord et du Pas-de-Calais* (Boulogne-sur-Mer, Dec. 19, 1896, p. 2c; ||4bhn7za3) said a cow calved an animal like a dog at Hardinghen, Pas-de-Calais.

Féchaine, Nord, France. L'Écho du Nord (Lille, Sep. 26, 1900, p. 3e; ||2hx6fknp) said a cow at Féchaine had calved a viable animal with the head of a bulldog, the teeth and body of a calf, and legs that were very short but shaped like a calf's. Five weeks old, it had no tail.

East Liverpool, Ohio. A report, about yet another viable canibovid, from the *Pittsburgh Post-Gazette* (Oct. 18, 1948, p. 15; ||y5olfm2c): "EAST LIVERPOOL, O., Oct. 17.—A dog-faced calf, resembling a bulldog, weighing 60 pounds and apparently normal, was born to a Brindle Jersey cow Sunday at the farm of R. L. Plumley, near here. The freakish bovine also has a pug nose and bowed forelegs."

* A 1943 Liverpool city directory lists an R. L. Plumley (||y6eztjlq).

Bristol, Rhode Island. From the *Morning Journal and Courier* (New Haven, Conn., Mar. 20, 1884, p. 3g; ||y9gcjkdz): "A strange freak of nature was reported at Bristol, R. I., today. A cow belonging to Benjamin R. Wilson gave birth to a calf having the head of a bulldog. The hoofs were shaped like a dog's paw. The calf was stillborn."

* Since the Bristol animal, calved by a cow, had not only a dog's head, but also dog's paws, hybridization is a far more plausible explanation than is random mutation. Also of note is the writer's bias in calling paw-like feet *hoofs*.

Canal Lewisville, Ohio. The *Somerset Press* (Somerset, O., Jan. 27, 1881, p. 1g; ||y97cvb8t) reported a viable canibovid with LIMB DEFICIENCIES.[3] "Mr. McGuire, a farmer living near Canal Lewisville [near West Lafayette] has an animal on his farm which is a curious monstrosity. It has the body of a calf, the head of a bulldog; has no legs below the knee-joints, is minus a tail, and its sex is indistinguishable. This apparent cross between the bovine and the canine species was calved a few days ago, is as lively as could be expected, and bids fair to live its natural lifetime. Still, it seems to us that we have heard of this monstrosity before, else one very much like it."

* Samuel T. McGuire (1853-1919), a farmer, died in West Lafayette (||y3auchbz).

Free State, South Africa. The *Rhodesia Herald* (Harare, Zimbabwe, Nov. 18, 1921, p. 11a; ||37n6fjvt) said a calf, recently born in the Free State, "has a perfect bulldog's head and tail, it barks and sits like a dog. It eats anything, meat, grass, etc."

Battle Creek, Michigan. A report in the *Livingston County Daily Press and Argus* (Jun. 21, 1888, p. 2b; ||2p9chms6) intimated a canine sire. "Charles [H.] Eberstein [1856-1942], of Battle Creek, has a cow which recently gave birth to a perfectly formed dog, with the exception of its feet, which were those of a calf, having hoofs. The head, mouth, teeth, legs and tail were perfectly canine. It was dead and was marked in black and white similar to a large dog owned by Mr. Eberstein."

Penn, Pennsylvania. The *Bloomfield Times* (Bloomfield, Pa., May 7, 1873, p. 4; ||y9jk87pj):"Mr. Schultz, of Penn Station [now called simply Penn], Westmoreland County, is the owner of a cow which gave birth to a singular monstrosity the other day. The head of the animal strongly resembled a bulldog, the body was that of a calf, though shorter, and the legs were very short and thick. The deformed bovine lived but a short time."

Savoisy, Côte-d'Or, France. Le Bien Public (Dijon, Dec. 9, 1925, p. 3d; ||uy7b25k9) said a cow calved an animal at Savoisy with a flat face like a bulldog, and short, twisted, hoofed legs. It was half the size of an ordinary calf.

Villiers-le-Sec, Calvados, France. Le Petit Haut-Marnais (Chaumont, Mar. 19, 1929, p. 3c; ||4nvkjun8) said, in translation, "a cow belonging to Mr. Corneux-Peilley has birthed a stillborn calf having a head like that of a bulldog and extremely short, twisted legs like those of a basset."

Pont-l'Évêque, Normandy. L'Éclair Comtois (Besançon, Feb. 22, 1923, p. 1f; ||bdmf9uun) reported the calving of a viable, tailless animal with a bulldog's head and teeth.

Edmond, Norton County, Kansas. The *Edmond Times* (Apr. 20, 1888, p. 3a; ||yyofkhdb) reported a stillborn canibovid: "Thursday last, A. R. Ford exhibited a strange freak of nature on our streets in the shape of a calf that had the nose, head, neck and legs of a bulldog but the body, hoofs, ears and tail of a calf. This monstrosity was dead at birth, but if Mr. Ford succeeds in preserving it, as he thinks he can, he will have one of the great curiosities of the times. This strange freak of nature was viewed by a large number of people on Tuesday, and all admit it the greatest curiosity they ever beheld." The *Times* later ran an update (May 4, 1888; ||y6m7dyf5), which read, "A. R. Ford has taken that dog-calf monstrosity spoken of in these columns two weeks ago and stuffed the same, and he has done it so perfectly that none of the peculiar features have been lost. It is a strange freak of nature and well worth seeing. The head, ears, neck, legs, and hindquarters of a bulldog; the eyes, hoofs and body of the cow, and the thick hide of the old ox. We understand that Mr. Ford proposes to exhibit this curiosity at the various public entertainments through the country during the coming season."

* An Allen Ransom Ford (1839-1911) lived in Norton Co. in 1880 (||y6yrmuhu).

Râches, Nord, France. Le Grand Écho du Nord de la France (Lille, Apr. 9, 1910, p. 3f; ||yxz6nhhu) said, in translation, "Saturday, April 2, a cow of Mr. Louis Malengé, a farmer and municipal councilman at Râches, had an extraordinary calf. This animal has a bulldog's head, a dog's paws and tail, and a calf's body." The use of present tense in describing this animal suggests it was viable (i.e., that it survived at least a week).

Asheville, North Carolina. A report in the *Asheville Citizen* (Apr. 16, 1920, p. 3c; ||yxe3f45s) about a canibovid mentioned that it was calved with its eyes closed like a puppy's, not open like a calf's.

> J. A. Reed [1885-1967], a dairyman of the Fairview section, reports the birth of a calf on his place which had a dog's head. The calf was dropped Tuesday by one of the fine imported Holsteins belonging to Mr. Reed, but evidently did not live but a few minutes, if at all, as it was dead when found in the pasture.
>
> Mr. Reed states that the calf had a head very much like a bulldog, with whiskers, wrinkled nose, dog's ears and neck and was born with its eyes closed, as dogs are. Calves are born with their eyes open. The news of the calf attracted much attention and scores of people, many from Asheville, went out to see it. It weighed about 100 pounds.

* Records list an Asheville dairy farmer Joseph A. Reed (||yynnhnkq, ||y2g9r7cs).

The Avesnes canibovid.

Avesnes, Pas-de-Calais, France. Le Grand Écho du Nord de la France (Lille, May 8, 1938, p. 7; ||bdhm7a6c) said a cow near Avesnes calved an animal (shown at right) with head like a bulldog's and doglike hindparts. Though viable, its had trouble walking and lacked a tail.

Sioux City, Iowa. The *Sioux City Journal* (Jun. 4, 1899, p. 4d; ||yyhy8dl8) gave the following account of a viable canibovid. It was unusual in giving a mature body weight for a canibovid.

> The midget cow is a mischievous little beast weighing but 280 pounds at her present age, two years. To some of the stockmen she is known as "The Dog Faced Calf." Surely, they are not amiss in giving this name to the strange animal. This freak is owned by M. E. Baggs, head cattle buyer for R. Becker & Degen, who named his monstrosity "Pansy Blossom." About one year ago this funny animal came to the yards in a load of cattle and Mr. Baggs purchased it for a mascot. It has the black and white markings of a Holstein and is as potbellied as a bullfrog. "Pansy's" face when peering through the boards of the pen very much resembles that of an English bulldog. She is exceedingly undershot and the white front teeth of the lower jaw glisten in the sun. She is as gentle as a lamb, being handled a great deal by inquisitive stockmen during her one year's residence in the yards.

* A Milton E. Baggs (1856-1934) lived in Sioux City in 1899 (||y6govn5s).

Flines-lez-Raches, Nord, France. Jardins et Basses-cours (Paris, Apr. 20, 1908, p. 75b; ||492h427n) said a cow owned by Charles Labbe, a butcher at Flines-lez-Raches, calved an animal with a bulldog's head and teeth and dog legs, but with cloven hooves. The report said it sat on its haunches.

Mézidon, Normandy. La Liberté (Paris, Mar. 3, 1931, 1g; ||3uw2r3j9) said, in translation, "LISIEUX, Mar. 2.—A freak calf has been born at Mézidon, on La Londe, a farm belonging to Mr. Mathuin Ballay. It has a large head like that of a bulldog, enormous eyes and a well-developed body, short legs, and neither tail nor genitals. Yet it has small breasts. The animal is very viable. Its body is red, the head, spotted."

Hamilton, Ohio. From the *Journal News* (Hamilton, Dec. 23, 1931, p. 3; ||yyme7kth): "B. E. Runtley, chief engineer of the Hamilton Milk Co., 55a North Sixth Street, has … a calf born last Sunday to a red cow owned by Paul Rahm, West Chester, R. R. 1. … it has little or no resemblance to a calf but does look very much like a large bull-dog. The freak was alive but died a short time after birth. The head is much rounded, like that of a bulldog, the hindquarters also have the shape of a dog of that breed. The legs are not more than half the length of a normal calf. The feet appear to be malformed, leaving but rudimentary hoofs. There is also a rudimentary tail. What there is of this appendage is somewhat suggestive of the screw-tail of some breeds of dogs. The hair is black with white markings."

* Hamilton is the seat of Butler County. A Paul S. Rahm (1906-1987), a farmer, was living in Butler County at the time in question (||7un6tvr3).
* The Hamilton animal was close to the dog end of the canibovid spectrum and thus might easily have been interpreted as a case of XENOGENESIS (see p. 1235).

Foster, Minnesota. The *St. Paul Daily Globe* (Feb. 19, 1878, p. 1d; ||y5sgrnmc) said, "A cow belonging to Henry House, Foster township, Faribault County, has given birth to a calf which is a complete monstrosity, having a head like a bulldog, low beetling brow, short pug nose, and teeth well extended in front, with a look of pure cussedness. Its body in front was shaped like a regular calf, while its hinder parts were like a greyhound's even to its tail."

* Farmer Henry A. House (1846-1900) lived in Foster in 1875 (||y5a2hwy2).

Guide Rock, Nebraska. The *Rock Island Argus* (Rock Island, Ill., Sep. 17, 1896, p. 7; ||kao6blq) describes "Taurus the dog-calf," a canibovid calved at Guide Rock. "There was a museum [that is, at the fair] which exhibited a freak of nature called "Taurus, the dog calf," [a] wonderful monstrosity given birth by a cow. This strange creature, which has attracted much attention … was bought by John Dunbar, who has had it mounted for exhibition. It … weighed 125 pounds at birth." Another report appeared in the *Times* (Bellows Falls, Vt., Jan. 15, 1896, p. 2; ||lvvbvt5): "C. O. Davis of White Cloud, Kansas, is a taxidermist, and a brother of H. B. Davis of this place. He recently mounted a monstrosity, half calf and half dog. … It was given birth by a cow belonging to Mr. Dunbar, of Guide Rock, Neb., and is part calf and part dog. It weighed 125 pounds and is covered with short dog hair. The head has a perfect bulldog shape with a calf's expression in the face. One ear is like a calf's, and the other is like a dog's.[4] The neck is thick and short. The shoulders are about the same size, but the right hip and hind leg are large like that of a calf, while the left one is small like that of a dog. The tail is smooth and covered with short hair, except a bunch of long hair right at the end. It has two tongues, one dog tongue and one calf tongue (the tongue is preserved in alcohol); the dog tongue has hair on it. It has calf's feet, with claws on the bottom."[5]

* A Charles O. Davis (1832-1913) lived in White Cloud in 1896 (||y6aaonhb). The *Kansas Chief* (Troy, Ks., Feb. 27, 1896, p. 3; ||32wt8zf4) described Taurus in detail. Also see the *Globe-Tribune* (White Cloud, Ks., Jan. 14, 1896, p. 4c; ||yxhrrpw5).

Mount Gilead, Ohio. From the *Cincinnati Enquirer* (Oct. 25, 1878, p. 1; ||y5x22wq2): "MT. GILEAD, October 24.—Another remarkable monstrosity hails from Morrow County, in the shape of a calf with the head of a bull-dog. The calf is owned by Mr. Thomas Tillay, living two miles east of Mt. Gilead, and would be a fit specimen for exhibition with the six-legged pig, two-headed calf, &c, all from Morrow County."

* A Thomas Tilay, not Tillay, aged 55, lived in Morrow County in 1880 (||22773upv).

Knoxville, Tennessee. The *Knoxville Sentinel* (May 15, 1895, p. 1; ||y6n3jtf6) said, "McCrary & Branson, the photographers, had a queer specimen brought to their studio this morning to be photographed. It was the head of a calf, and in shape and general appearance it exactly resembles the head of a full-grown bulldog. The mother of the freak is a thoroughbred Jersey cow belonging to 'Squire Joe Knott, of the Twelfth district. When the cow was about two months with calf, she was attacked by two savage bull dogs, and 'Squire Knott and some neighbors rushed to the rescue and saved her life. The dogs had torn her ears and nose, and she was in a frenzy of pain and terror. At full time she was delivered of this monstrosity, which died during the operation, as instruments had to be used. The calf has a true bulldog head and eye. The body was smaller in proportion than the head, and was the shape of a dog, but the feet and tail were those of a calf. The mother survived."

* A Franklin Bell McCrary (1849-1921) was living in Knoxville in 1895 (||yxgysvur; also see: ||yxsdumua). Enoch Lloyd Branson (1853-1925) was one of the two photographers at the studio mentioned in the report (||y2t4enxt, ||yxkeavdd). A Joseph L. Knott (b. 1855), a farmer, lived in Knox County in 1900 (||y6yy7f7x).

St. Petersburg, Florida. The *Morning News* (Savannah, Ga., Nov. 14, 1897, p. 14b; ||5aa8eums): "A cow owned by A. Welton of St. Petersburg recently gave birth to a calf that bore an exact resemblance to a Yorkshire pug. The resemblance is said to be almost perfect in size, color and formation, the body being terminated by the closely curled tail characteristic of this class of pug. The only point of departure is in the legs, which terminate in the regular bovine hoofs." The face of a pug is much like that of a bulldog.

Saint-Michel, Aisne, France. Le Réveil de l'Aisne (Chauny, Apr. 6, 1902, p. 3d; ||ysp2u6cw) said, in translation, "SAINT-MICHEL.—Two weeks ago, a cow belonging to M. Frognieux-Brunot gave birth to a calf with the head, paws and tail of a bulldog. This freak is still alive."

Newton, New Jersey. The *Allentown Democrat* (Allentown, Pa., Sep. 12, 1912, p. 7f; ||yyva9mry) said, "A bulldog-faced bull calf was the attraction at the baggage room of the Terminal Station yesterday. The freak animal is the property of James Smith, of Newton, N.J. … The animal has the face of a full-fledged bulldog and with the exception of one deformity, which is a split nose, would be taken anytime for a bulldog. The bull calf is two months old. It requires 18 quarts of milk to feed it with during 24 hours."

SOME BULLDOG-HEADED CANIBOVIDS CALVED IN FRANCE

1888	Belfort	||2s3cys69	fetus, cyclops, near term on July 7
1885	Falaise, Normandy	||yaxm4k4w	bulldog head, soon died, calved in Jan.
1895	Paris	||ydv9dhrt	specimen, bulldog head
1908	Venette, Oise	||yh8aw5da	bulldog head, calved in September
1911	Couchy, Oise	||3z8us8x4	soon died, calved in April
1912	Fontenelle, Aisne	||7ecwucp5	soon died, calved in Feb.
1919	Vulaines-sur-Seine	||59annjpv	viable, calved in June
	ditto	||59annjpv	calved in November
1922	Beurey, Aube	||4u4fawhk	ears on neck, calved in April
1923	Verteuil-sur-Charente	||mvy8asj8	soon died, legs on back, calved Dec.
1926	Brandon, Saône-et-Loire	||5xcnjzbu	viable, calved in March
1929	Crémille, Indre-et-Loire	||3dpfyckf	viable, calved in Feb.
	Valençay, Indre	||mvpy3trt	viable, tailless, calved in Dec.
1930	Hirel, Brittany	||3hddpxxa	viable, intersex, short legs, calved Mar.
	Saint-Aubin, Côte-d'Or	||2p8w3k2c	viable, misplaced nostrils, calved Mar.
1934	Talensac, Brittany	||2k6dvm98	viable, calved in Feb.
1935	La Fare-en-Champsaur	||5cn4vajw	3rd birth in locale, calved in Mar.
	La Salesse, Cantal	||vpjan3rk	eye near ear, calved in Jan.
1937	La Feuillée, Brittany	||3by98sfb	stillborn, short legs, calved in Nov.
1938	Romorantin-Lanthenay	||4f9ye4sd	viable, dog head, calved in May
1939	Talensac, Brittany	||3mrvhb9t	viable, tailless, male, calved in Feb.
1940	Romenay, Saône-et-Loire	||4zpz2vh9	viable, dog teeth, calved in Feb.
1943	Guingamp, Brittany	||mr3mfx7j	viable, dog teeth, calved in May
1947	Leignes-sur-Fontaine	||2bdae5sf	viable, intersex, calved in March
1948	Roanne, Loire	||2zmhhzw2	bulldog head and feet, calved in May

Sources: ||2bdae5sf: *La Croix du Nord,* Lille, Mar. 23, 1947, p. 1f; ||2k6dvm98: *L'Ouest-Éclair,* Mar. 2, 1934, p. 9d; ||2p8w3k2c: *Le Bien Public,* Dijon, Mar. 23, 1930; p. 3c; ||2s3cys69: *Le Ralliement,* Belfort, Jul. 8, 1888, p. 3b; ||2zmhhzw2: *La Croix du Nord,* Lille, Jun. 1, 1948, p. 1b; ||3by98sfb: *La Dépêche de Brest,* Nov. 18, 1937, p. 6f; ||3dpfyckf: *Le Tourangeau,* Tours, Feb. 3, 1929, p. 1a; ||3hddpxxa: *L'Ouest-Éclair,* Mar. 10, 1930, p. 6b; ||3mrvhb9t: *L'Ouest-Éclair,* Feb. 22, 1939, p. 9b; ||3z8us8x4: *Le Journal,* Paris, Apr. 21, 1911, p. 6a; ||4f9ye4sd: *L'Avenir,* Tours, Jun. 5, 1938, p. 4d; ||4u4fawhk: *Le Châtillonnais et l'Auxois,* Châtillon-sur-Seine, Apr. 5, 1922, p. 3b; ||4zpz2vh9: *Le Courrier de Saône-et-Loire,* Chalon-sur-Saône, Feb. 22, 1940, p. 2e; ||59annjpv: *L'Informateur,* Montereau, Nov. 28, 1919, p. 2e; ||5cn4vajw: *L'Express de l'Est et des Vosges,* Épinal, Mar. 28, 1935, p. 6g; ||7ecwucp5: *Le Guetteur,* Saint-Quentin, Feb. 16, 1912, p. 3b; ||mr3mfx7j: *La Dépêche de Brest,* May 27, 1943, p. 1b; ||mvpy3trt: *Le Moniteur de la Haute-Vienne*, Limoges, Jan. 1, 1930, p. 3d; ||mvy8asj8: *L'Avenir de la Charente,* Ruffec, Dec. 23, 1923, p. 2d; ||vpjan3rk: *La Croix de l'Aveyron,* Rodez, Jan. 20, 1935, p. 7g; ||5xcnjzbu: *Le Creusotin,* Creusot, Mar. 10, 1926, p. 2b-c; ||yaxm4k4w: *La Lanterne,* Paris, Feb. 1, 1885, p. 2d; ||ydv9dhrt: *La Nature*, Paris, Nov. 30, 1895, p. 54a; ||yh8aw5da: *Le Journal,* Paris, Sep. 20, 1908, p. 3e.

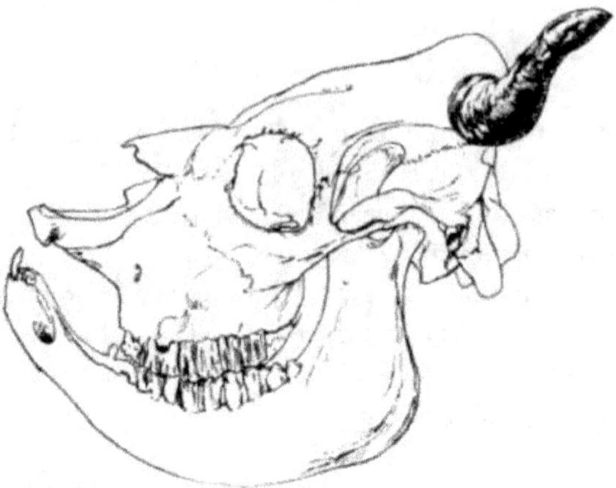

A niata skull (Galippe 1903). A normal cow skull is pictured on p. 530. Darwin (1845, p. 145) said he saw near Buenos Aires, "oxen of a very curious breed, called nãta [*sic*] or niata. … Their forehead is very short and broad, with the nasal end turned up, and the upper lip much drawn back; their lower jaws project beyond the upper and have a corresponding upward curve; hence their teeth are always exposed. … When walking they carry their heads low, on a short neck; and their hinder legs are rather longer compared with the front legs than is usual. … The breed is very *true;* and a niata bull and cow invariably produce niata calves. A niata bull with a common cow, or the reverse cross, produces offspring having an intermediate character." Niata skulls resemble (are?) those of canibovids. For a photo of such an animal, see ||yxdwzqh7.

Trézioux, Puy-de-Dôme, France. Le Bien Public (Dijon, Dec. 23, 1889, p. 3e; ||2amrv3hz) said, in translation, "A MONSTER.—A cow belonging to M. Gilbert Batisse, a landowner in the village of Tallay, in Puy-de-Dôme, has given birth to a male calf, which is properly formed but has the tail of a fox growing between its shoulders."

Saint-Longis, Sarthe, France. L'Ouest-Éclair (Rennes, Jul. 4, 1914, p. 5a; ||a5b7c924) said, in translation, "A cow belonging to M. Lambert, a farmer at Saint-Longis, has dropped a calf with a bulldog's head and dog paws. The young animal looks likely to live and is attracting attention."

San Antonio, Texas. The *San Antonio Daily Light* (Sep. 24, 1896, p. 4a; ||463dmrxt) said, "Gus Wagenfuehr, of 1524 South Presa Street, was the possessor of a freak of nature last Tuesday in the shape of a calf, which had a bulldog head and only one eye. The animal lived but a few minutes."

* A Karl Augustus Wagenfuehr (1850-1921) lived near San Antonio (||43tu9kta).

Aulnay, France. Le Petit Troyen (Troyes, May. 28, 1888, p. 3c; ||5n93rvnm) said that on May 13 a cow belonging to a M. Bequin Ferdinand, a resident of Aulnay, calved a tailless animal with the head of a bulldog, the jaw of a ruminant, and feet that were part dog and part calf. It lived three days.

A Bull-baiting by Samuel Henry Alken (1810–1894).

BULLDOG ORIGINS

Though it is often claimed that bulldogs were first bred for use in bull-baitings, depictions of such events, created by contemporary artists, do not picture dogs like the modern English bulldog. Moreover, a dog identical to an English bulldog is depicted on a bronze plaque dating to 1625 (Krehl 1901), which shows the breed existed before the time that it was supposedly bred. These facts and the fact that canibovids in photos do resemble English bulldogs suggest bulldogs were likely first produced by backcrossing canibovids to dogs and then selecting among the resulting variable offspring, a method much used by breeders, in connection with other kinds of crosses. Known hybrids are often given hyphen-assembled names indicating their parents (e.g., *wolf-dog*). This custom accounts for early occurrences of the name *bulldog* being hyphenated.

Another breed, the pug, is a likely offshoot of this cross. The pug and bulldog have similar faces, both resembling those of bulldog-type canibovids. Buffon (1791, vol. 4, p. 41) says the pug breed was first produced by crossing the bulldog with the "small Danish dog." The latter is pictured elsewhere (p. 791) in this book.

A canibovid calved in Orne, France, in 1932. In translation, the caption that accompanied this photo read, "A FREAK CALF WITH A BULLDOG HEAD. Nature sometimes produces unexpected freaks. For the last few days, a farmer in Orne has been the owner of a calf with a head that, in nearly lacking ears and having a flat muzzle, very much recalls that of a bulldog. And the body as well. Despite this conformation, the animal is quite viable." *Dimanche Illustré* (Lyon, Jul. 24, 1932, p. 1; ||wnu4m8vb).

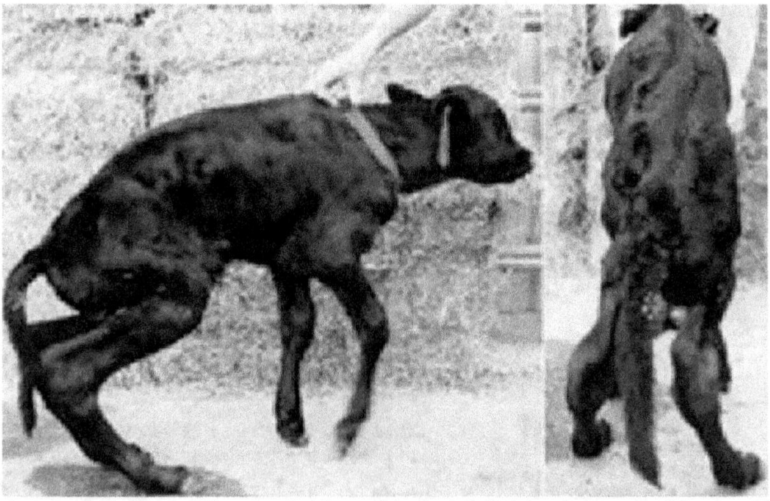

A supposed case of "contractural arachnodactyly" (also known as "fawn calf syndrome"). Such animals, calved by cattle, are canibovids because they have not only hooves and calf-like foreparts, but also dog-like hind parts and tails. (Image: Laurence Denholm).

Schenck's depiction of the Kalisz canibovid.

CONJOINED DOG-COW TWINS

Both conjoined twins and supernumerary appendages are common in canibovids About one human birth in 100,000 is conjoined. Among the 181 canibovids documented in this book, at least 22 were reportedly conjoined, which indicates the incidence is some 12,000 times the human rate.

Kalisz, Poland. Martin Cromer (1555, lib. ix, p. 238) recorded a canibovid, calved in 1269 at Kalisz, Poland, that had two dog heads and dog teeth, but seven hoofs like a calf. The larger of the two heads was in its normal place, he said, but the smaller, at the tail.[7]

Shelby, North Carolina. A *New York Sun* reporter (quoted in the *Rutherfordton Tribune,* Rutherfordton, N.C., Nov. 28, 1901, p. 4b) who had visited Shelby said that "a calf had but a few hours before made its advent there with no eyes or ears, but with the extra adornment of three tails, which it had the power to wag separately or all together."

Lyon. Le Petit Champenois (Reims, Aug. 2, 1924, p. 1f; ||5n7av8z8) said a cow of M. Jérémie Buisson in Saint-Priest, a Lyon suburb, bore an animal with four ears, four eyes, and a bulldog's head. It did not survive.

Watertown, New York. The *Gouverneur Free Press* (Gouverneur, N.Y., Apr. 26, 1911, p. 4c; ||kgvhk7w) said,

> Dr. G. H. Summerfeldt,[8] V. S. [i.e., veterinary surgeon], of Clinton Street, is today engaged in the work of mounting a monstrosity in the form of a calf the like of which was never seen in this section before. The freak was sent here by James Reed of Watertown and is a full-grown heifer calf from the breast of which appears the head and shoulders of another calf which has an almost perfectly formed pug dog head. The freak has seven legs and on one foot there are three toes, and the whole combination is without a doubt the rarest freak ever shown here.

San Pablo, California. The *Daily Alta California* (San Francisco, Jan. 16, 1871, p. 1b; ||y78twgpc) described a recently calved, viable canibovid.

> This calf, as it is called, for it appears to have no other claim to that name than it was born of a cow, belongs to Wm. Archer,[9] who resides near San Pablo, on that portion of the San Domingo Rancho known as Bay Creek. It was born last Saturday week, and as the owner states, "is a smart, lively calf," of the ordinary size. The face is flat like that of a bulldog and is ornamented with three nostrils. It has two lower jaws, distinct and separate in every particular, each of which is provided with a long, slim, pointed tongue, which the owner says it usually coils around his hand when feeding it. Under the tongue in each jaw is a tuft of hair, which appears to grow with the growth of the animal. The jaws are connected with the stomach by two throats which, like the jaws, are entirely distinct. One jaw projects prominently to one side, and in this a bold array of grinders is always visible. The owner says the animal cannot eat or partake of food without assistance. He has fed it with milk so far, and it seems to thrive admirably.

Ambérieu-en-Bugey, Ain, France. Le Cri du Peuple (Paris, Jun. 4, 1888, p. 2c-d, ||yc6c9sac) described a viable canibovid, a female with the head of a bulldog, three eyes and three nostrils, calved near Ambérieu-en-Bugey. Her two outer eyes, per the report, were dog-like, but the middle one, calf-like. She allegedly barked instead of mooed, and instead of an udder had the teats of a dog.

Tiel, Netherlands. The *Chronicon Tielense* (1789, p. 474) says a cow near Tiel had a calf with wolf hair, two heads and eight legs in 1420.

Dr. C. Garrod Ellison (1861-1939) and his wife Rachel Mahan Ellison (1865-1946).

Williamsburg, Kentucky. From the *Sun* (New York City, Mar. 28, 1897, p. 15g; ||y73m5rql): "WILLIAMSBURG, KY., March 22.—A cow belonging to Dr. Ellison of this place gave birth to a calf having eight legs, two tails, two bodies, and one head. The bodies come together at the shoulder, and the head is more in the shape of a dog's head than that of a calf."

* The existence of Dr. C. Garrod Ellison is well documented (||y3m7wp75, ||y4r6t7p3). He served two terms as mayor of Williamsburg.

Bridgeport, Connecticut. The *Bridgeport Standard* (quoted in the *Boston Evening Transcript,* Jun. 12, 1873, p. 2g; ||37zfrvwv) said,

> We saw today a very curious freak of nature … The nondescript was some six inches long, had a dog's head and a double body, or rather two bodies joined together on the breasts and forming one about halfway down and then separating into two, each part having two legs. There were also two forelegs attached to each body growing from the backs of the creature. The feet were cloven like a cow's feet, and it was wholly covered with a slight growth of very fine hair. One head, a body and a half, and eight legs formed the peculiarities of the creature. It was fished out of the water where it had probably been thrown by someone.

The Gué de Lauray canibovid twins.

Gué de Lauray, France. Puy (1655) reported a pair of conjoined twins calved June 13, 1655, at Gué de Lauray. Although his description does not seem to explicitly claim that this divided animal had canine traits, in his depiction the heads look dog-like.

McAlevy's Fort, Pennsylvania. The *Altoona Times* (Altoona, Pa., Jun. 22, 1889, p. 2; ||y5ovz65x) said, "Recently a heifer belonging to A. D. McCall, of McAlevy's Fort, gave birth to a calf that had two heads fully developed … Just back of the front shoulders and in the center of the vertebrae column, a third leg pointed up instead of down, and instead of a hoof … it was ornamented with a fully developed dog's paw. From the place whence this leg protruded from the back there were two distinct back bones, two sets of hips, two tails and four legs."

Paris. The *Journal de Seine-et-Marne* (Meaux, Sep. 6, 1903, p. 1b-c; ||3svu443f) said a cow at Gournay, a Paris suburb, bore a calf with a bulldog head, double above its three eyes. It died soon after birth.

Le Pouliguen, France. Le Guérandais (Guérande, Feb. 18, 1922, p. 2d; ||yc5w4je3) said a cow at Le Pouliguen bore a calf with a deformed head and six legs and that two of them, on one shoulder, were like a dog's.

Skálholt, Iceland. A stillborn two-headed calf with canine lower jaws and nine or ten legs was born at Skálholt in 1593 (Harris 1914, p. 5).

* Compare the Skálholt canibovid to the Rumilly CANICAPRID (p. 472).

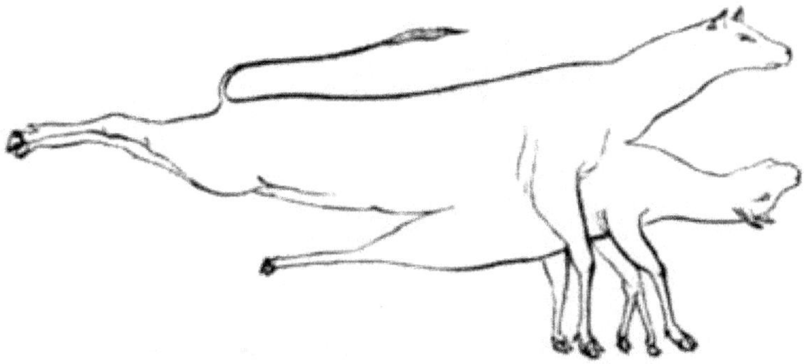

The Sharon canibovid twins (||y7m222ls). The report said the smaller twin was cyclopean and had a bulldog head, but this crude sketch fails to show it.

Sharon, Massachusetts. A report from Sharon published in the *Boston Medical and Surgical Journal* (Anonymous 1862, 66, p. 26; ||y7m222ls) said, "MESSRS. EDITORS.—The following case of 'freak of nature' in a calf, occurred in this place a few weeks since … The calf was double, or partially so, one being natural and of the usual size and form, the other imperfect and diminished in size. They were connected at the breast, by a union of about fifteen inches; the walls of the chest being perfectly united, forming but one cavity. Their necks were of the same length, but the head of the smaller (which was cyclopean), was somewhat shorter and broader, more resembling that of a bull-dog; the eye was imperfect, situated in the center of the forehead, and opened perpendicularly. One of the forelegs was normal in its form, the other double from the knee down, having one hoof on each bifurcation. These legs were reversed in their position corresponding in their direction with those of the other calf. The hind part extended about to the umbilicus of the other, having but one leg, diminished in length, and projecting directly back, and ending with one hoof. There was no anus, nor organs to distinguish sex. The accompanying rough outline, sketched from memory, may aid in describing the appearance of the animal."

Johnson County, Missouri. The *Butler Weekly Times* (Butler, Mo., Jul. 1, 1915, p. 5b; ||hbocasf) said, "The HOLDEN ENTERPRISE [published in Holden, Mo., a town in Johnson Co.] is responsible for the latest freak story. According to that paper a Johnson County cow last week gave birth to a calf with two heads, one of them a dog's head."

The Davos canibovid (||yw9chtd6).

Davos, Switzerland. A manuscript in the Wickiana collection (||yw9chtd6), accompanied by the picture above, claims that on May 6, 1571, a cow at Davos calved a canibovid with not only a calf head but also a dog head with three ears and four eyes.

Christian Gabriel Fischer

Stuttgart, Germany. Christian Gabriel Fischer (1743, p. 425) reported a canibovid with one head like a calf's and one like a dog's. He said it was calved at Stuttgart and had one doglike forepaw.

Richmond, New South Wales. The *Windsor and Richmond Gazette* (Oct. 13, 1894, p. 4a; ||y6n2zjq7) said, "A cow, the property of the H. A. College [Hawkesbury Agricultural College at Richmond], gave birth to a calf the other day having two heads,

> a double-jointed body, eight legs. If it had only lived, what sport the boys would have had with it. One of the heads of the calf was like a bulldog's.

La Flèche, Sarthe, France. The *Union de la Sarthe* (quoted in *Le Temps,* Paris, May 7, 1863, p. 2c-d; ||3hx2mdd3, also see: ||mr26pdcw) said a heifer belonging to a M. Roquet, a resident of La Flèche, had calved a repeatedly witnessed animal with a dog's head atop a calf's head, both horned, and that this creature, born April 5, 1862, was alive more than a year later. Supposedly, a dog tail, with dog hind parts beneath it, stood between the eyes of the dog's head.

MORE REPORTS

Elizabeth City, North Carolina. The Elizabeth City *Carolinian* (quoted in the *Progressive Farmer,* Winston, N.C., Nov. 8, 1892, p. 3a; ||jp5fyvl) said, "A live calf with a bulldog's head—a wonderful freak of nature—was one of the interesting curiosities at the Fair last week. The entire head had all the formations of the bulldog."

A calf with what Herzberg (1825, p. 5) said was the head of a dog. Six digits are visible on its right forefoot, but the left appears to be a cloven hoof. In this image, the face has been dissected.

Invercargill, New Zealand. The *Morning Herald and Miners' Advocate* (Newcastle, NSW, Apr. 30, 1896, p. 6a; ||y3o2mhna): "A monstrosity in the shape of a calf that has the head, tail, and skin of a dog, and the feet and teeth of its own kind, has been stuffed, and is now on exhibition in Invercargill (N.Z.)."

Palatka, Florida. The *Omaha Daily Bee* (Omaha, Neb., May 5, 1889, p. 1; ||y8xjyjxs): "A monstrosity of a calf with a dog's head was born at Palatka, Fla."

Memphis, Indiana. From the *Daily Star* (Cincinnati, O., Jul. 15, 1879, p. 1f; ||zujst4l): "JEFFERSONVILLE,[10] IND., July 15.—A calf with a genuine dog's head, recently born in this county, is still alive and growing.

> It has the voice of a dog and keeps up such an incessant barking that the neighbors are sorry it ever came around. It is owned by a farmer near Memphis.

Verviers, Belgium. Le Petit Haut-Marnais (Chaumont, Apr. 21, 1928, p. 3d; ||bd7t26kc) said, in translation, "At the Verviers abattoir, they killed a freak calf that had a head strangely like the head of a bulldog. Nothing was lacking, neither the ears, nor the fangs, nor the pendulous lips."

Licking River, Kentucky. From the *Grand River Times* (Grand Haven, Michigan, Jun. 25, 1856, p. 2c; ||yb6oyae9): "There is on exhibition in Cincinnati one of the greatest monstrosities ever known. It is a half calf half bulldog. It is but a few days old, born a few miles up Licking River, on the Kentucky side. Its head is round like a bulldog, its underjaw, the longest, its teeth pointed, and so irregular that it never can eat grass. It is short and bow-legged, has short, round ears, and a tail like a dog. It sucks its mother (which is a small two-year-old heifer) sitting on its haunches. It is constantly snuffling and smelling round like a dog after meat. It has hoofs, but no signs of horns."

Chivres, Côte-d'Or, France. Le Bien public (Dijon, Feb. 17, 1926, p. 4c; ||4p7rp9kx) said that a cow calved at Chivres a 10-lb. dog-headed animal she refused to nurse, that it was like a calf in other respects and soon died.

Ireland. The *St. Louis Republic* (quoted in the *United Opinion,* Bradford, Vt., Jan. 20, 1893, p. 2e; ||y7svdmu8) said,

> Mule footed hogs, "wonderful malformed sheep," live mastodons and sea serpents are all thrown in the shade by the reports which come from Ireland of a "bulldog" or "dragon" calf. The creature is the offspring of a 2-year-old Jersey heifer belonging to Colonel Doud, of Marcus Landing. A front view of the monstrosity, when its body is hidden from sight, instantly suggests the presence of a good-sized bulldog. It has the drooping red lips, the cropped ears and bony head to perfection. The fore feet seem to be a cross between claws, paws and hoofs. The hind ones are calf hoofs all right enough, but they are turned the wrong way, the result being that the creature appears to have the tail on the wrong end.
>
> A side view of the head, especially when the calf's mouth is held or propped open, reminds one of the pictures of the old oriental dragons, the enormous lidless, staring eyes adding to the resemblance. Some accounts of the freak say that it has the "wool of a dog on the forequarters and the regulation Jersey hair behind." There is no noticeable difference in either the color or texture of the beast's hairy covering. One odd thing is a double row of what are said to be "real dragon scales" along the belly.

Wallsee, Austria. Welt Blatt (Vienna, Jun. 24, 1881, p. 3c; ||y3netbot) said, in translation, "Today a live calf with a bulldog's head was purchased for the Veterinary Institute. It weighed 35 kg and came from Wallsee."[11]

St. Georgen, Austria. The *Neue Warte am Inn* (Jan. 16, 1909) said, in translation, "In St. Georgen, near Tollet, a heifer recently gave birth to an animal that was half calf, but more like a dog. The head of this monstrosity genuinely resembled that of a bulldog, only the lower jaw was defective so that the large tongue lolled. Also, the relatively short forelimbs were attached too far back in comparison to their natural position. The young calf, which was born dead, had to be cut in two, due to difficulties during delivery. When the rear half of the calf could not be extracted, the mother had to be slaughtered. The rear portion was more like that of an ordinary calf, but the legs were too short. Farmers will certainly find this most interesting news."

Beek, Netherlands. From the *Limburgsch Dagblad* (Sittard, Limburg, Sep. 9, 1938, p. 2b; ||yb8lrjdu), given here in English translation:

"Thursday, on the farm of Mr. G. Vroemen-Voncken on Burgemeester Janssenstraat, a calf was born that attracted many viewers on account of its monstrous form. This was a singular freak of nature. The unusually thick, blunt head of the animal has an extremely striking resemblance to that of a bulldog. Its short legs are no less remarkable, being much like those of a dog, and yet having normally shaped hooves. The other parts are formed like those of a normal calf but, overall, the animal is strongly reminiscent of a dog. Due to its monstrous form the animal was stillborn but thereby loses nothing of its interest. The creature has been preserved so that this wonderful freak of nature will remain available to the eyes of science and the world."

Armidale, New South Wales. The *Armidale Express and New England General Advertiser* (Apr. 8, 1881, p. 7; ||y32sld6s) said, "An extraordinary phenomenon, and one which seems to be without example in the annals of science, has just taken place in our town. A cow has given birth, three days ago, to a creature half dog and half calf. The head is that of a bulldog, with its cock-up nose, broad forehead, and ears short and straight. At the same time, the organs of mastication are imperfect. The greater number of the teeth are those of the dog, but those are large at the base and are only like a calf's at the back of the mouth. The hair is that of the calf, viz., tawny-red, but the hinder part of the creature is like a bulldog, short tail, and smooth legs. The foot resembles the ruminating animal, that is to say, the nails being separated by a biflex sinew, but they spread out something like a paw. What is most remarkable about the monstrosity is that it is a hermaphrodite. The female sex predominates and approaches the canine tribe. The teats are almost perfect. The creature seems healthy, and drinks milk with ease."

Wolcott, New York. From the *Pulaski Democrat* (Pulaski, N.Y., May 25, 1871, p. 2e; ||y9kfp9t4): "Wolcott has a 'freak of nature' in the form of a calf with a dog's head, or rather—say half calf half dog—and you have it. The head is dog, the ears are dog, the legs are dog; the body is calf, the hair is calf, the feet are calf; but the tail, the wonderful tail, is like Alva Rude's last egg, decidedly in doubt."

Bouée, Loire-Atlantique, France. Journal d'Indre-et-Loire (Tours, Aug. 26, 1899, p. 3; ||3eermkuv) said, in translation, "A most curious freak, which deserves to be reported, has been produced at Bouée (Loire-Inférieure). A cow belonging to M. Jean Belin, a farmer at Rochetannerie, gave birth last Friday to a strangely formed calf. It has the head and hindquarters of a bulldog, and the legs are those of a dog, with paws. Only the body is like that of a calf. Surprisingly, this animal is extremely vigorous and nurses admirably well, which it does while seated on its hind parts like a dog. Its cries are like those of a dog."

Pueblo, Colorado. Among various unusual animals exhibited at Pueblo, a reporter said he saw alive "a cow with a head that looked exactly like a bulldog's" (*Gilpin Observer,* Central City, Colo., Sep. 17, 1908, p. 4c; ||2p88nfpy).

NOTES

1. *Nova acta eruditorum* (1760, p. 409; ||hsxwwuc). Translation mine.
2. The report's "Moransburg" is here spelled *Moranburg.*
3. Appendage deficiencies occur at elevated rates in telenothians.
4. Asymmetric ears are reported for HOMINOSUIDS and a CANISUID (see p. 1210).
5. The Le Locle ursidequid supposedly had claws on the bottom of its hooves.
6. Vallisneri (1733, vol. 3, p. 453a). Translation mine.
7. Schenk (1609, p. 109, fig. 81), Liceti (1665, p. 18; ||y4fjq3zg), Irenäus (1584, citing "*Chron. Polon.,* lib. 3, cap. 56") and Lycosthenes (1557, p. 441) all cited Cromer's report. Though Cromer himself does not refer to the orientation of the legs, both Schenk and Lycosthenes pictured it with four legs turned toward one end of the animal and three turned toward the other.
8. In 1910, a George H. Summerfeldt, a veterinary, 52, lived in Gouverneur, and James Reed, 46, a farmer, lived in Brownsville, near Watertown (||y56tmt9y).
9. A William Archer, aged 43, lived in Contra Costa Co. in 1880 (||y4l4d2pw).
10. Jeffersonville is a town on the southern border of Indiana. Memphis is an unincorporated town, lying north of Jeffersonville in the same county (i.e., Clark).
11. The Veterinary Institute, Vienna (Thierarznei-Institut) was the forerunner of the University of Veterinary Medicine, Vienna (Veterinärmedizinische Universität Wien). Its Veterinary Pathological Museum houses the largest collection of veterinary pathological specimens in the world. The specimen may still be in their holdings. Wallsee-Sindelburg is a town in Amstetten, Niederösterreich, Austria.
12. *Colophon:* The Kalisz canibovid (Lycosthenes 1557, p. 441).

The Chinese canidovid (left), yeaned by the ewe at right.

CANIDOVIDS (DOG-SHEEP)

> *Monstrous lambs with the face of a dog are often found, the result of the mother having suffered some fright.*
> —ALBRECHT VON HALLER, *Bibliotheca anatomica*[1]

It is well known that the sort of mating that would be required to produce a dog-sheep hybrid does occur on farms. Various YouTube videos show dogs mounting sheep. Even the ancient Mesopotamians knew that such mating does sometimes occur (Freedman 2017, p. 6).

China. Photos on the Internet show a canidovid reportedly lambed on a sheep farm in China in 2011. The owner, a farmer named Liu Naiying, said he found it soon after birth. He was amazed, he said, to see it looked so much like a dog, despite being yeaned by a sheep and having a sheep-like fleece.

"I was herding the sheep and saw a sheep licking her newborn lamb on the grass," Mr. Liu said. "When I went up to check, I was shocked because it looked so weird, like a cross between a sheep and a dog. I was a bit frightened, as I've been raising sheep for 20 years and had never seen such a creature."

Oysterville, Washington. From the *Washington Standard* (Olympia, Mar. 16, 1888, p. 1f; ||z52ybqm): "A strange freak of nature presented itself last week among the sheep of R. H. Espy of Oysterville. A lamb was born bearing the face and head of a dog and the balance of it a sheep. The PACIFIC JOURNAL says it bore strong resemblance to a large Newfoundland dog, with eyes, nose, ears and the shape of his head. It only lived two weeks, but with care it might have grown to be of full size and proved a valuable curiosity."

* Robert Hamilton Espy (1826-1918) lived many years in Oysterville (||yxthheh7).

Summerfield, North Carolina. The *Asheville Daily Citizen* (Asheville, N.C., Mar. 2, 1893, p. 2b; ||gor8j5z) said, "The Summerfield correspondent of the WINSTON SENTINEL says that Mr. George Smith showed him a monstrosity that had a setter dog's head and neck, and a sheep's body with its hams covered with hair. It was accidentally killed."

* A George J. Smith (b. 1824), a farmer, lived in Summerfield in 1900 (||y3pby88d).

Tarter, Kentucky. From the *Adair County News* (Columbia, Ky., Feb. 20, 1901, p. 3d; ||hrmpm3y): "Mr. W. H. Wheat, who lives at Tarter, Adair County, is the owner of a fine white ewe which gave birth to twins last Sunday morning, one being a well-formed ewe lamb, the other a complete hybrid, the head, ears, tail, skin and hair precisely like that of a hound puppy; the legs and feet that of a lamb. Hundreds of persons have called to see the curiosity."

* A William Huffman Wheat (1884-1931) lived in Adair County in 1901 (||y29p5mt9).

Buenos Aires. The *Boletín Mercantil de Puerto Rico* (San Juan, P.R., Mar. 28, 1890, p. 3c; ||59c7hfnz) said, in translation, "FREAK.—A Buenos Aires newspaper says that on the farm of Mr. Juan López, a lamb was born recently with five legs and the head of a dog. It's blind, because it has, in place of eyes, two small horns in its eye sockets. The ears are long and shaped as in the canine race, but they are on its neck. ... Its getting very fat and promises to live a long time."

Cincinnati, Ohio. The *Cincinnati Daily Star* (Jan. 8, 1878, p. 1f; ||grxt2nk) stated that "a lamb with a double jaw and the countenance of a bull-dog" was on display in a local saloon.

* The report says the saloon owner was named John Birnbaum. A resident of that name appears in an 1880 census record for Cincinnati (||y38wd3ga).

The Chinese canidovid (see p. 465).

Newbury, Berkshire. From the *Reading Mercury* (quoted in the *Morning Post,* London, Jan. 17, 1838, p. 6a; ||bdfnsb5h): "A fine grown dog was exhibited at the White Horse Cellar, Newbury, on Thursday last, covered with a most extraordinary growth of wool, which resembles the fleece of a sheep, is of four years' growth, and lays on the dog in flakes of great weight; the proprietor has been offered a considerable sum for the animal, which he refuses to take."

Gap Creek, North Carolina. From the *Progressive Farmer* (Winston, N.C., Mar. 6, 1894, p. 3b; ||l817f4j). "Mr. A. D. Cowles of Gap Creek, Ashe County, has a freak of nature in the way of a yew [*sic*] lamb sheep, ten months old, which has three well-developed feet and legs. The left foreleg is shorter than the other and has a well-formed dog's foot with claws."

* An Arthur Duvall Cowles (1845-1902) lived at Gap Creek in 1894 (||y3hppvwx).
* The descriptions given in the Oysterville, Summerfield, Tarter, Buenos Aires and Gap Creek reports, all of which involve animals yeaned by ewes, recall the descriptions typically given of canibovids, in which dog-like characteristics tend to be concentrated in the foreparts, especially the face and head, while bovine traits predominate rearward. Canidovids, too, tend to show maternal traits toward the rear of their bodies.

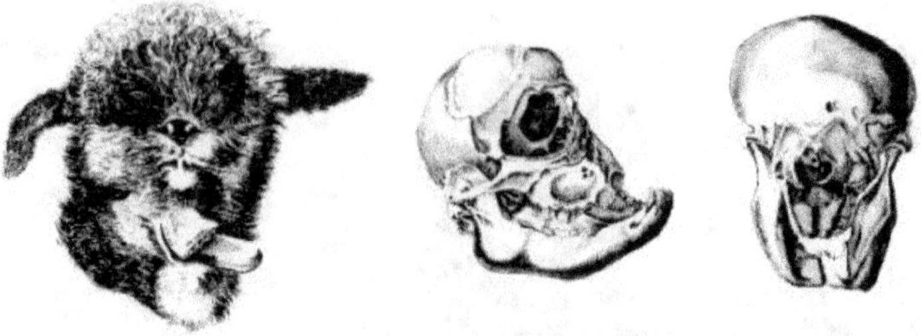

The face and skull of a cyclopean canidovid (Albrecht 1744, Tab. 9, figs. 1, 2 and 3; ||2uu8czw4). This animal was lambed in March 1742 at Rodach in the Duchy of Coburg. It died because it could not nurse. The irides were double.

Newcomb, New York. The *Enterprise and Vermonter* (Vergennes, Vt., Nov. 13, 1924, p. 7; ||h2kh7syy) said,

> One of the rarest animal freaks on record in the Adirondacks was recently exhibited at North Creek. It appeared to be a cross between a fox and sheep.
>
> As Harry Smith of North Creek and Harry Stanley of Warrensburg were motoring at night from Newcomb south, the freak appeared in the road before them. The car was slowed down, and the animal ran before it for some distance, then stepped into the ditch and stopped. The men jumped from the car and killed it.
>
> Upon examination the strange creature was found to have the paws of a fox, but black in color. The head also resembled that of a fox, but from the nose to the back of the head was jet black. The tail was shaped like that of a sheep, but larger. The body was covered with silklike wool nearly two inches long and dirty grey in color.
>
> When the men found what they had, they greatly regretted killing the creature, which they might easily have taken alive, as it offered no resistance. The freak would have been worth a lot of money exhibited alive. … The animal will be mounted and may be sent to the state museum at Albany.

Bendigo, Victoria. From the *Bendigo Advertiser* (quoted in the *McIvor Times* (Heathcote, Vic., Oct. 17, 1901, p. 2c; ||mefn2hz): "A FREAK OF NATURE.—There has been forwarded to Mr. G. A. Thomson, registrar of the School of Mines and Museum, by Mr. G. Groves, of Williamson Street … the dead carcase of an animal, which appears to be half sheep and half dog. The curious object was the foetus of a ewe that was recently killed by Mr. Scantleton, a butcher at Lake Charm. The carcase is nearly 2 ft long, and generally bears the resemblance to a lamb, being covered for the most part with wool, but has hindquarters and legs the shape of a dog, these portions being covered with hair, but the feet, or rather hoofs, are those of a sheep. The animal has no head, but an opening in the neck, which is terminated by a formation like two ears. The object forms an interesting study to the zoologist."

Pichi Richi Pass, South Australia. From the *Transcontinental* (Port Augusta, SA, Apr. 13, 1945, p. 4; ||lfgpaox):

> Mrs. H. C. French, of Pichi Richi Pass, possesses an aged ewe which has capped a production record by giving birth to a freak of unusual interest.
>
> Until this month the ewe had a highly respectable history.
>
> In April, 1943, she produced triplets (2 ewes and 1 ram lamb); February of 1944 saw her the proud mother of a beautiful pair of twins (a ewe and a ram). All were normal.
>
> Now fate has handed the bewildered animal a freak lamb. It was black with a pug dog's head, small ears, but only one eye in the middle of the forehead.[2] The lamb … was born dead.

Mundi Mundi Station, New South Wales. From the *Barrier Miner* (Broken Hill, NSW, Sep. 28, 1893, p. 2d; ||leztjmk): "From Mr. James Gemmell, manager of Mundi Mundi [sheep] Station, comes the account of a peculiar freak of nature in connection with a young ewe lamb, which was brought to light during shearing. The peculiarity is to be seen in one hind leg, from below the hock, from which point it develops into the natural paw of a dog. The wool has grown a few inches below the hock. Pure dingo hair succeeds, the paw being shaped and marked and having the three nails or claws as on a dog. The little thing is strong and lively and is a great pet with all. It is the intention of the manager to forward the lamb to the Adelaide Zoological Gardens."

Wimpfen, Baden-Württemberg. Johann Praetorius (1677, p. 297) said a ewe at Wimpfen yeaned a wolf with cloven hooves in 1645.

Stoverton, Ohio. From the *Springfield Daily Republic* (Springfield, O., Jul. 12, 1888, p. 2c; ||zyukzm8): "STOVERTON, July 12.—A most remarkable freak of nature is reported from the farm of George Sevingle, who lives near this place. He has among a flock of sheep a lamb over two months old, which has the nose and feet of a dog, but is in all other respects a sheep. Instead of having the ordinary teeth of the sheep kind, the animal's mouth, both upper and lower jaws, is filled with long and sharp wolf-like fangs, rendering it extremely difficult to pick grass like the other sheep. Consequently, it prefers to chew and masticate twigs and shrubs. It is regarded as one of the greatest curiosities ever seen in this neighborhood."

Nassiet, Landes, France. La Chalosse (quoted in *Le Mercure d'Orthez et des Basses-Pyrénées,* Jan. 11, 1844, p. 2b; ||5n875b2n) said a ewe at Nassiet had given birth to a viable animal like a wolf from the neck forward, with a wolf's head, hair, and teeth. From the shoulders back it was like a sheep except that its hind feet were like a wolf's, including the claws.

Huntingdon, Pennsylvania. The *Huntingdon Republican* (quoted in the *Nashville Union and American,* Apr. 5, 1871, p. 1b; ||jdbwc8m) said, "We have received two feet of a lamb from the farm of William Humphrey,[3] of this county, very much like the feet of a dog."

Lake Belt, Minnesota. The *Worthington Advance* (Worthington, Minn., Apr. 25, 1878, p. 1; ||yck2pkxj) said, "In the flock of Mr. Bottleson, of Lake Belt, … one lamb was born with the exact counterpart of a dog's foot in place of a natural one on the left hind leg."

York, England. The *Yorkshire Gazette* (York, Apr. 5, 1834, p. 2d) said, "A lamb, which was dropped in Castle Howard park, was brought to our market this morning, by a dealer in skins which a singular freak of nature has invested with a fox's face and four ears." It survived birth.

Miles City, Montana. From the *Anaconda Standard* (Anaconda, Mt., May 16, 1909, p. 3f; ||3u3wecwd): "You can depend on old Miles City to produce something unusual almost any time," said a resident of that city … "and the latest attraction we have to show is a seven-legged lamb. This freak was discovered on the old A. B. Clarke ranch, on the Tongue River, a few days ago. Four of the legs were in their normal position, two were about halfway of the body and the other was close to the tail. Instead of hoofs upon two of the pedal extremities they were shaped like the paws of a dog. The monstrosity has been carefully preserved, and it will be placed in Miles City's museum of curiosities." The *Billings Gazette* (May 14, 1909, p. 1f; ||yc4dvacm) said this animal would likely have lived but for a fatal fall.

Mount Eber, Kentucky. From the *Evening Bulletin* (Maysville, Ky., Jun. 2, 1893, p. 1e; ||grkphko): "MOUNT EBER, KY., June 2.—Farmer John Rollins has in his possession a genuine freak of nature that beats all curious things ever seen in this section. It is a lamb with the exact head of a dog and is as black as coal. Its forefeet are those of a dog, while the hind ones are like a sheep's. It is covered with fine silken hair halfway back and the balance is wool."

Metropolis, Illinois. The *Wyandot Pioneer* (Upper Sandusky, O., Feb. 11, 1859, p. 1f; ||z52ybqm): "Mr. Z. P. Posey, who has a farm near Metropolis, Illinois, gives us a description of a curious animal which he is raising at present. He writes: 'A very fine Merino ewe of mine brought forth, on the first of November, a curious creature which proves to be half dog and half sheep. Its head, eyes, teeth and ears are precisely those of a dog, while the rest of the animal is a sheep. It sucks like other lambs, and follows its mother about, barking like a dog. Its fore part is black as a crow, while its hinder parts are white and woolly like a sheep. Hundreds of people hereabouts have visited my farm to see this curious freak of nature.'"

Rochefort, France. Jean Cochon du Puy (1719), a physician practicing in Rochefort, said a ewe yeaned a nondescript with the pelt of "a wolf or mastiff" in 1715. It died at birth, he said, as the single opening in its face did not connect with the lungs. It had neither eyes nor nose.

Lipno, Poland. Rzaczynski (1721, p. 355) said an animal lambed at Lipno in 1705 had a dog's tail and head, and eyes on its chest, the face having another pair of eyes.

Leipzig, Germany. Wendler (1755, p. 604) said that in 1751, 1752 and 1753, a single flock yeaned "3, 4, or more lambs" with dog's heads.

Brandenburg, Germany. Sebald (1655, p. 366) recorded a headless, conjoined canidovid, with a double neck and hind parts like a fox, yeaned by a ewe at Brandenburg on April 16, 1640.

Rocky Ford, Colorado. The *Pueblo Chieftain* (*Pueblo, Colo.,* quoted in the *Daily Tribune,* Florence, Colo., Jul. 11, 1904, p. 3; ||yneejswm) said, "A passenger who arrived on Santa Fe train No. 603 yesterday afternoon brought to Pueblo one of the most peculiar freaks of nature ever seen around the depot. The freak was a half coyote and half sheep, the rear portions of the animal being like that of a coyote, while the front section is exactly like a sheep with horns. It is claimed the freak was captured in what is known as the sand hills near Rocky Ford. Mr. Trowbridge, who owns the freak, states that he saw two of them go into a hole in one of the sand dunes and as he could see but the rear portion he thought that some wolf had captured a sheep and taken it to the hole, on digging down, however, he came across two of the freaks and captured one of them, the other escaping. Mr Trowbridge states that the half sheep and half coyote was the fleetest little animal he has ever seen. This curiosity is about four or five months old and will be sent to the [St. Louis] world's fair to be placed on exhibition."

Screenshot from a video (||yy6pltgn). Present in 2007 at the Kitale Nature Conservancy, Kenya, this animal was described in the video as a "sheep with the back looking like a hyena." Such a composite can be termed a *hyaenovid*. The narrator said it was yeaned in Kenya in the upper Rift Valley. The Spotted Hyena (*Crocuta crocuta*) is present in that locale.

NOTES

1. Haller (1777, vol. 2, p. 437). Translation mine. Original Latin: "Monstra agnina frequenter reperiri, ore canino, a terrore matris nata."
2. Cyclopia is common in telenothians.
3. A William Humphrey was born (Apr. 2, 1812) at Huntingdon and lived nearby all his life (||y6wao354). For the story of Jacob's cattle, see Genesis 30:37.
4. A case similar to the Miles City canidovid is the Rumilly CANICAPRID (see p. 472).
5. *Colophon:* An ostensible cyclopean canidovid (Vrolik 1849, tabula 56). Vrolik received this specimen from the director of the Dutch Veterinary School at Utrecht.

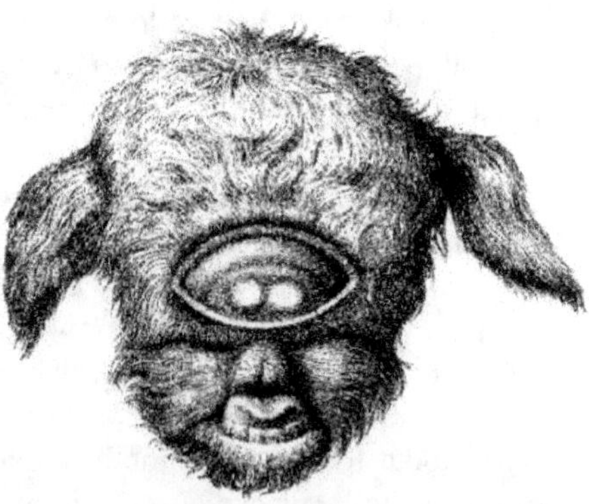

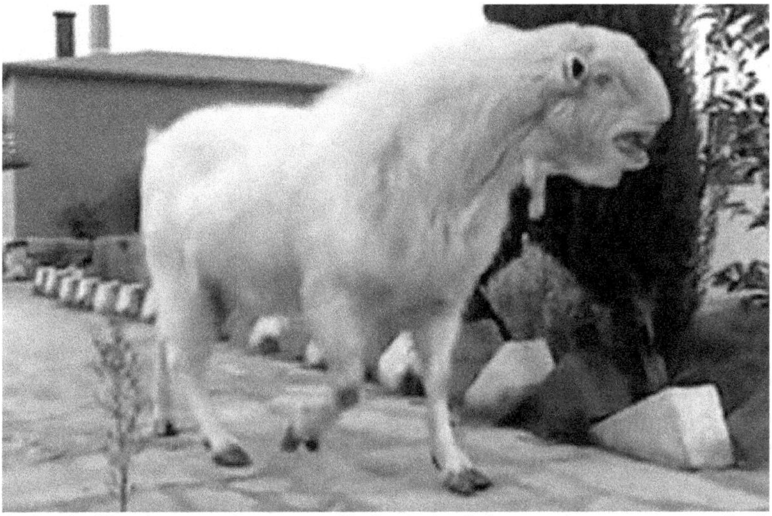

Animals like the one shown are likely derived from a dog-goat cross. Such creatures have been described as a breed, the Damascus goat.[1] The image is screenshot of a video (||34zfvcef), in which this animal can be seen to shake itself like a dog, behavior very unlike a goat.

CANICAPRIDS (DOG-GOATS)

Except ye see signs and wonders, ye will not believe.
—JOHN 4:48

Any animal combining the traits of a dog and a goat will be termed a *canicaprid*. That male dogs will mount nannies is documented on (e.g., ||y2xrmuum). The reverse mating, billy × bitch, also occurs, seemingly with less frequency. Goats sometimes nurse puppies (||y5sxwrkl), as do dogs, kids (||y4e5z7xl). Such fostering of young animals can result in sexual imprinting, that is, the tendency of the fostered animal, when mature, to seek mates of the same kind as its foster parent.

Reports about canicaprids are rarer than those for CANIBOVIDS, perhaps because cattle are far more common in developed countries than goats. Cattle in the U.S. outnumber goats by about 30 to 1. Indeed, the frequency with which canicaprids are reported seems to be even lower than that for CANIDOVIDS, and the reports are generally more antique. But given that the existence of CANIBOVIDS is well-documented by innumerable reports, the less numerous reports for canicaprids can be taken quite seriously despite their paucity.

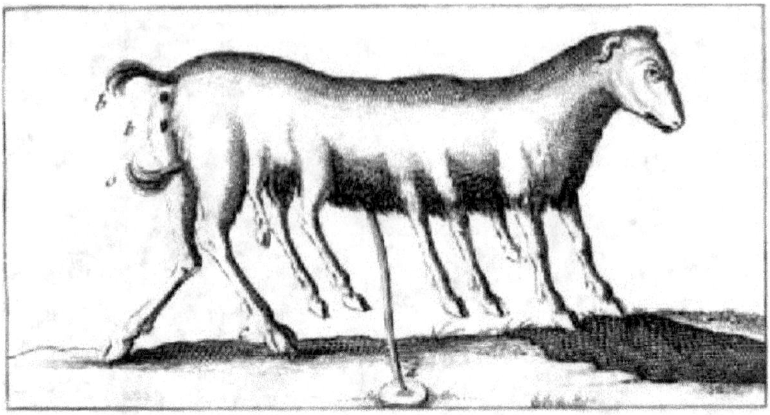

The Rumilly canicaprid. The rearmost right foot, labeled C, is the one the Merindol said was shaped like a dog paw. Backward legs have been reported for certain other telenothians.

Rumilly, France. In translation, the *Journal des Sçavans* (Jun. 14, 1683, p. 197; ||yaq6ze3u), the first European academic journal, said,

> *Extract from a letter from Rumilly,*[2] *Savoy, dated May the 19th, to the editor of the Journal, by Pierre Merindol, Professor of Philosophy.*
>
> On the 13th of this month [May 1683], a woman in a village near here, delivered a goat of a monstrous kid, of which I send you a drawing. The long umbilical cord, labeled A in the drawing, appeared first, and this woman, whom I questioned myself, said she forcibly extracted the fetus during delivery, but in doing so actually killed it. The end of the cord was filled with a matter that inflated it like a balloon. The kid had the muzzle of a dog, as well as one foot of a dog (labeled C in the drawing). The foot on the left, across from this dog paw, was like that of a goat, but reversed, as if to walk backward. These two legs were accompanied by eight others, all but two of which were shorter than the rest. The rest of the body was for the most part no different from that of other kids, except in that it was a little larger, and the hair was more like a dog's than an ordinary kid's.[3] [my translation]
>
> * The Miles City CANIDOVID (p. 468) resembled the Rumilly canicaprid.

Pitztal, Austria. Welt Blatt (Vienna, Feb. 22, 1910, p. 8b; ||y64w4hfn) said, in translation, "A MONSTROSITY. A farmer, Ehrenreich Sauteler, near Pitztal, in Tyrol, has delivered a monstrosity from a goat. Instead of having the form of a kid, it was like a dog, but it had hooves on its feet. The mother had to be slaughtered, and the monster also died."

Las Vegas, New Mexico. The *Las Vegas Optic* (quoted in the *Albuquerque Daily Citizen,* Feb. 20, 1902, p. 5; ||h67fqyd) said,

> Half dog and half goat. You could swear to that. So could anyone who saw the thing on the west side streets this morning as it limped, goat fashion, behind a rural Mexican's outfit.
>
> The creature was a wonder for combination. It had a goat's body, but a dog's head. The head had a dog's eyes and a dog's ears, but under the ears, from a lank goat throat, fell long strands of angora goat silk of unmistakable quality. The feet were those of neither dog nor goat. They were like the feet of a sheep. About the hocks fell long shaggy bunches of fine soft silk and larger, shaggier strands of a finer texture adorned the body and the hind legs. The tail was a white fluffed stump, more like no tail at all than anything else. The dog-goat might as well have been born a rabbit so far as caudal appendage is concerned. Except for the head, not a person alive would have guessed that this goat was a dog. A number of dogs came up to make friends, but they invariably decided that the acquaintance would not be acceptable. The poor beast evidently did not understand one word of dog talk anyway. An accident to its left hip caused it to limp.

Hartford City, Indiana. From the *Indianapolis Journal* (Dec. 13, 1903, p. 4; ||h67fqyd):

> HARTFORD CITY, IND., Dec. 12—A dog covered with the silky fleece of an Angora goat, from which it differs only in color, is owned by Hosh Mills, a bartender of this city. The little animal is a most beautiful creature and has many unusual characteristics.
>
> The mother of the dog is an Irish setter, and the sire is said to be one of the most beautiful specimens of the goat family to be found. A massive head of the St. Bernard adorns a short, thick body not over eighteen inches long, and its height does not exceed six inches. Its legs are short and the feet unusually large, while the stubby tail points skyward. The little freak is one of a litter of ten pups, and the remainder possessed no features similar to this one. It seemed more "pert" from the start than its little brothers and sisters, and the mother seemed to take special pride in caressing it in preference to them, especially when anyone is around.
>
> The cross in breeding is thought to be one of the most remarkable ever heard of in this country. The hybrid is covered with a fleece that is silky in texture and so thick that one is unable to pull it apart far enough to see the skin.
>
> Mills prizes the animal very highly, and, though numerous offers have been made for it, he has refused to part with it. So far as known, this is the only cross of this kind on record, and stockmen all over the country have become interested in the matter.

Buffalo, New York. The *Buffalo Commercial* (Feb. 22, 1887, Mar. 8, p. 3e; ||yyhay9uy) said, "The dog Jack, owned by the Buffalo Hat Manufacturing Company, 77 Seneca Street, attracts much attention. This animal has been pronounced to be one of the strangest freaks of nature. He was born December 10, 1885, in New Zealand, his father being a Cashmere goat, and his mother a New Zealand bitch, which accounts for his being half goat and half dog. His mother is now owned by a very extensive cattle merchant in New Zealand, and his father is chewing grass in the same country."

Hermsdorf, Germany. The *Teplitz-Schönauer Nachrichten* (May 1, 1875, p. 4; ||4f26mzn7) reported a creature with the body of a dog and the hooves of a goat born at Hermsdorf. The mother was not identified.

Mussey, France. The *Gazette de France* (May 6, 1768, p. 156a; ||k2ndjra; also see: ||mtho6oo) said, in translation, "We have from Joinville, attested by a procès-verbal [i.e., a detailed, authenticated account drawn up by a public official], the following particulars. A goat belonging to Maurice Feron, a vintner at Mussey,[4] Champagne, has borne an extraordinary animal, now two weeks old. Instead of the small, pointed ears, slender, bent legs and short face of a kid, this animal has pendulous ears about six inches long and its head resembles that of a hunting dog.[5] The two forefeet are flat and rather large; and it has a tail, about four inches long, like that of a dog. But its body and hindquarters are normally formed, as well as its hind feet, which resemble those of an ordinary kid. Its head is black, and a streak of that color extends down its back. The rest of the body is light gray. Its voice is at times like that of a kid, but at others, that of a small dog. The goat that bore this creature refused to nourish it and seemed to dread its sight. But in the end, she got used to it and has allowed it to nurse."

Plombières, France. Léopold Turck (1855, p. 603a), a physician who practiced in Plombières, stated that a goat there had borne two offspring "with the complete external form of a dog" ("qui avaient tout de la forme extérieure du chien"). He said the owner, frightened by this prodigy, at once gave them to a neighbor, who ate them. The description given amounts to an allegation of xenogenesis.

Lewarde, Nord, France. The Grand Écho du Nord de la France (Lille, Jun. 12, 1910, p. 3d; ||sfjep92r) reported that a goat at Lewarde yeaned a kid with the head of a bulldog.

Santa Maria Gaudalupe, Tecalitlán. The *Gazeta de Mexico* (Aug. 5, 1788, p. 1; ||mo6wgbg) said, in translation, "SANTA MARIA GAUDALUPE, TECALITLÁN, July 5.—On the Techachales ranch here, a bitch whelped a kind of goat, with the head and other parts of a goat from the belly up, but the rest like a dog, from which it can be inferred that the mother, one of the sheepdogs on the ranch, must have mated with a goat and that this monstrosity resulted."

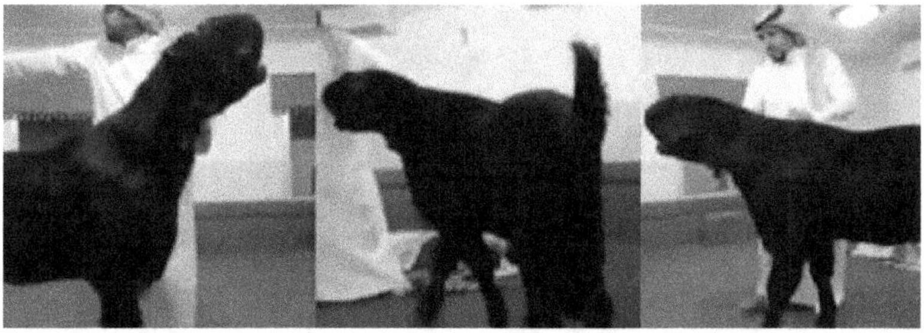

A series of screenshots from a defunct video ("Dog-Goat Hybrid Video - REAL"; ||34zfvcef). No description of the animal was given, but it was larger than a goat or dog, almost the size of a cow, likely a case of heterosis.

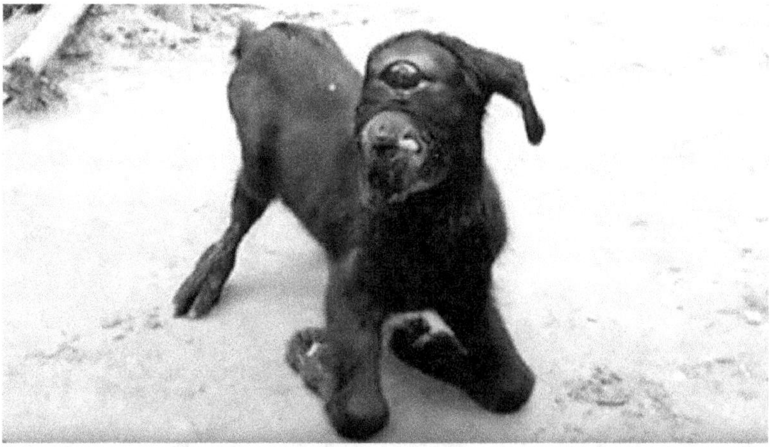

A cyclopean canicaprid yeaned by a goat in Assam, India. Note the doglike forepaws. The single eye is of the double-iris type, which are usually much larger than ordinary eyes (||2dc6wkvz).

NOTES

1. The Damascus breed is also known as the Aleppo, Halep, Damascene, Shami, or Chami. Damascus and Aleppo (Halep) are cities in Syria.
2. Today Rumilly is a commune of some 12,000 inhabitants in Upper Savoy.
3. Merindol goes on to say the animal had two tails and multiple excretory orifices, and that it would likely have lived if properly delivered. Biographical information about Merindol can be found here: ||y8lx9xod.
4. The village of Mussey has since been merged with Val-d'Ornain, a commune about 25 miles north of Joinville in northeastern France.
5. The French term used here, *chien courant,* seems equivalent to the English *scent hound,* that is, a tracking breed such as a basset or blood hound.

A reconstruction of the scene described in the report quoted on the opposite page, the 1887 branding of a horned, shaggy, bear-headed ursibovid in Arizona.

URSIBOVIDS (BEAR-COWS)

Fiction is obliged to stick to possibilities. Truth isn't.
—MARK TWAIN, *Following the Equator*

Animals combining the traits of a bear with those of a bovid will be termed *ursibovids*. There are fewer reports about ursibovids than about CANIBOVIDS, probably because bears are not domestic animals and therefore less common than dogs. Moreover, farm dogs come into frequent contact with domestic bovids (cattle, sheep, goats), while a bear would do so only under unusual circumstances. Nevertheless, numerous reports about ursibovids do exist. In such reports all the alleged mothers are bovids, and quite a few of the offspring have reportedly been viable.

Gila County, Arizona. From the *Clarion* (Clifton, Arizona, Nov. 9, 1887, p. 1; ||y39w8jux): "During the recent round-up in Tonto Basin [in Gila County], among the cattle driven in for branding was a strange monstrosity in the shape of a yearling heifer, bearing the head of a bear adorned with incipient horns. The animal was not over two feet in height, owing to the shortness of its legs, which terminated in peculiar shaped hoofs which were split like those of other bovine animals. Each hoof was about five inches long and was formed of two parts, the outer one being nearly two inches longer than the inner. In other respects, the brute seemed to be perfect, with the exception of the hair, which was unusually long and resembled closely in appearance that of a bear. The mother of the monstrosity was in no way remarkable and appeared to regard with the utmost favor her strange offspring. As the mother bore the brand of John V. Meadows, one of the leading stockmen in the Basin, the same iron was placed on the calf, which was again turned loose on the range."

* In 1887, both black and brown bears were present in Gila County.

Rhône, France. Le Réveil du Beaujolais (Villefranche-sur-Saône, Jan. 19, 1904, p. 3e; ||2zj4e5hc) said a M. Braillon, a vigneron near Corcelles-en-Beaujolais, had a cow that calved a freak with a bear's head. It weighed 65 kilos (143 lbs). Four years later, *Le Réveil* (Oct. 27, 1908, p. 2g; ||4pv3fmsa; and see: ||5n6m5vha) said M. Antoine Guerrier, another vigneron, had a "calf with the head, the legs and the tail of a bear" ("un veau ayant la tête, les pattes et la queue d'un ours") at Le Fond de Blacé. It was then five days old. In mountainous eastern France, brown bears were once common.

The Wildhaus ursibovid, with its mother, as depicted in a report preserved in the Wickiana Collection (||yfab8nch). It said it had the head and fur of a bear.

Wildhaus, Toggenburg, Switzerland. A manuscript in the Wickiana Collection (||yfab8nch), dating to 1578, states that on December 1, 1577, a cow at Wildhaus had a calf with both the head and fur of a bear (the same description as that given of the Gila County ursibovid).

Vitrac, Dordogne, France. La Croix de l'Aveyron (Rodez, Mar. 13, 1910, p. 4a; ||3nx3h334) reported that a cow at Vitrac had produced a bearlike calf. With care from its owner, it survived. Another deformed animal, calved in the same herd, died.

Westfield, Pennsylvania. The *Westfield Free Press* (quoted in the *Potter Enterprise,* Coudersport, Pa., Apr. 26, 1928, p. 2; ||y417hz3v) said,

> A two-year-old heifer in the herd of Harry Marsh, of Marshlands, gave birth to an animal one night recently that has attracted no little interest and attention. The creature, that should have been a Holstein calf, looked as little like a calf as one could imagine. In fact, it holds the greatest resemblance to a bear of any known animal.
>
> It has shiny black fur, its ears are set closely to its head, its legs are short and resemble those of a cub, the hind legs being entirely dissimilar from those of a calf. It has the divided hoof of a calf. But on each segment of the foot are claws such as are provided bears. It has the teeth of a bear and has so little

resemblance to its mother that the owner of the heifer when he first saw the creature could not for a time comprehend the occasion for the presence of a creature, the like of which there was not on earth, or in the waters under the earth.

The creature, like most monstrosities, was dead at birth but evidently of full birth age. Its appearance excited so much interest that several hundred people have been to see it and there have been theories stated, but no definite conclusion obtained. Some claim it is a marked object of birth, the result of a severe fright the heifer had from a bear during the period of gestation [see MATERNAL IMPRESSIONS]. The other theory being that the queer animal was the offspring of a bear and heifer. … Whatever be true, the heifer was pastured in a place that was sometimes visited by bears, but further than that no information is at hand.

At the suggestion of many, Mr. Marsh has decided to have the freak mounted, and it is more than possible that scientists will have some rare and spirited discussions over the skeleton of the animal that seems to be neither true bear nor calf.

Montpelier, Vermont. From the *Montpelier Evening Argus* (Apr. 29, 1918, p. 5d; ||yy6lt2d9): "A freak calf was born to a small two-year-old heifer owned by Mr. and Mrs. Robert Cook of the Jones Brook section nine days ago. It appeared to be part bear and part cow. It had a bear's body, head and tail and a calf's nose. It was small and black in color about a foot high. It had calf's feet, and horns were beginning to show through the skin. The mother of the beast was a small black heifer. … Two small bears have been seen in this section. The calf was well and healthy until yesterday when he was fed new cow's milk. It died Sunday afternoon. The calf, which was a great curiosity, had been visited by several hundred persons."

* A Robert Douglas Cook died at Montpelier in 1928 (||yy2dpos8)

Salorno, South Tyrol, Italy. *Bozner Zeitung* (Nov. 14, 1885, p. 4a; ||yvpkuaza) said a cow belonging to a Karl Franceschini, in Salorno, bore a calf with a complete bear's head ("vollständigen Bärenkopf").

Orny, Moselle, France. From *Le Courrier de Metz* (Mar. 1, 1911, p. 3b; ||mvrxw8ud): "ORNY.—Recently, M. Rollin, a farmer, received a freak in lieu of the calf he expected. The foreparts of this animal are of an extraordinary size and the hind parts, like those of a bear, are paralyzed. The monster cannot walk, but it is doing well all the same.

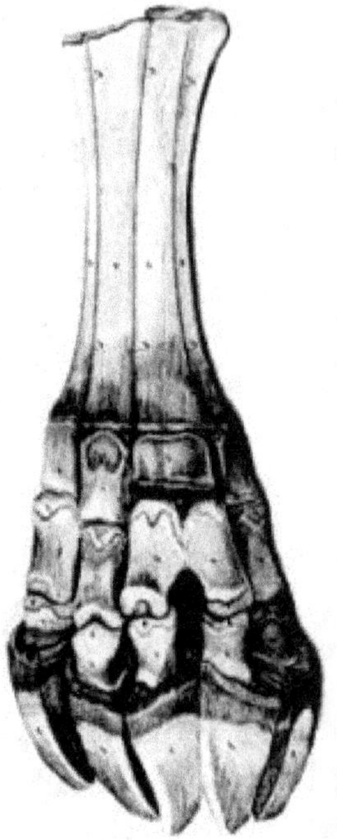

Hind foot bones taken from a Nuremberg "calf" (Trew 1735, Tab. 3).

Athol, Massachusetts. From the *Times* (Munster, Indiana, Oct. 29, 1923, p. 7g; ||y39zb8o8):

> ATHOL, MASS., Oct.—A freak calf with front feet like those of a bear is the property of H. Zack, of North Orange, a cattle dealer.
>
> The animal, which is three months old, has front feet almost exactly like a bear's, and its head also resembles that of a bear. When the animal walks it lopes like a bear.

* A 1920 Athol city directory (||y5jnbt77) lists a Hyman Zack (1874-1946), a cattle dealer, as living in North Orange, a place just north of Athol. Records (||y6bu8tls) list him as Hyman Chaim Zack.

Stowe, Vermont. The *Rutland Daily Herald* (Rutland, Vt., Mar. 13, 1916, p. 4; ||yyben8dv) said, "A freak calf of a few days old in the herd of Herbert Page on Edison Hill in Stowe is said to show a striking resemblance to a bear. It is about two feet high. Its legs are shaped like that of a bear, and although it has hoofs, it walks back upon its dewclaws like a bear. It has a thick, smooth fur and its ears and shape of its head are like those of a bear. When approached suddenly, it paws with its forefeet."

* A Herbert Willis Page Sr. (1858-1934) died at Stowe (||y6x3y26j).

Urs, Ariège, France. Le Journal du Midi (Avignon, Mar. 28, 1929, p. 4a; ||2crewkwk) said an animal calved in the Pyrenean village of Urs had bear ears and that it walked with its toes and metatarsals flat on the ground (as in the Stowe report quoted above), that is, it was plantigrade, keeping the heels down, as do bears and humans. Most locals thought it was a hybrid.

Wausau, Wisconsin. From the *Wausau Pilot* (quoted in the *Crescent,* Appleton, Wis., May 24, 1873, p. 4e; ||y2ullpnn): "FREAK OF NATURE!—On the premises of a gentleman named Joseph Odet about 10 miles east of this city, there has been born a hybrid calf-and-bear. The forequarters of this nondescript are said to be those of a calf, except that the head is held low like that of a bear, and it laps it's milk like a dog, while the hindquarters are those of a bear, behind, covered with bear's fur instead of hair, its tail short, and legs bending out with the unmistakable bear's feet and claws."

* A Joseph Odets lived near Wausau in Marathon County in 1864 (||yxopa9b6).

Obermühl, Kappl, Austria. The *Tiroler Stimmen* (Jun. 18, 1892, 3b; ||mrxy2v6v) said a cow slaughtered at Obermühl had a fetus taken from her that had a large head like a bear's, and bearlike hind parts as well, but front legs normal for a calf.

Winnemucca, Nevada. The *Daily Union* (Sacramento, Ca., May 7, 1874, p. 3e; ||y83uqvs3) said, "At Fairbank's ranch, says the WINNEMUCCA REGISTER, is a bovine monstrosity, which, in the hands of P. T. Barnum or Yankee Robinson, would be worth a fortune. It is a tailless calf with features much resembling the bear. From its general appearance, in one or two more generations it would become a complete biped. Fairbanks is taking great care to raise it ..."

* Farmer Randolph Fairbank (1836-1923) lived near Winnemucca in 1874 (||y35radcv).

Northampton, Pennsylvania. From the *Morning Call* (Allentown, Pa., Apr. 18, 1910, p. 5b; ||y4cq286b):

> A western cow, probably bred in the Rocky Mountain region, was purchased about a month ago at a public sale by H. H. Rice, of Fourteenth and Canal streets, Northampton. The cow, which appeared normal in every way, one week after its purchase, or to be exact, twenty-five days ago, gave birth to a calf, a freak with an extraordinary physical make-up. ... The freak appears to be a cross between a cow and a bear, separate parts of its anatomy embracing bones, flesh and hair similar to those characteristic to both the bear and cow. The lower portion of its four legs and the shape of its skull bones are distinctly those of a cow, while the upper portion of the legs, the hips and the shoulders incline to bear. The hipbones are not nearly so large nor prominent as those of a normal calf, in fact they are exactly in accordance with those of the cub. The shoulders, or upper portion of the forelegs seem larger than those of the ordinary calf, a condition characteristic to the bear. Indeed, it seems as though the legs are double jointed and that the joints are the dividing line of the different species.
>
> While the head favors that of the cow, the eyes are but half the size of those of a calf. Horns can be felt but they are broader and not nearly so pointed as normal horn sprouts, appearing more like a well-defined bump in the bones of the skull.
>
> The most prominent proof of the bear portion, however, appears in the fur of the skin. This is woolly and an exact duplication of the young bear skin. The tail of the freak is the tail of a bear, short and covered with bear skin and fur. Its back is rounded out more than that of the normal calf and it seems more active and stronger. Its flesh appears harder than the flabby flesh of a calf. To more clearly illustrate the contrast between the freak and a calf, Mr. Rice has a normal calf in the stable, two weeks older than the freak, with which to make comparisons.
>
> Vital organs are present to a number to indicate both the bear and the cow existing in this strange composite or hybrid animal.
>
> The freak is in a very healthy condition and every day added to its age brings out more prominently its exceptional anatomy. As its freakish nature gives it a monetary value far in excess of the per pound rate of veal,

notwithstanding the height to which the price of that luxury has soared, Mr. Rice has decided to sell the animal to a collector or exhibitor of freaks.

Accordingly, he inserted an ad in the "for sale" columns of the MORNING CALL and in the course of several days had two prospective purchasers, one representing a museum and the other a side show to Buffalo Bill's shows, which will exhibit in Allentown in the near future. $400 was bid for the freak by one of the callers while another asked Mr. Rice to refrain from selling until he could communicate with his people. The figure offered is slightly lower than Mr. Rice asks, and it is believed a sale will be consummated in the next several days.

A subsequent report about the Northampton ursibovid from the *Allentown Leader* (Allentown, Pa., Feb. 18, 1911, p. 11; ||y2ajnm5x) said,

A year or two ago H. H. Rice, former coal dealer and brewer at Northampton, purchased a cow for $28. It was said the cow came from the Rocky Mountains. Mr. Rice did not have the cow very long when she gave birth to a blind calf, represented to be half bear and half cow, and being both male and female. The freak attracted much attention and Mr. Rice refused $500 for the animal.

He has now sold the calf, according to report, for $1500. The owner will exhibit it around the county and expects to make a fortune.

Worth, New York. The *Watertown Times* (May 8, 1891, p. 8a; ||y9v3hogb) said, "In the town of Worth, this county, is to be seen a monstrosity, half calf and half bear."

Waupaca, Wisconsin. The *Ladysmith News* (Ladysmith, Wis., Oct. 20, 1906, p. 5b; ||4xj8psm7) said D. Fitch (David William Fitch, 1865-1923) had in his stables in Waupaca "a black ox which is half bear, with no seeming place for a tail."

Pottsville, Pennsylvania. The *Republican* (Pottsville, May 21, 1888, p. 1f; ||y2n64csq) claimed, "Over 400 people viewed the strange freak, half bear and half calf yesterday."

FOR SALE.
260 opera chairs for sale cheap. Also a calf, five months old, half calf and half bear, weighs about 325 pounds. H. H. RICE, Fourteenth and Canal Sts., Northampton, Pa. 11-6t

An ad offering the Northampton ursibovid for sale (*Allentown Morning Call,* Aug. 17, 1910, p. 12b; ||y4gbhk8p). Given that one article quoted above gives the birthdate of this animal as approximately March 24, 1919, the information in this ad indicates that this animal weighed 325 pounds when less than five months old.

Hersbruck, Bavaria. Welt Blatt (May 18, 1878, p. 9; ||y4b32er8) said, in translation, "FROM HERSBRUCK: Near here in Velden yesterday, the bailiff's cow could not calve … and could not be saved. And when they cut open the body, there came to light a thing no one in these parts had ever seen, which honestly had the following appearance: The head was like that of an unusually large pug dog, the neck, nonexistent. The body was covered with the long hair of a bear and, overall, was formed like that of a bear, with a bear's short tail. Only the feet were those of a cow. The creature is fully developed and would, perhaps, have lived if the mother had not been slaughtered."

Volders, Tyrol, Austria. Welt Blatt (Mar. 29, 1887, p. 5; ||yxat7u3a) said, in translation, "On the Riccabona farm in Volders, Tyrol, a cow has calved five stillborn offspring. Four were normal, but the fifth was a so-called bear-calf. The mother is doing well."

Alstadt, Duchy of Coburg. Blathazar Chilian (1683, p. 357) reported a cow at Alstadt calving a nondescript with a bearlike head.

Harzgerode, Saxony-Anhalt. Spangenberg (1572, 436b) said a cow at Harzgerode calved a bearlike offspring in January 1538 that had neither tongue nor anus.

St. Marys, New Brunswick. From the *St. John Daily Sun* (St. John, N.B., Apr. 25, 1904, p. 2a): "There has been born in St. Marys an animal which is causing considerable attention. Its mother is a cow and is half calf and half bear. Its head and neck are shaped similarly to a bear's, and it has two tails, one similar to a calf's on its side and the other similar to a bear's attached at the regular place."

Dexterville, Wisconsin. From the *Lake Geneva Herald* (Lake Geneva, Wis., Jul. 8, 1887, p. 7f; ||yy44clvs): "A 2-year-old heifer at Dexterville recently gave birth to twins, one a natural calf and the other being half bear and half calf. Its head, legs and feet are like a bear, the other parts, like a calf, color red and white. The bear-calf is perfect in development and good size."

XENOGENESIS

Buza, Slovakia. Welt Blatt (May 16, 1877, p. 7b; ||yxat7u3a) said, in translation, "On the 6th,[2] Herr Kaiser, the Košice veterinarian [i.e., *Kurschmied*] was called to the village of Buza where a pregnant cow of a farmer, named Glück, had been unable to calve despite all efforts to assist her … she had to be slaughtered. And what was inside! Instead of a calf, the assembled farm workers were amazed to find a fully formed bear cub weighing 60 Viennese pounds [~33.5 kg]. It had a long, shaggy coat and resembled a calf only in the upper portion of its face. [so, this is not quite a report of xenogenesis] There were no nasal bones, and it was a cyclops with a single eye in the center of its forehead."

Struthen, Thuringia. Praetorius (1678a) said that in 1671, at Struthen, four different cows, during a three-week period, calved bear cubs. Plausibly, a bear inseminated them all and bear-like offspring were reported as cubs.

Rust, Michigan. the *Crawford Avalanche* (Grayling, Mich., Aug. 21, 1890, p. 3c; ||y4jb9etf) reported a cow calving a bear cub at Rust.

AN URSIDOVID (BEAR-SHEEP)

Thundorf, Bavaria. The *Neue Warte am Inn* (Mar. 31, 1943, p. 7; ||y5k7nzgc) said, in translation, "A remarkable freak was found at Thundorf near Bad Reichenhall, a little bear-lamb. Just as cows sometimes calve bear-calves, so in Thundorf a sheep yeaned a lamb with a stocky body, short legs and a bearlike head. The creature lived a long time. Its mother had been wandering free."

MUSTELOVIDS (MINK-SHEEP)

Antioch, Illinois. The *Chicago Tribune* (Jun. 19, 1868, p. 2h; ||yyxjvjc7) said an entire flock of sheep yeaned minks: "Mr. Peter K. Blunt resides on his farm in the town of Antioch, and for years has raised sheep with varying success. He, like the neighboring farmers, makes calculations upon having his ewes give birth to their young about April or May. Three years ago, the present spring, … Mr. Blunt was the owner of about fifty fine, healthy ewes, and the animals, from their appearance, promised to add a lamb each to the flock of their owner. Instead of dropping lambs, however, each and every ewe gave birth to a little animal about the size of and resembling the mink—thickly covered with fine fur and bearing not the slightest resemblance to the natural progeny. They were, in every instance, dead when they came into the world."

A LUTRIBOVID (OTTER-COW)

Vulaines, Aube, France. Le Petit Troyen (Troyes, Jan. 26, 1934, p. 2f; ||3fzjttcj) said, in translation, "VULAINES.—A FREAK.—A calf belonging to Mr. Princez has borne a calf with the head of an otter. The strange animal, which was otherwise normal, was stillborn."

The Otter, or Ancon, breed.

LUTROVIDS (OTTER-SHEEP)

Dover, Massachusetts. Humphreys (1813, p. 88) reported that "Seth Wight, who possessed a small farm on the banks of the Charles River, in the town of Dover and State of Massachusetts, ... kept a little flock composed of fifteen ewes and one ram. In the year 1791, one of the ewes produced a lamb of singular appearance. ... he killed his former ram and reserved the young one for breeding. The first season, two lambs only were yeaned in his likeness. In the following years, a number more, distinguished by the same peculiarities. Hence proceeded a strongly marked variety in this species of animals, before unknown in the world. It has been called by the name of the Otter breed [aka Ancon]. This name was given from a real or imaginary resemblance to that animal, in the shortness of the legs and length of the back; by some supposed to have been caused by an unnatural intercourse ... otters were then sometimes seen on the banks of this river."

Ought an otter mount a sheep,
While said sheep lies fast asleep?
Ought an otter sire a son,
With a ewe his brother'd shun?
Ought an otter's daughter bleat?
Or have hoofs instead of feet?
Mother Nature might say no,
But who knows what's apropos?

—GENE MCCARTHY

Gwan-Go-Mock (Moves Swiftly), known as Bow String Jack, the Chippewa chief who in 1895 reportedly bagged a goat-bear in Itasca County, Minnesota.

CAPRIDURSIDS (BEAR-GOATS)

They observed many fruitful valleys, full of grass, and herds of very strange creatures feeding there.
—JOHN BARROW
The Life of Admiral Sir Francis Drake

According to a news report that appeared in many newspapers in 1895, numerous witnesses saw multiple goat-bears near Deer River, Itasca County, Minnesota. Though offered simply as news, this tale has elements of Greek myth—comrades on a quest, bizarre composite monsters, and a weird hairy being lurking in a cave. But the fleeces in this case were more ursine than golden. From the *Fort Worth Gazette* (Mar. 24, 1895, p. 1; ||yytyx8dy):

MYSTERY SOLVED BY A SHOT

THE WHEREABOUTS OF A MISSING MAN DISCOVERED

BY THE KILLING OF ONE OF HIS PECULIAR PETS

HE HAD BEEN FOR YEARS LIVING THE LIFE OF A HERMIT

DEER RIVER, MINN., March 22.—A mystery of long standing has been cleared up by the killing of an Indian, known here as "Bow String Jack," of a strange animal resembling to a marked degree a domestic goat, and yet retaining nearly all the characteristics of a common brown bear. It has solved the mysterious disappearance of Thomas Freeman, who fled from his home here some years ago, after getting into trouble with a neighbor over a young girl both were in love with and has brought to light the strange way in which a man can live when deprived of the refining influences of civilization.

A few days ago, the Indian hunter was roaming the woods in the northern part of this (Itasca) county, about 30 miles from this place when his attention was attracted by a peculiar looking animal a short distance away, which did not make any threatening demonstrations or try to run away on his approach. When within a short distance of the animal, Bow String Jack fired and killed the brute, putting a 44 Winchester bullet through its head. On making an examination of his catch he was surprised to find that the creature was in reality half bear, half goat, and had around its neck a broad leather strap, on which was a metallic tag bearing the name, Freeman.

The Indian brought the animal to this place and placed it on exhibition at the store of J. H. Fairbanks, where it attracted no end of attention. Alfred Stone, who is an old resident and knew Freeman, called attention to the strap on the animal's neck, and it was explained to the Indian that it was more than probable that Freeman could be found somewhere near the place where the bear goat was killed. Jack was willing to pilot a party to the place, and four men set out at once.

Bow String Jack led the party to a point near the shore of Long Bow Lake, a wild and almost inaccessible place at the extreme northern end of the county, where a white man is not seen once in ten years, so far as anyone here knows.

From the west end of the lake a huge cliff of dull gray rock rises abruptly to a height of about 300 feet, and here the Indian pointed to the place he had killed the bear. It was found that someone had visited the camp made by Bow String Jack during his absence, which was taken as conclusive evidence that there was someone living near the lake. The party at once made a careful search of the entire neighborhood, and after a long hunt their endeavors were crowned with success, for a cave was found in the rocks, which had been occupied but a short time before, and a fire was yet burning in a rude stove of stone. In one corner were several garments of fur, and beside them stood a rifle and other things, showing that the owner was a woodsman. All the furniture in the cave was of the most primitive nature, and not a scrap of paper or cloth could be found.

The members of the party determined to see the mystery through and so hid in the cave. Their wait was not a long one, for in less than half an hour the occupant of the cave returned, and it proved to be Thomas Freeman, who has been so long missing. At first, he was not recognized, as his hair and beard had grown to a tremendous length, and his skin was tanned such a dark color that he resembled an Indian more than a white man. When he saw the men in his cave, he started to run but was forced to stop and return to the party. It was a long time before he could be induced to talk, but when he did open his mouth, the words rolled out in a perfect torrent, and there was no stopping him.

He had not seen a white man before, he said, since he became a hermit, and, so far as he was concerned, he never intended to see one again. He had become disgusted with his own race when he had the trouble with a neighbor at Deer River, and it was his intention to remain in the wilderness so long as he lived. He was happy in his present condition, with his pets around him, and wanted nothing better than to be forgotten by the world, the world forgetting. He had long since ceased to take any interest in the affairs of the outside world and did not even know the name of the present governor of the state. He was singularly silent on the subject of his pets, and, as there was not an animal of any kind around his cave, the members of the searching party became very curious to know what sort of pets Freeman had. The hermit refused to answer any questions concerning the matter, merely saying that he had several, and they were in a safe place. He offered the further information that one of the pets had escaped from its pen a few days ago, and, although he had searched everywhere, he could not find it.

Freeman refused to accompany the party back to Deer River, but offered to accompany the men to the trail, which lay some four or five miles to the south. After going about a mile, the party heard strange noises coming from a small valley in the hills and insisted on investigating, much against the wish of Freeman, who insisted that it amounted to nothing. In the valley the party found a large corral of stone, inside of which could be seen a dozen or more peculiar looking animals, which, on close investigation, were found to be of the same variety as the one killed by Bow String Jack a few days before.

The leader of the party promised that no one should molest the corral if Freeman would explain what the animals were, and under this promise Freeman told a strange story, which would be hard to believe, but for the reliability of the members of the party, and the fact that one of the animals seen is now on exhibition in this city.

According to Freeman, when he left Deer River, he was disgusted with life and only wanted to get away from the haunts of men. He had taken with him when he left his home a female goat, which had been an old pet, as he wanted something for a companion in the wilderness. Arriving at Long Bow Lake, he had made his home in the cave, where he got along very nicely. On one of his hunting trips into the woods surrounding the lake he had killed a mother bear and captured her two cubs. The female cub had died but the male had lived and became almost as much a pet as was the goat. Soon the two animals became the best of friends and were inseparable. Where one went the other was sure to follow, and when the bear was about 1 year old the goat gave birth to a kid, which was a strange mixture of bear and kid. This was the beginning, and soon Mr. Freeman went into the goat-bear raising business, being somewhat of a naturalist.

He found that a male bear and a female goat would mate under proper conditions as readily as a male and female goat, but try as he would, he could not induce a male goat and a female bear to mate. The young at birth resembled a bear more than a goat—rough, round body, short, blunt head and small, sharp eyes. The most notable thing about them was the legs, which were small and slender, resembling those of the goat. As the little ones grew older, they lost something of the bear characteristics and took on many of the goat peculiarities—the fur became lighter and coarser, hoofs appeared, and two small horns appeared on the heads of the males. These peculiarities became more marked as the animals grew to maturity, and to see the body of a full-grown animal one would think it was a goat with an oddly formed head, while to see the legs one would swear it was a bear.

Efforts are now being made to induce Mr. Freeman to give up his hermit ideas and place his herd of animals on exhibition in the various cities of the country. A trust is being formed to offer him a large sum for the herd in case he will take no action looking to putting them before the public, and the plan will probably be successful, as Mr. Freeman confided to Mr. Stone that he was getting tired of living in the wilderness.

Note: The J. H. Fairbanks mentioned in the report may be the John H. Fairbanks Jr. born in Itasca County in 1846 (||y3e3cgf). Indeed, this seems likely given that the population of Itasca in 1890 was just 743 (||y2snpq3a). Elizabeth Flint, born Anna Elizabeth Canfield in Minnesota (||y4nuelur), wrote about her acquaintance with Bow String Jack, when she and her husband lived in Itasca County after their marriage in 1911 (||yybeb9fo; ||y4ssxqeh).

Lempach, Germany. The *Breslauische Sammlungen* (p. 325; ||y2te4vlv) stated that at Lempach, near Dresden, May 24, 1720, a goat kidded a nondescript with bear paws on its forelegs and a swollen mass on the back of its head.

* *Breslauische Sammlungen* is a short title for what has been called the most important German-language scientific journal of the early eighteenth century (||ya4h3769).

URSISUIDS (BEAR-PIGS)

An ursisuid reconstructed. (artist: @chromatophore)

Manor, Georgia. The *True Citizen* (Waycross, Ga., May 3, 1890, p. 4c; ||y2og87gw) said, "Mr. J. D. Ellis,[†] living near Homerville, is in possession of a most curious looking animal, half bear and half hog. The animal has the hoofs, nose and eyes of a hog, while its body is covered with hair like a bear. It often stands erect and walks in that position. It is three months old, perfectly gentle and will eat anything offered. It has teeth like a bear and makes a grunting noise which sounds like the noise made by a young cub. It weighs about twenty-five pounds. A sow owned by Mr. Steedley, who lives near the Okefenokee swamp, brought up two of these curious animals, a male and a female. The female has since died. Mr. Ellis proposes to place it on exhibition.

* In 1890, a Lawrence Peter Steedley Sr. (1836-1904), a farmer, lived in Manor (||y3shr2v9), a place near the Okefenokee, between Waycross and Homerville.

Roanoke Rapids, North Carolina. From the *Roanoke Rapids Herald* (Sept. 10, 1936, p. 33a; ||3rfu38pm): "Manager Bartlett of Auto Service Station at the Junction is the possessor of a freak pig which has excited much comment. The front part of the pig is normal, but the rear end and back legs are very much like those of a bear. Some observers say it is simply a deformed pig; that the hind legs are really those of a pig in a twisted condition. But there are indications of claws there. To these skeptics we also point to the tail. If that isn't a bear tail, we never saw a bear's tail. Higher up on the back is also a pig tail."

A hog-badger (*Arctonyx*) (Wikimedia, Rushenb ||yzsxwcdu)

Hog-badgers. The Hog-badger of southeast Asia, assigned to Order Carnivora, mixes traits of suids with those of a badger. In particular, Nowak (1999, p. 730) refers to its "long, truncate, mobile, and naked snout, which is often compared to that of a pig (*Sus*)." Nowak (ibid.) says it uses this appendage "in rooting" (as do pigs). It is therefore a CARNISUID, as are ursisuids and CANISUIDS.

The Boone County canisuid.

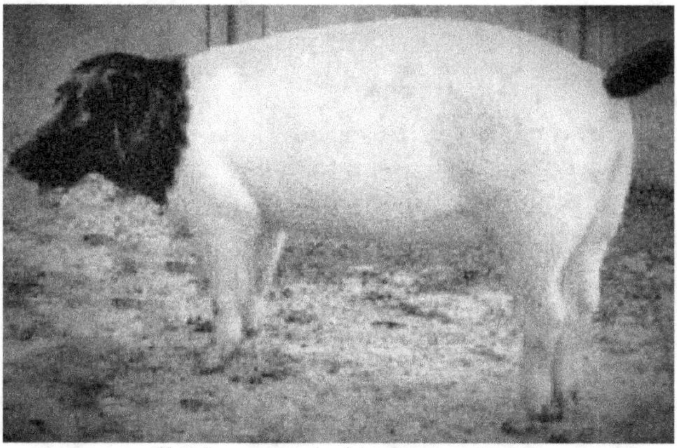
The Sayara canisuid (reconstruction).

The Eureka canisuid (reconstruction).

Canisuids (Dog-Pigs)

If a pig mounts a dog, the great will become small.
—A Mesopotamian birth omen[1]

London. The *Chattanooga News* (Oct. 17, 1903, p. 4; ||mue5an4d) said a London animal had a pig's head but a dog's body and bark.

Boone County, Missouri. St. Louis Post-Dispatch (Aug. 2, 1896, p. 29a; ||y43lsmw3) pictured a viable "dog-faced pig" farrowed in Boone Co.

Sayara, Colorado. The *Democratic Press* (Ravenna, O., Jul. 26, 1888, p. 1h; ||j6gy4aj): "A ranchman at Sayara, Colo., has a pig that has a perfect dog's head, with dog's hair covering the head and neck. Excepting this and a short and bushy tail, the rest of the animal is like a pig."

New Bern, North Carolina. The *Roanoke Beacon and Washington County news* (Plymouth, N.C., Jul. 20, 1934, p. 3; ||2fzaeyjv) said a farmer named J. S. Tingle, "reported two freak pigs born on his farm near New Bern a few days ago. The two animals had front feet similar to those of any ordinary pig, but their rear feet had claws and were like unto those of a dog. One of the animals possessed a natural head, but the other had a head resembling a dog. The two pigs or rare animals died two days after birth, it was reported."

* Hybridization better explains of two such animals in a single litter than does mutation (see, too, the next several cases).

Frankfort, Kentucky. The *Frankfort Roundabout* (Dec. 26, 1891, p. 2c; ||h2hlcks) said, "Mr. Richard [Johnson] Thomasson [1835-1907], of the Switzer neighborhood, this county, killed a sow about a week ago which had in her womb a litter of seven [lifeless] puppies. Some of them had one ear like a dog and the other like a hog,[2] all had dog's tails and were marked like a dog."

East Macon, Georgia. The *Macon Telegraph* (quoted in the *Fairfield Herald,* Winnsboro, S.C., Jun. 2, 1869, p. 4e; ||hncnsys9) said, "The most remarkable lusus naturæ that we ever heard of occurred on Sunday last in East Macon. A sow had a litter of pigs, stillborn, and they were actually half dog and half pig. The head and fore feet were those of a dog, the feet being supplied with claws and in every respect resembling those of a puppy. The balance of the body was that of a pig, the hind feet having the hoof and shape of the pig's hind feet." The *Telegraph* (Apr. 21, 1893, p. 3b; ||22t6zmtv) later reported another canisuid in East Macon.

Princeton, Kentucky. From the *Cincinnati Enquirer* (Aug. 6, 1892, p. 1; ||2tccb4d3): "PRINCETON, KY.—A sow on the farm of John Cook,[3] two miles south of Princeton, recently gave birth to a litter of nine pigs. Seven out of the number are on the usual pig order, but the other two are strange freaks in their class. One was born with only three legs, while the right forefoot of the other is that of a dog with sharp claws. They are both doing well."

Eureka, Nevada. From the *Abbeville Messenger* (Abbeville, S.C., Aug. 3, 1886, p. 3; ||n29xrpp): "A monstrosity in the way of a medium-sized dog, with the head of a hog, is the property of a Shoshone in Eureka, Nev. The hybrid generally goes along with his nose to the ground."

Welheiden, Germany. Johann Jakob Huber (1748, p. 6, §4) wrote, in translation, "Another monster borne by the same kind of animal as in the previous section, namely a sow, is not so deformed as the one just discussed, nor is it so unusual that it will require many words to describe. Near the end of last year, a female fetal pig was brought to me from the village of Welheiden, which lies about a mile from our city [of Kassel]. This fetus in part—the front feet, upper jaw and muzzle—was like a dog, as was its neck. Dissection of the internal anatomy revealed nothing unusual. Nor has this birth gone unnoticed among the peasants. The man who brought me the pig, said a certain dog had many times tried to engage in coitus with the sow mother, and perhaps succeeded, and that from this intercourse the fetus just described was produced."

Auburn, Maine. From the *Lewiston Evening Journal* (Lewiston, Me., May 12, 1915, p. 12; ||os538as): "Mr. [Edmond J.] Boucher [an Auburn taxidermist] tells of one freak which is positively the funniest. This animal lived to a ripe old age and undoubtedly attracted much attention as Mr. Boucher tells the story. 'I was driving through the country a while ago, when I saw an animal that acted like a dog and looked like a pig. It ran out of a house and gave a little grunt, then went back again. Naturally, … I went into the house to inquire. The animal proved to be a pig in everything but legs and feelings. His legs were those of a dog and piggy had a dog's soul, too. He lived in the house, wouldn't have anything to do with other pigs, and the other dogs wouldn't recognize him for one of their kind either. It was the queerest thing I ever saw, that pig, running around, lying under the stove and eating like a dog. He would run up to you just like a dog, too, and beg to be petted.'"

Perth, Scotland. Boece (1575, 17, p. 355) said a litter of dog-headed piglets was farrowed at Perth in the reign of King James I (1406-1437).

Palleau, Saône-et-Loire, France. Le Courrier de Saône-et-Loire (Chalon-sur-Saône, Dec. 15, 1890, p. 2c; ||bddh5by8) said, in translation, "PALLEAU.—Last Thursday, a sow belonging to Mr. Jacquetot farrowed a freak. Its body is that of a pig without hair. The head, big and squared, is that of a bulldog. It has only a single eye in the middle of its forehead. This strange animal died after two days.

Asheville, North Carolina. The *Asheville Citizen* (May 18, 1908, p. 8d; ||ms2wx9p4): "A peculiar freak was born on the place of J. T. Williamson who is a farmer near here. One of the puppies of a fine shepherd dog has feet like those of a pig, while all the others are perfectly normal. This odd puppy is perfectly healthy and does not seem to be inconvenienced in the least by its extraordinary pedal extremities. The feet are cloven, and as hard as those of any ordinary pig, but strange to say, the puppy has a growl like any other puppy and does not know how to squeal."

Hillsboro, Ohio. From the *News-Herald* (Hillsboro, Jun. 11, 1896, p. 8b; ||zrco5al): "S. P. Michael ... has a strange freak of nature. A pig with claws on three feet like a dog, the hind ones and one front foot. The pig is living and perfect, except the feet."[4]

Bernay, Normandy. Le Républicain du Gard (Nîmes, Sep. 7, 1928, p. 1; ||mp8p3ypd) said a cow at Bernay calved a little animal (50 cm tall) with a calf's body, bulldog's head, and a tail like that of a fox terrier.

Beaufort, South Carolina. The *Cleveland Gazette* (Aug. 3, 1889, p. 1) said, "There is at or near Beaufort, S.C., a monstrosity supposed to be a pig. One of its forefeet is a dog's and one a pig's, one of its hind feet is a dog's and the other a pig's; one ear is a dog ear, and the other is a pig ear;[2] it has two tails, one resembles a dog's, the other a pig's."

Williams Township, North Carolina. The *Alamance Gleaner* (Graham, N.C., Dec. 20, 1883, p. 2; ||yecm5x9a) reported that "P. M. Pearson, Esq., of Williams Township has a sow that recently gave birth to pigs whose mouths, noses and teeth were exactly like a dog's."

Thuringia. The Thuringian naturalist Johann Matthäus Bechstein (1807, p. 112) wrote, in translation, "I myself heard from a reliable person that his dog bore two pups with pig's heads and obvious pig snouts."

Graz, Austria. The *Grazer Tagblatt* (Graz, May 29, 1898; ||y6rd4e8t) ran an advertisement for a sideshow featuring "a living beast, half pig, half dog, ten months old, with four dog's paws" (translation mine).

Halle, Saxony-Anhalt. The German physician Johann August Unzer (1752, p. 445), a Halle native, described (in translation) a pig-dog hybrid: "The sire of this bi-natured creature was a mongrel dog, and the dam, one of those short-legged Indian pigs one sees from time to time."

Strasbourg and Gotha. Christian Gabriel Fischer (1743, p. 426) recorded two canisuids ("dog-faced piglets"), one in a Strasbourg collection, the other in the cabinet at Gotha.

Chevinay, Rhône, France. The *Hamburger Relations-Courier* (Feb. 9, 1758, pp. 1-2; ||mr5hn9cr) said, in translation, "In Chevinay, four miles from Lyon, a farmer, Jean le Brun, luckily killed, with many stabs of a spade, a dreadful monster that attacked his son. In the preceding two months the animal had killed 23 men, women and children and over 100 sheep and other livestock. This monster was easily six Parisian feet long, and in several parts like a wolf but in others like a wild boar."

* The Lyon cynanthropoid (p. 299) may have been this same animal.

Goulburn, New South Wales, Australia. From the *Sydney Morning Herald* (Jun. 13, 1931, p. 14h; ||ycrq6b38): "GOULBURN, Friday. On the Sydney road, nine miles from Goulburn, a man and two youths were cutting wood when a strange animal, apparently half dog and half pig, rushed at them from a creek. One of the party killed it with an axe. One local authority declared it to be a very big wombat, but others think otherwise. The animal's head resembled that of a pig. It had short, squat legs, with a heavy coat of soft silvery grey fur. The animal was more than three feet long and had no tail."

XENOGENESIS

San Juan, Puerto Rico. El Mundo (San Juan, Jul. 29, 1921, p. 6a; ||4ytpzwvd) said a dog in San Juan whelped seven viable piglets.

Guadalajara, Mexico. La Opinión (Los Angeles, Sep. 29, 1934, p. 6d; ||3f22xmft) said a dog in Guadalajara whelped five piglets.

Palúdzka, Slovakia. The *Breslauische Sammlung* (Fall 1724, p. 223; ||yxulnh5k) said a sow at Kis-Palugya (now Palúdzka) farrowed two puppies.

Prussia. In a manuscript listing of observations, the Prussian physician Israel Schmied (in Paullini 1688, p. 49; ||y635ks5e) reported that in 1632 a sow covered by dogs later farrowed five doglike offspring.

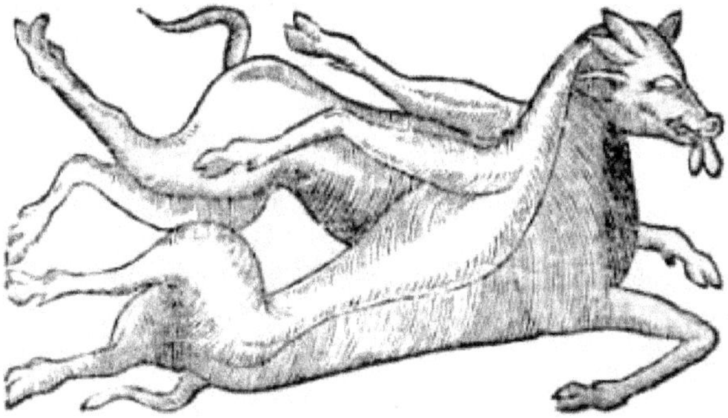

The Metz canisuid (Paré 1573, p. 397).

CONJOINED DOG-PIG TWINS

An abnormal repetition of body parts (e.g., supernumerary limbs, conjoined bodies, etc.) is commonly seen in telenothians (see COMPONENTS, p. 1197), especially in canibovids and canisuids.

Metz, Lorraine. The French surgeon Ambroise Paré (1573, p. 396) described a conjoined canisuid calved in 1572. In translation, he said, "In 1572, on the day after Easter, at Metz in Lorraine, at the inn called Saint Esprit, a sow farrowed a piglet with eight legs, four ears, and the true head of a dog. The rear parts were divided into two bodies as high up as the stomach, but the forepart was single. Two tongues hung from the mouth, which had four large teeth on either side, as many above as below. The sex was indeterminate, there being a single orifice under each of the two tails, which had the appearance shown in the figure [see above], which was kindly sent to me by M. Bourgeois, a doctor of medicine and a man of good judgment, who lives in the above-mentioned town of Metz and who is well acquainted with the case."

* Liceti (1634, p. 21) and Aldrovandi (1642, p. 615) also record this event.

Milledgeville, Georgia. The *Union and Recorder* (Milledgeville, Mar. 13, 1883, p. 4b; ||46mwpjhh) said, "A day or two ago, Mr. Robt. H. Miller [b. ca. 1844] of this county [Baldwin] brought to this office, a pig a few days old … It was perfectly black, his head shaped like a dog's, and without any hair on it. The pig had six well-formed legs. It was alive."

Hazelhurst, Georgia. The *Atlanta Constitution* (Jul. 21, 1915, p. 12; ||5cndvkb2) said, "D. S. Herrington, who lives about six miles from Hazlehurst, has … a freak pig … This pig has two bodies joined together, eight legs, two tails, four ears and two throats. Two of the feet are shaped in every respect like dog's feet, and it has dog's hair. The pig died when it was a few days old."

St. Louis. The *St. Louis Post-Dispatch* (Oct. 11, 1896, p. 29; ||msxy66cc) said, "E. Spore, who lives at La Salle and Montrose streets, brought to the POST-DISPATCH office a pup with seven legs and two tails. The father was a mastiff and the mother a Newfoundland. The mother presented her owner with this extraordinary freak ten days ago. It has two legs in front, four behind, where the body is double, and one on top of its back. The queer pup flourished for a time but got discouraged and died. … Another peculiarity about it was that its head and ears are shaped more like those of a hog than a dog."

St. Croix, Indiana. From the *Indiana State Sentinel* (Indianapolis, Aug. 2, 1893, p. 5e; ||mr3foz2): "ST. CROIX, July 25.—On the farm of Thomas Hornbeck near here this spring and summer have been born a calf without ears, a litter of seven pigs, four of which have six legs each, one of them having feet like a dog; also, a chicken with two heads and two necks. The chicken died, but the other freaks are living and prospering."

Bangor, California. The *San Francisco Chronicle* (Jun. 30, 1887, p. 8c; ||y7zz6beb) said, "M. Howser of Bangor brought to town yesterday a most remarkable freak of nature. It is a pig, or a pair of pigs, joined at the shoulders, with the remainder of their bodies entirely separate. The monstrosity has but one head, but on it are two sets of ears, one set being well formed, and eight legs and two tails. The head is very large and looks more like that of a dog than of a pig. It was born dead or died in a very short while after. C. J. Smith has it preserved in alcohol at his drugstore, and many have been to look at it."

Halsnaes, Denmark. Bartholin (1657b, p. 383) said, in translation, "I saw another marvelous specimen in Copenhagen, farrowed March 7, 1648, at Halsnaes, Zealand. First came five normal piglets. Next day another followed, misshapen birth composed of two piglets joined at the breast and having four feet on either side. There were two heads, pressed against each other. One had a dog's ears and snout, though the snout was like a pig's at the end. The eyes, too, were piglike. Long teeth jutted from the underjaw. The other head was not unlike the other except that the snout lay open so that the teeth showed."

* A conjoined nondescript with both pig and dog traits, born at Dubuque, is covered in the section on TRIPLIANS, as it reportedly also had human traits.

ADOPTIONS AND IMPRINTING

Many animals, when raised by a mother not of their own kind, will prefer, upon reaching sexual maturity, to mate with individuals of that same kind, a phenomenon called *sexual imprinting*. Pigs have often been raised by dogs, and vice versa. And the two sometimes choose to mate. No doubt, such behavior is often a manifestation of imprinting.

SOW ADOPTED PUPPIES.

ASHLEY, Ill. — (UP) — Albert Brown, a farmer living near here, reports that a sow belonging to him has adopted four motherless puppies and is nursing them along with her own litter.

Report about a sow adopting and suckling four puppies (*Index-Journal,* Greenwood, S.C., Mar. 31, 1933, p. 3; ||2m2ph85y).

My wife once went to a farm where the family had years before given an orphaned piglet to a Doberman Pinscher bitch to raise with her litter. She successfully suckled it, and when he grew up to be a boar, he would have nothing to do with other pigs and was treated as a dog by his owners. And he behaved like a dog. He spent his days and nights with the dogs of the farm and acted as if he were a watchdog when strangers arrived and even attempted to bark along with the dogs. One can easily imagine the kind of mates this boar would have chosen. The *Ward County Independent* (Minot, N.D., Apr. 5, 1906, p. 10; ||yym3wj9f) describes another such case of a bitch adopting a pig.

MOTHER DOG ADOPTS PIG

A farmer named Emmet arrived from Iowa last week with a car of goods, and with it an extraordinarily intelligent Shepherd dog, with two or three puppies. He also had a number of little Jersey red pigs when he started, but all but one died, and it looked to Farmer Emmet as though he might just as well knock this one in the head. The mother dog, however, thought different, for she took pity on the little orphan, if dogs ever do show such a feeling, and nursed the little pig alongside her own puppy. The pig began to grow stronger and now is as sleek a looking a little porker as one would ever care to see. After Emmet had reached Minot, he concluded that the pig would live all right, and took the little fellow away from the dog, but the canine would not have it so and was seen carrying the pig back to her nest by the nape of its neck. The strange sight of the pig and the pup nursing side by side, is quite unusual.

* A video (||y58mmxft) documents a dog adopting an entire litter of piglets.

The *Provemont Courier* (Provemont, Mich., Sep. 8, 1916, p. 4d; ||yb6938vt) reported the reverse, a sow adopting and raising a puppy.

> DOG DEVELOPS HABITS OF A PIG
>
> EVEN HIS BARK IS MIXTURE OF GRUNT AND SQUEAL
>
> GREENSBURG, IND.—Early last winter an automobile party passing the home of Seymour Cole, west of here, dropped a small bird dog pup. The dog made its way to the house, but as it was not welcome, was sent to the barn. Here it found an old sow with a litter of pigs. The pup attached itself to the litter and was made welcome.
>
> After being weaned along with the pigs, the dog continued to live with the hogs, roaming over the field by day and sleeping with them at night. It developed a taste for the same food, eats corn, etc., as if it were a hog and now its bark is changing in sound to resemble more a mixture of squeal and grunt than the bark of a dog. This story is vouched for by persons who have seen the dog and the hogs.

NOTES

1. Freedman (2017, p. 72).
2. Asymmetric ears have been reported for HOMINOSUIDS and CANIBOVIDS (see p. 1210).
3. A John Woodson Cooke (b. Jun. 12, 1845, d. Jan. 12, 1911), a farmer, was living in Princeton at the time in question (||5fv2ryvj).
4. Two men named Samuel Preston Michael, father (1816-1902) and son (1854-1932), lived near Hillsboro in 1896 (||yz4ftkdz).
5. *Colophon:* A dog-headed canisuid that Schmuck (1679, #5) said a sow farrowed at Pforzheim, Germany, Mar. 5, 1654.

The Le Locle ursidequid.

URSIDEQUIDS (BEAR-HORSES)

But he had the appearance of a monstrous animal whose ears and tail were those of an ass, and whose other parts were a bear's.

—VINCENT OF BEAUVAIS
Speculum Natural, 23:76

Le Locle, Switzerland. The *Schweizerische Beobachter* (1807, pp. 48-52; ||y3cqsmy9) reported an ursidequid foaled near Le Locle. The account included a crude drawing, reproduced above. The report began by suggesting that the strange birth was due to the mare having been frightened by a pair of dancing bears brought to entertain guests at a local inn (see MATERNAL IMPRESSIONS). After this speculative prelude, the article gives the following account (in translation):

> It was on May 16th, 1805, that this mare produced an unnatural and shockingly misshapen foal that indeed had the true voice of a bear. It was a male. The mother, far from caring for it in the same way she had watched over all her previous foals, sought to keep this one at a distance with kicks.
>
> The dairyman [i.e., owner Alexis Simonin] was away and his workers imagined, from prejudice or superstition, that such a monster would bring a curse upon the livestock, and they decided, especially given that they could not imagine how this monster could ever be of any use, to kill it. And so, after a few hours, they did, albeit with a great deal of trouble, as the beast turned out to be surprisingly strong. They then buried it.
>
> When all this became known in Le Locle two days later, residents came to investigate the story on the spot and had the beast exhumed

and taken back to town. There, for two days, it lay in a barn where everyone hurried to see it, and everyone avowed that Simonin's workers had exterminated a work of nature that should have been kept alive by any means possible. Its course of development would have been a matter of fascination for naturalists and might have cast light on the broader workings of nature herself, which are as yet shadowy and little known, and over which the opinions of naturalists are sharply divided.

One would have liked to have this monstrosity stuffed, but since that could not be accomplished, one must settle for the illustration accompanying this article [see p. 501], which accurately shows the nature of this animal, and the following description of the same:

The head: The upper half was fully like that of a bear, not only in its form, but also with respect to the eyes, ears and hair. The lower half was as in a horse.

The neck was round and much shorter than in an ordinary foal. The hair, except for the mane, lay in the opposite direction from that of the rest of the body, being turned toward the head.

The forelegs were natural as far down as the hooves, but these were bent forward like bear paws making nearly a right angle with the rest of the leg. One hoof bore five extremely hard claws, the other, four. The body and hind legs were like those of an ordinary foal, except for the hair and the claws, which were also present on the hind feet, although there were only two on one foot, and three on the other. Over the whole body the coat was gray and shaggy like a bear's.

The voice was, according to the eyewitness testimony of seven or eight people, a true bear growl and bore no similarity whatsoever to a horse's whinny.

When the live animal stood on its feet, it had an odd posture, in that it was very low in front but very high toward the rear. In general, it was of about the same size as an ordinary foal.

The great crowd of people who saw it will bear witness that all its various deformities obviously connected it to a bear.

Except for perhaps a few unintentional omissions, this report is taken word for word from the French language original, which was attested by several eyewitnesses as fully accurate and authentic.

* A report in the *Morgenblatt für gebildete Stände* (Stuttgart und Tübingen, Nov. 18, 1807, p. 1104 ||y4dvfn78) names some witnesses to the Le Locle birth: Alexis Simonin, the owner of the mare, Lt. Heugenin, in whose barn the creature was exhibited, F. Troz, the mayor of Le Locle, and "Auriet," the author of the report.

* The Westfield URSIBOVID, too, was reported to have had feet with claws. And reports say the Guide Rock CANIBOVID had hooves that bore claws.

* Brown bears (*Ursus arctos*) were present in Switzerland until the early 20th century.

PHOCOBOVIDS (SEAL-COWS)

The stars create new forms, and the heavens insert new faces.
—MARTIN DELRIO
Disquisitionum magicarum (1606, p. 382).

Combrée, Haut-Anjou, France. L'Ami du Peuple (Angers, May 26, 1940, p. 4c; ||49rmxvwf) said, in translation, "BEL-AIR DE COMBRÉE.—A FREAK.—Thursday, Mr. Aristide Segin, butcher in Bel-Air, had a freak calf born to a Cotentine heifer. This monster, stillborn, weighing no less than 80 kilos, had a seal's head, huge and hideous. Its four appendages, very short, were also those of a seal, except the digits were separate. The hindquarters were very large. The tail was on the back."

Warm Springs, Montana. The *Anaconda Standard* (Anaconda, Mt., Feb. 15, 1898, p. 3; ||y39l6n6w) said, "One of the most unaccountable and highly interesting freaks that has ever been seen in this city was brought here yesterday. ... The freak is a calf with the head of a seal and without legs. Mention was made a few months ago in the STANDARD of such a calf having been born down near Warm Springs [Mt.]. The animal was dead ... The freak is about two-and-one-half feet in length. The rear part of its body is that of a normal calf, being covered with a coat of white and red spotted hair. The tail is also perfect. Where the neck of a young calf usually begins the red and white hair on this freak begins to give way to a hair of brown color. From the shoulders the neck and head of the seal begins. The neck is long and has the graceful contour of the water animals. The head is that of a normal young seal pup. Even the tusks of the seal are seen in embryo. The tusks are covered by a coat of reddish skin, but they are perfectly visible for all that. ... it is entirely devoid of legs. Where the legs of an ordinary calf begin there are, on this freak, stumps, the bottoms of which are perfectly smooth and flat. These stumps are covered with white and red hair."

Rennes, France. Le Journal (Paris, Feb. 16, 1935, p. 3b; ||yvkzkc62) said Ferdinand Tessier, near Rennes, had a cow drop "a calf with the perfect head of a seal" ("vrai tête de phoque") and flipper forelegs. It was viable.

Coastal Scenery with Cows and a Milkmaid (Adolf Henrik Mackeprang). Seals are not the only mammals to frequent the seashore. Various animals have been known to mount recumbent cattle (see p. 424).

Seal Rock, Oregon. Clackamas County Record (Oregon City, Jul. 23, 1903, p. 1c; ||2s45zutz) said, "In the window of a local drug store is a freak calf. The animal is dead and has been stuffed. It is owned by C. B. Clement of Portland, who represents an insurance company. Mr. Clement saw the calf at Seal Rock and purchased it to exhibit. The calf has the appearance of a seal, with its legs doubled up like flippers. It lived two hours after birth. It is known that cows are susceptible to the eye influence [see MATERNAL IMPRESSIONS], and Druggist Jones says that probably the mother of the calf was in the habit of gazing at the seals and sea lions that abound in the waters of Yaquina Bay." The *Oregon City Courier* (July 24, 1903, p. 5a; ||4hzbs9b3) described what was probably the same animal: "a calf which bears every resemblance in shape and form to a seal. The calf was born in southern Oregon near the coast, where seals are numerous. The calf is of a red color, with legs about as long as a seal's legs and with a head very much the shape of those animals."

Virlet, Puy de Dôme, France. L'Écho de l'Allier (Montluçon, Mar. 22, 1883, p. 3a; ||4wf7ewdu) said a cow belonging to M. Alibert Jacques, a resident of Virlet, dropped a calf weighing 61 kg (134 lbs) and with a head like that of a seal.

No locus in quo. A video that first appeared on the Internet in 2024 shows a living animal with a seal's body and a cow's head (||2u26rcyb). It may be a fake generated by artificial intelligence.

* Coastal farmers allow their herds to wander onto the beach where seals might be encountered. Most of the reports phocobovids involve cattle mothers. So one can picture a cow lying down on the beach, mounted by a seal. YouTube videos show recumbent cattle being mounted even by small dogs. A male seal could presumably do the same.

AN OVIPHOCID (SEAL-SHEEP)

Netherlands. The Dutch physician and author Levinus Lemnius (1564, IV, vii, p. 319), in his *Miracvla natvrae occvulta,* reported the recent yeaning of a lamb with the head of a seal. The *Miracvla* was first published in 1559, which suggests that year as the most likely date for this event.

* Irenäus (1584) referred to what was likely the same case. He dates it to 1564, perhaps taking his information from the 1564 edition of the *Miracvla.*

(Aldrovandi 1642, p. 109)

The Tachov canidequid
(Adlersflügel 1703, pl. 26)

CANIDEQUIDS (DOG-HORSES)

If a mare gives birth to a foal with the claws of a dog, the nation will be diminished.

—A MESOPOTAMIAN BIRTH OMEN†

Reports about dog-horses are few but stretch far back in time. The birth omen quoted above (Lenormant 1875, p. 112) and others like it, date back three millennia. One such mentions "the birth of a foal with the tail of a dog" (King et al. 1910, p. 6). Another reads, "If a mare gives birth to a foal with the head of a dog, her mistress will be cursed, and the nation will be diminished."† Animals like those described in the three omens quoted above appear in the reports quoted below. For example, the Tachov report tells of a horse with the tail of a dog.

Tachov, Czechia. Georg Simon Winter von Adlersflügel (1703, p. 138) described and pictured a canidequid he said he'd seen at Tachov Castle in Bohemia in 1648. It looked, he said, like a horse with a long dog's tail. Adlersflügel was one of the leading equine veterinarians of his day.

White Earth, Minnesota. The *Progress* (White Earth, May 26, 1888, p. 4e; ||y49ewxqg) said, "A mare belonging to Mr. Bellefeuille gave birth to a monstrosity on Tuesday. The feet, tail and head of this singular freak bore a striking resemblance to that of a dog. It was lively and strong, and would, in all probabilities, have lived had not the owner unwisely put it to death."

Victoria, B.C., Canada. The *Victoria Colonist* (quoted in the *Daily Intelligencer,* Seattle, Apr. 3, 1877, p. 3b; ||y4wa3c7y): "The celebrated racing mare, lately owned by Frank Saunders, had a colt yesterday. The hindquarters, legs, tail, &c., are exactly like those of a Newfoundland dog. The strange creature was alive and frisky yesterday. Its head and forelegs are those of a horse and it barks almost constantly."

Highlands, New Jersey. The *Daily Standard* (Red Bank, N.J., Oct. 22, 1909, p. 1g; ||3v478wju) said, "Clarence VanKirk of Highlands has a curiosity in the form of a puppy dog with a horse's head and mane, and a dog's legs and tail. The freak was given to Mr. VanKirk by William VanNote, whose small black and tan mongrel gave birth to this and three other pups … This 'horse-pup' lived but two days."

†. Lenormant (1875, p. 112; ||y6sguegn).

Kanwaka, Kansas. The *Dodge City Times* (Dodge City, Ks., May 1, 1884, p. 1d; ||y3hrd9bl) said, "A most wonderful freak of nature has just developed on the farm of Adam Scheer, a German living in Kanwaka township. One of his breed mares gave birth last night to a colt with perfect body and limbs, but the head, ears, eyes, nose and mouth of an English bulldog. It is said by persons who have seen it, that the head of the colt is not deformed or disturbed, but it is a perfect, well developed dog's head."

Mount Vernon, Ohio. The *Cincinnati Enquirer* (1888, p. 4f; ||3c5zdfy9) reported a mare at Mt. Vernon foaling a viable horselike canidequid with dog-like, clawed forelegs.

Plymouth County, Iowa. The *Omaha Daily Bee* (Omaha, Neb., Sep. 4, 1887, part II, p. 12f; ||y4bfad5b) said, "John Hardie, of Plymouth County, ... has a blind mare that last spring gave birth to a colt that is almost as much dog as colt. The thing has neither mane nor fore top, has a tail exactly like a greyhound, long, slender, and turning up at the end; has a full set of teeth like a dog and will eat meat. When told to do so, it will run after and chase the calves and chickens just as a dog would. In other parts it appears to be more colt than anything else."

Mount Sterling, Ohio. From the *Cincinnati Enquirer* (Aug. 18, 1884, p. 5d; ||2p95dzay): "MOUNT STERLING, OHIO, August 17.—Mr. Norris, a farmer living two miles from here in Pleasant Township, this county, is the possessor of ... a colt-puppy—a beast half dog and half horse. The animal was foaled by a fine driving mare, the property of Mr. Norris, on Sunday, August 3d. The colt, or whatever it is, has been unable to stand since its birth, but is getting along finely. ... The head is that of a greyhound—long ears, nose and small round neck devoid of mane. The body is long and slender, covered with fine hair, as is the tail, which is long and slim. The legs are round and slender. As there is no long hair on mane or tail, it has the appearance of a dog. The hoofs resemble a horse's hoofs. It laps milk and also suckles."

Pietermaritzburg, Natal. The *Natal Witness* (Jan. 5, 1884, p. 5) said, "A monstrosity in the shape of a colt with a head like a bulldog's was born on the Point Road on Saturday morning. It died soon after birth."

Sauvat, Cantal, France. Le Réveil du Cantal (Mauriac, Apr. 22, 1908, p. 3a; ||yckv9h2p) said, in translation, "Mr. Tournadre, Mirandes Farm, Sauvat, has a mare that just foaled a filly with a poodle's ears and tail."

Palouse, Washington. The *Palouse City News* (quoted in the *Spokane Review*, Jul. 7, 1891, p. 6b; ||yeopfcj8) said, "Samuel Eslick has a colt foaled five weeks ago which is a curious freak of nature. It has no mane, and its tail is devoid of hair. Its ears look like those of a dog after being clipped. The lower jaw of the little animal is about an inch shorter than the upper. The colt is very slimly built and gives promise of wonderful speed. It is said that Mr. Eslick refused an offer of $250 for the freak."

* A Samuel Artis Eslick (1834-1894) lived at Palouse in 1891 (||yg439zap).

Laurel, Delaware. From the *News Journal* (Wilmington, Apr. 27, 1914, p. 10; ||bddapvmt): "One of the most peculiar freaks of nature ever seen in this section is a dog-faced colt, owned by Allison Riggin, a tenant on the E. John Hearn farm between Laurel and Bethel. The colt is normal in every way except that its nostrils, ears and tail are like those of a dog. Hundreds of people have gone to the farm to see the curiosity." The *Evening Journal* (Wilmington, Apr. 28, 1914, p. 8; ||yck8y3ku) said the Laurel animal barked.

* A Robert Allison Riggin (1877-1939) and an Elijah John Hearn (b. 1865), a farmer, lived at Laurel (||vkur7h84, ||bdzjvzat, ||2s4f49zu).

CYCLOPEAN CANIDEQUIDS

Fitchville, Ohio. The *Eaton Democrat* (Eaton, O., Mar. 11, 1886, p. 1e; ||mtbxx5k7) said, "A monstrosity from Fitchville Township, Huron County, was taken to Norwalk. The body is of a large well-proportioned colt with head and features of a dog. It has only one eye, and that in the center of the forehead. It was born dead."

Tullamore, Ireland. An ass at Tullamore foaled a nondescript with a double spine, one eye, and "tail, ears and skull resembling those of a dog" (*Freeman's Journal*, Dublin, Jan. 4, 1842, p. 4d; ||2p8mhyjt).

Pittsboro, North Carolina. A cyclops foaled near Pittsboro is classified here, tentatively, as canidequid, though the reports mention no canine traits. They do say the lower jaw was absent, a common feature of dog-ungulate composites. (*News and Observer*, Raleigh, Mar. 26, 1882, p. 4c).

GREYHOUND ORIGINS.

The reports above suggest greyhounds descend from dog-horse hybrids. Such ancestry would explain why canine corns, regions of hyperkeratinization on paw pads, occur in greyhounds and similar breeds (sighthounds) but not in other dogs.

INCERTAE SEDIS

The *Perrysburg Journal* (Perrysburg, Wood Co., O., Apr., 3, 1874, p. 3e; ||3ayj7525) briefly described a related birth.

In Hellenistic lore, the Chimera (shown here on an Apulian red-figure dish, ca. 350 B.C.) was a creature mostly like a lion but with an attached goat's head. Not only the head and neck but also the forelegs of a goat appear in some depictions. A snake or a dragon replaced its tail in some depictions. As shown here, with both a lion's mane and engorged breasts, the Chimera is intersex (as are many telenothians).

A felidungulate, as depicted by Schmuck (1682, #8; ||ebzjxekh). In translation, the caption read, "Reports from London say a calf with a cat's head was born, Feb. 10, 1681. It lived four hours, and many saw it with wonder."

FELIDUNGULATES (CAT-UNGULATES)

> *In this year of 1553, I myself saw a calf like a cat, revolting and horrid, with a swelling under its neck ... Its testes were uselessly small.*
>
> —CASPER PEUCER†

Silver Cliff, Colorado. The *Silver Cliff Rustler* (Dec. 9, 1908, p. 1; ||y32ppoca) said, "Come and see this curiosity, this strange freak of nature resembling both a domestic calf and a wild mountain lion. This calf was born on the Geo. Cress ranch in Wet Mountain Valley, so you can see for yourself that this is not a fake but a real, born animal. If you want to see a Curiosity, a Monstrosity, a Lion-Calf, a Deformed Calf or a strange freak of nature resembling both a domestic calf and a wild mountain lion on exhibition, it is free at the Novelty Store, Silver Cliff.—JOHN FIGGE" The *Cañon City Record* (Sep. 20, 1906, p. 12a; ||5fbhn9cz) later reported,

> A freak of nature in the shape of a lion-faced calf is on exhibition in a tent on Main Street, near Sixth, and is attracting a great deal of attention in Cañon City, as it has elsewhere throughout the Country.
>
> The lion-faced calf is a native of Colorado. It was born on a ranch near Silver Cliff in the Wet Mountain cattle country. After staying with it for two days the mother became frightened and abandoned the young animal. It would have starved to death had not men on the ranch taken care of it.
>
> The head of the freak animal is almost identical with that of a half-grown mountain lion, short ears, widely set eyes and tapering nose and mouth. Its body is long and sinewy, the hair a pale brown, while the legs are short and stocky. The feet, however, are of soft hoof formation and cloven.

* The *Santa Fe New Mexican* (Oct. 21, 1903, p. 1; ||msaxhc7w) also reported the presence of a "half calf and half cougar," likely the Silver Creek animal, at a local fair.
* John Figge's full name was Johann Heinrich Franz Koke Figge (||y5x7oy4s).

Siltcoos, Oregon. From the *Eugene Guard* (Eugene, Ore., Mar. 29, 1931, p. 12; ||46e45wk2): "SILTCOOS, March 28—Disowned by her mother, a heifer calf born on the dairy farm of G. P. Fish, Monday, is entirely unconscious of being different from any other calf. The baby has the heavy, broad shoulders of a cougar, with small ears pointing stiffly backward. Its face is broad and flat, with three long sharp fangs in the front of the mouth and two on the left. It gives the uncanny impression of a cougar's head and shoulders mounted on a calf's body. Mrs. Cow, although she felt so badly, she was bawling, would not succor the infant. It was born in the rain and there she left it. It was cold and wet when found. It is being kept for the time being as a curiosity."

†. *Commentarius de praecipuis divinationum generibus* (1553, p. 328). My translation.

Idaho County, Idaho. The *Idaho County Free Press* (quoted in the *Idaho Semi-weekly World,* Idaho City, Jan. 15, 1889, p. 4a; ||26ctbma5) said, "Geo. [Strutt] Alkire [1850-1915] had a cow which gave birth to a calf that had the head and legs of a cougar, while the body and hoofs were that of a calf. The freak died, and Alkire had it stuffed for future reference."

Camas Prairie, Montana. The *Anaconda Standard* (Apr. 3, 1896, p. 10; ||bdf8heuw) said, "Ranchmen from Camas Prairie tonight brought in a natural curiosity in the shape of a calf with a monster head which bears resemblance to a human cranium. The body of the monstrosity is like that of a mountain lion. It has no legs but short stumps. The creature is dead, having lived but one hour after birth." Resemblance of this animal's cranium to a human may well have been due to rounding effect of hydrocephaly, common in telenothians.

Skrzydlice, Poland. The *Vereinigte Frauendorfer Blätter* (Frauendorf, Germany, 1845, p. 398b; ||2hef2ep6) said, in translation, "We have a bit of news about a remarkable monstrosity calved by a cow, on February 23, at Kundorf [now Skrzydlice] near Löwenberg in Silesia. This animal had the head of a lion, in which the build of the face—the forehead, the mouth and nostrils—while it lived, lay the semblance of a lion before your eyes. Its tongue, which lolled from its mouth, was ring-shaped at birth. ... It was slaughtered six days after birth because of a constriction in its throat that interfered with its taking food. The head was sent to the anatomical cabinet in Breslau."

No locus in quo. A video (||d862vhms) shows a cow-calved animal with feline traits.

A FELIDEQUID (CAT-HORSE)

Valley City, North Dakota. The *Times-Record* (Valley City, N.D., Aug. 7, 1913, p. 8f; ||y77a26hy) said one Walter Emery had a "colt well developed in every way excepting that it has ears and a tail that resembles a cat. Those who have seen the curiosity say that it is a splendid well-developed animal with these exceptions and will probably live to be a full-grown horse."

XENOGENESIS

Aelianus (*Varia Historia,* 1.29) said, "a ewe in the pastures of Nicippus the tyrant yeaned, not a lamb, but a lion."

INCERTAE SEDIS

The *Fairfield News and Herald* (Winnsboro, S.C., Mar. 31, 1886, p. 4d; ||4sc8jj9j) said, "H. J. Thompson of Lancaster County, has a pig, now about three weeks old, which was born without eyes and has only three feet. One of its feet has four claws and resembles that of a cat or cooter." The river cooter (*Pseudemys concinna*) is a freshwater turtle in the family Emydidae. So this animal classifies as either a felidungulate or a chelonibovid.

Jumarians

And what about this well-attested jumart, born of a bull, that they say is now present at the veterinary school—the one at Charenton, I believe. I will have a hard time submitting myself to their existence, in view of the disproportion of the genitals. But when facts speak, it is well to hold one's tongue.

—ALBRECHT VON HALLER
Letter to Charles Bonnet, May 2, 1775

Members of Order Perissodactyla have one weight-bearing hoof on each foot, those of Order Artiodactyla, two. Since they compose two different taxonomic orders, artiodactyls and perissodactyls are usually considered distantly related, and certainly many biologists deem them too distant to hybridize. And yet, there are innumerable reports describing animals combining traits of the two, creatures we will call *jumarians* (joo-MAIR-ee-ənz/).

About twenty years ago, I had lodged a Navarran stallion in the high mountains of Beaujolais. This stallion, full of ardor, covered a cow and, as a result, a jumart was born.

—CLAUDE BOURGELAT

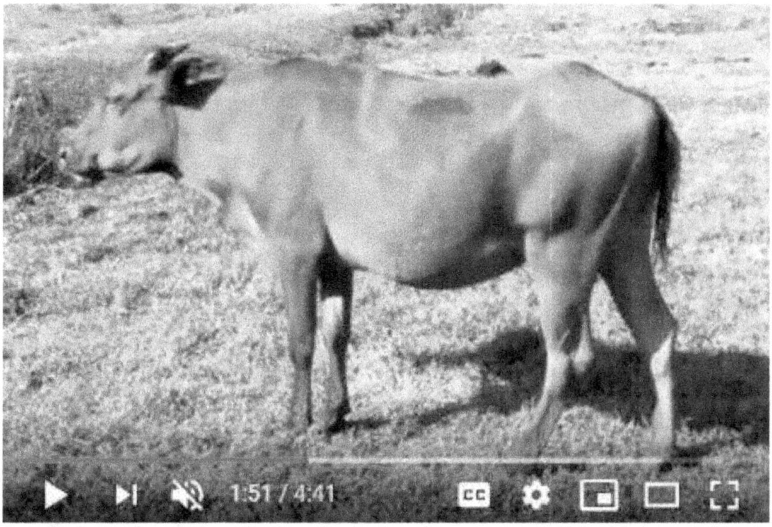

Screenshot of a video (||y54wqbrg) showing an ostensible jumart at Kitale Nature Conservancy, Kenya. Note the horse-like tail.

JUMARTS (COW-EQUIDS)

In India there are also oxen with solid hoofs.
—PLINY, *Natural History,* 8.31

Animals that combine the traits of cattle with those of equids are known by various names. But perhaps the most frequent is *jumart.*† Such composites constitute one of the most commonly reported kinds of telenothians. This chapter documents more than 140 separate cases. And no doubt many more reports, not listed here, exist. So there can be little doubt that such hybrids are a reality. For example, the picture above, a screenshot from a video of a living animal with a cow's head but horse-like hindquarters and tail, lends credence to their claims, in that it seems to confirm that cow-equid hybrids can in fact be produced—though they may be rare and/or underreported—as do the numerous reports, images and videos cited in this chapter.

†. The word *jumart* has been used in English since the eighteenth century, as evidenced by its presence in Johnson's *Dictionary of the English Language,* defined as "The mixture of a bull and a mare." John Locke (1706, p. 384) used the word. The words *jumart* and *jumarre* emerged earlier in French, apparently as a combination of the beginning of certain words meaning "mare" or "horse" with the end of words meaning "bull" (see p. 572).

A bull mounting a mare.

SEXUAL INTERACTION

It is well known that horses and donkeys do occasionally mate with cattle. Such intercourse is common on ranches and in other places where these animals meet. The knowledge is ancient. Akkadian cuneiform texts mention sexual pairings between ox and ass, and between ox and horse (Freedman 2017, pp. 13, 20, 22). And a biblical injunction (Deuteronomy, 22:9-10) shows the Israelites were concerned to prevent the production of donkey-cows:

> Thou shalt not sow thy vineyard with divers seeds: lest the fruit of thy seed which thou hast sown, and the fruit of thy vineyard, be defiled. Thou shalt not plow with an ox and an ass yoked together.

Writing in the first century A.D., the Jewish scholar Philo Judaeus (*Special Laws,* 3:8:46) recorded the fact that it was forbidden to allow cattle to cross with horses:

> And no Jewish shepherd will endeavour to cross a sheep with a he-goat, or a ram with a she-goat, or a cow with a horse; and if he does, he must pay the penalty as breaking a solemn law of nature who is desirous to keep the original kinds of animals free from all spurious admixture.

Even authorities who have denied the possible existence of jumarts, admit that equids do sometimes mate with cattle. For example, Buffon published an eyewitness account of such behavior (see p. 545).

EARLY REPORTS

The earliest reports about jumarts date back to ancient times. A famous example is Alexander the Great's steed Bucephalus, whose name meant, "ox-headed." Many classical authors seem to have taken it for granted that Bucephalus had the body of a horse but the head of a bull. Thus, Arrian (*Anabasis,* V, 18) wrote that "The mark by which he was said to have been particularly distinguished, was a head like an ox, from whence he received his name Bucephalus."

Bucephalus

As portrayed on ancient coins, Bucephalus did have horns like a bull. Chares of Mytilene, an historian who attended Alexander's court, wrote that Alexander's father, Philip, paid 13 talents for the animal (Rolfe 1946, 5.2), an equivalent of 312,000 sesterces in Roman money, which would be like paying a million dollars today. So, cow-horse hybrid or not, Bucephalus must have been an exceptional animal.

Some of the earliest references to probable jumarts are allegations of XENOGENESIS. Thus, the Roman historian Livy (23.31) told how a cow gave birth to a colt at Sinuessa in 215 B.C. Paullini (1686) published a list of curiosities compiled by monks at the Abbey of Corvey ca. 1200 A.D. Observation III of that collection (ibid., p. 186) states that in the year 874 the "monk Geroldus" witnessed a calf being foaled by a mare at Gadelheim, in Germany. Such claims likely represent cases in which a hybrid merely resembled either a calf or a colt but was reported as *being* a calf or a colt.

Ludwig Lucius (1624, p. 831) said a cow calved a jumart with the body and legs of a horse in Northumbria in the year 1215. Boece (1575, XIII, p. 279) also mentioned the Northumbrian animal and said (ibid., p. 355) that a cow bore a calf with the head of a horse in Scotland in the reign of King James I (1406-1437).

Gerolamo Cardano

In his *Contradicentium medicorum,* within the context of an argument he was making with regard to the hereditary content of semen Gerolamo Cardano (1548, pp. 444 & 448) cited the composite appearance of jumarts as an example. That Cardano mentioned the compound traits of jumarts, seemingly taking their existence for granted, suggests he was either aware of some still earlier primary report(s) or that the occurrence of such animals was common knowledge.†

Cardano was soon followed by Conrad Gesner (1551, p. 19), who, in his *Historia animalium*, wrote, in translation, "I sometimes hear of a peculiar sort of mule bred in France in the vicinity of Grenoble, born of a she-ass and a bull, and which in French is called a jumart." In the same place, he said (again, in translation) he had "learned from credible witnesses that in the Swiss Alps above Chur, in the village of Splügen, a foal was born from a mare served by a bull." Gesner, who was Swiss himself, was a professor at the Carolinum, the precursor of the University of Zürich. Elsewhere in the same book (p. 106), he refers again to the strange birth at Splügen. In translation, he said, "Here [in Zürich], we have heard the testimony of trustworthy men that they themselves saw a foal at Splügen in Rhaetia [that is, in what is now eastern Switzerland] that was produced from a bull covering a mare."

One source of this information was likely the physician Jakob Ruf (1505-1558), who, according to his German Wikipedia biography, lived as a young man in a cloister at Chur up to the time of the Reformation in 1526 but thereafter in Zürich, where he became the leading surgeon of the city. In his *De conceptu et generatione hominis*, he mentioned this same birth (Ruf 1587, p. 48) and gave a brief description of what was apparently the same animal.

> For in Switzerland, a horse served by a bull gave birth, at the expected time, to a foal with the hooves of a horse, but otherwise, in shape, hair and tail, like a cow. [translation mine]

†. The first edition of the *Contradicentium* was printed at Venice in 1545.

In *Exotericarvm Exercitationvm,* first published in 1557, Julius Caesar Scaliger, a resident of Agen in France, said, in translation, that a "hinnulus"

Julius Caesar Scaliger
1484-1558

> is the offspring of a bull and a mare. I have seen many and once had a pair myself. However, I now have just one, a young female. On her forehead she has two bony lumps each the size of half a walnut, traces of her father's horns [see photo of skull, p. 552]. In this type of animal, some say, the upper teeth are missing, and this flaw is indeed frequent, because the lower jaw is longer than the upper (as is seen in many fish†). They are called *befi* by the Gabali and Arverni.‡

Porta
1535?-1615

Likewise, the Italian polymath Giovanni Battista della Porta said, "I myself saw at Ferraria [i.e., Ferrara] certain beasts in the shape of a Mule, but they had a Bull's head, and two great knobs in the stead of horns."[1] And the Spanish Jesuit Juan Eusebio Nieremberg (1635, p. 111a) claimed to have seen in Madrid a horned horse belonging to Philip IV.§

Zacchias
1584-1659

Paul Zacchias (1651, p. 504), personal physician to popes Innocent X and Alexander VII, described a jumart, which he said was for many years in the possession of Cardinal Borghese. "And then, I will say a bit, about an ox-horse. It was a mule, as you might say, an animal born of a bull and a mare, with the

†. See note 4 on page 1210.
‡. Translated from the 1612 edition (p. 650). The Gabali were native to Aquitaine, and the Arverni, to the Auvergne. Venette (1732, p. 276) said, nearly two centuries later that such hybrids were still being bred in the Auvergne.
§. The king of Spain (reigned 1621-1665).

> face of a cow, but in other ways not unlike a mule, except that the legs were a cow's, though the hooves were a horse's. It was of ordinary size, or a bit smaller. The coat was thick and rough. Its teeth, both upper and lower, were so defective that it was unable to chew cud. Its voice was somewhere between a moo and a whinny, not the bray of a mule. Otherwise, it seemed mostly to resemble its mother. If I remember aright, it was the Cardinal de Comitibus who brought it [to Rome] from France, where these animals are often bred, and who gave it to Cardinal Scipione Borghese [1577-1633]. A powerful beast, it had a firm, pleasant gait. It lived 32 years and only died because it could chew neither grass nor hard food and had to swallow them whole.[†]

Johann Binhard (1613) described a likely jumart foaled March 15, 1600, at Wangenheim, Thuringia. Polish naturalist Gabriel Rzaczynski (1721, p. 357) stated (in translation) that in 1699, "A horse was shown in the town of Janow, in Lithuania, that had a horned head and cow hooves on its front feet." The Capuchin missionary Girolamo Merolla da Sorrento (1726, p. 3) described jumarts used as beasts of burden in Sardinia at Alghero in 1682 and said he had heard such beasts were also bred in the Cape Verde Islands.

John Locke
1632-1704

In a journal, John Locke (Lough 1953, p. 195) said he had seen a jumart in Paris. His entry for May 16, 1678, reads, "I saw today at Mr. [Salomon de] Foubert's [riding] Academy a beast got between a bull & a mare. About the head, especially the mouth, it has something of mixture in the make between horse & bull, & just at the foretop it had two little hornes which one might easily feele with one's finger. The taile also had a mixture

> of both species, & the teeth in the lower jaw overreached those of the upper. It was but little, but very hardy & laborious. They are never big. In Languedoc they are often to be found & they call them Gimar.

†. My translation. Also see Venette (1732, p. 276).

Right: A horned animal Bartholin (1678) saw in the stables of Frederick III, King of Denmark. Many authors say jumarts have broad muzzles. *Left:* One its deciduous horns. Jacobaeus (1796, p. 4) listed horns from this animal among the items in the collection of the royal Danish library.

In his account of the animal pictured above, seen in Copenhagen, an ostensible jumart, Danish anatomist Thomas Bartholin (1678, pp. 152-153) wrote, in translation,

> Leo Allatius, a most learned man, once saw in the stable of the Prince of Stigliano [of Naples] a thin but fierce stallion with a horn curving down from behind its right ear. Nicolaus Angelus Elisaeus of Naples, Regius Professor of the Gymnasium, told us how he saw the same or a similar animal. We ourselves have seen in the stables of His Serene Highness Frederick III, King of Denmark, a noble steed with horns growing from his ears. These were deciduous, always growing anew.

Patrick Blair (1720, II, p. 310), a Scottish surgeon, anatomist and botanist, and a Fellow of the Royal Society, informed his readers that

> from a Bull and a She-Ass is procreated a certain Animal, called Joumar, as I am credibly inform'd by the intelligent Dr. Sherard, who has often seen them in Turkey, where they are very frequent, and of great Use, as being excellent Beasts of Burthen, and of a quick Pace upon a March, a property not very incident either to Father or Mother. This Animal is a compound Mixture of both, and by being so, of a very unusual Shape.

Thomas Shaw
1692-1751

Thomas Shaw (1738, p. 239), who lived 12 years in Algiers (1720-1732), said he had seen there donkey-sired jumarts. "To the mule we may join the *Kumrah,* as, I think, these People call a little serviceable Beast of Burthen, begot betwixt an Ass and a Cow. That which I saw was single hoofed like the Ass, but distinguished from it in all other Respects, having a sleeker Skin, and the Tail and Head (excepting the Horns) in Fashion of the Dam's."†

François-Alexandre de Garsault (1787, p. 82), Louis XV's Master of Horse, gave an account of jumarts and mules. It said, in translation,

> Mules are monstrosities bred usually from a jackass and a mare, and rarely from a jennet and a stallion. Jumarts are equally monstrous, as they are derived from a bull and either a mare or a jennet, or from a jackass and a cow. These two types of animals are sterile, though they seem to possess all the requisite parts for reproduction.
>
> Mules are much more common than jumarts … A jumart is a small animal, a little larger than a donkey, but very strong. Its head rather resembles that of a bull, having a broad forehead and large muzzle, so much so that when one looks from the front, it seems almost like a bull without horns. Jumarts are common in the province of Dauphiné, where they are used as pack animals.
>
> To produce a mule, the breeder arouses a jackass with a jennet and then substitutes a mare in heat. Likewise, to breed a jumart, he excites a bull with a cow, or a jackass with a jennet, and then introduces the substitution. Jumarts produced by bulls breeding with mares or jennets differ from those produced when jackasses cover cows, in that the latter lack incisors in their upper jaws.‡

†. Shaw's kumrah was cited in the writings of various subsequent authors discussing jumarts, for example, by Réaumur (1751, vol. 2, p. 373), who said he planned to dissect a jumart, and by Buffon, who expressed doubts as to whether jumarts exist. Jean Chardin (in Harmer 1776, p. 445) said he had seen the same animal in Persia: "In this country there is also another animal of a mixed nature, begotten by an ass upon a cow." According to John Emerson, an expert on Persian history (quoted on Wikipedia), Chardin's "information on Safavid Persia outranks that of all other Western writers in range, depth, accuracy, and judiciousness." Chardin lived in Persia during the 1670s.

‡. Garsault seems to be the only author mentioning this effect of cross reversal.

Illustration showing how to mate a bull with a mare. (Georg Simon Winter von Adlersflügel's *Stuterey* 1703; ||jsrkk36).

G. S. W. von Adlersflügel

Indeed, eighteenth-century books on horse breeding gave explicit instructions on how to breed jumarts. Above is shown an illustration from Georg Simon Winter von Adlersflügel's *Stuterey* (1703), a manual on horse breeding. Adlersflügel was a master equestrian and the author of numerous texts on equine medicine. In connection with this illustration, he gave (ibid., pp. 128 & 130) instructions on how to get a bull to cover a mare, which in translation, read,

> One must keep the bull in a dark stall, get him used to wearing a rope halter and give him plenty of good feed. Rub him frequently with the same grooming powder used for stallions. Keep him in this manner for a few months, leading him about regularly by his halter, which should be done at night. And meanwhile, each day use a sponge to rub him well under his nose with the scent of a mare in heat. Then, on a quiet night, one can lead him to the mare and let him cover her. But it must be emphasized that one must do this repeatedly until the mare has conceived, at which time one need not, of course, lead him to her any longer. One must also place the mare in a hole dug out in the ground and restrain her head between posts so that the bull can easily come to her. Also, one should cover her head so that she will not see the bull.

The existence of such detailed instructions for producing horse-cow hybrids raises the question of why an expert on horse breeding should provide them if no hybrids ever resulted from such matings. In breeding ordinary mules, mares are often made to stand in a hole in just the same way, so that short-legged jackasses can reach them.

However, a correspondent of Suchetet (1890, p. 19) stated that farmers seldom produced jumarts on purpose because (1) relatively few couplings resulted in a hybrid, and (2) they feared the hybrid would be cloven-hoofed, as sometimes happened, in which case it would not make a good beast of burden and have to be destroyed at birth. For this reason, he said, most jumarts were born of fortuitous mating.

In late eighteenth-century France, the existence of jumarts seems to have been widely accepted. After the French Revolution, the new official calendar renamed the months, broke each month into three ten-day weeks, and gave the names of common animals to the 5th, 15th and 25th day of each month. The name of the 15th day of the new month of Messidor was *Jumart.* A French encyclopedia (*Cours d'études encyclopédiques ou nouvelle encyclopédie élémentaire,* vol. 6, p. 40), published during the revolutionary era, lists the beasts of burden as "the horse, the ox, the ass, the mule and the jumart." Voltaire, widely considered one of France's greatest Enlightenment writers, accepted the reality of jumarts as much as he did that of mules (see Voltaire 1792, p. 200).

Zirkle (1935, p. 34) says this cross "was cited as an authentic instance of hybridization for well over two hundred years." He also notes that many of the descriptions "seem to be independent eyewitness accounts from several different countries, and no one ever seems to have doubted the creature's existence." But this last statement is not entirely accurate; many people, for example Buffon (1749-1804, vol. 14, p. 398), did express doubts, given that cattle are artiodactyls ("even-toed ungulates"), while horses and asses are perissodactyls ("solid-hoofed ungulates"). In other words, he took it for granted that animals as different as bovids and equids simply could not cross. In his mind, it was an impossibility. He stated his argument thus (in translation):

> It is alleged that another kind of mule results from mating between the bull and the mare. ... In French they are called *jumars.* I brought one of these jumars from Dauphiné, I brought another from the Pyrenees, and by the inspection of the outer parts and by the dissection of the inner ones I found that both of these so-called jumars were simply

hinnies, that is, mules coming from a stallion and a she-ass. I believe myself justified, therefore, as much by this observation as by the lack of resemblance, to believe that this kind of mule does not exist and that the word jumar is merely an imaginary name with no correspondence to any real object. The nature of a bull is too different from that of a mare for them to reproduce together, the one having four stomachs, horns, and cloven hooves, while the other is hornless with a solid hoof and a single stomach. And the genitals being vastly different both in size and in proportion, there is no reason to think they could unite with pleasure, let alone with success. [Buffon 1771, p. 18]

But Buffon is just one witness, who by his own admission examined only two animals. Moreover, in a passage quoted below (see p. 545), he himself later described a bull that voluntarily covered a mare "three or four times a day," which clearly shows these animals do sometimes "unite with pleasure."

A jumart (Jacques de Sève ||y49ns9b5).

But in that same era, many did deem cow-horses real. Thus, the *Histoire générale des animaux* (Buc'hoz 1782-1788), an encyclopedia of animals, includes alongside images of other, more ordinary domestic animals a picture of a jumart (shown at right). The caption to the illustration read, This is a beast of burden produced from the union of a bull with a jennet or a mare, and from that of a stallion or a jackass with a cow. In the 1768 volume of *L'Avant-Courreur,* two such animals are mentioned. One, being the product of a bull and a mare, differed little from a small ordinary mule, except in that the upper jaw was much shorter than the lower. As for the other, bred from a bull and a jennet, it was only about 3 feet 2 inches tall. Its coat was composed of hair much like a cow's. The forehead had bumps in the places where cattle have horns. The upper jaw was also two inches shorter than the lower. It had the muzzle of a cow, of which its body also had the general length and conformation. And it was similar to a cow, too, with respect to its tail and in that it was knock-kneed. [translation mine]

Screenshot of a video (||yy6pltgn) showing an ostensible jumart at Kitale Nature Conservancy, Kenya, in 2007. The inset, which is more in profile, shows the animal's horse-like hindquarters and tail.

FOUR TRAVELERS BEAR WITNESS

Giuseppe Marc'Antonio Baretti
1719-1789

Giuseppe Marc'Antonio Baretti was one of four writers who had direct experience with jumarts. In his book, *An Account of the Manners and Customs of Italy* (1769, pp. 281-285), he says, "As for the carrying of burthens, we [Italians] make use of mules, and of another animal called *Gimerro* [In a footnote here, Baretti says, "Gimerro in English is Jumart."], especially throughout the mountains where horses would

> soon perish. Of mules we have great droves continually carrying merchandises, particularly over those parts of the Apennine that answer to the port of Leghorn; those of the Alps that lie between Italy and Savoy, Switzerland, and Tyrol; and those which geographers call the Ligurian Alps. Some of the muleteers of the Apennine draw even carts with mules; but those of the Alps never do, or at least I never saw any that did. Perhaps the greater height of the Alps and their unconquerable ruggedness causes the want of this convenience.

> It will not be improper to say something of the gimerros, as I find that no travel writer, of the many I have read, has ever mentioned them, and that they are but little known even to those of my English friends who delight in various and extensive reading. A gimerro is an animal born of a horse and a cow; or of a bull and a mare; or of an ass and a cow. The first two sorts are generally as large as the largest mules, and the third somewhat smaller. I have been told by some muleteers in several parts, that the sires of these animals are first shewn a female of their species just before the leap; then led forcibly to one of the species intended, which is kept at hand. The Alpine peasants assure us, that they might get a fourth kind between a bull and a female ass, but that they ordinarily prove sorry things. Of the two first sorts I have seen hundreds, especially at Demont, a fortress in the Alps (about ten miles above the town of Cuneo), that was much talked of during the last war between the French and the Piedmontese. There, many of these gimerros were used, chiefly in carrying stones and sand up to the fortress that was then a-building on a high rocky hill. Of the third species, I rode upon one from Savona to Acqui, so late as the year 1765. It was a sluggish beast, scarcely sensible of the bit and whip; but wonderfully sure-footed; and riding that way in January, as I did, in a most rugged by-road, the whole country round covered in a deep snow; many a mile in a narrow path, often on the brink of a precipice, and all the north sides of the frequent cliffs (over which I was to go) perfectly hidden under the hardest ice; going such a way, I say, I had really need of such a beast, that was very careful not to fall.
>
> The gimerros resemble the mules so much, that, if you are not told, you will scarcely ever think of the difference, which chiefly consists in the ears, not so long as those of the mules; in the parts of the head about the nostrils and mouth, which in the gimerros are generally rounder than in the mules; and in the middle of the back, which is sharper in the mules than in the gimerros. Those between a bull and a mare have likewise a fiercer aspect than the other two species; and the species of that on which I went that journey, have upper fore teeth remarkably more forward than their under [all other authors report the opposite]; yet they feed very well.

Thomas Sedgwick Whalley (1863, pp. 133-134) told how in a journey through Savoy from Allezon to Chamberry (today's Chambéry) in 1784, he traveled on a jumart, which in his journal he called a jumarre. "As I had journeyed thither [i.e., to Allezon] on my good little Prieur's own mule, and attended by his own servant, both of which were gone back to Belleveaux, the Prieur of Allezon had his favourite beast caparisoned for me to return to Chamberry and set me thereon. Now at first, I thought it was a sturdy, wicked-looking mule; but on eyeing him again and again, my belief was staggered, and I asked the Prieur's man who followed me what manner of brute it might be. 'C'est un jumarre,† mi Lor.' [It's a jumart, my Lord] 'Et qu'est ce que c'est qu'un jumarre mon ami?' [And what is a jumart, my friend?] 'O my Lor, c'est une béte qui est faite entre un taureau de montagne et une jument.' [Oh, my Lord, it is a beast begotten by a mountain bull upon a mare.] On I went, then, on my jumarre, examining him from head to foot, and above all things, having an eye to his vile dispositions to break my neck down the rocks, as we descended the rugged and sometimes almost perpendicular paths, that led to Chamberry. Did Vincent's mule dare approach it, a bite and a kick were his sure welcome. Did even the Prieur's man presume to come too near, his malicious posteriors were presently turned in a convenient attitude to have a fling at him. Thus, I crept down the mountains, sometimes with bodily fear, and sometimes worked up to spite myself, by the spite of my beast till landed, at length, safely in the vale, I jogged on with less apprehension, and with relaxed reins. All was well till we were crossing a little river, at the end of the town, when my strangely-begotten animal, resolving to avenge himself, with the true spirit of a Catholic beast, on his heretic rider, stopped short, gave himself one rude shake, and in an instant, and without time for prevention or even thought, I found myself in the middle of the river, and the gentle jumarre making repeated efforts to finish all my worldly cares by kicking out my brains. And his kind purpose would doubtless have been fulfilled, had not Vincent leaped into the water, and saved me from his charitable hoofs. Yet I mounted him again, drenched as I was, and ornamented as he was with a large silk net, fringed above and below, together with bells and with tassels—clapped my spurs in his sides, swore and rated, and went dripping and clattering and ringing through Chamberry, in all possible state.

†. Whalley commented, "This very remarkable hybrid animal is particularly described in the scarce history of the Vaudois Church, settled from time immemorial in three Alpine valleys of Piedmont, published in folio by Jean Leger, the Moderator, in the year 1669. He gives a drawing of the beast and mentions his travelling eighteen leagues through the mountains on one with greater ease than on horseback. The jumarre is a cross between the bull and the horse or the bull and the ass, the head and tail resembling the former, though without horns. As the upper and lower jaws do not meet properly, they can only feed where the grass is long enough to be broken off with the tongue. The editor, when he was formerly in the Vaudois valleys, found the animal was still known."

Bernard Eugène Antoine Rottiers
1771-1857

Rottiers (1829) tells of jumarts he saw while stationed in Russia. In translation, the article, which appeared in the French-language *Journal des Haras,* reads, "The question of whether jumarts exist, which has been so long debated among naturalists, seems to have at last been resolved, that is, unless one wishes to voice doubts about the word of an honorable man known for his wide knowledge and for the scientific research he has carried out during his long and difficult journeys. We refer to Colonel Rottiers, who was born in Antwerp and entered the service of Russia in 1808, where he was sent to Georgia at the beginning of 1812 to serve as head of state to Prince Orbelianoff. He had requested he be sent to that destination in order to study the little-known people residing in that region of the Caucasus bordering on the Black Sea. He resided there at Tbilisi, where he remained until 1818 when, leaving, he traveled to Constantinople via Asia Minor. During that journey he passed through Gori [the regional capital of Shida Kartli in eastern Georgia], there in Georgia, where he had the remarkable experience we here quote to our readers:

> Given that we found it impossible at Gori to obtain carriages like those we had travelled in as far as that place, we found ourselves obliged to content ourselves with some horses and mules of a peculiar kind, and in whose existence many people still refuse to believe. This hybrid born of the union of an ass and a cow, the jumart, is, to put it plainly, extremely common in the northern part of Persia [see footnote, p. 522]. I saw a great number of them in Georgia. It is larger, of better appearance, and less stubborn than either an ass or an ordinary mule but retains few characteristics of its cattle mother. I understand all that might be objected as to the impossibility of intercourse between two such extremely different species of mammal, but to such objections I respond: It does happen! It does! I bear witness to this fact, together with a host of other travelers. And I hope my testimony will tip the balance definitively in the favor of truth. [translation mine]

And Christophe de Villeneuve-Bargemon (1815, p. 89) described how he saw many jumarts during his sojourn in the Vallée de Barcelonnette (today's Vallée de l'Ubaye) in the French Alps. He said, in a letter written from that valley,

> Jumarts are common in this region; and people here think quite a lot of them because they unite the power of a bull with the patience and sobriety of an ass. They are bred by keeping a bull and a jennet by night in the same stable. [translation mine]

Villeneuve-Bargemon

So then, are these animals equid-bovid hybrids? A key to resolving this issue will no doubt prove to be the fact that specimens can be found. In 2007, two were at Kitale Nature Conservancy (pictured, pp. 515 & 526). By now they are likely deceased, but a skull is probably still present in a Paris museum (see at right, and p. 532). Since DNA can be extracted from bones, and as this specimen would presumably be from an F_1 hybrid, it would be easy to test this skull with molecular genetic techniques. Attempts could also be made to produce jumarts, via AI (or by such methods as those described on p. 523).

A jumart skull in the Musée Fragonard d'Alfort.

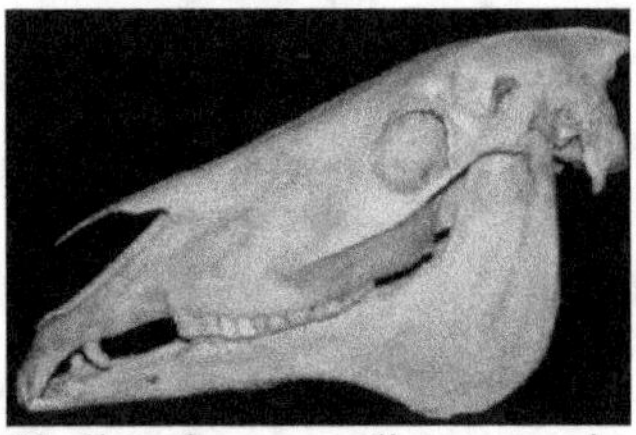

Skull of an ordinary mule. Hinny skulls do not differ in any consistent way from those of mules.

BOURGELAT

I was first led to the Paris specimen after reading Ackermann (1898, p. 74), who said there was a skeleton of a jumart in France in the collection of the Musée Fragonard d'Alfort, an anatomical museum associated with the

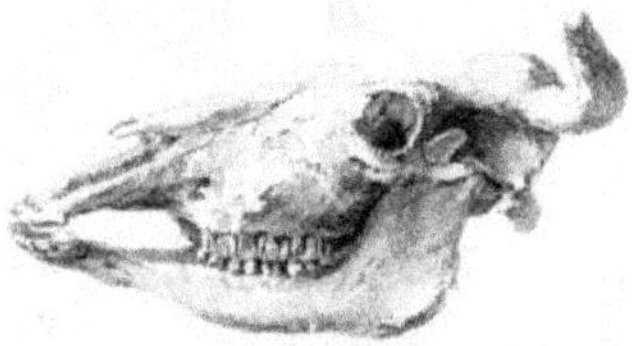

Cow skull

French National College of Veterinary Medicine at Alfort, located today in Maisons-Alfort a suburb of Paris. Ackermann's assertion prompted me to explore the museum's website, which did indeed turn out to have a page discussing jumarts (accessed Apr. 20, 2010), but it has since been deleted.[2] It included a photograph of a jumart skull (see the picture on the opposite page and the larger image, p. 532). The following is a translation of the original French of that now-defunct page:

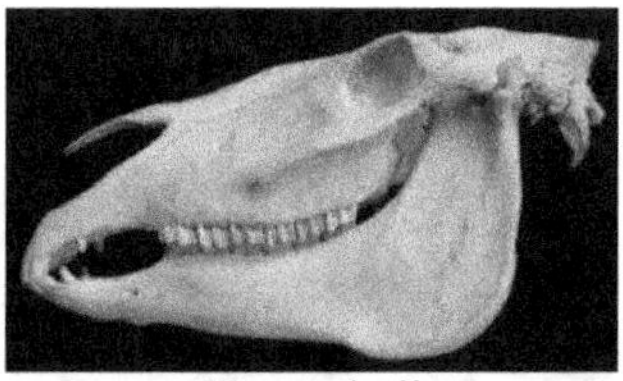
Horse skull

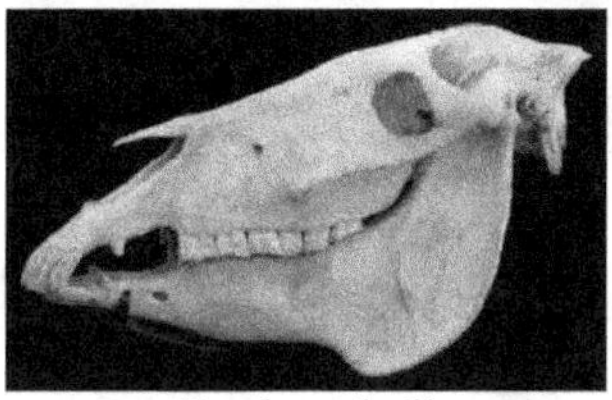
Donkey skull

> Monstrosities were little studied in the early years of the Écoles Vétérinaire, but one exception stirred longstanding controversy: the jumart. This animal, said to be the hybrid of a bovid and an equid, was reported by [Claude] Bourgelat, who used all his prestige to make the idea of this cross accepted. In his correspondence with the naturalist Charles Bonnet, he described it precisely. This jumart was unusually strong. Its forehead, muzzle, and lower jaw were those of a cow, but the teeth and uterus, those of a mare. Notably, it lacked a gall bladder. Elsewhere, Bourgelat said he had produced such a hybrid by mating a Navarre stallion with a cow. It only lived four months and was more like the mother than the father. The founder of the Écoles Vétérinaire [i.e., Bourgelat] noted that several cases had been presented to the Ecole de Lyon and that Dauphiné [a former French province] was certainly one region particularly favorable to their production. The jumart became the center of a huge controversy in which the major biologists of the era asserted their opinions. Buffon, in the *Supplément à l'histoire des quadrupèdes,* recognized that mares and bulls do copulate despite their differing genital organs. He reports seeing in 1767, at the home of his miller, two or three copulations a day without any birth having resulted. For Buffon, the jumart was a fantastical animal though he did admit other types of interspecific crosses are possible, as for instance that between wolf and dog. Garsault, Haller, and Huzard backed his opinion [But we have seen (p. 522) Garsault disagreed. Late in life, Haller, too, changed his mind (see p. 536)] and that opinion triumphed after nearly a century of controversy. The Musée Fragonard d'Alfort still has a jumart skull dating from the time of Bourgelat. The cranium is globular, the lower jaw, longer than the upper, and upper incisors are displaced toward the rear.

The Alfort skull (access a high-resolution image here: ||yy8cxayp). Given that the label, visible beneath the skull in the (enlarged) photograph, reads "Tête de jumard. Auteur: C. Bourgelat" ("Head of a jumart. Collector: C. Bourgelat"), this specimen must have been obtained from one of the animals Bourgelat claimed to have had had during the years he served as director and inspector general of the French National College of Veterinary Medicine at Alfort.[3]

The jumart skull mentioned in the just translated, quoted text was pictured on the same webpage (a larger photo of the same skull is shown above). If this skull is in fact that of a jumart, then clearly jumarts are possible, however imaginary they might be imagined to be. The skulls of mythical animals do not end up in museum collections. So, either this type of hybrid has been produced, or the skull pictured is a fake or that of some other creature. Note, however, that this skull, which the label in the Musée Fragonard's exhibit in fact describes as a "tête de jumard" ("head of a jumart"), differs markedly from that of a cow, horse or ass (see pictures above, pp. 530-531, comparing these skulls). Nor is it at all like that of either an ordinary mule (mare × jackass) or a hinny[†] (stallion × jennet). So, if it is not a jumart, what is it? And what are the animals shown above in the video screenshots (pp. 515 & 526)?

†. The picture of a mule skull (p. 530) suffices to represent also, the skull of a hinny since Allen and Short (1997, p. 385) say, "We still lack conclusive proof as to whether there is any consistent phenotypic difference between mules and hinnies."

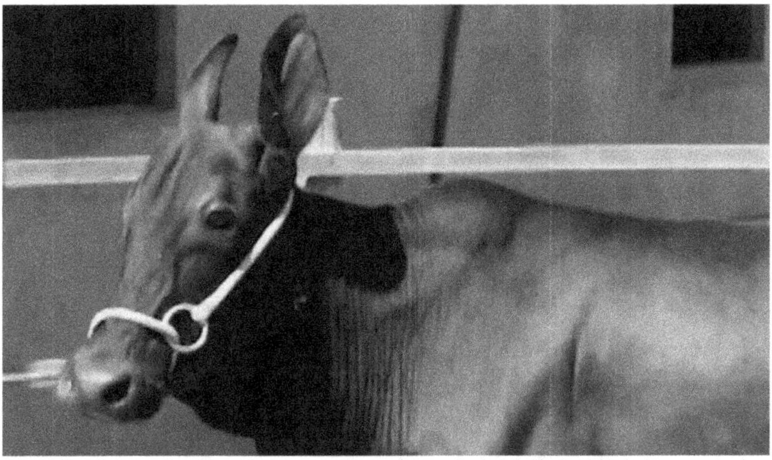

An animal described on Imgur as a cow-horse hybrid (||y4ctl58a). It seems that no further information about this creature is available.

The French college of veterinary medicine is one of the world's oldest. Its main branch at Alfort opened in 1766, and its museum, the Musée Fragonard, also located at that branch, opened in the same year. The founder of both,[4] the veterinary surgeon Claude Bourgelat (1712-1779), began the museum's now extensive collection. He was a member of both the French and Prussian academies of sciences. Many veterinary colleges in other countries were founded by his students, who took the French college as their model.

Claude Bourgelat

Charles Bonnet

In his *Oeuvres d'histoire naturelle et de philosophie,* Charles Bonnet (1779, pp. 349-352) described correspondence he exchanged with Bourgelat on the subject of jumarts. "Confronted with authorities of conflicting opinion, I greatly desired to be able to make up my mind on the interesting question of the existence

> of jumarts. In a magazine, I had seen the description of a jumart that M. Bourgelat, Inspector General of the Écoles Vétérinaires de France, had

> caused to be dissected under his direction at the Lyon branch of the school, but dared not trust the report of a journalist. Having to publish a new edition of my book, I directly contacted M. Bourgelat himself, and the expert and detailed letter I received in response from that celebrated individual leaves no room for doubt that jumarts are in fact very real. The author of the letter began as follows: "I believe in the existence of this particular kind of hybrid, known as the jumart, as much indeed as in my own existence. I have had several of them sent to me from the Haut Dauphiné by students at the veterinary school, which were born on their fathers' farms." Such formal testimony from such a renowned academician as M. Bourgelat, who is better able than anyone to judge the truth, would seem to carry the greatest weight. [translation mine]

Elsewhere, Bourgelat asserted that he himself had got a jumart from a cow that had been covered by a stallion. Bourgelat's claim, as quoted by Bonnet (1779, p. 547), reads,

> I can, by a fact to which I can bear personal testimony, even more willingly believe that a jumart could be produced from the coupling of a jackass with a cow. About twenty years ago, I had lodged a Navarran stallion in the high mountains of Beaujolais. This stallion, full of ardor, covered a cow and, as a result, a jumart was born. As you might well suppose, I prized this precious creature; and to inspire the horse's keeper to take good care of her, and I agreed to pay, as soon as she was weaned, a much higher price for her than for an ordinary colt, but she lived only four months. [translation mine]

Bourgelat's account is important because it cannot be explained by the argument that alleged jumarts are all hinnies (i.e., offspring of a jennet covered by a stallion). Those who have argued that jumarts are impossible have often used this argument to undermine the validity of reports about jumarts. But the offspring of a cow would, of a certainty, have a cattle component in its hereditary composition, and therefore could not possibly be an ordinary hinny, even if it happened to resemble one.

In the present context, it also appears significant that (1) descriptions of jumarts regularly mention that the upper jaw is much shorter than the lower (the usual distance mentioned is "an inch and a half"), and (2) that in the pictures shown above (see pp. 530 & 532), of what the museum website describes as the "skull of a jumart" ("tête osseuse de jumard") that the incisors in the upper jaw do, in fact, fall far short of those in the lower. This is not the case in a horse or an ass, and a cow does not have upper incisors. There seems to be no known animal with a similar mismatching of the incisors, nor one with a tooth and skull configuration like that seen in this putative jumart.

Leger (1669, p. 7) says this mismatch generally prevented the jumarts he had seen from grazing, except in fields with long grass, where they could crop by pulling it off with their tongues. Similarly, Paul Zacchias (quoted above, see p. 520) mentioned that an old jumart had to be put down because it could eat neither grass nor hard food.

Also, the Alfort subject had an obvious knob over each eye, a feature noted in many descriptions of jumarts. (These knobs are better visible in the smaller of the two images shown of the specimen, as the larger image was taken from an angle making it harder to see them.)

In the article quoted above, Bonnet (1779, pp. 351-352) includes an excerpt from Bourgelat's letter, describing a female jumart:

> Fear of altering the anatomical description of this jumart in abridgment has prompted me here to transcribe nearly word for word from M. Bourgelat's letter. In its external appearance, this animal has a mouth or muzzle like that of a cow.
>
> With its enveloping soft tissue removed, one finds the cranium much more rounded than in a horse. The frontal bone is larger; the nasal bones more depressed in their upper part; the orifices of the nasal cavities, and the cavities themselves, much narrower; the orbital opening, round, whereas in the horse it is oval; the palette, much longer and more concave; the upper jaw an inch and a half shorter than the lower, the former being, as in cattle, at least two inches wider than the latter. Each jaw contained twelve molars, six on each side; those of the upper jaw describe an arc posteriorly. … the region corresponding to the space between the canines and first molar was flattened relative to the horse, one and a half inches in length, and convex, whereas in the horse it is concave.
>
> This jumart had neither canines nor wolf teeth. The incisors, eight in number in the lower jaw of a cow, here numbered just six in each jaw. They were an inch and a half long. They were not vertical but inclined forward so that the upper jaw did not bear on the lower jaw except at the very tip of the left first incisor.
>
> The tongue was like a cow's. And its papillas or buds were as easily distinguished as in that animal. The throat was much larger in diameter than in a horse. … The stomach was single and shaped exactly as in a horse, but much larger. The spleen was of the same shape and consistency as that of a cow. The uninary bladder at its greatest dilation had a diameter of no more than three inches.

> The uterus was exactly like that of a mare or a she-ass. [description of uterus omitted] The ovaries were the size of a fava bean, soft and flaccid. [Hybrids often have small, soft gonads]
>
> It may also be mentioned that there was no gall bladder [A gall bladder is present in cattle, but not in horses.] and no difference in the structure of the remaining visceral organs, which resembled those of a mare in all respects. The musculature was similar to that of a horse in all respects. M. Bourgelat ended with the following words: "Since that time we have opened and dissected several jumarts, both male and female, and I can assure you that all had a single stomach and no gall bladder." [translation mine]

According to French zoologist Isidore Geoffroy Saint-Hilaire (1860, p. 146), Bourgelat's testimony also convinced the prominent anatomist Albrecht von Haller (1708-1777), though he had at first denied the existence of jumarts. Indeed, in a letter to Bonnet dated May 2, 1775 (Sonntag 1983, p. 1163), Haller wrote (in translation),

> And what about this well-attested jumart, born of a bull, that they say is now present at the veterinary school—the one at Charenton,† I believe. I will have a hard time submitting myself to their existence, in view of the disproportion of the genitals. But when facts speak, it is well to hold one's tongue.[5]

And the previous year Haller (1774, pp. 497-498) had spoken of jumarts in published work as if he deemed them real. Geoffroy (ibid.) also says the Italian biologist Lazzaro Spallanzani (1785, p. 219; also see p. 317) accepted their existence. In translation mine, he said that the "observations of a celebrated French naturalist [i.e., Bourgelat] leave no room for doubt concerning the occurrence of jumarts, hybrids from a cross between a jackass and cow, or between a bull and either a mare or a jennet, though Monsieur de Buffon has formally denied their existence."

Bonnet (1779, p. 547) remarked that he looked forward to further enlightenment from any new research by Bourgelat regarding jumarts. Bourgelat, however, died that same year.

†. Since Charenton, a larger community, was adjacent to Alfort, people spoke of the college of veterinary medicine at Alfort as if it were in Charenton.
‡. Spallanzani was a Fellow of the Royal Society.

A WATER-STORAGE ORGAN?

In an interesting sequel to the events at Alfort, the Swedish-born German naturalist Karl Asmund Rudolphi (1805, pp. 46-47), who visited the Musée Fragonard d'Alfort at the beginning of the nineteenth century, mentions that one jumart there had an unusual organ, which apparently served the function of water storage:

> The animal ... was at Alfort for a long time, and people were surprised that he was never seen drinking, and they concluded that he in fact never drank. For all attempts to make him drink were in vain. But then, all at once, he drank two buckets of water, which, given the small size of the animal, provoked new astonishment. But at autopsy they found a hollow organ on the upper surface of the stomach of about the same size as the water-stomach ["Wassermagen"] of a llama. [translation mine]

OTHER REPORTS

At about the same time as Bourgelat and Bonnet were exchanging letters, the German writer Johann Jacob Volkmann (1777, p. 240) reported that two living jumarts, a male and a female bred in Savoy, could be seen in the menagerie of the Landgrave of Kassel.[†] Wilhelm Friedrich von Gleichen-Rußwurm (1778, p. 30) recorded another German case he encountered at Bayreuth.[‡] In translation, he said, "The jumarts I saw at Bayreuth were bought at Arles in 1755 for more than a hundred pounds apiece. Though it had not been said to what species their father belonged, their conformation clearly indicated an ass, for which they were sometimes mistaken. They were not much larger than the biggest asses used by our millers. They were male and with dark brown coats, and their heads were wide, with broad foreheads. Their ears were quite long, their necks, backs, legs, hooves, and tails like a donkey's, and their voices, too, were almost like a donkey's. In addition, they were very lascivious, mean and stubborn, rearing if there was the slightest chance to assault the stableman. They were strong and dragged a heavy load of bread daily to the yard from the bakery, which was a quarter of an hour away. How old they were, I don't know. But they generally live to be 25 or 26 years old. Although these creatures are rarely larger than the largest donkey, a trustworthy friend of mine, who spent a few years in Madrid, assured me that in the stables of the king there he had seen a jumart larger than the largest mule, and which served as a mount."

†. A fact confirmed by Prizelius (1777, p. 165) and by Hartmann (1786, p. 376).
‡. Confirmed by Hartmann (1786, p. 376).

Isle of Ely, Cambridgeshire. A report about a jumart of a rather unusual nature, appeared in *Jackson's Oxford Journal* (Jun. 18, 1768, p. 1c; ||y2fq4com). From the description given, the animal in question may have been a bilateral gynandromorph, a rare condition in which one side of a hybrid is like one of its parents, the other, like the other. "We hear from Thorney Abbey in the Isle of Ely, that Mr. Patrick, an eminent Grazier of that Place, had a two-years-old Filly with a Foal, which produced on Saturday last a very uncommon Creature: One side of it is like a Calf, red and white; the other side like a Foal, quite black, except a few red Spots interspersed. It has a Head like that of a Calf, and two Horns already about three inches long; has Fore Feet like those of a Foal, sucks at the Mare, and continues in Health."

* Only a few telenothian bilateral gynandromorphs have been reported (see p. 1231).

Sarcey de Sutières. In an encyclopedia entry, agronomist André Sarcey de Sutières (1788, pp. 3-9) said, in translation,

THE JUMART OR HIPPOTAURUS

Jumarts are produced from the union of a bull with a mare or a jennet. A jumart is also produced by the mating of a jackass and cow, but they say that those with such parentage often differ from the others. There also exist, so they say, jumarts with cloven hooves, but I have never seen any.[6] I am, however, quite familiar with the others.

These animals are found fairly commonly in the vicinity of Chambéry, in Switzerland, and in the [former French] province of Dauphiné. They are bred quite commonly in Leforest, around Saint-Chamond, Saint-Étienne, and in the mountains around Pila, for in those regions their utility is well known.

A DESCRIPTION OF THE JUMART

I myself have owned three jumarts, all of whom have rendered me great service. The one I had the longest stood three feet three inches high at the withers and his hair was red with a deep brown cast. His legs were like those of a donkey, but he had knees like those of a cow. His chest was broader and more rounded that a donkey's, and his upper jaw, shorter than the lower. His muzzle and forehead were like those of a cow, and his eyes, too, were exactly like a cow's. But his ears were a little wider and shorter, like those of a donkey. His tail and backbone were like a donkey's, though in the conformation of his flanks he resembled a cow.

THE UTILITY OF JUMARTS

There are few animals from whom one can receive such great service as a jumart. And if the strength of these creatures is considered in proportion to their size, it would seem to be the greatest of any animal in existence. In fact, they can bear even the heaviest baggage and do so for longer than any other beast of burden. And when pulling a wagon, a jumart would sooner break the reins than pause for even a moment.

By pulling his little tumbrel, my jumart kept our home supplied with food, even where we live, atop the mountain across from Ance. Every day during the warm seasons he brought home, via steep roads covered with rocks, all the hay gathered in the vineyards. And in the cold seasons, he carried stone from the quarries. In 1770 or 1771, in the part of Beaujolais where we lived, there was such a flood that all the villages around the mountain were submerged, a disaster that caused shocking damage, but which was ignored because it was all so far from Paris. Many trees were uprooted, and vineyards, stones and earth were all swept away. A bit of pasture I have along Alix brook at Chatillon, was buried beneath 22 feet of stone and earth. However, level land being precious in this mountainous region, I resolved to clear it and to use my jumart to help me in the work.

We had to move all those rocks back into the little wood above the pasture, and that was no easy job because the road was steep and eroded.

This brave animal saw that the roots of the trees, which had been laid bare by the rains, crossed the road, and in climbing, he gripped them with his teeth to catch his breath or to help himself onward. As amusing as this behavior might have been, it is well that I loaded him accordingly. Otherwise, he would have killed himself by ripping out all his teeth. At any rate, he was the best of all my domestic animals, and the one that I could most rely upon.

This courageous animal was of a most docile nature, whenever his driver asked him to do anything. The other animals weren't so willing. They, sometimes, would refuse to move forward, or, after going forward a certain distance, turn and gallop back to the house, or lie down to throw their burden to the ground.

One year he took care of my wife whose presence was required on a farm of mine six miles away across the mountains where she had to feed the livestock. She had to go there two or three times a week. The jumart, because of the sureness and gentleness of his tread, was chosen to carry her there. She rode atop a saddle, and with a baby in her lap, and he was incredibly gentle with her. He stopped only in those places where his rider could dismount most conveniently. He kept to her side in the fear that she might take a misstep and moved forward at the mere prompting of her voice. Whenever he saw that she had the baby, he had no need of his bridle, as if he realized that she needed both hands to carry her child. The paths being very difficult, often in the descents the rocks, rolling underfoot, would make him stumble, but whenever he lost his balance, he would crouch so that my wife could slide off, then he would recover and go to a place where he could see that she could easily remount. These actions, so often repeated, convinced us without doubt that he was able to reason. And he quite expected his behavior to be acknowledged; as soon as he arrived, you had to give him a piece of bread dipped in wine, and if you delayed, he'd eat nothing else, never cease with what was his own peculiar species of braying.

He would let no one mount him except my wife. If someone else tried, he would bite and buck. And if that did not suffice, he would march forward quietly and then suddenly turn and gallop into a wall or tree. And if all that failed, he would simply lie down.

One might believe that this story is sheer fabrication, or at least exaggeration. But I could relate deeds of this animal that might astonish you, such as his ability to open doors and to undo the most complicated knots, to come to your call and to go back again on command, etc., etc. One sees animals that can do surprising things, but if this one had been trained when he was young, he could have been put on tour in shows. The other two jumarts that I had were intelligent but not like him.

I'm astonished such useful animals aren't bred in great numbers. Who knows whether, if in devoting oneself to the production of various hybrids, one might not succeed, with time and the assistance of certain combinations, in re-creating the various species of animals known to the ancients, and which have disappeared for lack of circumstances favorable to their continued existence.

Grognier. Louis-Furcy Grognier (1773-1837), a member of the Académie Française, was a professor at the École Nationale Vétérinaire at Lyon, and later its director. Like many veterinarians of his era, Grognier (1841, p. 83[7]) affirmed that jumarts exist, though he considered them rare. "Enlightened, trustworthy individuals claim to have seen these animals, but most naturalists deem them imaginary. However, whatever the differences may be between equid and bovine genitals, one cannot deny that a bull can cover a mare, and a stallion, a cow; both these things have often been seen. But from the fact that one observes a hundred such unions without progeny, does it follow that they are never fecund? It is an established fact that in the country, where males and females of all species are put together pell-mell on the pastures, hybrids are sometimes born that have the head of a calf, the tail of a cow, and knobs in place of horns, but having body and limbs built like those of a horse. At the school of veterinary medicine in

Lyons an animal was examined that in its general form was like a mule, but whose face and upper jaw resembled a bull. The tongue was covered with papillae as in the bovine species. This animal had neither a bull's bellow, nor a horse's neigh, nor the bray of an ass; rather its sharp, shrill cries recalled the bleating of a goat. I could supply additional examples but will be contented to say that, while I do not reject the existence of jumarts, I think they are rare." [translation mine]

Grognier (1805, p. 194) said one jumart was described to him by his colleague, Claude-Julien Bredin (1776-1854), who, like himself, served as director of the École Nationale Vétérinaire at Lyon.

> while at Sospel in the Maritime Alps department where [Bredin] was performing his duties as a veterinarian to the artillery, was shown an animal that was supposed to be the offspring of a bull and a she-ass. It stood 3 feet 5 inches at the withers. Its body resembled that of a donkey, but its hair was like that of a mule; its rump was much more rounded than that of a donkey; the stomach was distended, like that of a cow; the neck, very short and thick; the hooves, well formed and undivided; the cannon bones, slender and without the slightest indication of a groove in their anterior portion; the knees, extremely large; the mane was black and shaggy; the head, large and well-set; the forehead, prodigiously wide, and in its middle deeply depressed; there were two prominent knobs just above the eyes, but they had no trace of callosity, nor anything about them that resembled horns.
>
> The incisors of the lower jaw were properly placed but those of the upper, displaced to the rear. The first incisors were large, but the second and third, very small; the eyes were extremely large, and the end of the nose, quite slender, as in an ass. This bizarre animal was a male, and its genitals resembled those of an ass. [translation mine]

Brieude. French physician Jean-Joseph Brieude (1821, p. 391) concurred with Grognier's view that jumarts are rare. He said he had seen only two in his native Cantal. Given that these animals do seem to be rare, the attitude expressed by Grognier himself (1805, p. 194) is reasonable: "When a truthful and forthright person claims to have witnessed something extraordinary, one has no right to say that such a thing cannot exist simply because one has not seen it himself. Nevertheless, many authors reason as follows: 'We have never seen a horse-cow hybrid, therefore anyone who says he has is either mistaken or wants to trick us.' Why not say instead that animals of different species unite with difficulty and that, when they do, offspring very rarely result?" [translation mine]

Enniskillen, Ireland. In a report entitled HORNS IN THE HORSE, military veterinarian Charles Percivall (1828) described an animal he treated, probably at Enniskillen: "This horse had … excrescences … placed about two inches above the orbital processes of each frontal bone, in the centre of the forehead, two inches apart, projecting from half to three quarters of an inch from the surface of the frontal bones, very much resembling what is felt in young calves five or six months old."

Greenwich, England. And there are other, detailed and firsthand reports by veterinarians about jumarts. One such was authored by veterinary surgeon John William Gloag (1830). It reads,

> About eighteen months since, I was at Greenwich, and my curiosity was excited upon hearing that a monstrosity was then exhibiting, partaking of the peculiarities of the cow and horse. The animal seemed to be about fifteen hands high (a mare, if I dare call her so), of the commonest description, and three years and a half old. The most striking peculiarities were in the tail, hips, head, breast and off [i.e., right] foreleg. The off foreleg, from its singularity, claimed my first attention; and I was induced to examine it with great care, as, at first sight, I suspected it might have been the result of disease. The shoulder-blade and the humerus seemed to be shorter and more upright than usual, and the latter bone was thicker, and turned very much outwards. This leg was in every respect similar to a cow's; but the similarity was still more evident when I descended to the foot. There were two hoofs, with separate joints, and a natural [separation] between them. There was no sign of a frog, but the animal trod upon her heels and used the foot precisely as a cow. I walked her round the stable several times, and it was really ludicrous to observe the difference in the manner of progression between the two legs. The colour of the hair on the body seemed to present a strange mixture between that of both animals; but it was long, soft, and loosely attached. The head had a very curious appearance: it seemed to be broad at the poll, and the ears were very large, and rounded, and turned backwards and inwards and covered with long, shaggy hair. The lower part of the face projected very much on each side, giving a width very dissimilar to the face of the horse.[8] There were six incisor teeth in each jaw, but very unevenly distributed; and they seemed to partake much of the mixed character. The breast was wide and hung down, strongly resembling the dewlap [seen in cattle]. The body, neck, and near [i.e., left] foreleg, resembled those of the horse. The hips, tail, and [hind] legs above the hock, were exactly similar to those of a cow. The spinous and traverse process of the ileum projected very

much, and the pubis stretched backwards, giving the great prominence of the bones of the ischium always so observable in a cow. The sacral processes did not project, and the tail did not arch over the pubis, as in the horse, but seemed to fall more abruptly: it was shaved nearly to its tip, where was left a tuft of hair, which was frizzled, and very like that of a cow. The tail felt very fleshy and at its base the skin was loosely attached; and the vagina was without colour and hung down as in a cow. The patellae were very thick and large, and the hinder extremities, above the hock, were totally dissimilar to a horse. The mammae were not very large, nor was the udder unnaturally distended. At this time, I related the facts to many veterinarians, among others to Mr. Sewell.

A short time since I called upon Mr. J. Skilt, V.S., at Southwark Bridge, and he informed me that he had seen the skeleton of a very strange animal at the knacker's; and wished me to go and examine it with him. I did so; and was much surprised and pleased once more to have an opportunity of surveying the monstrosity I have just been describing; and I am enabled to add the following particulars to my previous description of her:—The occipital and pterygoid processes are large, and the former very broad and projecting, and differing from the triangular-shaped process in the horse. There are no signs of horns. Just above the three anterior molars on each side is a large projection, similar to specimens of disease wherein the teeth have still remained in their sockets. The three first molars are perfect cow's teeth. The ileum and pubis are laid nearly flat, as in the cow. The number of ribs and vertebrae are the same as in a horse. All the joints seem to be unusually large. The greatest peculiarity is, however, in the off fore foot; and I am obliged to be very concise in my account of it. Articulating with the lower head of the large metacarpal bone is another, altogether shorter and broader than the large pastern of the horse, and cleft nearly to its middle; and, articulating with each division, are the remaining small bones of the foot. There are two hoofs, separately joined, with a natural [separation] between them; but they have grown enormously since I saw the animal alive. The knacker … did not perceive anything in the stomach different from that of the horse. Any gentleman who wishes for more information can examine this, and many other curious specimens, at Winkley's Yard, in Fryer's Street, Borough. The only history I could learn of the animal was that the owner had purchased her from the breeder, who stated that her dam, when very old, had been put out on Finchley Common with a bull; that no horse could have access to her; and that this very singular animal was probably the offspring of the mare and the bull.

* An entry about Gloag in a history of British veterinary medicine (Smith 1933, vol. 4, pp. 39-40) reads, "JOHN WILLIAM GLOAG (1812-1886), an army veterinarian of great ability, singularly attractive personality and high ideals. He was all that is implied in the phrase 'a great gentleman.' In 1858 he was elected a Vice-President of the R.C.V.S. He was a careful and exact observer and furnished many papers to the *Veterinarian*."

Joly (1860) reported an animal at Toulouse he thought was a mule. Its two forefeet, he said, were "truly bizarre and of a form exactly like the hoofs of an ox" (translation mine). He said he'd seen two similar animals. One was the nine-month aborted fetus pictured above (Lavocat and Joly 1853).

Santa Fe, Argentina. Azara (1802, p. 226) said, in translation, "Reliable reports say a horse recently foaled in Santa Fe de la Veracruz had 4-inch-long horns like a bull's, upright and sharp, growing next to its ears, and that Don Juan Agustin Videla, who lives near Buenos Aires, brought one from Chile with sturdy upright horns three inches long."

Aberlour, Moray, Scotland. The *Elgin Courier* (Scotland, Aug. 3, 1855, p. 3): "Mr. Maconachie, Tombain, Aberlour, has among his herd a year-old stirk of the common breed, which is distinguished from all the others by being furnished with close-hoofs, exactly similar to those of a donkey! The creature has another peculiarity of invariably going down upon its knees when eating."

Struthers. Anatomist John Struthers (1863, p. 279) described an animal like a horse that had a single cloven forefoot. "In September 1859, I examined, on Ford Common, Northumberland, a two-year-old filly which had been born with the left fore foot cleft like that of the ox. Each of the two toes had its three phalanges, which could be made to move past the corresponding phalanges of the other toe, showing the complete division of the foot as far up as the fetlock … joint. The division externally was carried to the same extent as in the ox. The lower end of the great metacarpal (cannon) bone felt as if bifurcated like that of the ox, so as to give separate articular support to the two toes. Farther up, the great metacarpal, as in the other forelimb, presented the usual form of that in the horse. The two lesser metacarpal bones were felt to terminate at the usual place on the right side, and as if a little farther down in the left limb; but of this I could not be quite certain. The two hoofs were quite separate and complete, each having its own horny "frog" as in the ox. … The whole foot was much larger or more spread than the other."

NATURALISTS VS. VETERINARIANS

In the 1800s many veterinarians accepted the existence of jumarts, a credence apparently based on observations made in their practice. But many naturalists, who were not veterinarians, rejected their existence. This latter school of thought seems traceable in Germany to Blumenbach and in France to Buffon. In his *Handbuch der Naturgeschichte* Blumenbach(1779, p. 112) claimed that all alleged jumarts were in fact hinnies. The book was widely cited for decades.[9] And in an account also much cited by later writers arguing that jumarts do not exist, Buffon (*Histoire Naturelle,* 1829, vol. 16, p. 392) told how matings between a horse and a cow had failed to bear fruit.

> On my family estate, in 1767 and thereafter, the miller had a horse and a bull that had such passionate feelings for each other that whenever the mare was in heat, the bull, if allowed, never failed to cover her three or four times a day. These couplings, repeated many times over several years, gave their master high hopes of seeing offspring. But it never resulted in anything, and all the residents of the place were witnesses, both of these two animals mating—which was real and often repeated for several years—and of the lack of offspring. This fact seems to prove that, at least in our climate, a bull can produce no offspring with a mare, and it is this finding that makes me doubt, very legitimately, the existence of this first kind of jumart. I have no facts as certain that I can cite in contradiction of the possible existence of the second sort of jumarts, the type of which Dr. Shaw wrote and which he says comes from the donkey and cow. I admit that, even though the difficulties involved seem about as great in both cases, the positive testimony of someone as well educated as Dr. Shaw makes the existence of this second type of jumart more likely than that of the first. And with regard to the third type, that derived from a bull and a she-ass, I am convinced, despite the testimony of Mérolle [i.e., Girolamo Merolla da Sorrento (1726, p. 3)], that it does not exist anymore than a jumart from a bull and a horse. There is even more disagreement, yet more natural distance, between a bull and a she-ass than between a bull and a mare,[10] and from the fact, which I have mentioned—that a bull produces no offspring with a mare—we can infer that no progeny will be produced from the union of the bull with a female ass. [translation mine]

But reports kept on appearing. A *Placer Times* reporter (quoted in the *New-York Daily Tribune,* May 8, 1851, p. 6d; ||y5zalpdt) said he saw in Napa Valley, California, "a young colt with hoofs split crosswise, and each part separate like that of a cow." And the *True Northerner* (Paw Paw, Mich., Jan. 5, 1872, p. 3d; ||2th87rvk) said, "The latest freak of nature is an animal at Council Bluffs [Iowa], which is half horse and half cow."

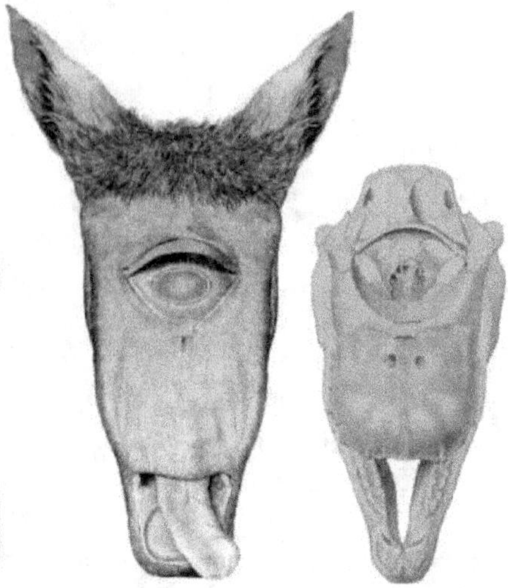

Left: A probable jumart foaled in Italy at Rovigo and described by Regnault (1775). It survived four months. At that age, a foal's nutritional requirements begin to force it to do more than nurse. So this death was likely due to an inability to crop, resulting in starvation. *Image: Moreau de la Sarthe (1808).* Ruben (1824, p. 12, obs. 15) stated that the Rovigo animal had a protuberance on each side of its forehead. *Right:* Head of a similar animal described and pictured by Ruben (1824), foaled in a village near Berlin. The eye, as in most cyclopses, was much larger than usual. It was single but had a double lens (ibid., p. 6).

Left: A cyclops foaled in or near London (Boyle 1665). In this case, the oft-reported rostral protuberances look like short horns. Boyle said the single eye was far larger than a normal one and bore two separate irides and corneas (a form of synophthalmia). The forehead was deeply depressed. *Right:* A cyclops foaled by a mare (Kitt 1894, p. 69, fig. 19).

Lincoln, Massachusetts. The *Boston Cultivator* (Jul. 21, 1860, p. 2d) described a "colt" foaled in Lincoln: "For the eyes, it had but one, very large, and in the center of the forehead. It had no nostrils, and the upper jaw tapered down thin, and was two or three inches shorter than the lower jaw. The foal appeared perfect in other respects. It lived about two hours." The opposite page lists other mare-foaled cyclopses.

NATURALISTS VS. VETERINARIANS

In the 1800s many veterinarians accepted the existence of jumarts, a credence apparently based on observations made in their practice. But many naturalists, who were not veterinarians, rejected their existence. This latter school of thought seems traceable in Germany to Blumenbach and in France to Buffon. In his *Handbuch der Naturgeschichte* Blumenbach(1779, p. 112) claimed that all alleged jumarts were in fact hinnies. The book was widely cited for decades.[9] And in an account also much cited by later writers arguing that jumarts do not exist, Buffon (*Histoire Naturelle,* 1829, vol. 16, p. 392) told how matings between a horse and a cow had failed to bear fruit.

> On my family estate, in 1767 and thereafter, the miller had a horse and a bull that had such passionate feelings for each other that whenever the mare was in heat, the bull, if allowed, never failed to cover her three or four times a day. These couplings, repeated many times over several years, gave their master high hopes of seeing offspring. But it never resulted in anything, and all the residents of the place were witnesses, both of these two animals mating—which was real and often repeated for several years—and of the lack of offspring. This fact seems to prove that, at least in our climate, a bull can produce no offspring with a mare, and it is this finding that makes me doubt, very legitimately, the existence of this first kind of jumart. I have no facts as certain that I can cite in contradiction of the possible existence of the second sort of jumarts, the type of which Dr. Shaw wrote and which he says comes from the donkey and cow. I admit that, even though the difficulties involved seem about as great in both cases, the positive testimony of someone as well educated as Dr. Shaw makes the existence of this second type of jumart more likely than that of the first. And with regard to the third type, that derived from a bull and a she-ass, I am convinced, despite the testimony of Mérolle [i.e., Girolamo Merolla da Sorrento (1726, p. 3)], that it does not exist anymore than a jumart from a bull and a horse. There is even more disagreement, yet more natural distance, between a bull and a she-ass than between a bull and a mare,[10] and from the fact, which I have mentioned—that a bull produces no offspring with a mare—we can infer that no progeny will be produced from the union of the bull with a female ass. [translation mine]

But reports kept on appearing. A *Placer Times* reporter (quoted in the *New-York Daily Tribune,* May 8, 1851, p. 6d; ||y5zalpdt) said he saw in Napa Valley, California, "a young colt with hoofs split crosswise, and each part separate like that of a cow." And the *True Northerner* (Paw Paw, Mich., Jan. 5, 1872, p. 3d; ||2th87rvk) said, "The latest freak of nature is an animal at Council Bluffs [Iowa], which is half horse and half cow."

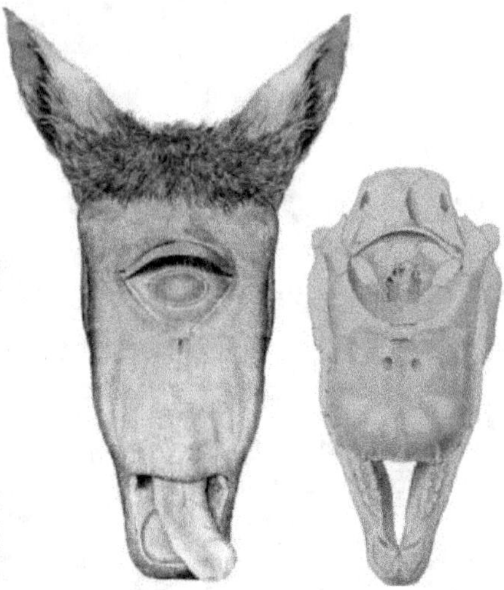

Left: A probable jumart foaled in Italy at Rovigo and described by Regnault (1775). It survived four months. At that age, a foal's nutritional requirements begin to force it to do more than nurse. So this death was likely due to an inability to crop, resulting in starvation. *Image: Moreau de la Sarthe (1808).* Ruben (1824, p. 12, obs. 15) stated that the Rovigo animal had a protuberance on each side of its forehead. *Right:* Head of a similar animal described and pictured by Ruben (1824), foaled in a village near Berlin. The eye, as in most cyclopses, was much larger than usual. It was single but had a double lens (ibid., p. 6).

Left: A cyclops foaled in or near London (Boyle 1665). In this case, the oft-reported rostral protuberances look like short horns. Boyle said the single eye was far larger than a normal one and bore two separate irides and corneas (a form of synophthalmia). The forehead was deeply depressed. *Right:* A cyclops foaled by a mare (Kitt 1894, p. 69, fig. 19).

Lincoln, Massachusetts. The *Boston Cultivator* (Jul. 21, 1860, p. 2d) described a "colt" foaled in Lincoln: "For the eyes, it had but one, very large, and in the center of the forehead. It had no nostrils, and the upper jaw tapered down thin, and was two or three inches shorter than the lower jaw. The foal appeared perfect in other respects. It lived about two hours." The opposite page lists other mare-foaled cyclopses.

Mare-Foaled Cyclopses: Likely Jumarts

<table>
<tr><td>1801</td><td>London</td><td>||bpa6tbnz</td><td>Likely viable</td></tr>
<tr><td></td><td>St. Buryan, Cornwall</td><td>||2p9v8uxk</td><td>NN</td></tr>
<tr><td>1841</td><td>Windsor, New South Wales</td><td>||2yf2n6e5</td><td>SM</td></tr>
<tr><td>1858</td><td>Richmond, Kentucky</td><td>||ar3dea53</td><td></td></tr>
<tr><td>1868</td><td>Brown, Mifflin, Pennsylvania</td><td>||8rm2k48t</td><td>SM</td></tr>
<tr><td>1871</td><td>Newport, Ohio</td><td>||yc5hsuea</td><td>SM, S, stillborn</td></tr>
<tr><td>1872</td><td>Horseheads, New York</td><td>||yc8rzk3x</td><td>SM, NN, viable</td></tr>
<tr><td></td><td>London</td><td>||3zbsb45s</td><td>SM</td></tr>
<tr><td>1873</td><td>Rowley Regis, England</td><td>||3tyfkn3b</td><td>NN</td></tr>
<tr><td>1876</td><td>Sandy Creek, Illinois</td><td>||y3sw6a66</td><td></td></tr>
<tr><td>1883</td><td>Monroe County, Mississippi.</td><td>||pabhja7x</td><td>SM, NN</td></tr>
<tr><td></td><td>Scandia, Kansas</td><td>||2ddb5kwm</td><td>SM, NN</td></tr>
<tr><td>1888</td><td>Buffalo, New York</td><td>||bddxj25n</td><td></td></tr>
<tr><td>1889</td><td>Omaha, Nebraska</td><td>||4cp2ypch</td><td>SM, NN, S</td></tr>
<tr><td></td><td>Pennsville Twp., New Jersey</td><td>||2saa4uhy</td><td>SM, NN</td></tr>
<tr><td>1891</td><td>Goodwater, South Dakota</td><td>||33hft87t</td><td></td></tr>
<tr><td></td><td>Richwood, Ohio</td><td>||4ksvskn3</td><td>NN</td></tr>
<tr><td>1893</td><td>Boston</td><td>||4pmdy42s</td><td>NN</td></tr>
<tr><td></td><td>Myersville, Maryland</td><td>||uuphjp3h</td><td>SM, ears far back</td></tr>
<tr><td></td><td>Palmyra, Missouri</td><td>||2w29ahre</td><td></td></tr>
<tr><td>1894</td><td>Eureka, Wisconsin</td><td>||59s2fk7v</td><td>SM, S, stillborn</td></tr>
<tr><td></td><td>Yukon, Oklahoma</td><td>||5y3ewm7p</td><td>NN</td></tr>
<tr><td>1896</td><td>Boone County, Missouri</td><td>||46s4mhy5</td><td></td></tr>
<tr><td></td><td>Raritan, Illinois</td><td>||3h7rnart</td><td></td></tr>
<tr><td></td><td>Rhinebeck, New York</td><td>||3swtapd8</td><td>Viable; S, SM, NN</td></tr>
<tr><td>1897</td><td>Elmwood, Ohio</td><td>||yc77ubec</td><td>SM; K</td></tr>
<tr><td></td><td>Springfield, Missouri</td><td>||ynnkvrs5</td><td>SM, K</td></tr>
<tr><td>1901</td><td>Fairmount, Missouri</td><td>||2u3s66fb</td><td>NN</td></tr>
<tr><td>1905</td><td>Bishopville, Ohio.</td><td>||4a27rkwd</td><td>SM, NN; lived a day</td></tr>
<tr><td>1907</td><td>Rutland, Vermont</td><td>||2p8tx49s</td><td>SM, S, stillborn</td></tr>
<tr><td>1911</td><td>Moline, Kansas</td><td>||3w4dsje7</td><td>SM, NN; born alive</td></tr>
<tr><td>1914</td><td>Marietta, North Carolina.</td><td>||ybbb8ks4</td><td>Lived a few hours</td></tr>
<tr><td>1915</td><td>New Castle, Pennsylvania</td><td>||2p94pctn</td><td>SM, NN</td></tr>
<tr><td>1922</td><td>Grant, Nebraska</td><td>||2p94pctn</td><td>SM, NN; K</td></tr>
<tr><td>1923</td><td>Montfort, Wisconsin</td><td>||35vhk5be</td><td>SM, NN, K</td></tr>
<tr><td>1927</td><td>Strasbourg</td><td>Schwarz</td><td>SM, S</td></tr>
<tr><td>1934</td><td>Greeley, Colorado</td><td>||5n84upj5</td><td>NN; lived a few days</td></tr>
<tr><td></td><td>Tissemsilt, Algeria</td><td>||nhzcy86j</td><td>SM, S; lived 48 hrs</td></tr>
<tr><td>1937</td><td>Cresco, Iowa</td><td>||4jzxxny3</td><td></td></tr>
<tr><td>1935</td><td>Spencer, Iowa</td><td>||8n6rw37u</td><td>NN</td></tr>
<tr><td>1938</td><td>North Platte, Nebraska</td><td>||3kfpsbhb</td><td>NN, S</td></tr>
<tr><td>1945</td><td>West Monroe, Louisiana</td><td>||y7f3k5hn</td><td>SM, NN; lived 24 hrs</td></tr>
<tr><td>1946</td><td>Gillette, Wyoming</td><td>||2w8d798s</td><td>SM, NN, S</td></tr>
</table>

Key: S: Synophthalmic. SM: short maxilla; NN: no nostrils; K: killed

* Other mare-foaled cyclopses include the Stevens Point CERVIDEQUID and the Milpitas HIPPOCAMELID, as well as the Fitchville and Tullamore CANIDEQUIDS.

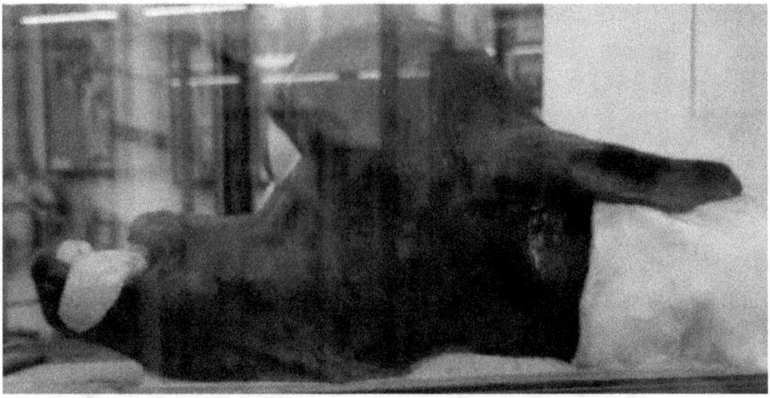

A synophthalmic cyclops (Musée Fragonard, ter.16.203; plaster cast prepared by Petitcolin in 1902, *see:* Lopez-Jugant 2010, p. 70). *Image: John Pelter.*

During the nineteenth century, scholarly opinion gradually shifted toward a rejection of the possibility of not only jumarts, but also of all interordinal hybrids. Thus, we find that Henry Harcourt Hyde Clarke (1867), in describing a jumart he saw in Turkey, felt obliged to mention that the actuality of such animals was doubted by naturalists. "The jumart, or hybrid between the bovine and equine race, is still believed in through all the southern countries. There was a reputed jumart at Seidekene, near Smyrna, in Asia Minor, during my stay there, and I heard of another. The jumart came into Smyrna several times, and I had made preparations to get a photograph, but it always escaped me. The description fully conforms to that given in books of natural history of the alleged jumart. This one was said to be the offspring of an ass and a cow; whereas the jumarts recorded in books are said to be the offspring of bulls with mares and she-asses. The existence of the jumart is doubted by most naturalists."

At least one naturalist, the French paleontologist and entomologist Paul Gervais (1855, p. 154), specifically stated this to be the case, "Naturalists," he said, "have accepted the opinion of Buffon, but such is not the case with veterinarians, who are far from convinced" (translation mine). It seems that naturalists were more influenced by theoretical notions about what was possible with hybrids, while veterinarians were swayed by what they had seen in their encounters with actual animals. Gervais went on to explain, in translation, why he himself did not believe in jumarts: "Sadly, these singular productions

†. Others have also reported ass-cows bred by Muslims (see pp. 521, 522 and 529).

have either not been described with the care they deserve or, in those cases where they have been, they are easily recognized as a few monstrous individuals, just as one sees in any species, and which can be explained without supposing that some foreign fertilization has occurred. Up to the present day at least, the doubts of many veterinarians on this subject have not been based on any authentic observation, and these doctors could not be under more obligation to provide evidence to resolve this important issue. It is nonetheless demonstrated today that jumarts do not exist in countries where they were reported, in Italy, Corsica and Algeria, for example, and moreover have not been observed elsewhere, in the way that we observe mules and hinnies. These monsters to which Mr. Grognier refers are never viable, and the stories, more popular than scientific in which they typically appear, seem no better founded than are the various tales people spout about various monstrous human births, about which modern teratology provides such exact knowledge."

But many of the reports cannot, as Gervais asserts, be satisfactorily explained "without supposing that some foreign fertilization has occurred." When Gervais says, "monstrous individuals," he seems to be referring to what today would be called mutants or mutations. But it is unclear how a mutation would transform, say, the right forefoot of a horse into the cloven hoof of a cow, as in the Greenwich animal (p. 542) and convert the hips and tail of that animal into those of a cow, as Gloag also describes. The many ostensibly mixed traits of that animal can be more economically accounted for by supposing it was a hybrid between a cow and horse. For it to have been a mutant, it would have had to have undergone not one, but multiple, independent mutations, which would all have just happened to be mutations that gave it traits like those of a cow. Such an event is unlikely, to say the least.

Gervais certainly had no basis for his claim that jumarts are never viable. Indeed, in a given hybrid cross both viable and inviable offspring are typically produced (McCarthy 2006). And in crosses that are more distant, the percentage of inviable hybrids tends to be higher than in closer ones. We have in fact seen, along with the many reports about inviable jumarts, many others about viable ones.

And the argument that jumarts are hinnies cannot hold in the case of mare-foaled animals with clear bovine traits or for horse-like animals calved by cows. Various reports do describe such composites.

The forefeet of the Norwich jumart (Pycraft 1926).

CALVED JUMARTS

Norwich, England. Pycraft (1926) pictured a jumart calved at Norwich, which he described as a "mule-footed ox." Its front feet bore solid hoofs like those of a donkey, but it otherwise looked like a calf. Pycraft said its anomalous feet could not be accounted for as a case of reversion to an ancestral state: because "none of the Ruminants ever had feet with solid, horse-like hoofs." But hybridization does provide a reasonable explanation.

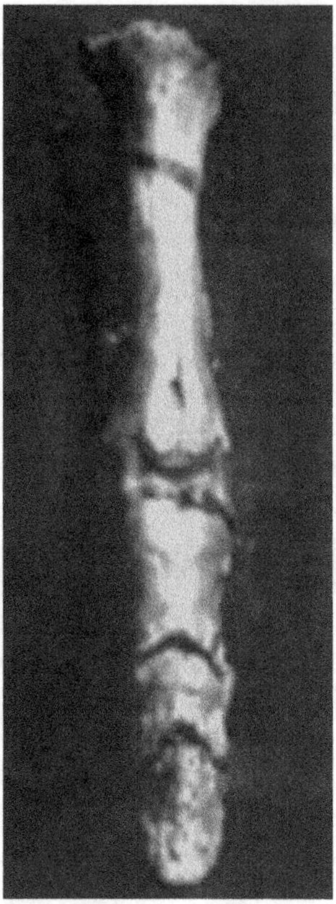

Osseous structure of one of the forefeet of the Norwich jumart (Pycraft 1926).

Nowata, Oklahoma. From the *Bartlesville Morning Examiner* (Bartlesville, Ok., Sep. 27, 1933, p. 4; ||5n7mkc64): "G. Durst, who lives on a farm three miles west and one mile north of Nowata, has a freak calf. It is seven months old and weighs 450 pounds. It has the neck and shoulders of a horse, no brisket, and six legs. The extra legs are above the shoulder on the right side. The freak has attracted considerable attention, and Mr. Durst expects to get a good price for it as a sideshow attraction."

Turin, Italy. The Italian veterinarian Longo Tommaso (1863) reported a cow calving an animal at Turin with a forefoot like that of a horse.

Michigan Bar, California. From the *Sacramento Record-Union* (Aug. 25, 1881, p. 3a; ||y5ofxbnr):

> Augustus Farenbocker, of Michigan Bar [a former mining camp], in this [i.e., Sacramento] county, has a calf which was born on the 25th day of May last, which has five naturally formed legs, and one, the foot upon the extra leg, is a perfect horse's foot in shape. The fifth leg is situated immediately in front of the right shoulder. It commences in a flat bone attached to the left shoulder blade and extends over to the other side and branches out from the lower part and front part of the right shoulder. The shoulder joint of this leg not only articulates forward and backward like its four legs, but also has the outward movement like the human arm. This leg is shorter than the others, so that it does not use it in walking, but moves it at will, and it is in all respects healthy and complete in formation. The foot is a perfect round hoof like a colt, turning naturally somewhat upwards and in. It certainly is a very strange malformation and an interesting one. It is a heifer calf and doing finely, and it is understood it will soon be placed upon exhibition in this city.

Brennilis, Brittany. Le Progrès de l'Algérie (Constantine, Mar. 28, 1888, p. 3a; ||33a64wff) said a cow at Brennilis calved a viable animal that looked like an ordinary calf except that from its shoulders emerged a fifth leg terminating in a single hoof. The Brennilis jumart must therefore have been quite similar to the one calved at Michigan Bar.

* The Michigan Bar and Brennilis jumarts are examples of xenomelia (see p. 1215).

Dornach, Switzerland. La Justice (Paris, Sep. 17, 1881, p. 3g; ||2s4jfjj9) said a cow calved a cyclops with ass ears at Dornach.

Poverty Bay, New Zealand. The *Poverty Bay Herald* (Jul. 18, 1893, p. 4a; ||y2ff7fcn) reported two viable jumarts. "A cow recently gave birth to a pair of singular animals. They resemble colts more than calves, although both possess rudimentary horns and hoofs of cattle, but in all other respects they seem to be young horses, having long, flowing manes and tails of colts, only these latter are unusually long and bushy. One is a male and the other a female, and both are well-developed, well-shaped animals. … The mother is a young Jersey of unmixed breed and a valuable animal. … The other cattle and horses on the farm alike refuse to consort with the strangers, and it has been found necessary to isolate them in a separate pasture. The colt-calves are playful little creatures, and seem to possess affectionate dispositions, coming to the bars at a whistle from the man having them in charge. Crowds from all over the country have been to see the curiosities."

HORNED JUMARTS

Fresnoy-la-Rivière, France. The *Journal de Senlis* (July 1, 1854, p. 1d; ||y6c3oe2s) said, in translation, "Everyone knows the mule is produced by crossing a jackass with a mare. But many people are doubtless unaware that man can unite a mare with a bull. The animal born of this union is called a jumart, according to Napoleon Landai, or a joubart, according to the author of the *Expert Blacksmith*. It has the form of a horse and the face of a bull. Usually, they have solid hooves, as do the former, but sometimes, cloven hooves, as in the latter. But, until now, Nature has refused them horns. A jumart is now present at Fresnoy-la-Rivière in the stable of the widow Dangicourt. He is unusual in being the offspring, not of a mare, but of a jennet. His father is a Berrichonne bull. And this is not his only distinction. Nature has given him what she has refused all his predecessors. His forehead is adorned by two very well-formed horns, entirely similar to those of a cow."

St. Louis. The *St. Louis Globe-Democrat* (quoted by the *Alexandria Gazette*, Aug. 10, 1878, p. 2f; ||y83g4jpg) said,

> A HORNED HORSE.—Mr. H. Howard, corner of Seventh Street and Clark Avenue, is the owner of a horse that has two rudimentary horns growing from the interior of the base of the ears, strongly resembling the horns of a buffalo. The horns are of equal size, about 6 inches long and curved like those of the bovine species. Inside the ears there is a soft, flexible ligature, covered with hair; the points are tipped with about two inches of hard, black horn, and project forward. When the ears are moved the horns become more rigid in the soft parts and stand out like the horns of an ox or a buffalo.
>
> The horns drop off every spring, about the 1st of May, and are succeeded by new ones of the same size. The horse is a bay mustang, about fourteen hands high, and rather poor in flesh at present. He is perfectly gentle, and will readily enter a house or other enclosure, and allow himself to be handled. Mr. John Grimsley, who knows something about a horse, examined his mouth yesterday, and pronounced him seven years old. He says the animal is the most wonderful thing in the equine line he ever saw, being a greater curiosity than Fremont's Woolly Horse, or the hairless horse exhibited in the city some years ago.

Westergate, Kent, England. Eustace (1903) read a paper before the Linnean Society of London in which he recorded "the occurrence of bilateral prominences on the frontal bones in two thoroughbreds."

A jumart that Aldrovandi (1642) called a *hippobos pedibus monstrificis* ("horse-cow with monstrous feet). He said (ibid., p. 573) it should have been named a hornfoot ("cornipes"), as it had eight horns in place of hoofs.

MISPLACED HORNS

Ypsilanti, Michigan. The *Jamestown Weekly Alert* (Jamestown, N.D., Jul. 17, 1902, p. 2d; ||yy5855r6): "Wm. Blake of his city bought a colt Saturday from a farmer named Newburger of Ypsilanti which is quite a curiosity. The colt has instead of the regulation horse's hoof on its left front leg two horn-shaped growths similar to the hoof of a cow in some respects but longer. One of the horn-shaped growths is used to walk on by the colt. The other grows out of the fetlock. The upper one is about 14 inches long and the lower is about 10 inches in length. The colt is a year and a half old, is healthy and walks with but little difficulty."

Paris, Missouri. The *Paris Appeal* (quoted in the *Shelby County Herald*, Shelbyville, Mo., May 1, 1895, p. 4; ||yyhv8gww) described a jumart with a horn attached in a seemingly random manner. "A mare belonging to Mart Grossball, living 4 miles north of Paris, gave birth to a colt on Tuesday night that is a wonderful freak of nature. Its head was that of a calf and one leg has a horn. The freak died when foaled."

Urana, NSW. The *Wagga Wagga Express* (Feb. 3, 1934, p. 3c; ||yabqtft9) reported a mare at Urana, NSW, foaling an animal with a horse's head and shoulders, calf hind parts, and its forelegs replaced by small horns.

SUCHETET

In 1890, André Suchetet (1849-1910) published an article in which he argued that jumarts were merely imaginary animals. An online biography (||y3sg6ohr) describes him as a politician and "big landowner" ("gros propriétaire terrien"). From 1898 to 1910 he served as a député in the French National Assembly (the equivalent of a U.S. representative). Although Suchetet was a member of the Zoological Society of France, he seems never to have been affiliated with an academic institution, nor to have held a doctorate. He did, however, publish some scientific communications, including his much-cited article dismissing the existence of jumarts (Suchetet 1890).

In his paper, Suchetet cites many authors who had reported seeing, or hearing about, jumarts, or who merely expressed their opinion about the possible existence of jumarts, yea or nay. However, in his tone, and even in the title of his article ("La Fable des Jumarts"), he makes it clear from the outset that his own opinion was that jumarts do not exist, have never existed, and cannot exist. But *why* did he, and the various other naysayers he cites, reach such a conclusion? Especially, one wonders why he would have chosen to ignore the reports of so many eyewitnesses. He even dismisses one witness who he himself says (ibid., p. 21) "had provided details that were sufficiently precise" ("a donné des détails assez précis"). It was the case of a farmer living at Sainte-Eulalie in the canton of Burzet, Ardèche, who when his stallion had refused to mount one of his mares, had replaced the stallion with a bull, who covered the mare. The mare conceived, according to the witness, though she was at no other time in the presence of a stallion, and, he said, she eventually produced a hybrid offspring (ibid., p. 19).

Suchetet's decision—and that of others—to reject the very possibility of jumarts can be accounted for in various ways:

- *Bias.* Simple bias is one likely factor. Among religious people there is a widespread notion that, by divine edict, the various kinds of animals and plants are kept separate through an inability to hybridize. The notion is that this prevents "confusion of the species." Anyone who held this general belief would obviously be more willing to believe that jumarts, specifically, are impossible. And, even among the non-religious, there is a prevalent belief that hybrids are somehow rare, unnatural and undesirable. Indeed, many biologists believe a loss of the ability to produce fertile offspring in matings with other populations is an essential step along the road to a population's becoming a separate species. Those who harbor such opinions are, again, more willing to accept unproven claims that jumarts simply cannot be, since to accept that they did exist would be to reject an axiom of their *Weltanschauung*.

- *Failed mating.* One argument often cited contra the idea of the possible existence of jumarts is that mating between equids and bovids has sometimes failed to produce hybrids. For example, in a case cited above, Buffon made much of the fact that mating between one bull and a single mare had produced nothing. But evidence of this kind is of limited value. For, as everyone knows, not every mating leads to a pregnancy, even when hybridization is not in question. Indeed, as any infertile couple will affirm, in some cases even insemination many times repeated can fail to result in impregnation. And when hybridization is in question, more inseminations are typically required than in unmixed mating (McCarthy 2006).
- *Difference.* Another argument is that equids and bovids are too different in terms of their anatomy or in terms of evolutionary relationship to produce a hybrid. But the parents in a hybrid cross *typically* differ in anatomy, and it is unknown just how much two animals can differ anatomically and still to produce hybrids. As to evolutionary relationship, in that case too, it is unknown how distantly related two hybridizing animals can be (indeed, this book explores that very question). But those who do believe that "distance" precludes hybridization will be likelier to believe jumarts are simply deformed hinnies (since horse × donkey is a much closer cross).
- *Genitals.* Another, rather antique argument, which no one would take seriously today, is that the genitals of equids and bovids are too different to allow mating to occur. Haller, for example, mentioned this reservation with regard to jumarts in his correspondence with Charles Bonnet (Sonntag 1983, p. 1163). This notion is completely negated by the fact that both donkeys and horses have been many times observed copulating with cattle. Indeed, cases of hybridization are on record between parents differing greatly, not only in the size of their genitalia, but even

with respect to body size. Male Steller sea-lions more than five times as large as female California sea-lions. And yet natural hybrids occur (DeLong 1982; Gorodezky 1995; Miller 1996, pp. 471-472).

- *Gestation periods.* It is sometimes argued that horse-cow hybrids are impossible because the gestation period of a horse (48 weeks) is much longer than that of a cow (39 weeks). During my research into hybrids, I have often met with this assertion. It is, in fact, an age-old claim, dating back at least as far as Aristotle (*On the Generation of Animals,* 2.4). Counterexamples invalidate this line of argument. One well-established case is that of the cama, the hybrid of a dromedary camel and a llama. A dromedary has a gestation period of some 56 weeks, whereas a llama has one of 48 weeks. Another is the wolphin, the hybrid of a false killer whale, in which gestation lasts 67 weeks, and a bottle-nosed dolphin, in which it lasts only 50 weeks. In the latter case, the lengths of the parental gestation periods differ by four months, whereas the difference between horse and cattle is only about two. The domestic pig has a gestation period of 115 days, whereas that of the babirusa averages 153 (150-157 days), a difference of 33 percent, about the same, then, as that between dolphins and false killer whales (34%). And yet, they produce hybrids (Thomsen et al. 2011). Elsewhere in this book, numerous reports about likely horse-dog hybrids (CANIDEQUIDS) have been cited. The gestation period of a horse is more than five times as long as that of a dog. In the former, it averages 342 days in the latter, 63. It seems, then, that this claim about gestation periods is just an old wives' tale, repeated by scientists and non-scientists alike.
- *Chromosomes.* One reason that might be cited for the impossibility of producing jumarts is that cattle, donkeys and horses have different chromosome numbers ($2n = 60$, $2n = 62$ and $2n = 64$, respectively). But obviously, horses and donkeys produce hybrids together, and they differ no less with respect to chromosome count than do cattle and donkeys. Moreover, well-documented hybrids are known from other types of mammalian crosses in which the parents differ more with respect to chromosome number than do cattle and horses or donkeys. A few examples among many are: horse × zebra, sheep × goat, and Indian muntjac × Chinese muntjac.

More Reports

Theodor Kitt (1885), a professor of veterinary medicine in Munich, described two separate animals with four hooves like those of a horse, but in other respects like calves (his cases I and II). Another looked like a calf, except for its soliped right forefoot (his case IV).

The converse configuration to Kitt's cases I and II was reported in the *DeSoto Times* (Hernando, Miss., Jun. 12, 1890, p. 2b; ||yysg4srh). The brief notice reads, "A freak is reported from Jefferson County, Miss. A colt, otherwise perfect, has cow legs from the knee down, the hoof being split."

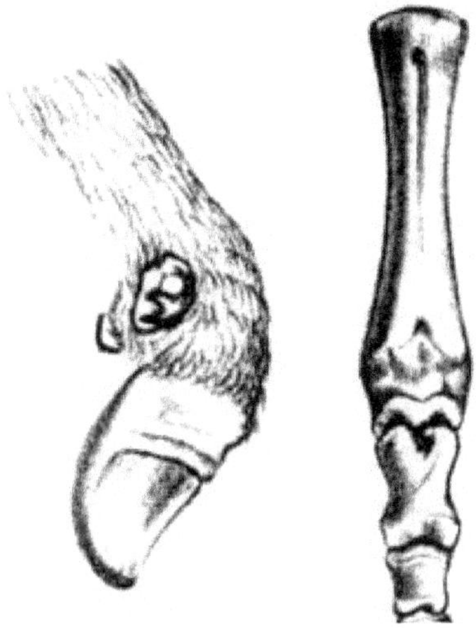

Left: One of the four solid-hoofed feet of an animal that otherwise looked like a calf (Kitt 1885, fig. 3). *Right:* Bone structure of the leg of a jumart that had only one solid hoof (Kitt 1885, fig. 2).

Magdalena, Sonora, Mexico. From the *Tombstone Epitaph* (Jan. 25, 1890, p. 3f; ||y7vbbdz6):

> Ventura Martinez lives in Sonora, about six miles from Magdalena, and is a prosperous rancher. Among his other possessions is a curious freak of the animal kingdom in the shape of a three-year-old stallion which has hoofs like a cow and a horn about four inches in length above each eye. In other respects, the animal is a perfect horse, and Mr. Martinez has been offered large sums of money at different times for his curiosity but says that money will not purchase it. It is probable that the freak will be brought to this city in a few months.

Alfort, France. Describing the solid-hoofed right forefoot of a bullock in the Alfort collection, Barrier (1884, p. 490) said, in translation, that "Altogether, the digital member of the subject individual is quite similar, both in form and anatomic structure, to that of a horse."

* The birth locale is unknown for this case, as Barrier lost the specimen donor's letter.

Fournels, Lozère, France. Le Journal du Midi (Avignon, May 6, 1892, p. 3a; ||3wyp6yc6) reported another French jumart, a calf born at Fournels in Lozère with a perfect mule's leg with a solid hoof.

Byaduk, Victoria. The *Hamilton Spectator* (Sep. 30, 1879, p. 2f; ||mxw3vda) reported an inviable horse-cow: "A MONSTROSITY.—A lusus naturae, in the shape of a foal with a calf's head and feet was born on a farm near Byaduk last week. With the exception of the head and feet, all other portions of the body resembled those of a horse. Its mother was a fine mare. Mr. Christie, who supplied the information, says he never saw anything like it, nor do we suppose anybody else ever did, and had it not been stillborn, it would have been worth hundreds of pounds to a showman."

Fabius, Michigan. The *Advertiser Record* (Constantine, Mich., quoted in the *Breeder's Gazette,* Chicago, Jul. 2, 1885, p. 19a; ||2w3w5t7k) said, "On Wednesday night of last week a mare belonging to Mr. Fred Hagenbuch, of Fabius, gave birth to a male colt, well-formed and perfectly symmetrical in all respects except that one of its feet is cloven and hoofed like the foot of a cow."

* A subsequent issue of the *Gazette* (July 16, 1885, p. 90c; ||2p6tmh4y) quoted N. C. Woolf, responding to the Fabius report: "My neighbor, Mr. D. M. Hall has a two-year old colt that exactly fills the above description. For a few months Mr. Hall has taken great pains in shoeing and thinks he will succeed in making a pretty good hoof."

Poland. Joseph Drzewiecki (1891, p. 530b), a Polish physician, reported seeing a jumart. "My father … related to me that on his farm a peasant had a horse, whose face greatly resembled that of a cow's. … I took the first opportunity to see the animal and must acknowledge that its face presented a striking resemblance to a cow. Only the horns were wanting. … in Poland the peasants keep the horses together with the cattle in the same stable."

Williston, Vermont. The *Herald and News* (West Randolph, Vt., Jun. 18, 1891, p. 1d; ||r7c85vk) reported a colt-like JUMART foaled in Williston with six legs, one with a hoof like that of a cow.

Sonoma County, California. The *Indianapolis Journal* (Jul. 23, 1893, 2, p. 16b; ||yc6xtmtj) said,

> A colt with a head like a calf and having but one eye and no mouth, was foaled on the Fair ranch in Sonoma County. It lived two hours after birth. The dam was a large draft mare.

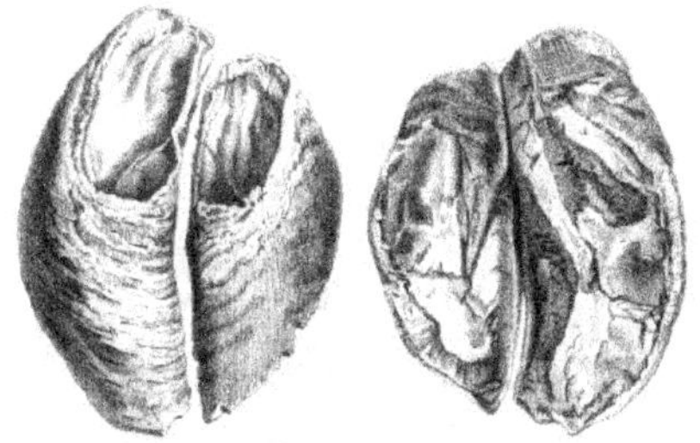

Cow-like hoofs from the right forefoot of an animal otherwise looking like a horse (Arloing 1867, Plate 2; ||y5m94qa5) *Left:* superior view. *Right:* inferior view. Arloing (p. 69) says this individual lived five or six years, serving as a draft animal, before being sacrificed due to age.

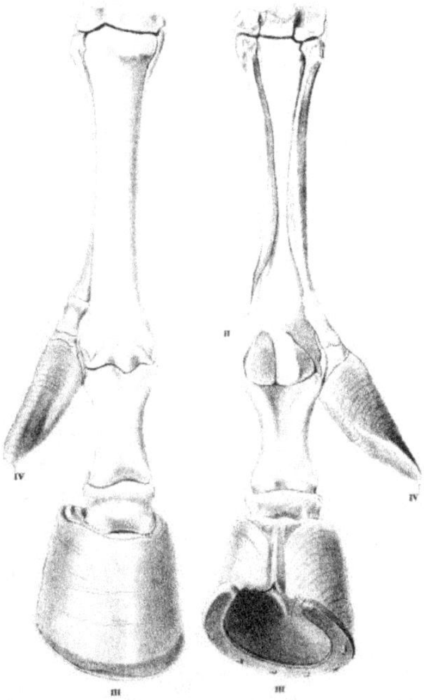

Forefoot of an otherwise horse-like animal with a supernumerary digit bearing a cow-like hoof (Wood-Mason 1871, Plate 1; ||y2uzlpnq). Arloing (1867, Plate 1; ||y2m5lrme) pictures an almost identical case.

Saturnin Arloing (1846-1911), a well-known French veterinarian and professor of anatomy, the son of a blacksmith (more about Arloing: ||y4oaroxo).

James Wood-Mason (1846-1893), an English zoologist who served as director of the Indian Museum at Calcutta (more on Wood-Mason: ||y5ek8cmx).

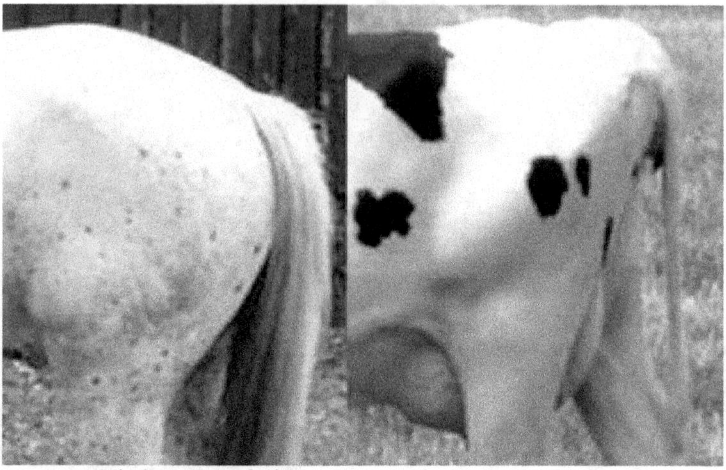

Hindquarters of a horse compared with those of a cow.

Leeds, North Dakota. The *Courier Democrat* (Langdon, N.D., Jul. 13, 1893, p. 3c; ||6cjs9jfs) said, "A singular freak of nature is a colt belonging to Jonathan Matthews of Leeds that was born with a fully developed udder and has to be milked. There has been an increase of milk ever since it was born."

* The *Journal de la Manche et de la Basse-Normandie* (May 26, 1909, 4e; ||ykrpf7jp) reported a filly, foaled at Sottevast, Normandy, that produced abundant milk.

Elizabethport, New Jersey. The *Broad Ax* (Salt Lake City, Aug. 3, 1901, p. 2d; ||yy22gef7) said, "Veterinarians are interested deeply in a freak cow-horse, which is in the possession of Mr. William S. Hugo of Elizabethport, N.J. At first glance the animal looks like a mare of natural size, but on approaching her hindquarters the formation of a cow is discovered in the hip bones, which are level with the backbone. She measures [~60 cm] from one hip bone to the other. The mare has natural shoulders and head, but when traveling has the peculiar stride of the cow. ... The mare can get over the ground in lively fashion, while not appearing to be going fast. In the stall the animal chews her cud, as does a cow or bull, and if watched closely many of the attributes of the bovine can be observed. When swishing flies her motion is the same as that of a cow. She can gallop, but in a clumsy fashion only."

Plestin-les-Grèves, Brittany. L'Ouest-Éclair (Rennes, Jul. 11, 1901, p. 3; ||mryk8su6) said a cow at Plestin-les-Grèves bore, on July 1, a calf with 5 legs, one of them bearing a solid hoof like a horse's.

Puttelange-aux-Lacs, Moselle, France. Le Patriote Lorrain (Metz, May 18, 1924, p. 3e; ||yc8cr8mk) said, in translation, that a mare at Puttelange foaled a newborn animal that had, "indeed, three forefeet. The 'spare' foot is a horse hoof. The other two are cow hoofs." It soon died.

Burr Oak, Iowa. From the *Ottumwa Semi-weekly Courier* (Mar. 7, 1901, p. 7f; ||y4hzqxf9): "DECORAH, March 5.—On the farm of H. W. Stockman in Burr Oak township this county, there was foaled a colt which is a freak. Its hoofs were like a cow's. Its eyes were side by side in the center of the forehead. It had no nostrils, its lower jaw was turned up and the mouth, which was small, was higher up than the nostrils should be."

* The *Union* (Princeton, Minn., Oct. 8, 1903, p. 1; ||y9pl5w7u) reported a sideshow animal, that it described as a colt with cow hoofs.

Daubeuf-près-Vatteville, Eure, France. In translation, the *Journal d'Évreux et du Département de l'Eure* (Évreux, Jul. 27, 1901, p. 3b; ||22y63wcc) said, "DAUBEUF-PRÈS-VATTEVILLE.—FREAK.—M. Narcisse Jérôme, a farmer at Daubeuf, owns a 15-day-old calf with a head like that of an ass. This freak seems to disprove the claim that freaks never live. It is growing and fattening in the normal way."

* Also see: *Le Petit Républicain du Midi,* Nîmes, May 21, 1906 (p. 3b; ||52fxmuhw).

Mountain Rest, South Carolina. The *Keowee Courier* (Pickens Court House, S.C., Feb. 22, 1905, p. 5a; ||ycz2vw2q) said, "C. W. Hunt, of Mountain Rest, reports a curiosity, monstrosity, malformation or phenomena at his farm in the shape of a combination calf—half calf and half mule—which was born last week. Its feet are those of a mule, small and shapely, and have no sign of the calf's cloven hoof. Its head and ears also are quite mulish, bearing a decidedly closer resemblance to a mule than a calf. The remainder of the body of the creature is that of a calf. At three days of age the little freak was alive, apparently healthy and gave every promise of growing to maturity."

Fremont County, Wyoming. The *Sheridan Post* (Sheridan, Wy., Oct. 12, 1906, p. 6; ||4py4v489) said, "W. L. Snider, a ranchman of Fremont County, has one of the most remarkable freaks of nature ever seen. The monstrosity is in the form of a colt which has but one eye, this being located in the middle of the forehead, and one foot of a calf. The animal is sound and perfect in every other respect."

* Three months later, the *Deadwood Pioneer-Times* (Dec. 7, 1906, p. 1b; ||bdzbf9jc) reported what was likely the Fremont animal, alive, in Pinedale, Wyoming.

Saint-Girons, Ariège, France. Le Réveil du St-Gironnais (Aug. 12, 1906, p. 2a; ||44u3cneb) reported the foaling of an animal with a calf's head and five legs. It scarcely had a tail but seemed to be entirely viable.

Cedar Junction, Kansas. The *De Soto Eagle Eye* (De Soto Ks., quoted in the *Stayton Mail,* Stayton, Or., Oct. 25, 1907, p. 1; ||2e7a8bty) said, "Henry Lee of Cedar Junction has a heifer calf born Saturday October 5th. It has four perfect well-formed legs and attached to the shoulder blades are two additional legs which turn up and fold over the back more like wings. These legs are also well formed and perfect in every respect except that one bears a mule hoof. The calf weighs about 140 pounds."

Louisville, Kentucky. The Union Flag (Jonesborough, Tenn., Mar. 20, 1868, p. 2; ||mk4fvabz) said that "on Main Street, Louisville, Ky., may be seen … the hide of a calf which had two bodies, eight legs, with hoofs like those of a mule, four ears, one eye two tails and no head, the ears and eye being where the neck should have sprung from the body."

St. Louis, Missouri. "Snyder," a 2,700-pound animal exhibited at the World's Fair in 1904, had two horns and was billed as the biggest horse in the world (*Oregon Daily Journal.,* Jan. 30, 1904, p. 9b; ||25baythn).

Albany, Oregon. The Herald (Pullman, Wa., Apr. 25, 1908, p. 4, col. 4; ||y88czdn3) said, "ALBANY, ORE., April 21.—A freak colt has been born on the farm of E. Hartsock, in Benton County, near this city. One of its front feet is a cow's hoof. The colt is now a week old and apparently is not hampered by its strange foot. It runs about and will probably grow up."

Nevada, Iowa. From the *Monroe City Democrat* (Jun. 11, 1908, p. 8a; ||jzztq72): "NEVADA, IA., June 5.—A freak colt was foaled here yesterday. A Shetland pony mare belonging to Orrin Lash, a retired farmer … gave foal to a colt, and three of the hoofs of the bay Shetland are as perfectly formed cow's feet as one ever looked upon. The fourth hoof, the right hind one, is naturally formed. The colt is strong and healthy, chases about the barnyard with the mother and is natural in all other ways. … At first Mr. Lash was inclined to kill the little fellow, but as the colt was strong and apparently able to make his way through the world all right, his neighbors prevailed upon him to save the little one. A great many people have called to see the freak."

* The *Evening Times-Republican* (Marshalltown, Ia., Nov. 13, 1908, p. 9e; ||yeyrsmjm) reported what was likely the Nevada animal.

Stratford, Connecticut. The *Bridgeport Evening Farmer* (Nov. 14, 1910, p. 8e; ||y4uccbpr) said, "A horse with cow blood in its veins is the queer freak brought from Oregon by L. W. Warner and now housed in his barn in

Stratford. The creature plainly exhibits the mixed characteristics of horse and ox. Mr. Warner intends to place it on exhibition."

Colby, Washington. From the *Ranch* (Seattle, Mar. 1, 1911, p. 11a; ||y4y2km62): "I have a heifer that is somewhat of a "freak of nature," she has a tail like a horse; but what interests me most is that she will soon freshen, and only one side of her udder is developing. What can I do to make the other side develop?—A. V. R. Colby, Wash."

Webb City, Missouri. The *Dispatch* (Seneca, Mo., in the *Ripley County Democrat,* Aug. 15, 1913, p. 1c; ||y76pmja8) said, "A Webb City man paused in Seneca long enough to show the editor of the DISPATCH a colt with two cow hoofs in front."

Larimore, North Dakota. The *Bismarck Daily Tribune* (Apr. 29, 1915, p. 4d; ||4322z4es) said, "A freak colt was born at the John Williams farm, near Larimore, which had two distinct hoofs on one of its front legs. The colt, however, did not live and Dr. H. M. Elsenlohr has the hoof."

Pontivy, Brittany. La Dépêche de Brest (Jan. 1, 1924, p. 3f; ||3jymruue) said two fetuses in a slaughtered mare both had a calf's hooves and tail.

Kent County, Ontario. The *Windsor Star* (May 20, 1924, p. 1f; ||yxvpwhyf) reported a foal with a cloven hoof born in Kent County.

Friedingen, Germany. The *Bregenzer Tagblatt* (Bregenz, Mar. 17, 1915, p. 3; ||ya8cec4v) said, in translation, "In the stable of the restaurateur Werkmeister in Friedingen, a calf was born that had a perfect horse's head. This monstrosity was a male and lived 28 hours."

Neuhof, Germany. The *Pilsner Tagblatt* (Pilsen, Jun. 4, 1917, p. 3; ||y9xasdeh) said, in translation, "A MONSTROSITY. In Neuhof, near Buchau, a cow belonging to farmer Albrecht Huml bore a female calf with the hind legs and tail of an ass. The birth of this vigorous monstrosity, which is still alive, created a great sensation."

Neudorf, Austria. The *Grazer Vorortezeitung* (Graz, Oct. 28, 1917, p. 6; ||ya9k3zkb) said that after a 44-week pregnancy, a cow belonging to Alexander Hügel in Neudorf bore a calf with a horse's hind legs and tail.

* The gestation period of a cow is about 39 weeks, a horse's, 48 weeks, or more. The 44-week pregnancy reported in the Neudorf cow, then, was shorter than a horse's but longer than a ordinary cow's. A mammalian hybrid typically gestates for a length of time intermediate between the gestation periods of its parents. Similarly, avian hybrids have an intermediate incubation period.

Rhine Falls, NSW, Australia. The *Bombala Times* (Nov. 22, 1912, p. 4a; ||ybtkv8ty):

> An extraordinary freak of nature made its appearance at Mr. William Lynch's, Rhine Falls, ten days ago, in the shape of a piebald foal which has a hoof like a cow's on one fore foot, while the other is misshapen and turning up so that the poor little creature walks on the fetlock. The tail resembles that of a cow rather than a horse. Mr. Lynch intended destroying it but has been advised not to do so, and it is probable it may be used for exhibition purposes later on.

Glen Elgin, NSW. The *Inverell Times* (Feb. 3, 1922, p. 4; ||y9f9dg3x) said,

> Mr. A. H. Woods, of Glen Elgin, has a freak filly foal running on his property. On one leg, from the knee downwards, the formation is exactly the same as a calf's, even to the cloven hoof. The animal, which was foaled about five months ago, is being poddy [i.e., bottle] reared.

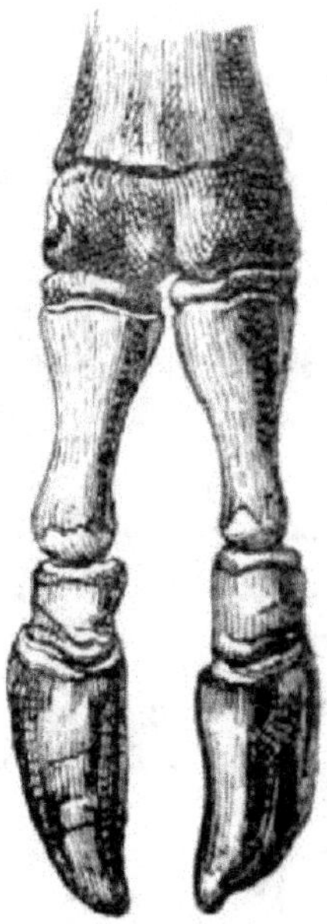

Forefoot of a mare-foaled animal (Danish College of Veterinary Medicine; Boas 1882, p. 277 & pl. XI.9).

Singleton, NSW. From the *Muswellbrook Chronicle* (Oct. 2, 1925. p. 8; ||ybtyfeed): "One of the strangest freaks of nature in the animal kingdom is in the possession of Mr. George Brooker, Maitland Road, near Singleton. It is a heifer, 15 months old, with some of the characteristics of a horse. For instance, it has high withers, with a dip in the back; the near hind leg is straight and shaped like that of a horse, while the short bushy tail heightens the grotesque appearance of the freak. The horns are just commencing to grow; but when the animal has its head down, and when one is standing at the rear of it, the heifer is for all the world like a poddy foal. When it runs the action of the hind legs—the near one being slightly shorter than the other—is similar to a horse cantering. The heifer is [supposedly] the progeny of a Jersey bull and a Jersey cross cow, and was bred by Mr. L. L. A. Reeves, of Mooral Creek, near Wingham, from whom Mr. Brooker has bought it. The animal is healthy and will eat almost anything."

Medford, Oregon. The *Medford Mail Tribune* (Jan. 21, 1931, p. 1g; ||2s4bersp) reported the birth of a calf with equine hooves on its front feet.

Paris. An editorial in *Le Journal* (Paris, Feb. 13, 1940, p. 5a; ||4rj595pe) extolled the racing abilities of a certain Mireille, a mare with the head of a calf ("jument à la tête de veau").

LIMB DEFICIENCIES

Valence-sur-Baïse, Gers, France. Le Midi Socialiste (Toulouse, Apr. 26, 1929, p. 1; ||mrxrhwu8) said, in translation, "VALENCE-SUR-BAÏSE (GERS), Apr. 25.—At the home of Mr. F. Castex, proprietor of the Tambourial estate, a cow has borne a freak calf with a donkey's head and without any trace of a tail. This strange animal … soon died."

La Graverie, Calvados, France. L'Ouest-Éclair (Rennes, Jul. 22, 1938, p. 7; ||3wa7nuwz) said a cow at La Graverie bore a viable jumart with equine hindquarters and only one front leg. The report offers a picture.

Tuntable Creek, NSW. The *Northern Star* (Lismore, Sep. 23, 1946, p. 7e; ||y9s962zp) said, "A calf born at Tuntable Creek on a farm owned by Mr. D. Cronan, was born with only three fully-grown legs. The fourth was a small undeveloped leg, growing backwards with a straight hoof similar to that of a horse. The calf is normal in every other respect."

Sealand, Flintshire, Wales. The *Cheshire Observer* (Mar. 20, 1858, p. 7e; ||yyrfd2uy): "On Friday se'ennight, a mare belonging to Mr. Davies of the Mill Farm Sealand dropped a foal, which exhibited a curious instance of malformation. The foal was perfectly made in all parts but the forelegs, which were most singularly shaped. The near foreleg bears a perfect resemblance to the leg and foot of a calf, and has a cloven hoof formed from the fetlock joint. The off leg terminates at the knee, to which there is a small pendulous bag attached, while from the back of the arm, a small leg bone about six inches in length, grows upwards, and strange to say has also at its extremity a very small cloven hoof. The mare it appears had carried the foal for the natural time, and dropped it during the night, but when discovered, it was dead."

CONJOINED JUMART TWINS

Jumarts with abnormal duplications include the Williston, Vermont, Saint-Girons, Nowata, Plestin-les-Grèves, Puttelange-aux-Lacs, Louisville, Cedar Junction, Michigan Bar, and Brennilis animals. In addition, there is at least one other case:

Limoges, France. L'Est Républicain (Nancy, May 24, 1947, p. 4d; ||n9u9yfek) described a 13-month-old jumart with a split spinal column and duplicated hind parts, bovine on one side, equine on the other. On exhibit in Limoges, it had five legs and, on its back, a third horn.

REPORTS DECLINE

Civilization is nature's worst enemy.
All wild things vanish when she comes.
–CHARLIE RUSSELL
On the Death of Edgar Paxson

As the twentieth century unfolded, reports about jumarts became increasingly scarce. This decline was probably due, at least in part, to a decrease in the rate at which such animals were produced. Barbed wire was first patented in the United States in 1867. The increasing use of this cheap fencing during the late nineteenth and early twentieth centuries drastically decreased the number of free-ranging animals worldwide. With the demise of the open range, animals became more segregated on farms and ranches, a trend that would reduce the opportunity for serendipitous matings between bovids and equids.

Also significant was the advent of the internal combustion engine. Horses and donkeys were more and more rarely used as sources of power, and they therefore became rarer on farms. By the end of the thirties, very few of these animals remained on working farms, at least in developed countries, where they had all been replaced by tractors and other machinery. Consequently, there was now little chance even for them to mate with cattle, let alone to produce jumarts.

Another factor, which may have been just as important as any actual decline in the rate at which jumarts were born, was a shift in attitude toward abnormal people and animals. The word *freak* is a shortening of the older term *freak of nature*. In the early nineteenth century, *freak* lacked the heavy negative connotation that it carries in the English language today. At that time, it simply meant "whim" or "caprice." The idea was that nature was doing something for no real reason. Indeed, *freak of nature* is an English rendering of the older term *lusus naturae,* meaning a "game" or "sport of nature." The notion was that a personified Nature, in creating abnormal organisms, was taking a break from the humdrum of her usual task of producing the so-called normal creatures of the natural order.

However, this idea of a frisky Dame Nature, playing around with her creations, was gradually supplanted during early twentieth century by the idea that "abnormal" individuals are the victims of mutations in their hereditary material. As such, they became objects of pity, and their public display in sideshows and dime museums came to be viewed as cruel and immoral. Likewise, under this new view, it was judged insensitive and even disgusting to speak of them in newspapers and other sources of information. Where freaks had once been family entertainment, appropriate even for the eyes of children, they were now something near obscenity, abhorred by all. With these changes in mentality, *freak* came to have the negative connotation that it still carries today.

This shift in attitude brought legislation placing outright bans on the public exhibition of freaks in many places (Fordham 2007). For example, at the outset of the twentieth century, an Indiana paper, the *New Richmond Record* (Oct. 25, 1900. p. 2b; ||yaxyjxob) was advocating an outlawing of freak shows.

> One of these days there will be legislation making it a misdemeanor to exhibit freaks such as some are exhibited at the public fairs and museums of the cities. There were shown at the recent Indianapolis street fair and in this city last week two black children that ought to have been confined in an idiot asylum. As it was, two or three besotted men lived off the money wrung from a curiosity-seeking community ever on the scent of something horrible or grotesque. Away out in California, a state a full quarter century farther from civilization, it is a penal offense to exhibit in public a monstrosity of any kind and the same sort of a law should be engrafted on the statute book of Indiana.

Increasingly, such laws were indeed enacted. By 1908, the *Cincinnati Enquirer* (Apr. 13, p. 2) was writing, "In most civilized countries there are now enacted laws forbidding the public exhibitions of monsters and revolting deformities. A more refined and more human popular taste now frowns upon such exhibitions, and they are less profitable to their promoters." The *Sheboygan Press* (Sheboygan, Wi., Feb. 12, 1923, p. 11) noted that "Fairs, including the State Fair, have cut out the freak tent, and carnival companies that once searched the world for them are also beginning to cut them out." With time, and in the absence of venues where people could witness living evidence to the contrary, prevailing notions of what was possible became less and less tempered by observation. It became the "educated" attitude to dismiss the aberrant as impossible, though such claims are more comfortable than rational. In the *Lord of the Rings,* Galadriel says,

> And some things that should not have been forgotten were lost. History became legend. Legend became myth.

Old reports about jumarts have been forgotten. The sideshows have largely disappeared. Journalists and scientists, in their wisdom, now deem horse-cows myth. And they never mention them, except with a certain bemused regret for the benighted ideas of a day gone by.

The Boston jumart.
Image: Boston Standard (Boston, Lincolnshire).

Because biologists and reporters now see jumarts strictly as the stuff of legend, any jumart born in our day will rarely reach the public eye, especially given that the general topic of freaks is now so widely judged in bad taste. But a few reports do evade the net of dismissal, disgust and disdain. Thus, an animal in a relatively recent news report appears to be a jumart. It belonged to Frans Buitelaar, who was a meat dealer in Lincolnshire during the 1960s and 70s, and is pictured in an online article in the *Boston Standard* (||y4re3xzn). According to the brief comments accompanying the picture, a local vet was consulted at the time, and he pronounced the animal "bovine." But might it not have been equine as well? The *Standard*'s image is shown above. Traits visible in the picture that seem consistent with the idea that this animal has equine inheritance are: (1) its very horse-like tail; (2) the rounded hindquarters region, which does not at all have the pointed shape characteristic of a cow; (3) absence of an udder; (4) saddle in the back, unlike the straight back of a cow; (5) long, thick, horse-like neck. Though Buitelaar is dead, his son, Adam Buitelaar, contacted in 2016, said he remembers seeing the living animal and confirmed it did look like the creature shown in the image above. And he remembered it having a horse's tail. He said, too, that it was a female, which makes its lack of an udder peculiar indeed—*if* it were a pure cow. Moreover, the animals shown in videos cited above have a similar structure, and videos are harder to fake than photos.

A jumart (Houël 1766).

VARIABILITY

The Buitelaar animal differed substantially from the creature shown in Jean-Pierre Houël's eighteenth-century illustration (above), supposedly drawn from life at the Ecole d'Alfort in 1766. Note that the Alfort animal has a cow-like tail and hindquarters, unlike the animal pictured by Buffon, the Buitelaar animal, and two of the animals shown in the videos cited above, which all have tails resembling that of a horse.

But an animal in one video (||ybu27b7p) does have a cow-like tail, which may simply reflect high variability in offspring from this cross, as is the case with certain other telenothians. Some authors do say that jumarts, specifically, are variable. In an old encyclopedia of agriculture, François Rozier (1785, p. 106b, s.v. *jumart*) states that not all jumarts are the same, with some exhibiting more bovine traits, others, more equine. And the descriptions of animals given in the reports quoted in this chapter describe animals that do vary widely.

Thus, on the one hand, the structure of the Buitelaar animal is consistent with the brief descriptions of the Neuhof, Neudorf and Colby animals, in that the equine similarity is primarily restricted to the hindquarters and tail. The same can be seen in two of the Kitale animals shown in video screenshots above (pp. 515 & 526). But the Mountain Rest, Elizabethport and Friedingen animals were reported to have the reverse configuration, that is, they were described as horse-like anteriorly and cow-like posteriorly. And then, some of the early

French cases and the Poverty Bay animals, from their descriptions, seem to have been of an equine character overall, but had horns and cloven hooves. There are also the two animals born in the U.S. in 1908 (in Albany, Oregon, and in Nevada, Iowa), which supposedly looked like ordinary colts, except that one had a single foot like that of a cow—as did Gloag's animal—while the other had three. Other reported jumarts had extra legs. Many were cyclopses.

In most kinds of crosses, hybrids produced by the first cross (F_1 hybrids) are markedly uniform from one individual to the next, because each receives the same genetic input from each of the two parents. For this reason, variability usually occurs only in later hybrid generations. So, when variability is observed in hybrids, it generally implies that the hybrids have at least some fertility. But these hypervariable jumarts seem to be F_1 hybrids. So, the variation observed in them cannot be the result of meiotic segregation. Perhaps some other, unrecognized developmental mechanism produces the observed variability?

Some of the reported variability is likely the result of crossing different parents. One jumart may be from a jackass and a cow, a second, from a bull and a jennet, a third, from a stallion and a cow. The particular breeds crossed would also differ in different cases. But a more broadly satisfying and specific explanation of the variation seen in these animals is epigenetic mosaicism, as explained elsewhere, in the section entitled COMPONENTS.

The hypervariability of jumarts seems to account for the many stillborn animals of this kind. Some variants will be more viable than others. As Cuvier (*Le Règne Animal*, I: 5-6) wrote,

> Since nothing can exist that does not fulfill the conditions rendering its existence possible, the different parts of each being must be coordinated in such a way as to render possible the existence of the being as a whole, not only in itself, but also in relation to its environment. [translation mine]

Seen in this light, the ability to survive under a given set of conditions is not a consequence of environmental influences. It is a necessary attribute, demanded of each animal from the moment of its inception. According to reports, the facial structure of some jumarts was such that they could not nurse; in others the teeth were such that they could not crop grass. A jumart that had its hooves replaced by horns, as pictured by Aldrovandi (see p. 553) would surely have had trouble getting about. None of the cyclopean births seem to have survived long. Such animals failed to meet Cuvier's conditions and therefore ceased to exist.

CONCLUSIONS

Genetic testing of the alleged jumart skull held in the Musée Fragonard's collection would help to resolve the question of whether hybridization between cow and horse is possible. Given that it would all presumably be an F_1 hybrid, such testing would be straightforward. Such a study could have potentially important implications, in that any definite verification that the specimen was derived from a hybrid between an equid and a bovid would demonstrate beyond doubt that the limits of mammalian hybridization extend well beyond what biologists generally believe possible. At present, in the minds of many, it seems the limit extends only to intergeneric crosses (Savriama et al. 2018), whereas a cross between an equid and a bovid would be interordinal (Perissodactyla × Artiodactyla), as has been repeatedly mentioned.

But given even the evidence collected here, it seems probable not only that it is feasible to produce hybrids between cattle and equids, but also that such births have in fact occurred. One might conceivably take the position that the many reports about these animals are all merely mistaken, or even somehow faked. But where there is so much smoke, it does seem that a certain amount of combustion must be present.

The situation may well parallel that seen in other, better studied, distant crosses, in which many inseminations are required to produce a single viable hybrid. For example, when turkeys are artificially inseminated with rooster semen only a tiny fraction of the inseminations lead to the production of a mature hybrid (about one insemination in a thousand). And from the standpoint of taxonomic classification, a cross between an equid and a bovid would, if anything, be even more distant than that between chicken and turkey, so the number of inseminations required might be even higher. And in fact, Suchetet (1890, p. 19), as has been mentioned, said one of the reasons farmers hesitated to attempt to produce this cross was a low conception rate. Anyone who chooses to investigate this cross by means of artificial insemination or experimental mating should keep this fact in mind.

NAMES

The word *jumart* was taken into English from French around 1700. It seems likely that *jumart* and various words cognate to it (see list below) were constructed by adding the first part of various cognate words meaning "mare" or "beast of burden" (e.g., *jum*ent, *gium*enta, *ium*entum, etc.) to the last part of various cognate words meaning "bull" (e.g., t*aureau,* t*auro*, t*oro*, t*aure*, etc.). This method of constructing names for hybrids is still used today (e.g., *liger* or *cabbit*). For example, the word *iumarus,* meaning "jumart," occurs in both Renaissance and New Latin. This name could have been constructed at an early date by taking the *ium-* from Latin *ium*entum ("beast of burden") and combining it with the *-aurus* in *taurus* to form *iumaurus*, which then shortened with time to *iumarus.* Latin consonantal *i* was regularly replaced with *j* in French in the Middle Ages, and the inflected endings of Latin (e.g., *-us*) were dropped. The result of such a process would be the spelling *jumar.* The -d and -t endings, seen in *jumard* and *jumart*, may have been added under the influence of the suffixes *-ard* and *-art* which are used in French to form nouns and adjectives, often with diminutive or pejorative meanings.

Many names have been used, in various languages, to refer to animals of this sort:

- ENGLISH: *jumart, ox-donkey, ox-horse.*
- FRENCH: *jumart, jumar, jumard, joumar, joumard, gemar, gemard, gimar, hippobus, hippotaure, onotaure, baf,*[13] *bif,*[13] *juméry.*
- VALÉIAN: *jumerre.*[14]
- OCCITAN: *jhimêrë, jhimêrou,*[15] *jumeri, gimere, gimerou, gimeri.*[16]
- LATIN: *iumar, iumarus, hippotaurus, onotaurus, hippobos.*
- PORTUGUESE: *jumardo.*
- SPANISH: *hippotauro, onotauro.*
- ITALIAN: *giumarro, giumerro, gimerro, ippobo,*[17] *bosmulo.*
- GERMAN: *Ochsenpferd, Ochsenesel, Maulochs, Gemar,*[18] *Jumar.*
- DUTCH: *Muilpaard.*[19]
- ARABIC: *kumrah, koomrah, berdoor.*
- RUSSIAN: юмарь (pl., юмарья).

An Italian asinovid (Aldrovandi 1642, paralipomena, p. 123).

ASINOVIDS (SHEEP-DONKEYS)

Paris. Gesner (1551, p. 906) stated, in translation, that “In the palace of the King of France, in 1549 [Henry II], was a monster, I hear, sent from Italy. Its foreparts were like a donkey’s, its hind ones, like a sheep’s”

Ohio. The *Jeffersonian* (Stroudsburg, Pa., Nov. 10, 1853, p. 1e; ||zn6ep6h) reported a 16-year-old asinovid at the Ohio State Agricultural Fair.

CAPRIDEQUIDS (GOAT-HORSES)

The *Meridional* (Abbeville, La., Apr. 15, 1899, p. 3; ||hbqh4ph) said, “What promises to prove an interesting freak of nature is now the property of F. P. Beauxis, of this town, and will be on exhibition at his store after tomorrow. It is a goat-colt, if we may use the term, being an animal with the head and mane of a goat, and the body of a colt. It makes a noise very much like a goat.”

AN ONOCAPRID (GOAT-DONKEY)

Paullini (1686) published an abridgment of a compilation of curiosities collected ca. 1200 A.D. In it (XLVI, p. 209), the monk Alexander Insulanus claimed that a nanny goat, kept with donkeys in his parents’ stable, kidded a nondescript with the ears and tail of a donkey. But in other ways, he said, it resembled its mother. Insulanus grew up near Hildesheim (ibid., p. 180).

An ovidequid (possibly Barnum's?).

OVIDEQUIDS (SHEEP-HORSES)

Reports about ovidequids are scant. Block (1902) reported that, in Germany, at Mecklenburg a mare foaled an ovidequid with a sheeplike head in 1901 (shown at right). A cyclops, its lower jaw was longer than its upper. *The Post* (Middleburg, Pa., Apr. 7, 1881, p. 1e; ||mswcdf6r) described an animal, in Philadelphia, covered in a sheeplike fleece. From the front, it reportedly looked like a horse, from the rear, like an overgrown sheep. P. T. Barnum (1856, p. 136), too, had a woolly horse. "The entire body and limbs," he said, "were covered with a thick fine hair or wool curling tight to his skin. He was foaled in Indiana" (in 1844).

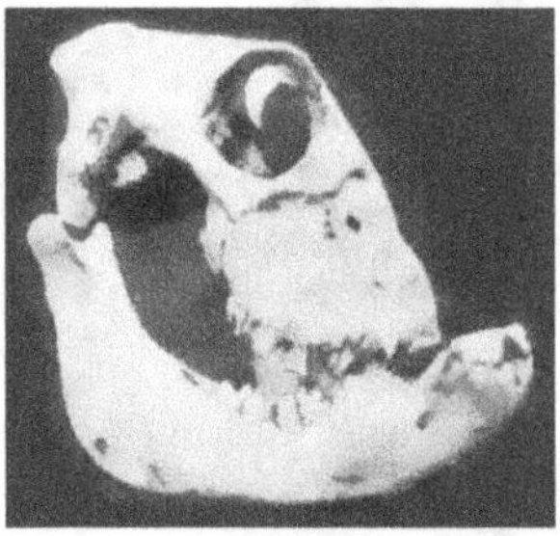

The Mecklenburg ovidequid. *Above:* lateral view of skull; *Below:* Face with a short upper jaw and a single large eye.

* Obsequens (*Prodigiorum liber,* 43) said two lambs with horse hoofs were yeaned in Lucania, Italy, in 104 B.C., one with the head of an ape.

* A breed of horse, the Bashkir Curly, plausibly derives from a sheep-horse cross.

"PRINCE IMPERIAL"—HALF BUFFALO, HALF HORSE

An alleged bisonequid (*San Francisco Examiner*, Jul. 24, 1892, p. 15).

BISONEQUIDS (BISON-HORSES)

The *San Francisco Examiner* (Jul. 24, 1892, p. 15; ||4d6xn5ra) reported the capture, in a herd of buffalo on the Great Plains, of a bisonequid, soon dubbed "Prince Imperial" and put on exhibit. In part, the report read, "The head, ears, nose, shoulders, haunches and legs were those of a horse. The dull, sleepy eyes, the shaggy coat and thin tail and the hoofs belonged to the buffalo species." It having been found in a bison herd suggests a bison mother. Another bisonequid, a maned animal reported as a "Half Horse, Half Ox," calved in 1865, is described in the *Indianapolis Journal* (quoted in the *Columbia Democrat and Star of the North,* Bloomsburg, Pa., Oct. 31, 1866, p. 1f; ||wa8pn5rc).

Notes

1. See della Porta (1658, ch. 9, p. 39; ||y5nw5brz).
2. Translation mine. For the original French, see: ||ycyrbpyd.
3. Rudolphi (1805, p. 46), 26 years after Bourgelat's death, described the head and neck of a jumart in the Alfort collection, which are apparently the source of the jumart skull pictured here, as Goubaux (1888, p. 479, footnote) said that, although this head and neck were not recorded in the museum catalog, he had seen the specimen, that it had been boiled, due to poor condition, and that only the skull is preserved.
4. Bourgelat also founded the original branch of the college at Lyon (January 1, 1762).
5. Original French: "Qu'est-ce-que ce jumar bien averé, né d'un taureau, qu'on pretend être à l'école vétérinaire—je pense, à celle de Charenton. J'aurai bien de la peine à me soumettre à leur existence, vu la disproportion des parties genitales. Mais si le fait parloit, il faudroit bien se taire."
6. Suchetet (1890, p. 19) says jumarts with cloven hooves were considered undesirable as beasts of burden and, therefore, not raised.
7. Also see Grognier (1805, p. 188).
8. Bourgelat, in his anatomical description of a jumart, also mentions this widening of the maxilla in comparison with a horse.
9. Blumenbach (1781, pp. 15-16) offers other arguments contra jumarts.
10. Taxonomically, this is not the case. Both crosses would be bovid × equid.
12. Probably, Henri II (reigned 1547-1559). This animal would likely have been kept in Henri's menagerie at the Chateau de Madrid, his palace in the Bois de Boulogne.
13. The *Dictionnaire de l'Académie Française* (Suppl. 3, 1836, Paris, pp. 73 & 91; ||y73xqpl9) says the French word *bif* signifies a jumart derived from crossing a bull with either a mare or a jennet, whereas *baf* means one from a cross between a cow and either a stallion or a jackass. However, Leger (1669) and Valmont de Bomere (1775, vol. 3, p. 512, s.v. *jumart;* ||ycyauoew) say that a baf results from crossing a bull with a mare, but a bif, from a bull with a jennet. So there appears to be some confusion on this point.
14. Valéian is the dialect of Occitan spoken in the Ubaye Valley.
15. See ||yanqphyl.
16. See ||yyxp3r9o.
17 See ||ycnku9zb.
18. See ||y85ntdjq.
19. See ||y8841486.
20. *Colophon:* A cyclopean jumart in the royal cabinet of natural history in Madrid, as pictured by Joan Baptista Bru de Ramon (1786).

The New Forest cervidequid. Reportedly, it chewed cud and had a deerlike tail. The caption reads, "Remarkable hybrid, from the New Forest, drawn from life by George Landseer."

CERVIDEQUIDS (DEER-EQUIDS)

It must be satisfying to be quite certain which way to look at it.
JAMES HILTON, *Lost Horizon*

Though both are jumarians, cervidequids are far more rarely reported than are JUMARTS, perhaps because cattle, as domestic animals, are more often found in the same context with horses than are deer. The best documented cervidequid, pictured above, was foaled in England in 1848. The report accompanying this picture, taken from the *Illustrated London News* (Dec. 9, 1848, p. 364; ||3ym9d5wn), said,

> REMARKABLE HYBRID
>
> This remarkable filly (seven months old) was found a short time since in the New Forest, and is evidently of a mixed breed, between the horse and the deer. The mother (a pony mare) was observed to associate with some red deer stags in the New Forest for some months, and, at last, this foal was seen by her side. The nose shows a proximity both to the stag and the horse; her forehead is round, like that of the deer; legs slender and distinctly double; hoofs pointed, and partly double; colour brown, lighter under the belly [as in a deer]; tail like a deer.

This extraordinary animal is the property of T. G. Attwater, Esq., of Attwater, at the village of Bodenham, three miles from Salisbury. Dr. Fowler, of that city has inspected the hybrid, and is quite satisfied with the correctness of the preceding statement; and Colonel Buckley (a Keeper of the New Forest) has likewise seen that animal and is of a similar opinion.

Richard Fowler
1767-1863

* Richard Fowler was a physician and a Fellow of the Royal Society, who occupied a leading position in Salisbury for many years. Colonel (later General) Edward Pery Buckley was Queen Victoria's Equerry (stable master) and, as such, had extensive knowledge of horses. He was also a member of Parliament, his position as a Keeper of the New Forest being only one of many roles he played in public service.

Edward Pery Buckley
1796-1873

The *Sherborne Journal* (quoted in the *Bradford Observer*, Bradford, West Yorkshire, Nov. 9, 1848, p. 7; ||49exttat) said,

> Mr. Attwater, of Bodenham, near Sherborne, has a mare which had been some time grazing in the New Forest, and which some five or six months ago, gave birth to an animal half deer and half horse! It is alive and well. Its head resembles that of a deer; its legs are slender, but hoofs are not divided; the mane is very curious, and almost baffles description; the colour is a bright fawn; the hindquarters are like those of a horse, but the tail is of the deer tribe. The animal, on the whole, is one of great curiosity, and one that chews the cud.

* Horses do not chew cud, nor do any other perissodactyls. But deer do, a trait seen only in artiodactyls. The fact alone of its chewing cud would place the New Forest animal, born of a mare, firmly in the jumarian camp even if its other cervid traits were ignored.[1]

Physician Frederick Hollick (1902) described the New Forest cervidequid as "one of the most remarkable and best-attested instances of hybridization, between animals very remote from each other, that I have ever met with, and proves that with due care, crosses may be effected to a greater extent than is usually supposed possible."

Frederick Hollick
1818-1902

Williston, North Dakota. From the *Devils Lake World* (Devils Lake, N.D., May 24, 1922, p. 8c; ||wgyhmpt):

> According to Joe Mendro of Williston, he recently saw and examined a colt in a farmer's pasture near this city, which is perhaps one of the queerest freaks ever seen in this section. The colt, Mr. Mendro says, had a head not much different from that of an ordinary buck deer, and as nearly as he could tell, he believed there were indications that one or more horns might presently grow out. Another peculiar fact about the animal was that the hoof of one foot was cleft into two divisions, while the remaining hoofs were entirely natural.
>
> * The Williston cervidequid's having one hoof cleft recalls reports about JUMARTS in which both cloven and solid hoofs were present in the same animal.

Osseo, Minnesota. The *San Bernardino News* (San Bernardino, Ca., Apr. 29, 1914, p. 1; ||yyhdefmv) said, "OSSEO, MINN., April 29.—Chris Neuman, a farmer who lives near here, has a colt which is said to be part horse and part deer. Its head resembles a deer, and the body and legs that of a horse. However, the animal is slightly deformed, its lower jaw being two inches longer than the upper,[†] making it necessary to feed the colt with a bottle. Apparently, it is thriving on cows' milk and is as frisky as a real colt and as agile as a fawn. Neuman plans to develop it into a racer."

Schweinfurt, Germany. Johann Lorenz Fehr (1688), a Schweinfurt physician, reported a local cow, which had been in the company of a buck, calving a nondescript that looked like a calf except for the presence of a deerlike tail.

†. Recall that a short upper jaw is common in JUMARTS.

* If the Williston, Osseo and Schweinfurt cervidequids were hybrids, then the only potential deer sire present locally, and large enough to easily mate with a mare, would be *Cervus elaphus*, known as Elk in America and as Red Deer in Europe. Videos show *C. elaphus* bucks mounting mares (e.g., ||wo9p8ln).

St. Joseph, Missouri. From the ST. JOSEPH GAZETTE (quoted in the *Western Reserve Chronicle,* Warren, O., Aug. 10, 1870, p. 4c; ||wmgdmbq): "Mr. Antiff Hyde, who resides some three and one-half miles south of this city, has something in the way of horse flesh that is a natural curiosity. It is a three-year-old colt with horns resembling those of a deer, protruding out straight about two and one-half inches above each eye. These horns made their appearance some two or three months ago, have grown with remarkable rapidity, and have now attained considerable length The colt is a very finely formed animal, about sixteen hands high. This singular freak of nature creates considerable of a sensation in the neighborhood."

* JUMARTS, too, have occasionally been reported as having horns.
* Elk were present in Missouri in 1870 but later went extinct there, though they have since been reintroduced. White-tailed deer have always been present.

Stevens Point, Wisconsin. The *Stevens Point Journal* (Jul. 8, 1936, p. 3d; ||4hz4shhk): "A freak male colt with one eye and a head shaped like that of a deer was born Monday to a buckskin mare at the farm of William Ordway, Bancroft, Route 1. The colt is normal in other respects."

* Cyclopia is common in JUMARTS.

Berlin. Another cervidequid, the supposed offspring of a mare and a stag, was reported in the *Edinburgh New Philosophical Journal* (1827, p. 390). The brief account of this animal, entitled, "Extract of a Letter to M. de Férussac,† dated Berlin, 27th January, 1827," reads, "There is here at present an animal produced between a stag and a mare. The authorities of the place have attested the phenomenon. The appearance of the creature is very singular; the fore part is that of a horse, the hinder part that of a stag; but all the feet are like those of the latter animal. The same stag has covered another mare. The king has purchased the hybrid for the Pfaueninsel,‡ where there is a menagerie."

†. Baron André Étienne Justin Pascal Joseph François d'Audebert de Férussac (1786-1836), a French naturalist. Other accounts (e.g., Ackermann 1898, p. 61; Anonymous 1827a, 1827b) describing the Berlin cervidequid trace to the letter written to Férussac.
‡. The Pfaueninsel is in the Havel River near Berlin. King Frederick William III (reigned 1797 to 1840) turned it into a model farm and created there a menagerie modeled on the Ménagerie du Jardin des Plantes in Paris, in which exotic animals were kept.

The Georgia cervidequid

Georgia. In a now-defunct page of a hunting and fishing forum, an anonymous participant posted the picture shown above. When accused of faking the image, he or she responded as follows: "I personally took the picture myself. The deer was killed in [Georgia] on a dog/hunting club around Christmas. It is the only deer we have ever seen like this. Even the hooves had white on them. So I guess I will take this as a personal attack that you are calling me a liar. I do not care for being called a liar. I guess that everyone who posts something on here needs a certificate of authenticity? This is not a photoshop. PERIOD. Believe it or not. I didn't post this picture or join this forum to be ridiculed and called a liar. Lower down on the same page, he or she continued, "So I took it that I was being told that when I said it was not photoshopped, that I was not believed. I should have been clearer that I was the camera man. It was taken on a raggedy phone. The deer had been skinned and quartered. I don't know how y'all do it in Nebraska, but we clean them ourselves at home and take the remains to the woods. That was just before he hauled it off. He wanted to show a few people the legs, etc., so they stayed with the hide. I would have mounted it. That guy didn't have the $ so he didn't keep it. I am sorry I have no other pics and can only offer you my word that what you are seeing is the real thing. I think it is the coolest wild deer I have ever seen in my life."

* White-tailed deer (*Odocoileus virginianus*) are the only cervids native to Georgia.
* Since the Georgia cervidequid was apparently shot in the wild, it may have been the offspring of a doe.

Colmberg, Brandenburg. In his book *Stuterey* (i.e., *Stud Breeding*), Georg Simon Winter von Adlersflügel (1703, p. 136), pictures a horse at Colmberg that he says he rode and trained himself, that had cloven, deerlike hooves protruding from the pastern on both front legs just above the ordinary hooves. Adlersflügel said a farmer had put the mother mare out to graze and that she was covered by a stag in rut. The animal in question was a red deer (*Cervus elaphus*).

* Adlersflügel was a master equestrian and the author of numerous books on equine medicine.
* Compare Colmberg case in which a horse had supernumerary deerlike hooves to Wood-Mason's depiction of a horse-like hoof with a supernumerary digit like a cow hoof (p. 559).

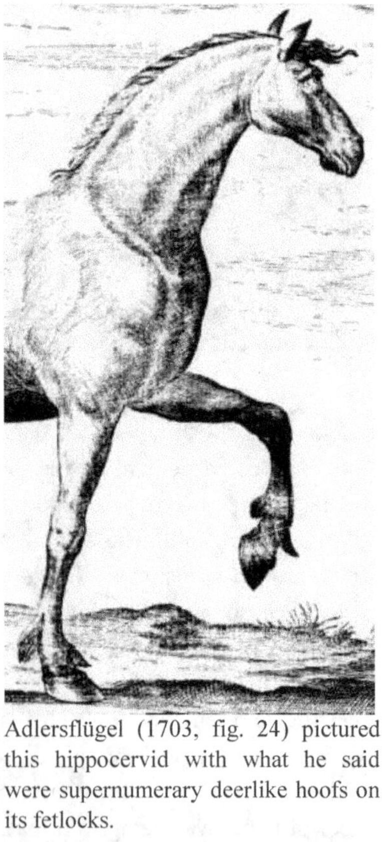

Adlersflügel (1703, fig. 24) pictured this hippocervid with what he said were supernumerary deerlike hoofs on its fetlocks.

EARLY REPORTS

France. The Swiss physician Jakob Ruf (1587, p. 48) said a cervidequid was given to the king of France.

> In fact, in France, a mare impregnated by a deer bore a foal that resembled a deer in its posterior portion and that no other horse could equal in speed. King Louis himself accepted it as a gift from its owner. [translation mine]

* The Louis mentioned by Ruf was presumably Louis XII (reigned 1498-1515). No other French king named Louis reigned during Ruf's lifetime.

Louis XII

Lhanhamelach, Wales. The twelfth-century monk Gerald of Wales said (in Hoare 1806, p. 32), "It appears also from the ancient and authentic records of those parts, that during the time Saint Elwith led the life of a hermit at Lhanhamelach, the mare that used to carry his provisions to him was covered by a stag, and produced an animal of wonderful speed, resembling a horse before, and a stag behind."

REPORTS COMPARED

In several of the older accounts about cervidequids, the front part of the animal was described as equid, the posterior, cervid, which is the same configuration as the London CERVASINID (see p. 584). But the alleged structure of the New Forest animal (see p. 577) was equid overall but, here and there, cervid.

A NON-REVERSIBLE CROSS?

In all reports specifying the parentage of a cervidequid, the mother was an equid, which suggests a strongly directional, perhaps non-reversible cross. Many hybrid crosses are directional, that is, either mating doesn't occur when the cross is reversed (in the present case the reverse, or reciprocal, cross would be stallion with doe) or such couplings yield no hybrids. To a certain extent, then, the situation seen in cervidequids seems to parallel that for jumarts, where reports mentioning equid mothers preponderate.

The London cervasinid (Adlersflügel 1703, plate G).

CERVASINIDS (DEER-DONKEYS)

London. Adlersflügel (1703, p. 124) described a deer-donkey hybrid present in London's Royal Menagerie in 1682. It had been seen, he said, by his good friend and patron Herr von Spiegel.

Tunis, Tunisia. Sir Grenville Temple (1835, p. 228) said he saw at the Bardo, the Bey's residence outside Tunis, "a curious little animal which died but a short time before I quitted Tunis; the fore part of whose body resembled the ass, whilst the hinder was like the deer's."

BREEDING

Adlersflügel (1703, p. 128) gave instructions which might be of interest to any would-be breeders of cervidequids and cervasinids. Thus, he said, in translation, "Begin by keeping a stag and a pregnant mare, who will foal in August, together in a park throughout the spring and summer, through the time of her delivery, so that they will get used to each other's company. There must be no other beast present. In this way, when the mare, having foaled, comes into heat [in September], the stag, when he enters rut without access to does, will mount and service the mare. And she, being lusty and in heat, will readily accept his advances."

Bambi, the Saint-Jogues alcidequid. Note the mooselike, down-turned muzzle.

ALCIDEQUIDS (MOOSE-HORSES)

In every respect the burthen is hard on those who attack an almost universal opinion.

—JOHN STUART MILL
The Subjection of Women

An animal combining the traits of a moose with those of a horse—that is, an *alcidequid*—is a type of CERVIDEQUID. Moose have in fact been observed mounting mares, but it seems that no hybrid of this type has been confirmed by genetic testing. Such reports tend to confirm the CERVIDEQUID mating pattern in which male cervids have often been observed covering mares, but never stallions covering does.

Saint-Jogues, Quebec. On May 5th, 2006, at Saint-Jogues, a rural locale about 20 kilometers north of the coastal town of Paspébiac, a mare foaled a colt that its owner, François Larocque, insisted was the result of her being covered by a bull moose.

Larocque claimed the mare in question had no access to a fertile stallion during the time that she would have had to conceive. He also pointed out that her offspring had unusually long legs, long woolly ears, a muzzle that curved downward (as in a moose), and two bumps on its head at the points where antlers might be expected to emerge (recall that JUMARTS, too, have many times been reported as having two such bumps on the forehead). Most of the similarity to moose in this animal seems, then, to have been from the neck up. As Larocque later told the newspaper *Le Soleil:* "When the mare gave birth, my sisters said: 'It has a moose head.'" Many telenothians exhibit a similar demarcation between the head and the remainder of the body (see CERVICAL DIVISION, p. 1199).

This alleged moose-horse hybrid, which Larocque named Bambi, provoked a great deal of discussion. Local biologists and veterinarians were sufficiently intrigued to request that it be genetically tested. But if such tests occurred, their results seem not to have been published.

Bambi, the Saint-Jogues alcidequid, standing beside her mother.

According to information on a now-defunct website forum, Larocque saw a bull moose prowling the rear of his property during September 2005. Moose come into rut in September. On the same website, it was stated that he owned two geldings only, and that there were no other male horses in the vicinity.

The gestation period of a moose is eight months, that of a horse, eleven. The diploid chromosome number of horses is 2n=64, while that of North American moose is 2n=70.

From the pictures shown here, as well as other photos shown on the Internet, it's clear that this animal, if it is a hybrid, is much more like a horse than a moose. But the mere fact that horse characters seem to predominate does not in itself preclude the possibility that it is derived from hybridization between horse and moose. It has long been recognized that in certain crosses the hybrids much more closely resemble one parent than the other.

Jay, Maine. From the *Daily Reformer* (Brattleboro, Vt., Sep. 9, 1913, p. 3c; ||j2gpt5n):

"MOOSE MARE" WAS A REMARKABLE FREAK

STRANGEST EQUINE PRODUCED IN NEW ENGLAND HAD

TROTTERS ALL STOPPED

SIRED BY BULL MOOSE

The famous "moose mare" of Maine, so called because her owner believed her to have been sired by a bull moose, was perhaps the most remarkable equine freak that New England has ever known. Her prowess on the track in the forties and fifties was widely known, and horsemen of the period who timed her for exhibition half-mile heats refused to believe their senses when she covered the distance in 1:10 to 1:12.

The animal, a nondescript creature with large, uncouth head, lapping ears and a scant mane, whose tail was a stub hardly larger than a goat's tail, worked during the forties on a stage and freight team that ran between Jay and Readfield, Me.

She was called the "moose mare" because of her resemblance to the monarch of the woods. She was foaled the property of Benjamin Goodnow of Jay from a fine Morgan mare, which he had at pasture the year before the birth of the freak, away from all other horses. The pasture was distant from any traveled road and had to be approached through a fenced field.

The mare was not taken from the pasture from May until late in October. Frequently a large bull moose was seen running in the great woods near the pasture during the mating season, and his bellowing was often heard. In the spring the mare gave birth to the most remarkable colt which farmers of the vicinity had ever seen, but she showed great affection for her strange offspring and guarded it with jealous care.

Mr. Goodnow was advised by his neighbors to destroy the foal but refused to do so.

"It is a Messenger," he said, "and they are the homeliest things in creation until they are about 12 years old."

He was sure the colt was sired by the bull moose and was curious to see how it would turn out.

When the creature was about four years old, he put it to work on a stage route which he operated, and it did constant service there for more than 20 years. She proved a cheerful worker, a fast walker and an intrepid roadster. She never refused to pull in all her life. Hardly a horse could be found that could equal her, and for years she worked against two horses day in and day out.

Her prowess was a matter of record in that part of Kennebec county, where she was well known. She could out-walk, out-run and out-haul any horse ever matched against her, and Mr. Goodnow placed many wagers on her at cattle shows and musters, backing her against any other animal in a feat of speed or strength.

In those days there were no railroads in that part of Maine, and all the winter traffic was done by horses. Augusta was a central point for stage routes, and many trotting races were held there. Ice races on the Kennebec were also common at Hallowell, and horsemen from all over the state would bring their speediest steeds, for buyers from Boston and New York were always in attendance.

Mr. Goodnow went to Hallowell with his moose mare one afternoon to see the racing and suggested that his animal be entered. None of the horsemen who had choice trotters would enter them against this queer-looking beast, however, and he finally became indignant.

He offered to wager any sum, from $1 to $1,000, that his mare could out trot any horse in Maine for any distance, from half a mile to 100 miles. A stranger accepted his challenge, for he had just purchased a trotter that could step a half mile in very fast time.

The race was a half mile, with $500 a side put up on the result. It was scheduled for the next afternoon, best two in three heats. Each man had a judge at start and finish, and these chose a third judge.

The moose mare won the first and second heats in a walk, the prize trotter stepped the best he knew how. After the purse had been awarded Mr. Goodnow, the moose mare did an exhibition half mile timed by many in from 1:10 to 1:12, a clip so fast that the timers could hardly credit it.

Mr. Goodnow kept the mare until she died at the age of about 30 years.

* The all-time official record for a trotter in a half mile is 1:51, some 40 seconds slower than the moose-mare. A horse trots at about 8 mph, a moose, at 20.

Ross River, Yukon, Canada. From the *Dawson News* (Dawson City, Yukon, quoted in the *Cordova Daily Times,* Cordova, Alaska, Nov. 1, 1922, p. 2b; ||uhmlpqy):

> Ross River Indians recently arrived at Selkirk exhibited the head and antlers of an extraordinary freak moose, which had been shot in the wilds of the Ross River. The animal looked about thirty years old, with stunted antlers that stuck out of its gnarled head like knobs, twisted and misshapen. It had a white-spotted hide, a long tail, and hoofs like a mule, not cloven like the rest of its tribe.
>
> The Indians said that the animal must have been very old, and had been wandering about the forests alone, avoided by its kind. Perhaps a devil moose, they said.

Nalibaki, Lithuania. In his *Hippica,* a text on horses first published in 1603, Krzysztof Mikolaj Dorohostajski (1861, p. 46) said that on an estate near Nalibaki, a village in Lithuania, now in Belarus, surrounded by forests and wild beasts, a mare foaled a colt like a horse, but with the head, form and hair of a moose.[2]

Hol, Norway. An animal, living in 2023 near the mountain village of Hol was a likely moose-horse hybrid (||4xhwz2sf).

NOTES

1. Dunglison (1850, p. 468) also discussed the New Forest cervidequid.
2. Original Polish: "samo wszytko jako koń, lecz głową, kształtem i sierścią łosiowi podobne." Rzaczynski (1721, p. 357) cited this case.
3. *Colophon:* Adlersflügel (1703, plate G).

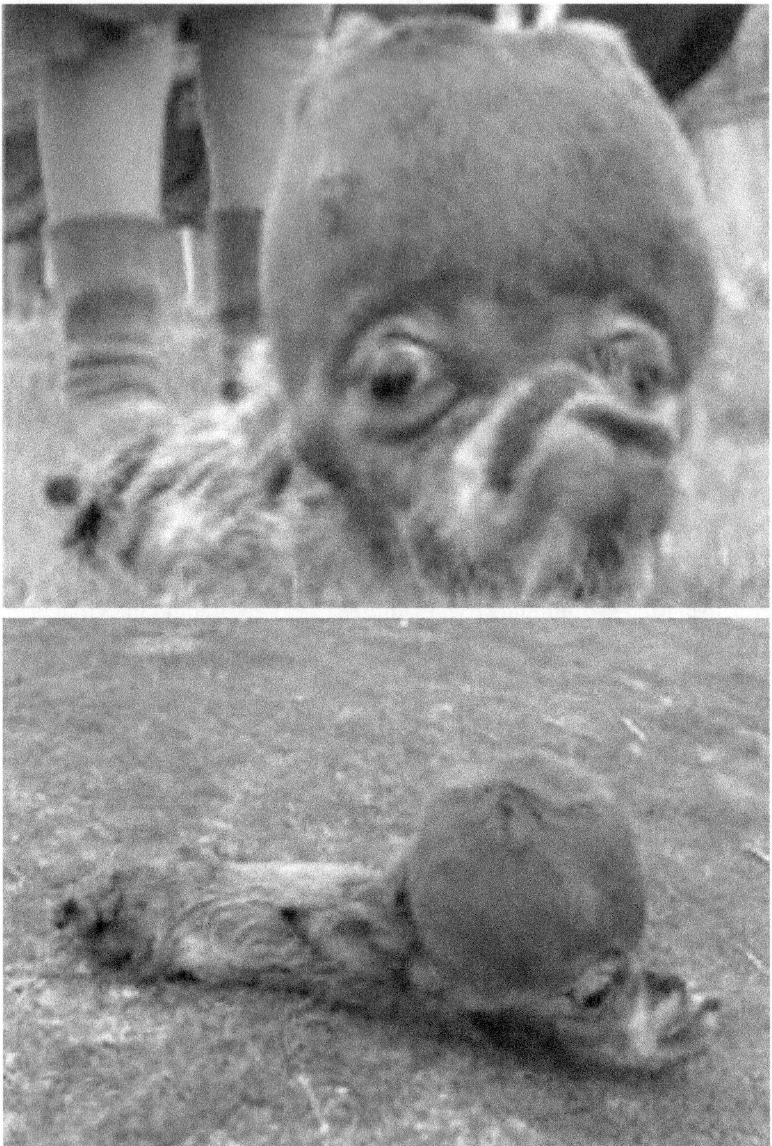

Severed head and neck of the Zărand hippocamelid, foaled by a Romanian mare in 2014. Note the short upper jaw, a frequent feature in jumarians. Hydrocephaly, apparent here, has been reported for many telenothians.

HIPPOCAMELIDS (CAMEL-HORSES)

Hippocamelus: a fabulous animal, half horse and half camel.
—LEWIS AND SHORT, *A Latin Dictionary*

Relatively few reports exist about hippocamelids. But the plausibility of those that do is increased by the many cases documenting the existence of other types of JUMARIANS, especially JUMARTS. Two camels, the Bactrian and the Dromedary, are treated as species.

Kanawha, Iowa. The *Kanawha Record* (quoted in the *Evening Times-Republican,* Marshalltown, Ia., Jul. 3, 1901, p. 5b; ||29b6tcda) said, "W. L. Channer, the butter maker at the Bruce creamery, informs us of a freak colt he owns, now about ten days old. It has characteristics of a camel, having a short tail, bare of any hair, no mane or foretop, a beautifully curved neck, and a very pretty animal throughout. Its ears are about 1½ inches long, white face, blue eyes, legs below the knee, round. Mr. Channer states that the dam of the colt is an Arabian mare, bought of a passing show, who sold her on account of her being lame."

* A 1900 census record shows a William Lewis Channer (1857-1935), a butter-maker, was living in Norway Township, 20 miles from Kanawha (||3w6jzpv7).
* The Kanawha hippocamelid may have been the animal advertised a few years later in the *Barre Daily Times* (Barre, Vt., Oct. 5, 1906, p. 8; ||m9sn6w7): "Strange freak, half horse and half camel, absolutely the only living one; puzzles scientists."

Relizane, Algeria. L'Hippique Algérienne et Tunisienne. (quoted in *La France Chevaline,* Paris, Jun. 14, 1899, p. 2a; ||4yntkcc9) said, in translation, "A FREAK (HORSE × CAMEL).—This is a freak of which the renown will soon sweep the earth. Read, readers, and you will see that we do not exaggerate: Mr. Michel Hueras just purchased at Relizane a 25-day-old Barbary mare, the issue of a mehari and a mare. [A mehari is a type of fast-running dromedary camel.] And that's not all that's bizarre. She has the head, the ears, the eyes, the muzzle, and the tail of a camel—and a bit of the neck, as well. The hump is replaced by a small bulge that seems to be growing. Her neck and tail are naked. On the forehead and hump, her hair very much resembles a dromedary's, as does her gait. The hindquarters, the legs and the rest of the body are like a horse. Another detail: she's a ruminant! Finally, and this will catch the eye of sportsmen, this freak has extreme agility. She makes extraordinary leaps with the speed of a gazelle. Expert horsemen who have seen her run affirm that a five- or six-month-old thoroughbred of the finest breed would not be able to keep up with her. And voila! A most appropriate breed for Algeria!"

Zărand, Romania. On June 29, 2014, in Zărand, a mare foaled a stillborn hippocamelid. As can be seen in the photos (p. 590), its skull was deformed by intracranial pressures, which also bulged its eyes. Reports say it had the elongated legs of a camel, as well as a camel's long, slender neck. The coat color and face were also like a camel's. In translation, one report (||vzvlhqu) said,

> People in the Arad town of Zărand are in shock and praying in church after a mare gave birth to a colt there that looks more like a camel with bulging eyes. Locals believe it's a bad sign, coming before Easter, and say they've never seen such a scary being.
>
> "I have something here that I've never seen in my life," said the owner of the animal, according to arq.ro [a local online news source].
>
> The mare gave birth last night, dying in labor. The colt lies in the owner's farmyard, people standing about, in shock at the way it looks. The owner says he'll ask the local veterinarian, who was not present during the delivery, for compensation. The vet, who is supposed to be responsible for Arad County, is still nowhere to be found. Meanwhile, local women are going to church to pray, believing the mutant colt is an evil omen.

Abnormal births have been considered ominous since ancient times. Even in the earliest civilizations such births inspired fear and religious awe. Indeed, the word *monster* comes from Latin *monere,* meaning to show, warn or indicate. Like comets, eclipses and other rare portents, hybrids have been taken as harbingers warning of evil things to come. It seems the residents of Zărand hold similar beliefs.

Milpitas, California. From the *San Francisco Call* (Mar. 29, 1902, p. 3e; ||rjaa2qg): "SAN JOSE, March 28.—A curiosity in the shape of a colt with a camel's head and one large eye in the center of the forehead was born this morning at the ranch of Samuel Parker, near Milpitas. The animal otherwise is a perfect specimen of a thoroughbred colt and well developed. A nose was lacking and as it could only breathe through its mouth it lived but a short time. The jaws and lower part of the head are those of a camel. The lower jaw is very long, while the upper one and the lip are short [as is the case with many jumarians]. On the upper lip is hair the same as on a camel's lip and the mouth resembles that of the humped animal. Above the mouth is an abnormal eye, as large as two ordinary ones. This optic is open. Above on either side of this cyclopic member are two other eyes. They are covered with skin but can be plainly felt. The ears are those of a horse but are turned around more than ordinarily. The mother of the colt was at the racetrack this winter while Norris & Rowe's [Circus] show was in quarters there and the monstrosity is supposed to have been due to a prenatal influence."

Montfoort, Netherlands. Anatomist IJsbrand van Diemerbroeck (1672, p. 279), a native of Montfoort, said that there that city, in 1626, a pregnant mare, frightened by a dromedary, later foaled a hippocamelid. "That mare gave birth to a colt," he said, in translation, "whose whole right foreleg was entirely similar to the leg of a dromedary, with the broad, round, cloven hoof, and the knee as thick as a dromedary's. The foal grew up into a sturdy horse, which we saw for many years, pulling plows and wagons with the other horses."

* Also see Haller (1774, vol. 1, p. 562).

* Note the attempts made to explain both the Milpitas and Montfoort animals in terms of the discredited theory of MATERNAL IMPRESSIONS, which was widely accepted, even among scholars, until the early twentieth century.

San Luis Obispo, California. An ad for Porter's Freak Animal Show described one of the animals on exhibit as a "colt with camel's head, one eye in the center of the forehead" (*Tribune,* San Luis Obispo Jul. 13, 1914, p. 4; ||zdc8z5fu). It would seem that this animal was distinct from that foaled at Milpitas, despite the similarity of the descriptions, as reports about the Milpitas hippocamelid say it died soon after birth.

Memphis, Tennessee. A humpbacked supposed scion of the famed American standardbred Axworthy was described in the *Washington Times* (Washington, D.C., Feb. 26, 1905, 9b; ||2kv43p32). This animal had such a humped back that it was nicknamed "Dromedary."

* The *Evening World* (New York City, Aug. 18, 1922, p. 10f; ||pva3uy96) mentioned what may have been the same animal on display at Coney Island.

Will horses and camels mate? An early writer, the scholar and physician Julius Caesar Scaliger (1612, p. 649), noted that mares sometimes allow camels to mount them. Though many horses are afraid of camels, when accustomed to each other, these animals can be on friendly terms (||vzan7lw). Buffon (1778, p. 19) said female camels will nurse foals and young horses. Upon reaching sexual maturity, such animals would likely prefer camels as mates (due to sexual imprinting).

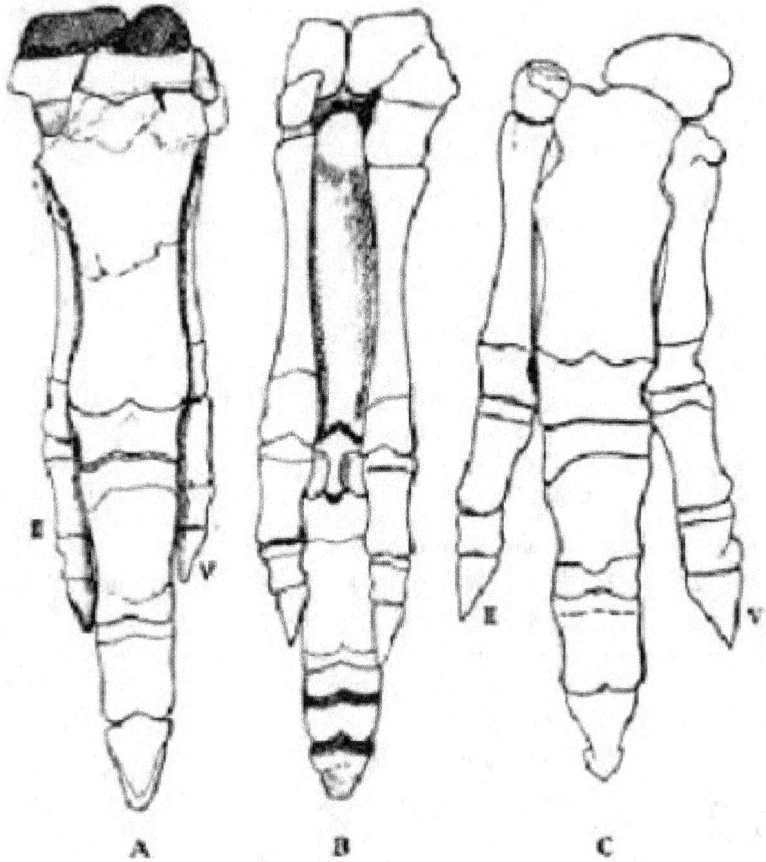

Feet of the Alfort hipposuid. *A:* left manus from the front; *B:* the same from the rear; *C:* left pes from the front. *Image: Bateson (1894, p. 387, fig., 120).*

HIPPOSUIDS (PIG-EQUIDS)

Solid-hoofed pigs resemble horses more than other pigs do.
—FRIEDRICH BLUMENBACH

The hipposuid most often reported is the mule-footed or solid-hoofed pig, in which the two weight-bearing digits are fused and have a single hoof, a trait of equids not of pigs and other artiodactyls.

Alfort, France. Barrier (1884, p. 491). described the hind and fore foot of a "mule-footed pig" in the Alfort collection. In place of two separate weight-bearing digits, as in a normal pig, this animal had a single digit with a single hoof, while the other two digits were normal.

Plymouth, Illinois. Wade (1880), a resident of Plymouth, reported that "Last month a sow in this vicinity gave birth to nine pigs, six of them being mule-footed (solidungulate). Three of the litter possessed the usual number of toes. All the pigs died."

Foot of a "mule-footed pig," reportedly one of a herd of such animals that once existed near Baton Rouge, La. (Auld 1889).

* Insemination by an equid would account for an ordinary sow farrowing six offspring with equid hoofs. Mutation does not, because mutations are independent events.

Washington, D.C. In a list of specimens in the United States Army Medical Museum, Billings (1884, p. 24) described specimen no. 1218: "Four malformed feet of a pig (*Sus scrofa*); the front feet are clubbed, with the hind toes abducted externally: the hindfeet are well developed, the front toes being joined together, forming a single hoof."

Cuba. Alfred Henry Garrod (1877), Prosector of the Zoological Society of London, discussed two pigs farrowed in Cuba that had been donated to the Zoological Gardens in 1876. The hoofs of all four feet in these animals, "instead of being cloven," he said, "are solid, much resembling those of the Ass." He went on to say that "The sow gave birth on November 15th to a litter of six, three males and three females. Of these, the hoofs were solid (like those in the parents) in three, namely in two males and one female. In the remaining male and two females the hoofs were double, as in the animal under ordinary circumstances." Two of these, a male solid-hoofed and a female cloven-hoofed specimen died within a few days of birth.

* Detlefsen and Carmichael (1921) reported that a solid-hoofed boar mated with cloven-hoofed sows consistently produced solid-hoofed offspring.

Munich. Theodor Kitt (1885, p. 64), a professor of veterinary medicine in Munich, described two solid-hoofed pigs. One had four solid hooves that Kitt said were extraordinarily similar ["ausserordentlich gleichsehend"] to a tapir's. (Most biologists deem tapirs close relatives of horses.) The other had only two, also tapir-like.

Rannoch, Scotland. Struthers discussed a large herd of solid-hoofed pigs that, circa 1840, belonged to Sir Niel Menzies in Rannoch. When kept separate from ordinary pigs their numbers increased into the hundreds. After the separation ceased, however, they died out.

* Menzies' once numerous solid-hoofed pigs may have declined due to impaired hybrid fertility, which resulted in their being swamped out of existence once they were allowed to interbreed with ordinary, fully fertile pigs.

Bexar, Texas. Elliott Coues (1878) deemed a population of solid-hoofed pigs established in the vicinity of Bexar, a new breed.

* Various modern reports speak of the mule-footed pig as having been popular in the early 1900s. But hog expert Foster Dwight Coburn (1916) provided data indicating they were uncommon even at that time. Thus, he states (p. 31) that a census of the swine in Indiana in 1907 counted 24,632 Poland Chinas, 26,729 Duroc Jerseys, and 7,834 Chester Whites, but only 229 Mule Foots. Solid-hoofed pigs were known even to the ancients, thus Pliny (11.106), who died in 79 A.D., noted that their presence "in certain places in Illyricum." Aristotle also mentions them (*History of Animals,* 2.1).

Hof, Bavaria. Hahn (1865, p. 18) described a pig sent to him from Hof which had soliped forefeet that he said were like those of an equid ("Einhufer") but without a frog. The hind feet, too, exhibited more fusion than normally seen in a pig.

Poland. Pallas (1766, p. 19) commented that the feet of pigs "always bear four digits, except that playful Nature sometimes produces solid-hoofed animals like domestic pigs with the two major digits fused. These, especially from Poland, are well-known to German butchers.

Schrottendorf, Lienz, Austria. The *Tiroler Volksbote* (Bolzano, Mar. 20, 1904, p. 11a; ||2rjmsujv) said, in translation, "Last week a sow in Schrottendorf farrowed a litter of ten, among which was one remarkable monstrosity. It was entirely like a donkey, except that it had cloven hoofs. It also had a pig's tail. The rest was exactly like a donkey—head, torso and legs. Otherwise, it was the size of a piglet, but stillborn." A donkey-like offspring farrowed by a sow can be reasonably explained only as a hybrid.

The Dodge City hipposuid (reconstruction)

TWO MARE-FOALED HIPPOSUIDS

Dodge City, Kansas. From the *Globe-Republican* (Dodge City, Feb. 11, 1892, p. 5a; ||24bvtxjm): "One of Stubbs Bros.'s high-bred mares gave birth to a monstrosity this week. It was a perfectly formed colt in all respects except the head, which has a striking resemblance to that of a hog. This portion of the freak is on exhibition at the grocery store."

* The Stubbs brothers were racehorse breeders who had a farm two miles east of Dodge City (*Globe-Republican,* May 23, 1892, p. 4b; ||y2bj8zcb).

Seneca Falls, New York. A report about a viable hipposuid appeared in the *Morning Post* (Camden, N.J., Jul. 19, 1934, p. 6g; ||r4hhdu2).

> SENECA FALLS, N.Y., July 18—A new freak for the sideshow world, best described as half pig and half horse, became the talk of the town here today as Dr. W[illiam]. L[lewellyn]. Clark [1885-1940], local veterinarian, placed the animal on exhibition at his hospital.
>
> The freak, born to a mare on a farm here, has the head of a pig, breathes and snorts like a pig,[†] but has the long legs and body of a colt. The animal is black with white marks on its head and nose.

†. A hybrid with a given kind of animal's head usually has that animal's voice.

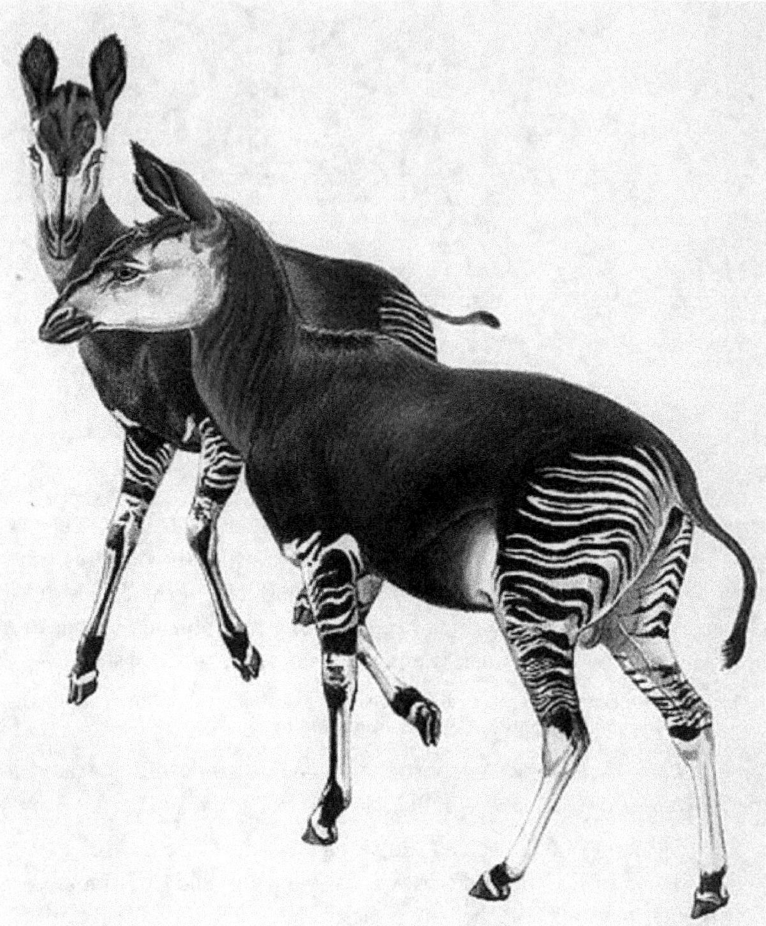

Okapis combine the traits of giraffes with those of zebras (||32aadmda).

On their rumps and legs, okapis have stripes much like a zebra's.

OKAPIS (ZEBRA-GIRAFFES)

The rule is perfect: In all matters of opinion, our adversaries are insane.
—MARK TWAIN, *Christian Science*

This chapter, like the one on monotremes, argues that a well-known kind of animal, treated as a species, the okapi (*Okapia johnstoni*), probably had a hybrid origin.

An okapi's head, horns and teeth are all giraffe-like.

This animal mixes traits seen otherwise only separately, in either giraffes or zebras. This fact would ordinarily constitute strong evidence that okapis are giraffe-zebra hybrids.[1] But in this case, few biologists would be persuaded. One might well wonder why.

Zebra (left), okapi (center) and giraffe in silhouette. The general conformation of an okapi is intermediate with respect to body, neck and leg length, as well as the slope of the back from withers to rump. Okapis are about the same size as zebras. Giraffes (not shown to scale) weigh 2-3 times as much.

The apparent reason is that scientists rely more on theory in this case than on observation. In trees representing accepted notions of evolutionary descent, giraffes and zebras are placed on widely separate branches, so it is generally believed that the two are simply too far apart to produce hybrids. Thus, it's no surprise that there are no reports from about attempts to create such hybrids. Experimentation does not occur, as theory says it is useless to try.

Okapi tongue (Image: Kaelin, Wikimedia; ||wssutbg).

Giraffe tongue

Scientists long ago assigned the okapi and the giraffe to the same taxonomic order (Artiodactyla) because they both have cloven hoofs, and to the same family (Giraffidae) because they share certain distinctive features: Both have large eyes and ears, thin lips and an extraordinarily long, extensible tongue that allows them to lick their entire faces, even the ears; their backs slope upward from rump to withers; they also share the same dental formula: (i 0/3, c 0/1, pm 3/3, m 3/3) × 2 = 32. Both okapis and giraffes, unlike other mammals, have molars with rugose enamel and bony horns that remain covered with skin throughout life (Nowak 1999, vol. 2, p. 1085). The two are therefore firmly linked.

And yet, the rump and legs of an okapi are covered with black-and-white stripes exactly like those of a zebra. And an okapi is about the same size as a Burchell's zebra.

The chromosome count of an okapi is also like that of a zebra, to which it is not supposed to be related, and unlike that of a giraffe. Giraffes have 30 chromosomes (Taylor et al. 1967; Hösli and Lang 1970; Koulischer et al. 1971), whereas okapis have a variable chromosome number of 44-46, depending on the animal in question; most seem to have 2n = 45 (Ulbrich and Schmitt 1969; Hösli and Lang 1970; Koulischer 1978). The chromosome number of Grevy's zebra is 2n = 46 and plains zebras have 2n = 44 (Benirschke and Malouf 1967). Variation in chromosome count is itself unusual among mammals, but common in hybrids.

Okapis also produce high levels of abnormal sperm. Thus, Penfold (2007) reports that 52% percent of the spermatozoa produced by these animals are morphologically abnormal. As those authors state, "okapi semen collected by electroejaculation routinely contains high numbers of non-motile and plasma membrane-damaged spermatozoa, apparently unrelated to season or the length of time since the male was housed with a breeding female." Hybrids produce abnormal sperm at far higher rates than do non-hybrids (see citations below[†]).

Various studies report that okapis have poor reproductive success (Benirschke 1978; Leus 2004; Loskutoff 1997; Rabb 1978; Raphael 1988; Raphael et al. 1986; Van Puijenbroeck and Leus 1996), which is also the case with many kinds of hybrids.

Giraffes and zebras exist in mixed herds in various parts of Africa and therefore are in potential breeding contact (these regions include those where okapis occur).

†. Researchers reporting abnormal sperm in hybrids: Anonymous (1979b; 2007, p. 57); Basrur (1969); Chandley et al. (1974); Dawley et al. (1985); Döring et al. (1972); Eastham and Nicholls (2002, p. 387); Fedyk and Krasinska (1980, p. 211); Guyer (1900, pp. 40-41; 1902, pp. 11-12; 1912); Hamaguchi and Sakaizumi (1992); Ishishita et al. (2015); LaChance (1984); McCool (1990); Park et al. (2011); Podevin (2011, p. 11); Röll and von Düring (2008); Santiago-Moreno et al. (2016); Shrigley (1938, 1940); Smith (1912, pp. 160-162); Smith and Thomas (1913, p. 40); Spangenberg et al. (2017). Stolla and Gropp (1974); Swan and Christidis (1987); Vasil'ev et al. (2003, p. 72); Wishart et al. (1988); Xiao-Lin Luo et al. (2014); Yamashina (1941, 1943); Zhao et al. (2016).

However, zebras are much smaller than giraffes, which might lead one to deem them physically incapable of mating. But hybrids sometimes occur between animals where the disparity in size is much greater. Male Steller sea-lions (*Eumetopias jubatus*) often mate, and sometimes even successfully hybridize, with female California sea-lions (*Zalophus californianus*). The former average around 1100 pounds, while the latter weigh only around 200 pounds, a ratio of 5.5:1 (the female often dies in such encounters). Such large disparities are nothing unusual in hybrid crosses. Florio (1983) reports a case of a lion father who weighed 550 pounds (250 kg), siring a hybrid on a leopard mother weighing a mere 84 pounds (38 kg), a ratio of 6.54:1.

In the case of a full-grown male giraffe, which typically weighs some 2,600 pounds (1,180 kg), with a female zebra, which would weigh about 770 pounds (350 kg), the weight ratio is only 3.4:1, that is, the difference is less disparate than in either of the two crosses just mentioned. And the difference would be even smaller with the cross reversed, that is, with a giraffe cow and a zebra stallion. Giraffe females weigh nearly a thousand pounds less than males, while zebra stallions weigh a bit more than females, which would yield a ratio closer to 2:1, not at all unusual in a hybrid cross. And with a young female, the disparity would be even less.

The bray of an okapi sounds just like part of a zebra bray. It apparently contains no giraffe-like sound, however, perhaps because giraffes are famously silent (though a recent study reports that giraffes do make low-frequency sounds on the lower bounds of human hearing). Many kinds of hybrids produce vocalizations that are combinations of sounds heard separately in their two parents' vocalizations.[†]

Another fact consistent with the idea that okapis might be giraffe-zebra hybrids is their rarity, which is typical of sporadically produced and reproductively deficient hybrids. The IUCN rates the okapi as endangered, although it also states that "there is no reliable estimate of current population size."

†. Thus, Marler and Tenaza (1977) say the song of a female gibbon hybrid (*Hylobates lar* × *H. muelleri*) followed this pattern. Jung et al. (1994) describe a junco × sparrow hybrid (*Junco hyemalis* × *Zonotrichia albicollis*) which, too, sang a song mixing those of its parents. Likewise, in studying natural hybrids between the Smooth Frog (*Geocrinia laevis*) and the Victorian Frog (*G. victoriana*), which occur in southeastern Australia, Littlejohn and Watson (1976a, 1976b) found that hybrid calls combine elements of the parental calls.

Comments on the Internet are frequent to the effect that okapis look like a giraffe-zebra cross. And many people even say okapis are obvious giraffe-zebra hybrids. This speculation is due to okapis combining, as has been mentioned, the traits of giraffes and zebras. And this belief has been widespread ever since the okapi first became known to science. Writing in the first years of the twentieth century, only four years after the discovery of the okapi, the British naturalist Richard Lydekker (1904, p. 474) commented that

> the absurd idea that the okapi is a hybrid between a giraffe and a zebra still appears to be current. Apart from the fact that hybrids between such widely different animals do not occur in nature, the okapi is essentially a giraffe in structure and fully a dozen specimens are known.

So Lydekker is implying that if the okapi were a hybrid, multiple specimens would not be known. But he was writing at a time when little was known about natural hybrids, and it was widely supposed that they never occur. Today, a wide variety of natural hybrids, from many thousands of different crosses, are known. And many reports describing a similar hybrid, the JUMART, are on record, as well as ones about other jumarians (CERVIDEQUIDS, HIPPOCAMELIDS and HIPPOSUIDS). So why shouldn't there be multiple giraffe-zebra hybrids?

Richard Lydekker
1849–1915

Sir E. Ray Lankester (1905, pp. 163-165), director of the natural history departments of the British Museum, concurred with Lydekker's opinion that the okapi could not be a giraffe-zebra hybrid:

Sir E. Ray Lankester
1847-1929

> As they live in these immense dark gloomy and damp forests, they are very difficult to shoot or to catch, and moreover they are not

> abundant. The natives cut the striped skin into girdles and bands for ornament. Two of these were sent home by Sir Harry Johnston before the animal was known and were described as coming from a new species of zebra, which was named *Equus johnstoni* by Dr. Sclater. Some people, on account of the Okapi being striped somewhat like a zebra, whilst it has the double hoofs of giraffes and also paired horns, have supposed that it might be a hybrid or 'mule' between a zebra and a giraffe. This is, however, a supposition which every naturalist knows to be quite out of the bounds of remotest probability. It is a fact that mules or hybrids never are produced by animals living in their natural conditions, except in a few rare cases among aquatic animals whose eggs are fertilized in the water after they have been laid. And no one has ever produced, even in captivity, a hybrid between any creatures so unlike each other as a double-hoofed and a single-hoofed mammal.

Today, Lankester's assertion that "It is a fact that mules or hybrids never are produced by animals living in their natural conditions" seems laughably inaccurate. It is now well known that thousands of different kinds of crosses occur naturally among birds, mammals and a wide variety of other animals.

And as to his allegation that "no one has ever produced, even in captivity, a hybrid between any creatures so unlike each other as a double-hoofed and a single-hoofed mammal," it seems abundantly clear, from the chapter on JUMARTS, that such hybrids have been produced in captivity, often serendipitously, on innumerable occasions. Moreover, elsewhere in this book it has been documented that many other, equally distant, or even more distant, combinations certainly exist.

And there is certainly nothing in the structure of a hoof, *per se*, that prevents interbreeding. After all, there are breeds of pig ("mulefoot pigs") that have solid hooves like those of horses, donkeys and zebras. And yet, they can produce offspring with ordinary cloven-hoofed pigs.

Even in his own day, not everyone agreed with Lankester. As he later wrote (Lankester 1920, pp. 132-133),

> A very eminent person whom I was conducting some years ago round the galleries of the Natural History Museum, declared as we stood in front of the specimen of the Okapi of the Congo Forest,[2] that it was clearly a hybrid between the giraffe and the zebra. He

insisted that it was obvious that such was its explanation and pointed to its striped haunches and legs, and its cloven hoofs and giraffe-like head. I failed to change his opinion.

But you don't have to be a "very eminent person" to see that the okapi does in fact look quite a lot like what one would expect in a giraffe-zebra hybrid.

Nevertheless, only experimentation and further study can better establish the true status of the okapi. For example, it would be interesting to see the results of repeated artificial insemination of female giraffes with zebra sperm, and vice versa. Genomic analysis would also be of interest.

NOTES

1. Moreover, old news reports (e.g., ||hbd374s) say peoples native to the region where the okapi occurs believed it to be a hybrid.
2. Okapis are not known to occur outside the Democratic Republic of Congo, north of the Congo River. Since lions do not occur in that region, it may be the only place where okapis with their poor reproductive abilities can persist as a self-sustaining population.
3. *Colophon:* Lundeberg and Seymour (1910, p. 217; ||wuoj4lc).

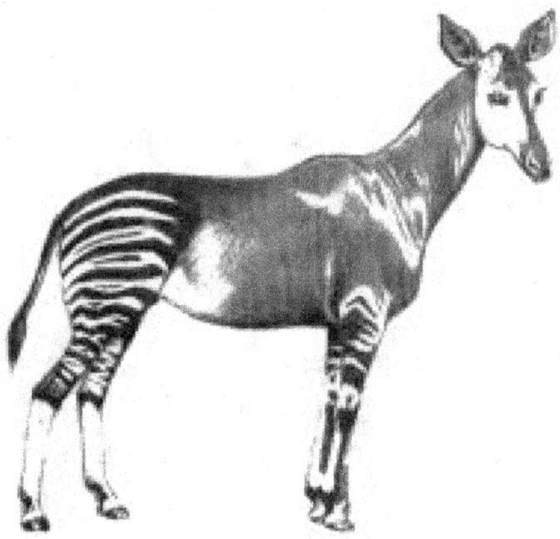

ELEBOVIDS (COW-ELEPHANTS)

Rohrsburg, Pennsylvania. The *Times Leader* (Wilkes-Barre, Sep. 15, 1903, p. 30c; ||5yurpvub): "John Gillaspy, of near Rohrsburg, has a monstrosity in the shape of a calf. This animal has the head of a calf while its body has all of the appearance of being that of an elephant. The body is full and clumsy and with all the characteristics of his larger herbivorous friend. The tail is also the very image of that of the beast that carries his trunk with him, and at first glance at the animal, not happening to see its head, one's first impulse would be to call the brute an elephant. The animal is now quite large, being fully three months old … All residents in that district have been to see the elecalf.

The Rohrsburg elebovid (reconstruction).

* A John Wilson Gillaspy (1845-1938), a farmer, lived near Rohrsburg in Greenwood Township, Columbia Co., Pa. (||2p9x58sh).

Willard, Georgia. The *Atlanta Journal* (Jun. 14, 1927, p. 9) said a cow calved an animal with a hairless, tough elephant's skin near Willard.

Grantham, Lincolnshire. The *Grantham Journal* (July 22, 1871, p. 2a) said a cow calved a "calf, which had two perfect heads on one neck, with four perfect ears, the other parts resembling those of an elephant."

Camelford, Cornwall. The *Western Mail* (Cardiff, Apr. 20, 1882, p. 3; ||3uewbue6) reported the calving at Camelford "of a calf with the perfect head of an elephant, including a trunk six inches in length."

Montbéliard, Doubs, France. La Croix (Paris, Dec. 23, 1906, p. 4; ||bdtupevj) reported a strange animal calved at Montbéliard. In translation, it said, "The young calf had an enormous head, entirely like that of an elephant, the trunk and tusks being very well formed. The owner has had this freak stuffed. It died during birth."

Teleglirids

Sir Robert Moray related that he had heard from Dr. Hinton of a copulation of a male rabbit and a female cat, which produced monsters whose foreparts were like a cat and the hinder parts like a rabbit, and that those monsters had reproduced more complicated monsters.

—THOMAS BIRCH
The History of the Royal Society of London[1]

A *glirid* is any animal assigned to either Order Rodentia or Order Lagomorpha. A *teleglirid* is any animal combining traits of a member of one of these orders with those of a member of some other order. For information about human-glirid composites, see the section on GLIRHOMINIDS (p. 391).

Photo from the *Daily Eagle* (Brooklyn, N.Y., Dec. 12, 1922, p. 31; ||ttzgqs4) The caption read, "This hybrid, said by its owner, Mrs. Edmond Devos of Omaha, Neb., to be a cross between a cat and a rabbit, was on display at a cat show recently held in Omaha. Its tail and method of locomotion are those of its rabbit father."

Man holding a pet feleporid (Utagawa Kunisada, Japan, 1858; ||mwt64xy3).

FELEPORIDS (RABBIT- AND HARE-CATS)

"We learn to think about everything," he said, "and then we train our eyes to look as we think about the things we look at."
—CARLOS CASTANEDA, *A Separate Reality*

Screenshot of a video showing the 2011 Tucuman feleporid (||wnlshzb). Note the plantigrade hind feet.

The mind learns early to classify, and thenceforth there is an urge to fit everything we see into our system of classification. Those in the thrall of the conventional way of seeing assign every organism—even animals with mixed features clearly marking them as hybrids—to a "species." Regarding birders tagging hybrids in this way, Sibley (1994, p. 164) noted "the human brain has a remarkable capacity to ignore glaringly inappropriate features while categorizing a bird based on other, more familiar, features. My experience is that an unexpected hybrid does not stand out at first glance." Such is surely the case with *feleporids*, animals mixing the traits of a rabbit with those of a cat. Many people would call the animal shown at right a cat, despite its rabbit-like tail and hind legs, and even though it hopped like a rabbit.

Feleporids are popularly termed cabbits or, less often, rabcats. It seems that virtually all the innumerable reports about such animals say the front half of the body looks just like that of a cat, while the rear half is glaringly identical to a rabbit's. And numerous videos showing feleporids are available,[2] in which they hop when in rapid locomotion but walk more like a cat when moving slowly. That buck rabbits will mate with female cats is amply documented on YouTube (e.g., ||rul8x4s, ||qrkyo2z), although the reverse mating seems not to occur.

There is ample evidence for the existence of such animals. Thus, the creature shown below (and on p. 609) was widely publicized in 2011. Found in Tucuman, Argentina, this animal is clearly shown in videos (e.g., ||wnlshzb, ||wnkldjm). It is an actual feleporid, not an impossible or imaginary one. Such evidence suggests, at the very least, that this issue should be investigated further. If feleporids are hybrids, it should be easy to verify the fact via biomolecular analysis of their DNA, given that many such animals would likely be F_1 hybrids.[3]

Animals adopted by a female not of their kind tend later to choose mates of that kind. Cats have been repeatedly reported to foster rabbits. Thus, the *Independent* (Raymond, Ill., ||bdfukfv9) said, “J. R. Chapman has a cat that found a young rabbit in the field a few days ago and carried it home, placing it with her kittens. And now the rabbit is just like one of the family and seems to be as happy as any of them. The mother cat nurses it just the same as her kittens.” Videos (e.g., ||yx2lsjhu) amply document cases of cats nursing rabbits.

Screenshot of a video showing the plantigrade position of hind feet of the 2011 Tucuman feleporid (||wnkldjm). The dashed line indicates the point where the phenotype of feleporids is typically divided. Forward of the line they are catlike, back of it, rabbit-like. Try blocking out each side of this picture and identify the animal you see. This demarcation typically applies even to such details as the claws (the foreclaws are curved and sharp, the hind claws, blunt). The fur forward of the line is catlike, behind it, rabbit-like.

REPORTS

Over the years, reports about feleporids have been so numerous that, in the material quoted below, no effort has been made to be exhaustive.

Sarcoxie, Missouri. The *Sarcoxie Record* (quoted in the *Cedar County Republican,* Stockton, Mo., Jun. 20, 1912; p. 1b; ||wwuwb94) said,

> A peculiar freak of nature bearing all the seeming of a cross between a cat and a rabbit is in the possession of M. A. Dunphy and Walter Woods,[4] who have bought the hybrid for the purpose of selling it to some show dealing in freaks. The animal has the head and front teeth of a cat, but its hindquarters, hind legs and tail are unmistakably those of a rabbit. It rears itself on its hind legs like a bunny, seems to fear cats and subsists on a mixed diet of grass and meat.
>
> The animal is not at all pretty to look at. Its movements show neither the grace of a cat nor the speed of the rabbit. … The rabbit belongs to the rodent family of animals, a family preyed on by the feline species to which the cat belongs. That two animals so antagonistic should combine to produce a hybrid of this character is a very remarkable thing, so remarkable as to be almost past belief. The young cat-rabbit, however, shows all the characteristics of its dual parentage and must be accepted for what it appears to be.

Trois-Rivières, Quebec. The *St. Maurice Valley Chronicle* (Trois-Rivières, Jan. 20, 1938; p. 3; ||r3gjzq9) said,

> We are indebted this week to a charming lady secretary for the story … we venture to assert, may yet make the name of Three Rivers and Cape Madeline resound throughout the world.
>
> It seems that a lady resident of the Cape has a cat which, not long ago, added to the feline population in a big way. Amongst the number of her progeny was one animal whose front section is reported to be catlike in every respect though its rear portion is pure rabbit.
>
> Moreover, this strange animal does not miaouw like a real cat (though it purrs) and it has seven toes on each foot. It shows a fondness for hay as an article of diet and has a fuzzy rabbit-like tail which indicates that in this or preceding generations one of its ancestors contracted a mésalliance with an enterprising jackrabbit.

Spartanburg, South Carolina. From the *Newberry Weekly Herald* (Newberry, S.C., May 31, 1921, p. 8d; ||teokzjq):

> Mr. J. W. Hodge,[5] driver of a transfer auto in Newberry, has shown us a living curiosity in the shape of half rabbit and half cat, which is a freak of nature.
>
> The "thing" came from Spartanburg, and if you don't believe it is what we say it is (it is hard to believe offhand) get Mr. Hodge to show it to you. Being so unusual, it is an interesting object to look at.
>
> The rabbit cat or cat rabbit has puzzled us since seeing it, and you don't know what to call it—but call it by any other name and it is not as sweet as a rose, although some girl wanting it for a pet might call it cute. It is a strange looking animal, from its "bunny cotton-tail" and rabbit like hind legs to its cat like fore part and head. …

Truemans, Pennsylvania. From the *Expositor* (Yale, Michigan, Oct. 28, 1910, p. 3a; 2pvx8m69): "W. H. Kisinger of Truemans, Pa., is the owner of a most peculiar animal. It was born of a cat mother, but has the body, legs and tail of a rabbit, with the head of a cat. It runs like a rabbit, by jumping up and kicking with its hind feet. It has cat claws on its hind feet but has both rabbit and cat claws on its front feet. She has had three litters of kittens, but none resembling her. She is sitting up most of the time and stands up and walks around when looking for anything that alarms her. She catches mice, frogs and bats. When necessary, she can jump six feet in the air and catch a bat with her mouth."

* The Truemans report is one of many that speak of feleporids producing offspring.

Marshallton, Delaware. The *Evening Journal* (Wilmington, Del., Apr. 20, 1909, p. 3c; mwmmn5rf) said,

> William W[atson]. Bacon [1879-1945], of Marshallton, has a freak cat. It is half rabbit and half cat with one yellow eye while the other one is blue. Two kittens appeared on Sunday last and one of the kittens resembles the mother by having no tail while the other kitten is fully endowed along that line. Many persons in Marshallton who know that cat called to see the young freaks.
>
> "Anybody not believing this can call and see them at any time," says Mr. Bacon.

Tamaqua, Pennsylvania. From the *Manning Times* (Manning, S.C., Feb. 27, 1901, p. 1b; ||j5b6tof): "A dispatch from Tamaqua, Pa., says: Richard Miller of Hauto has a curiosity in the shape of an animal that is half cat and half rabbit. The front portion of the creature, with the exception of its red eyes, is that of a cat, while the rear half is that of a rabbit. One half of the animal's body is covered with the white hair of a cat, while the remainder of its body is covered with the reddish-brown fur of a rabbit, ending in a short, bushy tail. It moves about with half run and half hop and is very tame. It lives on vegetables and milk and has no use for meat. It is about one-half the size of a full-grown cat."

Titusville, Pennsylvania. From the *Greencastle Herald* (Greencastle, Ind., Apr. 16, 1914, p. 3c; ||yjtcwuwr):

> TITUSVILLE, PA.—Dr. and Mrs. J. M. Wald have on exhibition at their home an animal that is apparently a cross between a rabbit and a cat. It was found in the woods of the Mohawk reservation in Canada by Louis Deer … The animal is without a tail and is white and yellow in color. Its hindquarters are like those of a rabbit, its head like that of a cat and it purrs like a kitten when pleased. When it walks its movements are like those of a bear; when it runs it jumps like a rabbit, and when it plays about it assumes the appearance of a cat.

Cowansville, Quebec. From the *Post-Crescent* (Appleton, Wis., Jan. 29, 1907, p. 8f; ||3j24prxn): "COWANSVILLE, P. Q.—A wild rabbit, with a head like a cat and the rest of its body like an ordinary hare, has been brought in here by Hugh Ford and is an object of much curiosity. Ford got the hybrid while running rabbits with a hound. As it jumped along in the ordinary way he noted nothing peculiar about it until he picked it up. He then saw that he had a real curiosity. An examination of the animal's stomach showed that it had been eating flesh, the remains of a bird being found. Its teeth, like those of a cat, accounted for its peculiar appetite. As the hybrid was fat and strong, it appears to have been able to get all the food it needed. Probably birds, mistaking it for a rabbit, failed to get out of the way. A cat that can jump 20 feet isn't liable to go hungry, in any case."

France. Blumenbach (1813, p. 14) described an animal with cat foreparts and rabbit hind parts, the alleged offspring of a cat impregnated by a buck rabbit. A French diplomat gave it to Frederick William III, King of Prussia.

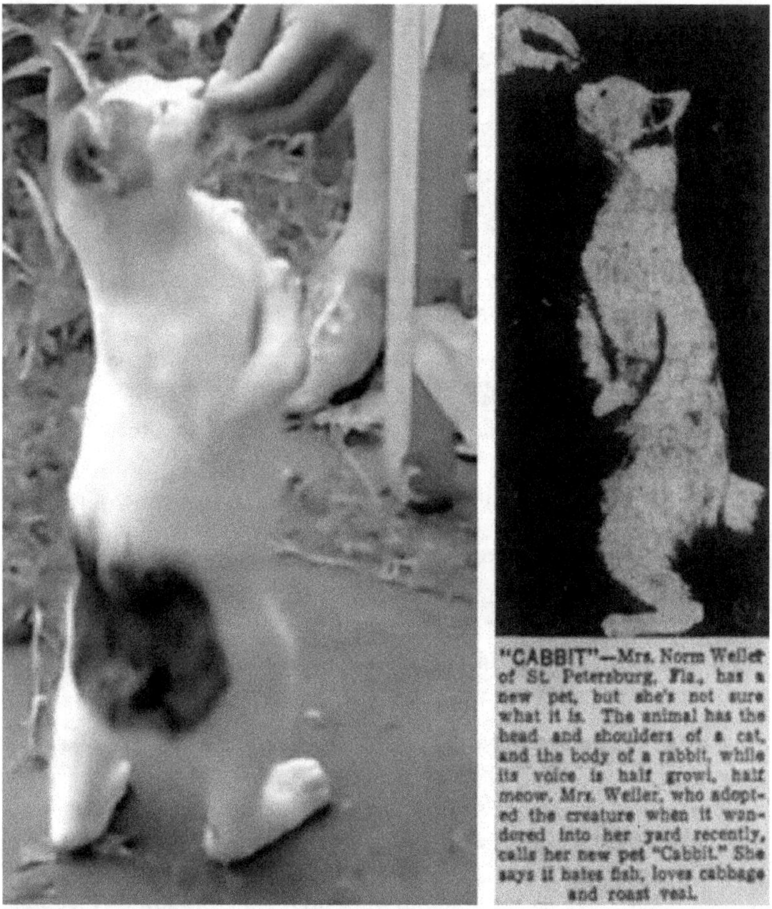

Two feleporids begging. In both cases, the posture is erect, and the hind feet are plantigrade. Ordinary cats do not have plantigrade hind feet. *Left:* screenshot of 2011 Tucuman animal (||rhn6vkq). *Right:* St. Petersburg animal (*Coloradoan,* Fort Collins, May 22, 1953, p. 9; ||vdg3kcn).

Lake Mills, Iowa. From the *Omaha Daily Bee* (Omaha, Neb., Jan. 1, 1901, p. 5d; ||zp8qx4s): "The authorities of the Iowa State College have had put to them a hard question by a farmer near Lake Mills, who has captured a strange animal which he wishes the college scientists to name. The animal looks like half cat and half rabbit. The head and shoulders exactly resemble a cat, but it has the long hind legs and short cotton tail of a jackrabbit and in traveling it takes the gait of a rabbit. The animal eats meat and drinks milk, and also loves vegetables and roots and corn. The farmer, William McLean, captured the animal as it was running wild."

Cambridge, Ohio. From the *News* (Mansfield, O., Sep. 29, 1892, p. 4e; ||ug5hyc5): "CAMBRIDGE. O., Aug. 26.—The town has a freak in the way of a half rabbit and half cat. The forepart of the animal is like an ordinary house cat, while the other half is shaped like a full-fledged white rabbit."

Milton, North Carolina. The *Milton Chronicle* (quoted in the *Southerner*, Tarboro, N.C., Nov. 19, 1875, p. 3; ||ta5h58c) said, "A CURIOUS CAT—Our neighbor Jones, the Postmaster, has quite a singular half-grown cat; it is half cat and half rabbit. The head part is cat, and the rear half is the shape of a rabbit"

Sayville, New York. From the *Salt Lake City Tribune* (May 5, 1904, p. 3b; ||y476p483): "Walter Kessings says it's a half cat, half rabbit. Mr. Kessing is a farmer and a man of repute. He went into the woods with a friend from the city with the intention of getting material for a rabbit potpie. There was a rustle in the underbrush, and an animal hopped out and darted for the road. There it halted, and, to the intense astonishment of Kessing and his friend, set up an able-bodied yowl, the like of which is heard only at an Egyptian concert on a city back fence. "Well, by gum!" exclaimed the farmer, "It lopes like a rabbit, it's got the cottontail of a rabbit, but by jiminety! it talks like a tomcat!" Dropping his gun, Kessing leaped forward and caught the strange creature in his arm. He took it home and has it on exhibition. It is a sure enough half cat half rabbit, with sharp claws in the forefeet and the stumpy toes of the hare."

* Feleporid reports often mention sharp foreclaws and dull hind ones.

Athens, Georgia. From the *Savannah Morning News* (Savannah, Ga., Jan. 30, 1901, p. 8a; ||u6ch4kcp): "HALF CAT, HALF RABBIT.—Mr. J[ohn]. B[Barber]. Wier [1852-1935], manager of the Athens Oil and Manufacturing Company, is the owner of an animal that is half rabbit and half cat. The head and forelegs of the freak are those of a cat, while the rest of the body is that of a rabbit. The fur, body, hind legs and short tail are exactly like a rabbit's. ... It is an excellent mouser, catching many mice about the house, but hops about like a rabbit."

Savannah, Georgia. From the *Morning News* (Savannah, Dec. 8, 1898, p. 3d; ||yrzcahsx): "John Ohsiek, who runs a grocery at Bay and West Broad streets, has a kitten that is something of a curiosity. It is small and black and at first glance would be taken to be just like any ordinary kitten. A closer examination will show, however, that it is minus a tail, nature having failed to provide it with a caudal appendage. A still closer examination will show that the hindquarters of the kitten are shaped like those of a rabbit. The kitten has no claws on its hind feet, and they are differently shaped from the front feet. The kitten … squats on its hindquarters like a rabbit and jumps like one."

Lambertville, New Jersey. From the *Warren Sheaf* (Warren, Minn., Mar. 11, 1897, p. 7b; ||hoafnaa):

> Nature cuts many queer capers, but no stranger example of her oddities can be seen in this section than a pet which is owned by a barroom keeper in Lambertville, N.J., says an exchange. The animal is half cat and half rabbit and has some of the habits and nature of both animals, while its anatomy consists of a peculiar commingling of each.
>
> The freak from its head to the middle of its body looks like an ordinary kitten, the head and front paws being perfectly formed. The hindquarters of the animal are those of a rabbit, with a typical "cotton tail." The front feet are armed with sharp, curling claws, while the hind feet are equipped with long, straight claws for use in running. The fur is white, gray, black and yellow, in spots, and in quality is in places soft, in others harsh.
>
> To see it run one would think it a rabbit, except for its short ears. When not in motion, it squats on its hind legs just as a rabbit does, and if a piece of cabbage or apple be given it, it will take it in its front paws and nibble it contentedly. This strange animal, if disturbed or maddened evinces at times all the nature of a cat, while again it will run away and hide behind any object of shelter. It is now about three months old.

Cold Spring Harbor, New York. From the *Sun* (New York City, Dec. 22, 1895, p. 30b; ||mr44zxya): "COLD SPRING HARBOR, L. I., Dec. 21.—Sylvester Howard, a resident of this place, is the owner of a Maltese cat which gave birth to five kittens a few days ago. One of the youngsters has a head like a cat and body, legs and tail like a rabbit. It has long hind legs, which it uses like a bunny. The soles of the feet are shaped exactly like those of a rabbit, and it has a short, white, cottony tail. The freak is now at the home of Eugene Howard, at West Neck."

Camden, South Carolina. From the *Omaha Daily Bee* (Mar. 15, 1884, p. 2; ||szoyxjs): "Mr. J. L. Brasington, of Camden, S.C., has a curiosity in the shape of a cat-rabbit or rabbit-cat, being one half cat and the other half rabbit. The head and shoulders are those of a cat, while the hindquarters are those of a rabbit, the legs and feet also being shaped precisely like those of a rabbit. The tail is also short and white like a rabbit's. In moving about, it jumps and squats like a rabbit and has often been chased from the street because it was taken for one. It cannot climb a tree because only the forepaws have claws on them. It will devour anything that a cat or rabbit will eat. The monstrosity is now about six months old and is very docile in its habits. An old cat is the mother of this curiosity. Only one other kitten was born at the time, and it is a full-blooded, perfectly formed cat."

Caldwell, Idaho. From the *Caldwell Tribune* (Nov. 17, 1900, p. 4b: ||zpdd6mc): "I. L. Harader, whose reputation for truth and veracity has never been blemished, reports an interesting study in nature at his home on the west side of town. A few days ago, an old cat drifted into his possession from the adjacent brushy lands that lie to the south of his orchard, bringing with her a family of four kittens. About this there is nothing strange, but when it is asserted that these kittens are half rabbit half cat, the fore-quarters being that of the kitten while the hind-quarters is that of a rabbit, that they have the bunny cotton tail of the rabbit, and jump instead of walk; our fusion friends will receive a great shock in the fear that the cat-rabbit trust will destroy the value of the jack-rabbit as a substitute for beefsteak in the event of a democratic administration of national affairs. But seriously, there is no joke about the strange family at Elder Harader's. The mongrels are worth seeing."

Cloverport, Kentucky. From *Breckenridge News* (Cloverport, Jun. 7, 1893, p. 4b; ||t5xvqgb): "There is a monstrosity at the stables of Col. Crit Davis that ought to be sent to the Columbian Exposition. We are quite well acquainted with the old-time mule, a cross between the horse and the donkey; we have seen a cross between the Guinea-fowl and chicken, but the cat-rabbit or rabbit-cat is a curiosity never seen or heard of in this locality, until Mr. W. Shy "raised" this one. In the language of our friend G., it is "'alf and 'alf." The fore part, head, forelegs and chest is that of a well-developed cat, while the other parts are that of a rabbit, even to the little white "molly cotton" tail. The hair of the cat part is like that of a cat and the rest like that of a rabbit, and what is stranger still, is that one part crawls while the other part hops, and by the conflicting movements locomotion is to a great extent frustrated. It eats both meat and bread as well as grass but catches no mice or rats. If anyone doubts this strange story, he can go out to the stables and there see for himself."

Bamberg, South Carolina. From the *Bamberg Herald* (Jun. 20, 1901, p. 3a; ||w6glnd7): "Mr. N. H. Williams[6] says he has a curiosity at this house in the shape of an animal which is half rabbit and half cat. The fore part is like a cat, and the hind part looks like a rabbit. It is about six weeks old, and he supposes it to be the result of tame rabbits and cats breeding together."

Baltimore, Maryland. The *Baltimore American* (quoted in the *Baltimore County Union,* Towsontown, Md., Jan. 25, 1902, p. 3f; ||5a9z9mcw) said, "Probably one of the most interesting as well as uncommon of animal freaks is a beast—half cat, half rabbit—that belongs to an old German shoemaker named Klinefeldt, who occupies a small frame shop and dwelling on the Reistertown Road at Garrison Forest Lane."

Lawrenceburg, Indiana. The *Plymouth Tribune* (Plymouth, Ind., Jan. 6, 1910, p. 3g; ||3xxbfr85) said, "Harry Fryen, George Fryen, Arthur Fryen and Karl Fryen, who have been spending a few days hunting near Lawrenceburg with friends, have returned to their home in Indianapolis with several valued trophies of their trip. They bagged 120 rabbits, 61 quail, a red fox, a small black eagle and a queer freak of nature, which is believed to be half cat and half rabbit. The freak was shot on the farm of Mrs. Mary A. Harrison, in North Hogan Creek, by Harry Fryen."

Tiffin, Seneca County, Ohio. The *Tiffin Advertiser* (quoted in the *Bucyrus Journal* (Bucyrus, O., Nov. 6, 1891, p. 3); ||ywbdxnz6) said, "Dr. Jeffrey, of this city, has a domestic pet which as a freak of nature is without rival. It is a live animal, half cat and half rabbit. It is white with dark markings. It has the tail, eyes and legs of a rabbit, but the claws, ears and head of a cat. It hops about and crouches like a rabbit but purrs and submits to caresses like a cat."

Owensville, Indiana. The *Plymouth Tribune* (Plymouth, Ind., Feb. 24, 1910, p. 3g; ||uytm9n4) said, "William Ned Harmon, living west of Owensville, has offered a reward of $5 for the capture of a strange animal that has been prowling through his premises. The animal is half cat and half rabbit. It first appeared at the home of James Land. The animal has a head resembling that of a common house cat. It has cat claws, but its legs and tail are like those of a rabbit. The strangest thing about this creature is that it runs like a rabbit and makes a noise like a cat."

Oakland, California. The *Dillon Tribune* (Dillon, Mt., Jun. 16, 1883, p. 1c; ||z9yewkf) said, "West Oakland, California, has produced a nondescript, which is apparently a combination of cat and rabbit. The forelegs are those of a rabbit; it has a rabbit's tail, conspicuous for its brevity and bushiness; it squats down like a rabbit, jumps like a rabbit, but mews like a cat. It is a black and white color, and very playful, and is inclined to scratch. The half cat and half rabbit is three months old. The mother of this ill-behaved little cross-grained animal is of the regular feline species."

South Carolina. Naturalist John Bachman (1850, p. 80) told how "An animal appeared in Beaufort, and two others at Columbia, S.C., which were asserted by whole neighbourhoods to be hybrids between the common cat and the gray rabbit (*Lepus Sylvaticus*). Some of these accounts found their way into the public prints of America, and we perceive were republished in London. The hybrids were represented as having the fur of the rabbit with its short tail and long hind legs, sitting on its tarsus, and leaping in the manner of the hare. They were moreover said to be so wild they seldom came to the house; after some correspondence and frequent disappointments, one of the animals was sent to us from Beaufort. The teeth soon told the story: it was a true cat, born, however, with a short tail, with long hind legs, and claws that could scarcely be called retractile. It may prove a second edition of the Isle of Man cat. If, however, these strange animals should be excluded from an intercourse with other cats we are inclined to believe that the variety might be perpetuated, and it would be regarded, as it is now considered by many, a feline-leporine hybrid."

Palm Beach, Florida. From the *Eagle* (Reading, Pa., Aug. 17, 1947, p. 32; ||nlycp6m):

> WEST PALM BEACH, FLA., Aug. 16 (AP)—Stewart Morgan has a cat—at least it's a cat when viewed from the front, but take a look from the rear, and it's a rabbit.
>
> Morgan says the animal, promptly named "Cabbit," is jet black, has the head, shoulders and forelegs of a cat, but just below the shoulder joint the spinal column does a squads left and—presto—the cat becomes a rabbit. The hind legs are like those of a rabbit and operate kangaroo-fashion.
>
> Even the fur, said Morgan, is only cat fur about halfway, and the rest of the body, covered with fur like that of a rabbit.
>
> The tail is nothing but a boneless tuft of black fur.
>
> Neighbors suggested that the creature was nothing but a Manx cat, but encyclopedias, while acknowledging the similarity, don't say a word about the hindquarters hopping around like a rabbit.
>
> The "cabbit" has a normal feline appetite. He eats meat and drinks milk but ignores such rabbit delicacies as lettuce and carrots.
>
> In fact, he might pass off as a cat, if he'd stop the kangaroo-jumping with his back feet.

Indianapolis, Indiana. Robert L. Hoover, a photographer for the *Indianapolis News* (Jul. 15, 1947, p. 30; ||yevfuw4b) snapped a photo of two typical feleporids (cat foreparts, rabbit hind parts). They were from a litter of five such, produced by a mother cat. He noticed them when he stooped to pet one and it leapt away like a kangaroo.

Elkton, Maryland. The following notices, all about animals from the same litter, are all from Elkton's *Cecil Whig*.

> [Aug. 25, 1883, p. 3b; ||v94wyx9]:
> An Elkton man has in his possession a curiosity in the way of a freak of nature, in shape of a young one, a cross between a rabbit and a cat. He owns a buck rabbit and a tabby cat, and in pussy's last family there were specimens bearing resemblance to both a cat and a rabbit. these kitten rabbits should be sent to the Smithsonian Institute.

[Jan. 12, 1884, p. 3a; ||s6bv5x5]:
The cross between a rabbit and a cat owned by Mr. George D. Bennett,[7] of this town, mentioned in the WHIG some time ago, is about to be added to the Zoo collection at Druid Hill Park, Baltimore.

[Mar. 22, 1884, p. 3a; ||ql6dvlw]:
The celebrated cat, half cat and half rabbit, owned by Mr. George Bennett, of this town, and which had attracted considerable attention, died suddenly in a fit last Tuesday. The cat astrophe is the greater from the fact that he was about to send it to the zoological gardens.

[May 3, 1884, p. 3b; ||s9f5swy]:
The half-cat half-rabbit that escaped from Mr. George Bennett several months ago, now makes its home in some hole or crevice about Partridge's warehouse, near the railroad station, and is frequently seen hopping around. It is very timid, however, and will not allow a person to approach it.

West Corydon, Indiana. From the *Indianapolis News* (Jul. 16, 1895, p. 2; ||veh6t8d): "CORYDON, IND., July 16.—Dr. M. C. Popham, of this place, owns a curious malformation in the shape of an animal, which is half cat and half rabbit. Its dam was a cat and its sire a white rabbit. The fore parts are those of a cat, but it has hind legs and tail like a rabbit, and it hops about instead of walking."

Fresno, California. From the *Fresno Morning Republican* (Jan. 15, 1929, p. 7; ||qsej8f8):

It purrs like a cat; it hops around like a jackrabbit; from the front it looks like a well-behaving kitten; the back view makes one change his mind and take it for a pet bunny.

What is it? Nobody's sure, but the only safe answer seems to be: Half cat and half rabbit.

Whatever it is, it's the playmate of Alice Mink at the H. H. Mink home at 245 North U street, and since Alice likes both cats and rabbits, she likes this animal friend just twice as hard.

Bob, that's his name, has the quiet, dreamy eyes and the cute little head of an ordinary kitten. His soft, silky fur is just like a kitty's, too, but there are patches of tannish coloring distinctly rabbitish, if you get the point.

> What makes him look like a rabbit more than anything else, though, are his long hind legs, which fold under him, and his very short, bushy tail. And when you see him go hopping jumpety-jump, for all the world like a jackrabbit, then no doubt is left that he's something more than a pussy.
>
> Alice says she doesn't know how old her pet is, as it was given her around New Year's as a present from the Verne Goodrich ranch at Hays and Shields avenues. It was learned that the kitten's mother had a litter of five, and that Bob was the only freak among them. There was another that resembled him but was normal and qualified as a feline without the shadow of a doubt.
>
> Yes, there are many rabbits around the Goodrich ranch and—well, one can't help being suspicious.

Northampton, England. Cleric John Morton (1712, p. 445) gave an account of an alleged cat-rabbit hybrid, in which he said, "That Two Brute Animals of different Species do sometimes join in Copulation and that a Mongrel or spurious Beast partaking of the Nature of both the Parents, is begot betwixt them, everybody must own, who knows anything of the Generation of the Mule; but then it is to be observed, that this happens only between Animals of Kinds near of Kin to each other, as, e. gr., the Horse and the Ass, between whom the Mule is engendered. For this Reason, I suspected the Story rife in almost every bodies Mouth about Two or Three Years ago, that at the Cross-Keys Inn in Northampton, they had a Creature in the Fore-part of it a Cat, in the Hinder-part a Rabbet, that came of a She Cat, which had coupled with a Buck-Rabbet at a neighbouring House. The Rabbet and Cat do indeed agree in this, that they both are multifidous [i.e., an obsolete word that meant having feet with multiple toes]; but in other Respects, especially in the Shape of the Head, in the Fashion and Number of their Teeth, in their Manner of Living and Temper, do so much differ that 'tis scarce to be imagined there should ever be any such Intermixture. But however, that I might satisfy myself more fully in the Matter, I went to view this so much talk'd-of Monster, which to me after all appear'd to be only a common Cat with a bobb'd Tail, and somewhat more bushy than ordinary, and with blunter Claws! Things that may easily be accounted for. And it had the same way of squatting down upon its Tail, that Rabbets have, which no doubt it was taught. The

Vulgar may still, if they please, believe that Relation; but I would not have it impose upon Persons of better understanding."

But could a "person of better understanding" explain how an ordinary cat would come to have multiple traits of a rabbit? And why is it that accounts given by witnesses, widely separated in time and space, agree in assigning the same rabbit traits to different animals? It seems that random mutation, given that it is in fact random, would not consistently produce the same traits.

Greenbush, Wisconsin. From the *Superior Times* (Superior, Wis., Nov. 30, 1907, p. 2a; ||3yxcvs7x): "PLYMOUTH.—Two hybrids, half cat and half rabbit, are on exhibition in a local grocery store here. The animals were brought in by H. R. Roehring, of Greenbush. One of the kittens has a head and forefeet of a cat, and haunches of a rabbit, the front part being white and the back part gray. The other freak is coal black. When the animals sit up they move about like a rabbit, but they drink milk and eat like a cat."

Budapest, Austria-Hungary. The *Pester Lloyd* (Budapest, Dec. 29, 1894, p. 6a; ||t6cbs3t) said, in translation, "A rare animal is now in the zoo's holdings. It is a freak produced from a cross between a female housecat and a buck hare, half cat, half hare, which was given to Archduchess Klotilde by a woman living in Ofne. The archduchess in turn gave the animal to the zoo."

Kempten, Germany. The *Österreichische Land-Zeitung* (Krems an der Donau, Austria, Mar. 24, 1906, p. 27a; ||t24eg7e) said, in translation, "HALF HARE, HALF CAT—Near Kempton, a farmer lost his cat. Later, however, she returned to the house and brought with her two young ones that were half hare and half cat. In their foreparts they were like cats, but in their hind parts they were just like a hare. This singular beast should be sold to some museum."

Joelton, Tennessee. The *Nashville Banner* (Sep. 2, 1934, p. 9; ||ue7yptc) said, "On the State Quail Hatchery premises, near Joelton … there is a creature which is half cat and half Belgian hare. … It has a short tail, hind legs like a rabbit and a rabbit hop … also it has a rabbit nose."

Genoa, Crete. From the *Herald Democrat* (Leadville, Co., Aug. 22, 1911, p. 3c; ||vuqkpyh):

> LONDON, Aug. 21. A curious animal is reported from Genoa, Crete, by a correspondent, who says that a person of his acquaintance some months ago lost his cat, a female. While walking in the country the other day he came across a hare accompanied by his cat, who was carrying in her mouth a charming little kitten-leveret, which she laid at his feet and which he took home with him.
>
> The animal has the head and paws of a cat, but all the rest of the body is hare, not only in form but in quality and color of the fur. The mixture of the father and mother in its walks and leaps is most remarkable. The animal also possesses a character of its own. In its capacity of hare, it prefers the garden to indoor life, but its cat nature brings it regularly to the dining room at mealtime, when it expresses by a soft purring its pleasure at being fed and caressed.

The Ogden canileporids.

CANILEPORIDS (RABBIT- AND HARE-DOGS)

Reports about canileporids are far rarer than those about FELEPORIDS.

Ogden, Utah. Regarding the picture above, the *Spokesman-Review* (Spokane, Wa., Feb. 15, 1935, p. 7; ||2p9bkfdj) said, "Half dog and half rabbit seem these 3-month-old-puppies, Toots on the left and Trixie on the right, the pets of Mrs. Thomas Dillard, 610 S. Grant. Toots has three normal legs and a stump, while Trixie has but two normal legs, on which she hops, after the manner of a kangaroo. … In addition, each has normal dog's hair about the head, neck and forward part of the body, but the hair over the back resembles the fur of a rabbit [as in the case of most FELEPORIDS]. The rear feet of Trixie are shaped more like those of a rabbit [again, like FELEPORIDS] than a dog. Mrs. Dillard brought the pets here from Ogden, Utah, when she came here to marry this month. … Mrs. Dillard said the mother had a habit of lying on the roof of a rabbit pen, and she thinks that is responsible for the strange freaks of nature."

* Bitches that frequent rabbit pens are certainly more likely to be mounted by rabbits.

The Lyon canileporid

Lyon, France. Aldrovandi (1642, p. 528) pictured this animal, which he described as a harelike dog ("canis leporini"), born with two legs. He said he received it from Lyon, but didn't say when (so, prior to his death in 1605).

Beauvais, France. In an alleged case of XENOGENESIS, *Le Petit Journal* (Paris, May 20, 1925, p. 2d; ||2bspm3yn) reported that a dog belonging to a M. Allain at Beauvais whelped three puppies and three rabbits, two of which died. But the third was supposedly fully viable.

Montreuil, France. The *Journal des Sçavans* (Anonymous 1684, p. 381) said, in translation, "A gentleman in Montreuil, who was out hunting, had one of his dogs bring him an eyeless leveret, about 15 days old, that had in place of a mouth only a small hole. It had four ears, two in the usual place, behind which there emerged two little legs, and, opposite them, two other larger ears resembling those of a hunting dog. Behind these ears were two other, reversed forelegs pointing upward. From there back, the leveret was separated into two bodies arranged side-by-side, each composed of the usual number of parts."

Burrowa, NSW. From the *Burrowa News* (Dec. 16, 1932, p. 5; ||hn2q5th):

> An affection between a half-bred fox terrier and a thoroughbred female Angora rabbit led to the birth of a peculiar little creature, half dog and half rabbit, according to the owner of the animals, Mr. Victor Morehead, well-known businessman of Minnamurra Avenue, Earlwood.
>
> The freak died, however, a few hours afterwards and Mr. Morehead has preserved it.
>
> More like a tiny puppy than a rabbit, it is black and white and has rabbit paws, legs and ears, but a dog-like head, tail and body.
>
> Mr. Morehead said it was impossible for anyone to have played a joke on him, and there had been no other rabbits near the mother rabbit for more than 12 months.

Prussia. In a treatise on lagomorphs, Christian Franz Paullini (1691, p. 89, §41) said, in translation, "a certain Prussian surgeon kept hares, and these impregnated one of his dogs. But due to her small birth canal, she could not give birth. So he delivered via cesarean section, two offspring like puppies and one like a leveret."

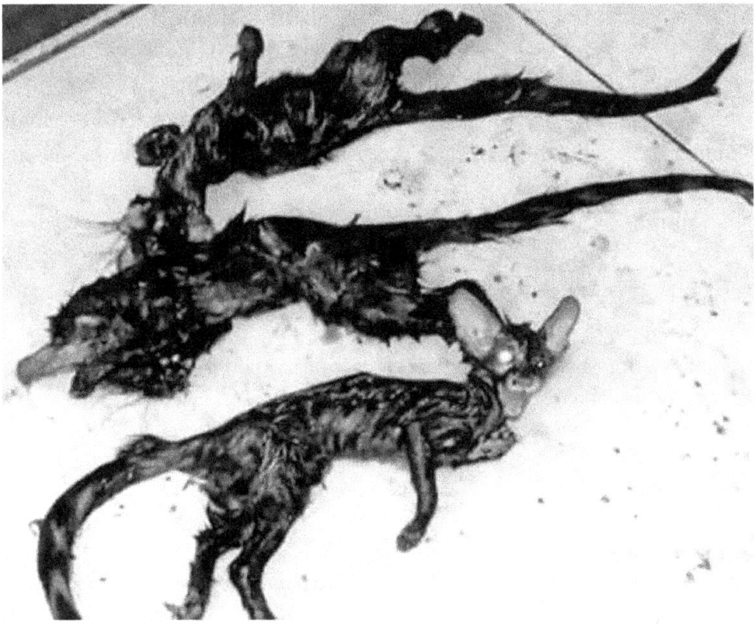

Three of the Lethabong felimurids.

A CANIMURID (RAT-DOG)

Saint-Quentin, France. Le Guetteur (Saint-Quentin, Mar. 7, 1875, p. 2b; ||44mrwm56) said, in translation, "A freak, dead almost as soon as it was born, came into the world at Saint-Quentin. A dog whelped a little monster with the head of a rat, an eye in the middle of its forehead, a closed nose, whiskered lips, and two teeth and ears of a dog."

FELIMURIDS (RAT- AND MOUSE-CATS)

Lethabong, South Africa. Internet reports say rat-cat hybrids were found in a house in Lethabong in 2017. According to the reports, there were six, four that were killed—as is often the case with strange hybrids—and two that were taken by the SPCA for testing.

Forbes, New South Wales. The *National Advocate* (Bathurst, NSW, Apr. 26, 1933, p. 3; ||2mkk84ec) reported a half cat half rat found at Forbes.

London. Philosopher John Locke (1819, p. 479) wrote, "I once saw a creature that was the issue of a cat and a rat, and had the plain marks of both about it; wherein nature appeared to have followed the pattern of neither sort alone, but to have jumbled them both together." Jacques-François Dicquemare (1778, p. 213), mentioned what was likely the same case: "The female cat covered, in London, by a big rat, bore young resembling both cat and rat, that were raised in the king's menagerie" (translation mine). And English physician James Drake (1727, vol. 1, p. 186) claimed, "There have been abundance of Observations of Mixtures of Animals of more remote kinds such as between … a Cat and a Rat, of which, however improbable it may seem, there is an honourable Family that can produce abundance of Witnesses."

* Cats have been known to adopt rats. Such rats would likely become sexually imprinted on cats and so, as adults, prefer feline mates. Cases, too, have been reported of cats giving birth to rats, but such accounts may, at least in some cases, be explained by assuming the cat in question adopted a rat.

* Cats exist having extra, rat- or mouse-like ears (||2xreethp, ||xbxua8jx).

A Myomustelid (Mustelid-Gopher)

Wood Lake, Minnesota. The *Wood Lake News* (quoted in the *Redwood Gazette*, Redwood Falls, Minn., May 12, 1914, p. 4c; ||y3tp9zkd) said, "Alfred Murray, who has been working the Henry [i.e., Heinrich] Braasch [1871-1917] farm near here, brought into town a queer specimen of wild animal the other evening and there has been much discussion as to what to name it. It has the stripes and spots of a common field gopher, but these can be seen only in the sun or other bright light. Its fur is shiny black. The bushy tail and the head resemble that of a common gopher, but the body has more the resemblance of the weasel or mink. Mr. Braasch has the animal at his farm now and will keep it, hoping the freak will develop and that he will be able to guess its pedigree when it is full grown. Mr. Murray caught the animal in a trap, but it is not injured."

A Lagolotrid (Rabbit-Raccoon)

Churchtown, Pennsylvania. The *Daily Evening Express* (Lancaster, Pa., Aug. 19, 1869, p. 2f; ||hkjb5z3a) said, "A Strange Animal.—A correspondent writes us that a strange animal was caught last week by J. A. Styer, on his farm near Churchtown. The head is shaped like the head of a rabbit, while the body and feet resemble those of a raccoon. The monstrosity will be on exhibition at the festival at New Holland, on Saturday the 21st inst."

The Cottonwood lagobovid
(reconstruction)

LAGOBOVIDS (RABBIT- AND HARE-COWS)

Cottonwood, Idaho. From the *Camas Prairie Chronicle* (Cottonwood, Feb. 26, 1904, p. 1e; ||vpyhvfu): "A large crowd of people viewed a freak of nature here in the shape of a one-legged calf with a head like a rabbit … The calf, which was born in a pasture near here, had one perfect hind leg and two short front legs, each about 5 or 6 inches long. One had a small foot resembling a rabbit foot and the other had neither foot nor hoof. The calf's nose was round and the lips full, making its head resemble that of a rabbit much more than that of a calf. It looked quite strong in spite of its odd condition and with proper care might live to occupy a place in a freak museum."

Colmey, Meurthe-et-Moselle, France. L'Indépendant (Metz, Feb. 25, 1867, p. 2d; ||bdcmpd5f) said, in translation, "COLMEY—A freak as rare as it is curious was recently born at Colmey. A cow there bore a calf with, except for the ears, a head shaped like a rabbit's. … This animal, not being able to suckle, due to its split upper lip, would have certainly perished, had not its owner had it killed a few hours after its birth."

Bühl, Haut-Rhine, France. Le Courrier de Metz (Mar. 9, 1907, p. 3; ||5376t64h) said an animal had been calved at Bühl with the nose ("mufle") and tail of a hare, and two tongues, as well.

The Dalles, Oregon. The Hood River Glacier (Jan. 23, 1903, p. 5b; ||mr2edee5) reported "a calf with a rabbit's head" that soon died.

Troy, Missouri. From the *Troy Herald* (Jul. 30, 1873, p. 5; ||uo8l4ad): "A WHAT IS IT.—A cow belonging to Capt. Colbert gave birth last Thursday night to a monstrosity. It was a red calf fully matured, but of a most singular malformation. The head is large, face prominent and rounding from forehead to tip of nose, presenting in general view the appearance of that of a rabbit, as does in fact the whole body. Its length from forehead to tail is fifteen inches. The spine at the shoulders is considerably elevated. The caudal spine is closely bent back upon the lumbar and measures only half an inch in length. The ears are set far back, and the head rests closely upon the shoulders. The legs are small, scarcely an inch in diameter, the fore ones being very short. The hoofs are not cleft. It was evidently born alive but was dead when found."

Bedford, Indiana. From the *Washington Herald* (Oct. 26, 1914, p. 11; ||4ksv44ku): "BEDFORD, IND., Oct. 25.—William Wilson, a farmer, is the possessor of a freak calf. The animal, apparently healthy, is ... restricted to an erratic jumping motion. It was born without a tail, the nearest approach to such an appendage being a hairy growth, similar in form and size to that displayed by a rabbit."

Washington Court House, Ohio. From the *Fayette County Record* (Washington Court House, quoted in the *News-Herald,* Hillsboro, O., Aug. 18, 1901, p. 1b; ||y6m7rf66): "Mr. Otis Boncutter, living out the Columbus Pike, has a cow that on August 3 gave birth to a freak calf, which he says is perfectly developed in every way with the exception that it has no tail, the end of the backbone stopping near the middle of the back, where the hips are joined on, without any part of a tail bone or stub. It has curved hind legs, is a female, is very pert and hearty, and is of a very wild nature. It leaves its mother after nursing and hides in the weeds. It has the actions of a rabbit in way of travel, can walk very little and gets around by hopping."

The Nesslau lagobovid.

Servance, Haute-Saône, France. La Franche-Comté à Paris (Paris, Apr. 30, 1911, p. 3c; ||3bc7enzx) said a woman at Servance had a calf with the paws of a rabbit ("un veau qui a les griffes d'un lapin").

Nesslau, St. Gallen, Switzerland. A report in the Wickiana collection (Ms F 32; ||mt7d3ppe) says a cow calved a lagobovid with the whiskered face of a hare at Nesslau on March 10, 1584.

Nescopeck, Pennsylvania. The *Columbian* (Bloomsburg, Apr. 4, 1901, p. 4; ||3yw82eyu) said, "Frank Myers, of Nescopeck, is the possessor of a freak in the shape of an animal, and since the creature came into the world, about three weeks ago, his barn has been visited by hundreds who desire to look at the monstrosity. It is a calf, the offspring of a full blood Alderney cow. The one and only leg is where the hind legs should be. It has a short stub tail, similar to that of a rabbit, and when standing resembles a kangaroo. Notwithstanding its many and unusual deformities, it is full of life, and gets about easily, if not as quickly as though it possessed four legs."

Manchester, Vermont. The *Weekly Messenger* (St. Albans, Vt., May 2, 1907, p. 2; ||y6kdefnu) said, "J. N. Thompson, of Manchester, has a freak calf. Its two forelegs are curved in toward the body and the animal rests on its hind legs the same as a rabbit does. It is also minus a tail."

East Fallowfield, Pennsylvania. The *Altoona Tribune* (Altoona, Pa., May 8, 1879, p. 1e; ||bdhfsuh7) said, "A few days ago, a cow belonging to Mr. Job Keech, of East Followfield [*sic*] township gave birth to two calves, one of which died shortly afterward. The other lives and has one of the most remarkable tails of any calf known. It is shaped exactly like that of a rabbit, is the same color, and size proportionately."

East Shrewsbury, Vermont. The *Vermont Tribune* (Ludlow, Vt., Apr. 30, 1897, p. 1e; ||43p45x9d) said George Ducharme Butterfly (1847-1915), of East Shrewsbury, "has a calf … which was three days old about a week ago, and has no eyes or tail, and the hind parts are shaped like those of a rabbit."

La Quintinie, Charente, France. La Charente (Angoulême, Apr. 8, 1938, p. 3g; ||uw2sfrue) said a tailless cyclops calved at La Quintinie had the hind legs of a hare, that it was a month old and in good health.

Montbronn, Moselle, France. The *Excelsior* (Paris, May 24, 1925, p. 2e; ||yvp2dd2t) said a fetus, taken at Montbronn from a slaughtered cow, weighed 180 pounds. In translation, it said, "One side of its enormous head bore a curious likeness to the head of a hare. One of the eyes was in the middle of the forehead."

Windsor, New South Wales. The *Times* (Gosford, NSW, Oct. 5, 1922, p. 13d; ||y8x8cuu9) said, "An extraordinary freak was seen in Windsor last week from the property of Mr. S. Marker, of Magrath's Hill. It was a calf that one of his cows had dropped. The little animal had the body, hair and feet of a bovine, but the head was that of a hare and the teeth and upper lip those of a foal, while it had a rabbit's tail. It had no nostrils, and the two eyes were set close together in the centre of the forehead, protruding in a sort of a bull's eye setting. One of the hind legs was turned the wrong way. The thing was alive when Mr. Marker found it, but it did not live long."

Summerau, Austria. The *Linzer Volksblatt* (Linz, Austria, Nov. 23, 1911, p. 5c; ||vm5lqdo) said, in translation, "At Pfarre Rainbach bei Freistadt, a farm in Summerau, a cow recently gave birth. And, therewith, a remarkable double being entered the world: The front half of this curiosity is that of a calf, the rear, that of a hare. If this beast steps forward with its forefoot, the rear half hops after. Also, the tail is exactly like a hare's. This monster is already now 14 days old, so it's completely viable."

A BOVOCASTRID (BEAVER-COW)

Tillamook, Oregon. The *Tillamook Headlight* (Jun. 8, 1923, p. 1f; ||2s3rdfac) said an animal calved there "resembles a beaver more than anything else, and just below the center of the forehead was one eye … The teeth … were chiseled like those of a beaver, and the tail was shaped like the same animal." It had no nostrils.

BOVOSCIURIDS (SQUIRREL-COWS)

Indianapolis, Indiana. The *Sentinel* (Feb. 1, 1885, p. 8; ||5n6ju8ds) said the womb of a cow slaughtered in Indianapolis contained a hairy mass with "a perfect squirrel's head and neck, with eyes, mouth and ears in a very perfect state."

Middletown, Pennsylvania. The *Wilmington Commercial* (quoted in the *Times,* Reading, Pa., Jun. 1, 1869, p. 1g; ||2utpwr2j) said, "A cow belonging to Eli Yarnell [b. ca. 1820, Pennsylvania], of Middletown [Pa.], recently gave birth to a calf with a head resembling that of a seal or ground-hog with a body like that of an ordinary calf, and with a tail of a ground-hog. Its legs were but six inches long, and the feet were like those of a calf. This curiosity has been stuffed for preservation. "

No locus in quo. Photos exist of another groundhog-like animal calved by a cow at an unknown locale ca. 2009 (||2fkc9z3m).

A LEPIDEQUID (HARE-HORSE)

Thrace. In his account of the passage of the Persian host over the Dardanelles in 480 B.C., Herodotus (7.571) reported a case of xenogenesis wrote, "When the whole army had crossed, and the troops were now upon their march, a strange prodigy appeared to them, whereof the king made no account, though its meaning was not difficult to conjecture. Now the prodigy was this: a mare gave birth to a hare."

A GLIRIDEQUID (DORMOUSE-HORSE)

Kozliczky, Poland. Rzaczynski (1721, p. 358) said that in 1718 a mare at Kozliczky near Malbork, foaled a colt with a small ears and a tail like a dormouse's.

SUILAGIDS (RABBIT- AND HARE-PIGS)

Westphalia, Germany. Paullini (1691, p. 114) said a sow in Westphalia farrowed leverets with pig snouts in the summer of 1690.

Five suilagids with rabbit-like forepaws, reportedly farrowed in Le Mars, Iowa *Des Moines Register* (Aug. 29, 1937, p. 72; ||yf37wcz7).

Indian Castle, New York. The *Buffalo News* (Buffalo, N.Y., Dec. 2, 1910, p. 28b; ||34d7adc8) said, "Three freak pigs saw the light of day in one litter a short time ago at the Gen. Lansing farm in Indian Castle. One of the pigs had only three legs. Another one had legs shaped exactly like those of a rabbit, and the third one had only two legs, both of them front ones. The first two pigs have died, but the one with the front wheelers is still alive, after two months."

Dunn, North Carolina. From the *News and Observer* (Raleigh, N.C., Jun. 5, 1927, p. 31; ||yc6pyyuh): "DUNN, Jun. 4.—L. C. Blackman, who lives near Dunn, has recently discovered a new species of the animal kingdom. At least he was an animal which nobody has been able to name. It appears to be a cross between a rabbit and a hog. The eyes are like those of a rabbit, the ears like those of a hog, the front feet are like those of a rabbit and the back feet like those of a hog, and the animal carries a tail like that of a hog. The hair on the front half looks like that of a rabbit and that on the rear half looks like that on a hog. The mouth is shaped like that of a rabbit, but the teeth are of the usual hog type. The animal represents a perfect combination of a hog and a rabbit."

Ulm, Germany. Ulm physician Peter Rommel (1684) reported that after a sow in Ulm farrowed hare-mouthed offspring, they were unable to nurse, and their owner cast them into the Danube.

Lüchtringen, Germany. Meinhart of Lüchtringen recorded piglets with hare's ears born in 1212 A.D. (ibid.).

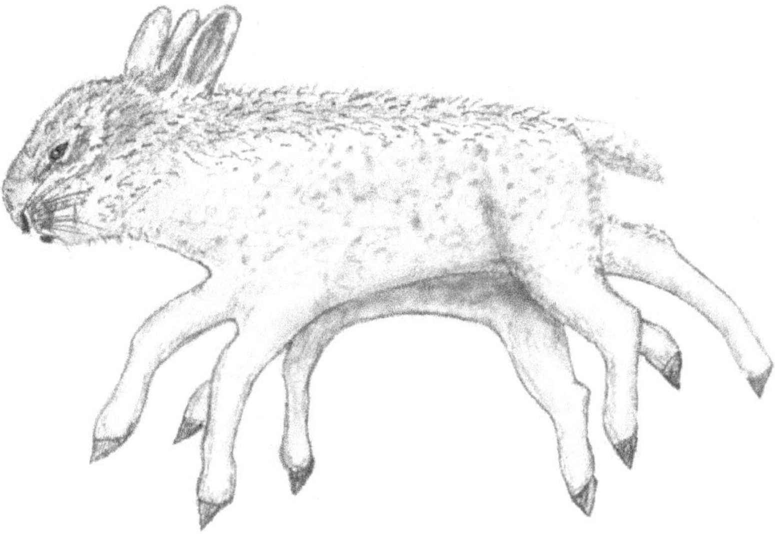

The Lappion lagovid (reconstruction).

LAGOVIDS (HARE-SHEEP)

Lappion, France. Chanvalon (1727) said, in translation, "In January, at Lappion, two leagues from Notre Dame de Liesse, a ewe recently yeaned a stillborn monster. It had the bodies of two lambs—one male, the other female—but one head, a hare's. The nose, whiskers, teeth and eyes were unmistakable. The ears, too, were a hare's, but there were three, two in their right places, a third between. The two bodies, with their four forelegs, were joined at the neck. The bodies were otherwise separate, each with its own belly, tail, and two hind legs, all of the proper size and proportion. The neck and shoulders, as well as all the back to the tail, were covered with hare fur. But wool clothed the eight legs, two tails and lower parts."

Upwey, Dorset, England. The *Sherborne Mercury* (Sherborne, Feb. 27, 1855, p. 3e) said a ewe at Upwey "gave its owner a nondescript half like a lamb and half like a hare; and it is not decided whether the creature (which is still alive …) is a hybrid or merely a freak of nature."

Jakobsberg, Germany. Paullini (1686, obs. 46, p. 209; 1691, p. 63) listed a lamb with hare ears, yeaned in 1174 A.D. at Monte Iacobi (now Jakobsberg) in the Diocese of Corvey.

La Longeville, Doubs, France. L'Éclair Comtois (Besançon, Jul. 8, 1906, p. 2d; ||d9b4u2mn) said a ewe, belonging to a farmer in La Longeville, yeaned a viable lamb with ears like a hare's.

Sidmouth, Devon. The *Western Times* (Exeter, Apr. 22, 1843, p. 3) said, "There is now in the possession of Mr. Snell, of Sand Farm, in the parish of Sidbury, a lamb, having five distinct legs, one of them being attached to the outer part of the animal's shoulder, and differs from the others only in so much that its extremity somewhat resembles the foot of a hare." This is an example of XENOMELIA (p. 1215).

A RODOSUID (RAT-PIG)

Bismarck, North Dakota. The *Bismarck Tribune* (Apr. 16, 1924, p. 8e; ||yepym4p8) said, "A freak pig, born on the state prison farm, has been taken to J. D. Allen, taxidermist, at Mandan, for mounting. The freak is described as a pig with a rat's nose and head, rat's front feet and rat's claws, especially … on the front feet."

A LEPICAPRID (HARE-GOAT)

Taufers im Münstertal, Italy. The *Tiroler Bauern-Zeitung* (Apr. 4, 1912, p. 4c; ||5466h6wb) said a goat at Taufers im Münstertal yeaned an animal that had no hind legs below the knee and that had hanging from its hide a paw entirely like ("volkommen ähnlich") a hare's. This is an example of XENOMELIA (p. 1215).

AN OVOMYID (SHEEP-GOPHER)

Hanford, California. The *Hanford Sentinel* (Jan. 25, 1923, p. 5; ||y4nhcv9l) said, "An unusual freak, a gopher-headed lamb, was on exhibition at the Hanford Veterinary Hospital yesterday, having been brought from the ranch of Mrs. Frank Lavinge in the morning by Dr. Anderson … The animal was born yesterday morning and was alive when found … It was born with no lower jaw, and in place of a stomach has a well-developed craw beneath its neck, similar to that of a chicken. It also has no tongue, and its teeth, just breaking through, are those of a gopher, while its nose is also gopherlike. Further the esophagus is missing. Otherwise it is normal and well developed."

* *Craw* is an obsolete word for *crop*. Pocket gophers have a storage pouch, linked to the oral cavity, that might well be interpreted as a crop.

A horned hare (Hoefnagel 1575-1580: plate 47).

HORNED LAGOMORPHS

Old reports refer to horned rabbits and hares, for example, Clauder (1688b, fig. 73, p. 368), Gassendi (1655, p. 56), Ray (1738, p. 24), and Schreber (1792, pp. 875, 928). Although such creatures may seem to be artiodactyl-lagomorph composites, another explanation is possible: Lagomorphs infected with Shope papilloma virus (SPV) often grow hornlike carcinomas (Shope and Hurst 1933).

* More about horned lagomorphs: ||wexzkoa, ||skqgcxg, ||svrm433, ||wwo2zhp, ||uwagqr9, ||y7m6hl3l, ||vtmrjyl, | |wgom7rm.

INCERTAE SEDIS

Saint-Quentin, France. The *Messager de Gand* (Ghent, Feb. 12, 1837, p. 3c; ||r3redk5p) said a rabbit at Saint-Quentin kindled a fetus with a frontal proboscis in 1837.

London. Lowne (1872, p. 45, no. 173), in an account of specimens in the collection of the Royal College of Surgeons, described a hare-kindled fetus with a frontal proboscis.

Garden City, Kansas. From the *Kansas City Daily Gazette* (Mar. 9, 1888, p. 3; ||2aue4sbj): "A Garden City (Kan.) man is the proud possessor of a jet-black rabbit which has, instead of an ordinary nose, a protuberance shaped like an elephant's trunk, which it uses sometimes to convey food into its mouth."

LAGOGLIRIDS (RABBIT-DORMICE)

Jean Baptiste van Helmont
1580-1644

Brussels. The Belgian chemist Jean Baptiste van Helmont (1652, p. 240b) said that among seven hybrids produced by crossing a doe rabbit with a male dormouse only their dormouse-like tails betrayed the identity of the sire.

* Rodents and rabbits are, respectively, assigned to two separate orders, Rodentia and Lagomorpha.

CAVILAGIDS (RABBIT-GUINEA PIGS)

James Conway Walter
1831-1913

England. The Country-Side (Aug. 11, 1906, p. 189a; ||yt9m6jyp), an English periodical, published a letter from naturalist James Conway Walter, attesting hybridization between between a rabbit and a guinea pig: "THE QUESTION OF HYBRIDS.—Sir,—In your issue of June 30th (page 117, first column), Mr. J. Bowey, of Gateshead-on-Tyne, gave a case of a cross between rabbit and guinea pig, and doubts were implied of the fact. I am able to testify to such a cross having occurred within my own knowledge. Some three years ago [ergo, ca. 1903], Mrs Stodthart of the White Hart Inn, Horncastle, Lincolnshire, had a rabbit and guinea pig which lived together, and the result was the birth of a young one, in the form, generally, of a very large guinea pig, but with the longer ears and fur of a rabbit. I have myself seen this specimen [in a separate letter published in the Aug. 25, 1906 issue of The Country-Side, James Conway Walter says that other members of the Lincolnshire Naturalists' Union also saw the animal] and, having had some experience in hybrids, I thoroughly believe it to be a genuine case.

* Walter was vice-president of the Lincolnshire Naturalists' Union, and the author of numerous books, among them *Letters from the Highlands* (1859), *Forays among Salmon and Deer* (1861), *Literæ Laureatæ* (1890) and *Records of Woodhall Spa and Neighbourhood* (1899).

England. Apparently, the occurrence of this cross has been known for a long time. In his book *Anthropologia Nova* (1727, vol. 1, p. 186), the English physician James Drake, a Fellow of the Royal Society, stated that "besides the vast variety of Mongrel Dogs, begotten betwixt Sires of different species, there have been abundance of Observations of Mixtures of Animals of more remote kinds such as between … a Buck-Rabbet and a Guinea-Sow, of which lately I have had a very credible Account."

James Drake
1667-1707

Sydney, New South Wales. The *Encyclopedia Britannica* (1910 ed. p. 27b; ||ycxnfwda) stated that "Crosses have been made between the common rabbit (*Lepus cuniculus*) and the guinea pig (*Cavia cobaya*), and examples of the results have been exhibited in the Zoological Gardens of Sydney, New South Wales."

No locus in quo. A video (published 2012, accessed Apr. 4, 2024; ||bdfuz65t) shows an alleged guinea-pig/rabbit hybrid born in a local pet shop. The video itself was crude and the newborn it pictured could have been the young of almost any animal, but the comment is of interest: "I had a long-haired guinea pig sharing with a lavender Dutch rabbit when I was a kid. I found the guinea pig dead one morning having partially given birth to a very elongated baby with lavender Dutch markings. I didn't realise the significance at the time and was more upset by the death of the female, so I extricated the enormous (single) baby and buried them both. Often wondered about it since, but no evidence now. I know biologically it should be impossible."

Overgrowth of womb. In mammalian crosses where the mother is smaller than the sire, hybrids may grow to large for the womb, which will often kill the mother, if the offspring is not delivered by Caesarean. In such cases, the passage is too small for the birth of the oversized offspring. Guinea pigs are smaller than rabbits. This phenomenon is also seen, for example, in the mouse cross *Peromyscus maniculatus* × *P. polionotus,* and in the bovid cross *Bison bison* × *Bos taurus.*

NOTES

1. Birch (1754, p. 393). Sir Robert Moray (1609-1673) was a founder and the first president of the Royal Society. Sir John Hinton (1603-1682) was physician to both Charles I and Charles II. At about the same time, Sir Christopher Wren and his friend William Aubrey, also members of the Royal Society, reported a feleporid and, unlike other reports, said the alleged parents were a doe rabbit and a tom cat (minutes of the Society's June 3, 1680, meeting; ||o8pewtt).
2. Some videos of feleporids: ||so62bkr, ||r9d2nhr, ||ulz4cxy, ||swdmw9e; ||tlv7bzy, ||tcabq9l, ||waceccq, ||rb4c5qy, ||vke48ub, ||wnlshzb, ||w4qaxmo, ||vjxg4xg.
3. As a rule, only later-generation hybrids are hard to analyze in this way.
4. A Walter Woods (b. ca. 1885) lived in Sarcoxie in 1910 (||ufvp665). A. Michael Andrew Dumphy (1883-1963) lived in Jasper Co., Mo., in 1907 (||t5euzqs).
5. A John Wesley Hodge (1889-1949) lived in Newberry in 1921 (||tw5ettc).
6. A Ninons Henry Williams (1851-1926) lived in Bamberg in 1901 (||wko5aha).
7. A George Bennett (b. 1862) was living in Elkton in 1880 (||wkj9rml).

Virginia Opossum (*Didelphis virginiana*).

Thyloplacentals

The floor of our familiar room has become a neutral territory, somewhere between the real world and fairyland, where the Actual and the Imaginary may meet, and each imbue itself with the nature of the other.
—NATHANIEL HAWTHORNE, *The Scarlet Letter*

A marsupial is any member of the major division Marsupialia of Class Mammalia. Many, but not all marsupials, have a fold of skin or pouch (marsupium), containing the mammary glands, which holds the young during early development. The other major division, Placentalia, encompasses all other mammals, with the exception of MONOTREMES (platypus and echidnas). Call any mammal mixing the traits of a marsupial with those of a placental a *thyloplacental*.

A didelfelid, that is, an animal combining the traits of a Virginia opossum (*Didelphis americana*) and a cat. A feral domestic cat gave it birth in North Carolina in 2015, along with a litter of otherwise normal offspring. Such a litter could be produced by the mother first mating with a domestic tomcat, which fertilized most of her eggs, and then with an opossum, which fertilized one. The recently recognized Lykoi breed is likely derived from this same cross. More pictures: ||yckz3faj.

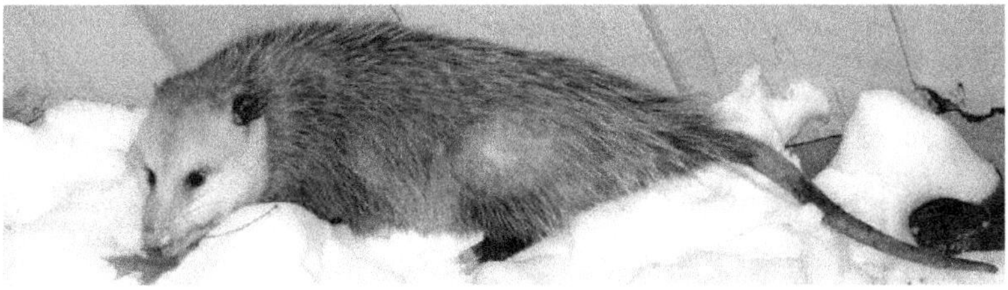

A Virginia opossum (*Didelphis virginiana*)

DIDELFELIDS (OPOSSUM-CATS)

Rafinesque
1783-1840

Lexington, Kentucky. Constantine Samuel Rafinesque (1821, p. 113; also see Rafinesque 1820, p. 85) seems to have been the only scholar to have reported hybridization between a cat and a Virginia opossum (*Didelphis virginiana*), a cross he described as "well-attested" ("bien avéré"). "A cat was left in a cabin," he writes (in translation) "in the Kentucky woods, abandoned for several months. This cabin was perfectly isolated,

> being several leagues from any other. There were no cats within eighteen miles. Upon his return to the cabin, the owner found his cat nursing a litter of five little monsters, resembling cats in body and fur, but having the head, the feet and tail of a Virginia Opossum (*Didelphis virginiana* to naturalists). The animals lived and were shown as a curiosity in all the region; but they died young and without propagating themselves.

* Cats sometimes adopt and nurse opossums (e.g., *Topeka State Journal,* Jul. 15, 1898, p. 4b; ||2p8sdm2e). Such nurslings, having reached maturity, would likely prefer feline mates.

Du Pont, Clinch County, Georgia. From the *Morning News* (Savannah, Ga., Nov. 30, 1899, p. 9a; ||2ddv9pem):

> The latest of nature's freaks, seen at Dupont, was a would-be house cat, and owing to his maternity, is forced to adopt the name of cat, while from its general appearance it would seem a sheep, or an opossum. The animal has ears, eyes and feet like a cat, while its head, body, etc., are very much

> like those of an opossum. For a long while after the cat's advent into the world its body was devoid of covering, but as it grew up there appeared upon its body a covering of very fine wool. The animal is now nearing maturity, and, strangest of all, it hasn't a tooth in its head. The cat is the property of Mr. Jerome Corbett, residing near Dowd post-office, near Dupont.

Christiana, Delaware. The *Wilmington Evening Journal* (May 22, 1912, p. 5c; ||ycytxzwt) reported that "Irvin Leach [of Christiana] has a freak in the form of a kitten with feet shaped like an opossum."

Another view of the North Carolina didelfelid. Note the naked opossum-like tail and the long pink toes.

Brush-tailed Possum (*Trichosurus vulpecula*)

TRICHOFELIDS (POSSUM-CATS)

The Brush-tailed Possum (*Trichosurus vulpecula*) has the widest distribution of any Australian mammal. Nowak (1999, p. 93) comments that "Because of its fecundity and adaptability to a variety of conditions. *T. vulpecula* has been compared to the Virginia Opossum (*Didelphis virginiana*) in North America." In the case of both *T. vulpecula.* and *D. virginiana, opossum* is often shortened to either *'possum* or *possum.*

Adelaide, South Australia. From the *Adelaide Times* (Nov. 7, 1853, p. 18; ||lcz9yzx): "CURIOUS PHENOMENON—At the Anderson's Arms Inn, situated on the Mount Barker Road, where it emerges from Glen Osmnod, and on the ground of which the Bachelor's Picnic Party was held, may be seen a litter of Kittens, half Cat and half Opossum. They have the tail, the fur and fore claws of the latter animal."

Burtundy, New South Wales. A letter to the editor from the *Australasian* (Melbourne, Jun. 2, 1877, p. 24; ||mhezv36) said,

> "Sir,—Did you ever hear of a cross between the domestic cat and the opossum? A cat here has produced six kittens at her second litter. And four out of the six are covered with opossum fur. At the first litter she had two kittens only. They are both here alive and well, and do not differ from other cats in any way. Of the mother herself, she was born on the ground; and both her father and mother, sisters and brothers, are

all ordinary cats. Whence come the furry young ones? They are at my feet as I write, soft-wooled, large frilled, short-legged, and broad between the eyes, with short ears. There are no Angora cats in the neighborhood; besides, I never saw a woolly cat. What do you think? An old boundary rider spotted the kittens at once as being opossum cats. He had seen similar cases upon the Murray [River].
STOCKING. Burtundy, N.S.W., May 19. *Stawell, Victoria.*

Ballarat, Victoria. A subsequent letter from the *Australasian* (June 16, 1877, p. 24; ||lcz9yzx) reported a fertile trichofelid.

Sir,—In reference to the above subject, I would wish to remark that some years since I saw a specimen of what appeared to be a cross between these two animals. It was dead, and was left by the owner at a Mr. St. John's, then a taxidermist of this town, with instructions to have it stuffed. The owner stated that, when living, it caught and ate mice like a cat; also, that at other times, it would make a good meal off a cabbage in her garden. I took particular notice of it at the time. The forepaws were those of the opossum, while the hind ones were like the cat's. The fur of the opossum was distinctly observable on the surface, whilst underneath the fur or hair of the cat could be seen. The general colour was that of the opossum, but faint lines could be seen like those of a tabby cat. The most remarkable thing about this creature, however, was that, according to the statement of its owner, it had on several occasions had kittens, which would tend to show that it was not a mule [i.e., a hybrid], unless the law affecting mules is upset for, I believe, the first time in this instance. [Despite the reader's belief, many hybrids have in fact produced offspring.] I think Mr. St. John is still in Victoria, and if so, would doubtless remember the specimen above described. Would not a natural history column add interest to your valuable paper?
PLATYPUS, Ballarat.

Stawell, Victoria. Another letter in this series (*Australasian,* Jun. 23, 1877, p. 25; ||lpvp8lc) reported an additional fertile trichofelid.

Sir,—Allow me to contribute my little story. About nine years ago, on a station in the Wimmera district, I saw a tomcat with a decided opossum look about him—a long bushy tail, pointed nose, large round eyes, without the contraction of the pupils, sleepy in the daytime lively at dusk, furry coat mixed with hair, colour white with a brown shade. While stroking the animal another wild puss darted across the floor, having all the characteristics of Tommy [i.e., of a cat], supposed to be his progeny.
NATURALIST. Stawell, June 18.

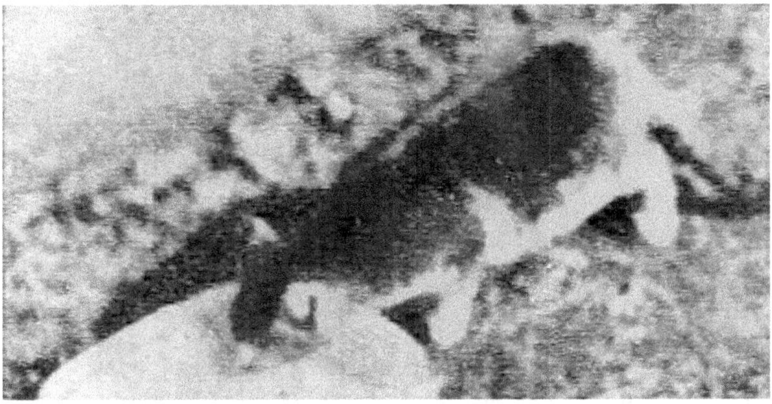

The Kingaroy trichofelid.

Kingaroy, Queensland. A letter in the magazine *Pix* (vol. 18, No. 6, Aug. 10, 1946, p. 2; ||mr25548h) reported a hybrid of a domestic cat and a brush-tailed possum. It read, "Believe it or not, this cat's grandfather was an opossum! The claim is made by the cat's owner, a farmer of Kingaroy (Qld.). It certainly has all the signs of its strange ancestry, for it is an outcast in an otherwise happy cat family, has an opossum-shaped body of a blue color, and has a ringed tail which tends to curl. It also is inclined to take everything into its paws before eating. Bushmen say opossum-cat crosses are not uncommon."

Albury, New South Wales. The *Sydney Mail and New South Wales Advertiser* (Nov. 3, 1877, p. 555c; ||mqvd955), in a column offering news from Albury, referred to a "lusus naturae which turned up at a township on the other side of the river, in the shape of a cat. This animal has fore paws exactly resembling those of an opossum, whilst the hind feet are precisely similar to the feet of a kangaroo. In other respects this monstrosity displays no special departure from the ordinary feline type."

Campbell Town, Tasmania. A letter to the author of a children's column, "Mr. Flamingo," as it appeared in the *Launceston Examiner* (Launceston, Tasmania, Mar. 25, 1899, p. 2; ||m8ykkkk) said, "My Dear Mr. Flamingo,—I am greatly interested in your children's column, and read the stories every week. I think you would like to hear about an opossum cat which I have. She is a curious little creature and knows me so well that she follows me about everywhere. At night, when I am in bed, she finds her way to my room and coils up on my pillow; but the poor little thing is never allowed to stay all night with me. She does not love my brothers at all, because they torment her so. We also have an opossum, which is very amusing. Sometimes we put it up the pear tree, and it pulls the pears off, and throws them down at us. If we run away, he is not long coming down the tree after us.—Your constant reader, Frances Ruth Pitt (aged 10 years).—Campbell Town, March 22."

Axedale and Geelong, Victoria. The *Argus* (quoted in the *Bendigo Independent,* Jan. 14, 1907, p. 2; ||ltm7yzd) said, "I notice in THE ARGUS of last Friday [Jan. 11] a letter from one of your correspondents, "T.T.," from Hamilton, about a hybrid cat and 'possum, and also your note to the effect that there is a scientific assurance that cats and 'possums will not interbreed. With all due respect for science, it seems strange that there should be so much evidence to the contrary. I can bear out the evidence of your Hamilton correspondent, for some years ago, while on a visit to Axedale, the constable in charge of the local police station had a peculiar-looking cat. The animal, which was of a dingy black, had fur like a 'possum, and heavy hindquarters. This animal was one of a litter found in the hollow of a tree. The mother, a domestic cat that had gone wild, was shot. The owner, who would not part with the animal on any account, was satisfied it was a hybrid between a 'possum and a cat. I saw a similar specimen at Macedon about three years ago." Two correspondents [other than "T.T.] write stating that hybrids between a cat and a 'possum are now in the possession of Mrs. Hansom, of Blyth Street, Geelong, and Mr. Duff, at the Queenscliff end of the Nepean Road [near Geelong]."

Sussex Inlet, New South Wales. From *Smith's Weekly* (Sydney, Jan. 24, 1920, p. 18; ||lcoxn8n): "Down at Sussex Inlet … lives a quaint old Irishman. He … centres his affection on "Sinn Fein," a weird hybrid, half-'possum, half-cat. This curious creature possesses the characteristics of both parents mixed. He has one cat's-eye, the other a 'possum's, and his fur is nondescript grey. He loves the rain, climbs a tree with the ease of his 'possum ancestors, walks on four legs cat fashion, lives on fish, and has never been known to caterwaul or purr under the gravest provocation."

Port Pirie, South Australia. From the *Recorder* (Port Pirie, Sep. 1, 1934, p. 2; ||mr3786h3):

> When Alex Swensson, Jr., began talking about a freak cat he had at home, his friends put it down to a rough Saturday night and let it go at that. … Yesterday afternoon I had a look at the animal,

> which is half cat and half opossum. The head and tail are all cat. The body and legs 'possum, particularly the claws, which are built for fast and high climbing.
>
> Tempted with a piece of meat the freak will sit up and cry for it in approved 'possum style and seldom is heard the homely "meow." Its run resembles a pacing horse, as much as anything, with the front portion moving rhythmically and the hindquarters rolling.

Kumbia, Queensland. Another report about a possum-cat hybrid, which produced offspring, appeared in *Queensland Country Life* (Brisbane, Aug. 31, 1939, p. 3; ||kfjxbc2):

> It is generally admitted by bushmen that cats and opossums will occasionally cross, although there is some disagreement whether the progeny is sterile.
>
> The accompanying photograph† is an animal claimed to be the grandchild of a hybrid.
>
> The first cross, the owner, Mr. Wason, of Kumbia, said, showed many signs of its opossum father, having the inside of its tail clean, and the fur and shape of its body being opossum-like.
>
> Amongst the many cats at Mr. Wason's place, this animal was always an outcast. She, however, had one kitten which was the mother of the present two. One of these shows little of the opossum blood, but the subject of the photo has many traces. The feet and head are those of a normal cat, but in the shape and coloring of the body the opossum can be plainly seen. The tail, though well furred, is ringed and has a tendency to curl at the tip.
>
> When handed food this animal will take it in its paws and carry it to its mouth. Sometimes it will even pick food up off the floor with its paws.

* Note that the Kumbia report gives the direction of the cross (possum ♂ × cat ♀). It also says a female F_1 hybrid backcrossed to a cat and produced a female B_1 hybrid, and that this B_1 female backcrossed again to a cat and bore two progeny. So it would seem that the female hybrids are sometimes fertile (which is consistent with Haldane's Rule).

Drysdale, Victoria. A report in the *Argus* (Melbourne, Nov. 10, 1933, p. 7; ||17e7c3f) stated that another possum-cat, at Drysdale, also produced "kittens" in a backcross to cat.

†. This photo not shown here due to the poor quality of the digital image accompanying the archive version of this article.

Carisbrook, Victoria. The *Argus* (Melbourne, May 14, 1936, p. 4; ||lbtrtqw) published a letter from a Robert Downing of Carisbrook: "When we were walking home from church on a recent night we heard a series of growls, hisses and barks, and my father's torch, flashing along a fence, revealed our cat and a possum-cat with backs up hissing at each other while our dog, a four-months-old staghound, was barking at both and apparently thinking it all great fun. … I picked up the possum-cat (being scratched while doing so), and dad called the dog off. Since then I have seen the dog and the possum-cat making friends while our white cat turns up her nose at them. Now the possum-cat is scared of the white cat and is thoroughly chummy with the dog."

A DIDELOTRID (OPOSSUM-RACCOON)

Newtown, North Carolina. From the *Perrysburg Journal* (Perrysburg, Wood Co., O., May 1, 1914, p. 6; ||2p8hkab4):

> NEWTOWN, N.C.—A freak opossum which was captured fifteen years ago and released after being marked was killed here by Gus Price and Walter Rowe … The animal is a cross between an opossum and a raccoon. It has stripes running around its body, is 5 feet 3 inches long, measuring from the tip of the tail to the end of the nose; and weighs 21 pounds.
>
> O. C. Boston., an old-time hunter declares it is the same animal he captured in Alexander County in 1899. At that time he marked it slitting its ears and released it. … Citizens of this town have been at a loss for some time to know what sort of animal was making nightly depredations on hen houses. Although many tracks of the strange creature had been found, no one had seen it until its discovery today. Old hunters here declare they have never seen anything like the unusual creature.

A BOVOMACROPID (COW-KANGAROO)

Botany Bay, New South Wales. The *Westport Times* (Westport, N.Z., Jan. 5, 1883, p. 4; ||2p83ar55) said, "A wonderful freak of nature is on exhibition in the premises near the Exchange Hotel, Shortland [a suburb of Newcastle, NSW]. It was, our informant says, calved by a cow in Botany Bay, on March 10th last. It has one head, that of a kangaroo, and a formation representing the place nature apparently intended for another, eight legs, four ears, three eyes, two of which resemble a calf's, the other being unlike them, but, it is said, similar to that of a kangaroo, and two tails. … the ears are under the eyes."

HOMODIDELPHIDS (HUMAN-OPOSSUMS)

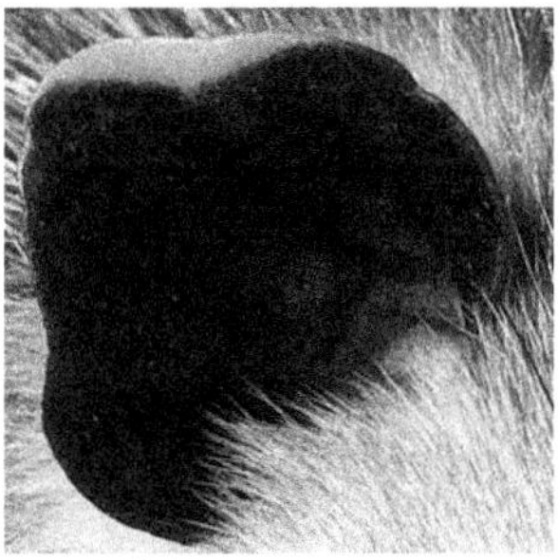
An opossum ear.

Lusby's Mill, Kentucky. Physician T. J. Denny[1] (1871) reported a woman at Lusby's Mill giving birth to a stillborn fetus with opossum-like traits. "On the night of the 9th of June Mrs. S. was delivered of a most singular monstrosity. It consisted of a fœtus, a seven months' gestation. Its head, body and upper extremities were natural, with the exception of the ears, one of which was an

> imperfect human ear, the other an exact resemblance of the ear of an opossum. ... Its hips were perfectly natural, but it had no legs, nor any sign of any lower extremities [see DEFICIENCIES] except an appendage about eight inches long, resembling the tail of an opossum. It also had a nail, similar to a fingernail, on the end of its tail; it had no anus or external genitals. It was alive when delivered but died in a few minutes."

* Thomas J. Denny lived at Lusby's Mill in 1880 (||2p955m3v).

Hannibal, Missouri. J. B. Bolton[2] (1881, p. 438), a Hannibal physician, reported a woman birthing a homodidelphid that "had one ear like that of an opossum" but looked otherwise like an ordinary human being.

* Physician J. B. Bolton lived in Clay, seven miles from Hannibal, in 1880 (||32j53m58).

Brunson, South Carolina. The *Dispatch* (Lexington, S.C., Dec. 1, 1909, p. 2; ||y2naqh38) reported a case of near-XENOGENESIS. "BRUNSON, November 28.—Special: One of the most unique products of humanity on record is at present in evidence in the town of Brunson. Very recently a negro woman, a resident of this town, gave birth to an object or being decidedly, in appearance, much more animal than human. The mother is an unmarried woman of about 35 years of age; black, of good health and strong physique. The child or product is in appearance wholly animal, and is in shape, size and general make-up a facsimile of an opossum, minus the tail, and furry skin. In the place of the tail, there is a stub of about an inch in length, the end of it suggesting amputation of the caudal appendage. The shape of the four legs and feet, with claws; the shape of the head and size of the body, limbs and head are all 'possum-like or animal-like. Only the eyes are unlike the 'possum's. They are larger, rounder and further back on the sides of the head. This monstrosity is not living but is carefully preserved in alcohol in the office of Dr. J. L. Folk, resident physician. Dr. Folk will probably produce it at the next meeting of the State Medical Association for inspection and diagnostic theory or hypothesis."

Habersham County, Georgia. The *Boston Medical and Surgical Journal* in giving an account of a report, about three cases of congenital deformity, by Georgia physician Thomas Gibbs Underwood (1857), in a now unavailable copy of the *Nashville Journal of Surgery and Medicine*, "The first case is quite extraordinary. A lady in Habersham County, Georgia, was delivered, at the end of the seventh month, of a child whose left leg was destitute of a foot and ankle, the stump appearing as if the limb had been amputated by a dull instrument, a short time previously; the bone protruded about an inch and a half and appeared to have been broken off obliquely. The right leg "resembled in shape the foreleg of the opossum. The foot was turned inwards, and the toes were grown together to near their extremities, and were destitute of natural nails, but had claws precisely like those of the opossum. The ring finger of the right hand was off at the first phalangeal articulation and very much resembled the toe of the above-named animal, and on the extremity was a claw, like those on the toes."

* In connection with the missing appendages reported in the Habersham case, see DEFICIENCIES.

* Records show that a Thomas Gibbs Underwood was living in Habersham County in 1850 (||y85aw2hx).

A HOMOMACROPID (HUMAN-KANGAROO)

Logan County, Kentucky. The *Cincinnati Enquirer* (Dec. 7 1893, p. 9e; ||ye22rykb) said, "Several days ago a white lady living in Logan County, southern Kentucky, gave birth to a child which, in almost every respect of form and feature, resembled a young kangaroo. The human freak lived only a short time but has been preserved in alcohol by a physician. The only way in which the appearance of this unhuman looking child is accounted for is that some time before its birth the mother read a book relating to kangaroos and afterward appeared much enthused on the subject."

* The explanation offered in the Logan report refers to the theory of MATERNAL IMPRESSIONS.

* Russellville, the seat of Logan County, is about 30 miles from Clarkesville.

TELENOTHIANS CONTINUES IN VOLUME II

TinyURL Keys

Throughout the book, to save space, Internet links are provided as TinyURL keys. Each such key is preceded by a pair of vertical bars. For example, the key ||3w7u6c98 corresponds to the TinyURL http://tinyurl.com/3w7u6c98. To use that key type “http://tinyurl.com/” into the URL bar of your browser and append the key.

www.ingramcontent.com/pod-product-compliance
Lightning Source LLC
LaVergne TN
LVHW050344210926
848481LV00017B/1452